Noise Control

in

Building Services

Noise Control

in

Building Services

Sound Research Laboratories Ltd

Edited by

ALAN FRY

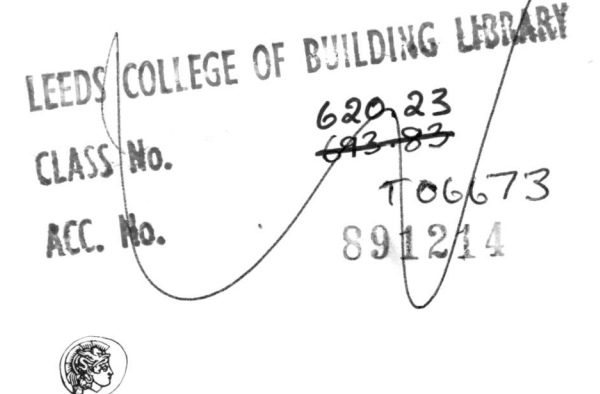

PERGAMON PRESS

OXFORD · NEW YORK · BEIJING · FRANKFURT
SÃO PAULO · SYDNEY · TOKYO · TORONTO

U.K.	Pergamon Press plc, Headington Hill Hall, Oxford OX3 0BW, England
U.S.A.	Pergamon Press, Inc., Maxwell House, Fairview Park, Elmsford, New York 10523, U.S.A.
PEOPLE'S REPUBLIC OF CHINA	Pergamon Press, Room 4037, Qianmen Hotel, Beijing, People's Republic of China
FEDERAL REPUBLIC OF GERMANY	Pergamon Press GmbH, Hammerweg 6, D-6242 Kronberg, Federal Republic of Germany
BRAZIL	Pergamon Editora Ltda, Rua Eça de Queiros, 346, CEP 04011, Paraiso, São Paulo, Brazil
AUSTRALIA	Pergamon Press Australia Pty Ltd., P.O. Box 544, Potts Point, N.S.W. 2011, Australia
JAPAN	Pergamon Press, 5th Floor, Matsuoka Central Building, 1-7-1 Nishishinjuku, Shinjuku-ku, Tokyo 160, Japan
CANADA	Pergamon Press Canada Ltd., Suite No. 271, 253 College Street, Toronto, Ontario, Canada M5T 1R5

First edition 1988

Library of Congress Cataloging-in-Publication Data
Noise control in building services
edited by Sound Research Laboratories Ltd.
1. Soundproofing 2. Noise control. 3. Buildings—Environmental engineering. I. Sound Research Laboratories.
TH1725.N66 1988
693.8'34—dc19   88-23842

British Library Cataloguing in Publication Data
Noise control in building services.
1. Buildings. Noise. Control measures
I. Sound Research Laboratories Ltd.
693.8'34
ISBN 0-08-034067-9

Printed in Great Britain by A. Wheaton & Co. Ltd, Exeter

FOREWORD

SRL was formed in 1969. Those of us involved in setting the new venture on the road had studied, trained and practised in the field of building services so it was natural that the company should develop quickly into a recognised authority on noise control in this specialised branch of engineering.

Although the company has expanded considerably since into other fields SRL has maintained its strength of expertise in the foundation activity. The laboratories, which were added in 1976, were primarily designed for acoustic measurements in the presence of air flow and are probably still unique in their scope in the corporate sector.

With many years of experience behind them, the senior engineers in the company have combined to write this volume in order to pass on some of their knowledge to those of other disciplines who have only found time previously to gain a glimpse of the state to which the technology has now developed.

R. I. WOODS

Chairman, Sound Research Laboratories Ltd

 SRL is a member of the Salex Group of Companies.

CONTRIBUTING
AUTHORS

R. D. Bines
BTech, CEng, FIMechE, MIOA

P. R. Dunbavin
MSc, MIOA, MinstSMM, MSEE

A. T. Fry
BSc, ARCS, MinstP, FIOA, MASHRAE, MCIBSE

T. J. Hickman
BSc, MSc, PhD, MIOA

P. R. Hobbs
MIOA

M. S. Langley
MSc, PhD, CPhys, MinstP, FIOA

T. L. Redmore
BEng, MSc, PhD, MIOA

D. F. Sharps
CEng, MIMechE, MIOA, MBIM

G. M. Tomlin
MIOA

T. K. Willson
FIOA

CONTENTS

Chapter 6   Vibration Isolation

Chapter 7   Ductborne Noise — Transmission

Chapter 8   Ductborne Noise — Flow Generated

Chapter 9   Ductborne Noise — Breakout

Chapter 10   Noise from Room Units

Chapter 11   Noise to the Exterior

Chapter 12   Construction Site Noise and Vibration

Chapter 13   Laboratory Testing

Chapter 14   Other Building Services Plant

Suggested Further Reading

Index

INTRODUCTION

This book is aimed at Building Services consultants and engineers who find they have a need to be involved in matters of noise and vibration control, or at least require to be conversant with such matters. Hence, by the very nature of its collected material and formal presentation, it also lends itself as a supplement to student courses in Engineering Acoustics — it is an applied engineering book.

It has been written by the senior engineering staff of Sound Research Laboratories, who come from a broad background of skills and experiences. Hence, they have been able to supply a contrast of expertise and evolve a balance for the experienced Building Services Engineer and the Acoustics Engineer alike: for example, the background to sound and its measurement/presentation for the Building Services Engineer, and the more detailed equipment sections for the acoustician, such as "Noise from Room Units".

Following on from the acoustics introductions, the core of the book covers the duct distribution system with the fan as the key generator of noise, and many natural and bespoke items as the loss and absorption mechanisms for attenuating this noise. However, the all-important awareness of noise generated en route by the flow of the conditioning air — often labelled 'regeneration' — has not been ignored and a complete chapter addresses this difficult subject.

The consequences once the resultant noise has reached the conditioned space have been considered early in the book, and most importantly a connection between the scientific objective noise measurements and the very relevant subjective reactions is established by various criteria and design targets.

Away from this duct-borne noise but staying within the building, the problems of noise control by sound insulation between rooms — more especially building services plantrooms — and the control of structure-borne vibration induced noise have been covered in separate chapters. Naturally, the vibration isolation chapter also covers the lower frequency vibrations, which are felt rather than heard.

Outside of the building both the early days of construction site noise and the more permanent problems of noise radiated to the surrounding environment have each received a full chapter.

Testing and accreditation of equipment for its noisiness are considered by acoustic consultants to be a very necessary feature of improving prediction quality and techniques for the future, and consequently we have included a substantial chapter on current test methods and the links to National and International Standards.

All this has been achieved with a range of complexity from rather chatty chapters to the more mathematical nature of, for instance, the flow generated noise contribution. Despite this, we hope the image of the texts remains clearly aimed at

INTRODUCTION

engineers who appreciate a necessary degree of understanding, but above all require graphs, tables, nomograms and formulae to assist them either in their own tasks of noise and vibration control, or more especially in understanding and keeping a watching brief on others.

As much of the source material comes from in-house and proprietary experience blended with more established publications, a conscious decision has been made not to include specific references but simply include a list for further reading and reference. We apologise now to those readers — and reviewers — who feel strongly on such matters.

All this apart, we expect the book to make a worthwhile piece of reading and become in its own right a reference book of useful engineering guidance and data.

ALAN FRY

CHAPTER 1

"A SOUND BASIS"

1.1 INTRODUCTION

Although it is not necessary to be a qualified mechanic to drive a car, most people find it easier to learn to drive — and thereafter to make the most of their motoring — if they have at least a general idea of what goes on "under the bonnet". Similarly, the environmental engineer will find that a basic knowledge of the fundamental theory of sound is of great help in appreciating the problems of noise in heating, ventilating and air-conditioning systems. Any introduction to the subject must inevitably include some material which may seem elementary and some which may appear to be of mainly academic interest. Nevertheless, the fundamental principles have a great bearing on practical acoustic problems and a knowledge of fundamentals can help the engineer to avoid the more common acoustic pitfalls. It is the purpose of this chapter to provide the background to the physics of sound as a basis for the consideration of specific installation problems in the remainder of the book.

1.2 WHAT IS SOUND?

1.2.0 The Question

Subjectively, sound is something we hear as a result of vibration in the air. These vibrations yield pressure fluctuations which can be measured on a sound level meter. Vibrations are more usually associated with mechanical devices such as a mass vibrating up and down on a spring. Air also has mass and like a spring it has stiffness — that is it resists compression as it does in the pneumatic tyres of a car. As sound travels through the air, this air is locally compressed and expanded. The air tries to regain its normal equilibrium state; a region of compressed air will try to expand and in so doing acts on the adjacent region compressing it in turn and so the sound wave spreads and is propagated. Figure 1.1 shows a representation of a sound wave generated by a piston reciprocating in a tube.

As the piston moves forward into the tube it compresses the air in front of it and this compression travels down the tube as a pressure wave — sound. On retraction the piston creates a rarefaction and this then follows the compression down the tube. They travel at the "speed of sound". There are two mechanisms for generating sound that will be considered.

1

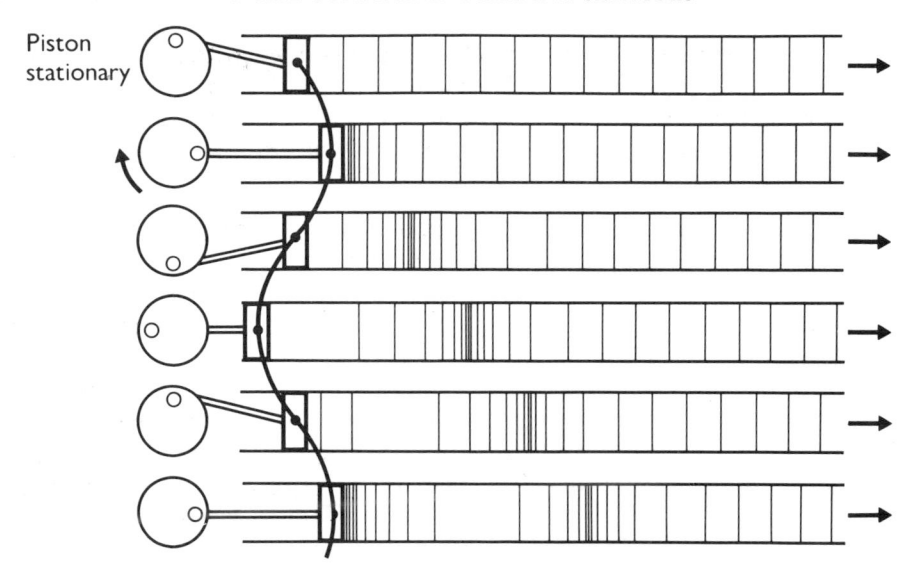

Piston
stationary

Figure 1.1   Sound Wave in a Tube

1.2.0.1 Any moving surface will cause disturbance of the air around it. If it vibrates very slowly or if the air can travel easily from the front to the back, then all that happens is air movement to and fro. If the surface is very large or vibrates at a high frequency or if, as in a loudspeaker, the air is prevented from moving to the back of the vibrating cone, then the disturbance moves away from the surface as a sound wave. The vibrating covers of the machine in Figure 1.2 can be very effective sound sources.

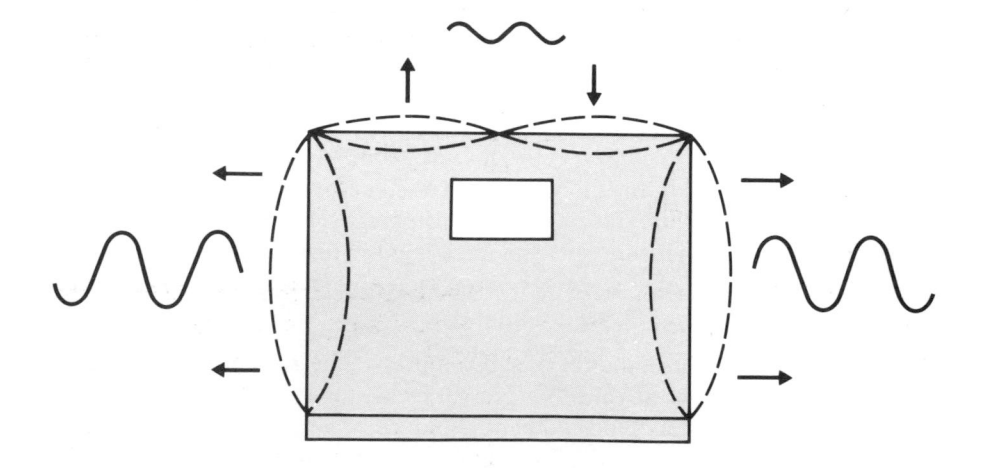

Figure 1.2   Sound Generated by Vibration of Machine Covers

The sides are shown to be simply vibrating and totally contributing to pressure changes and hence sound generation. However one half of the top cover is travelling up whilst the other half travels down, which allows for some to and fro air movement. Hence a partial cancellation occurs and as a result much less sound is generated by the vibration of the top cover.

1.2.0.2 The second mechanism is a change in the way air or gases are moving, for example, if moving air is stopped or changed in direction or gases suddenly expand. An example of sound generated this way is wind blowing on a building or through the sails of a windmill. In an air conditioning system the blades of a fan accelerate and noisily change the direction of air flowing through, whilst it is the sudden drop in pressure that accelerates air escaping from a compressed air line which results in sound. The sound of an oil burner originates from the tiny oil droplets exploding with a sudden expansion of gases. A similar sudden expansion of gases occurs in an internal combustion engine, much of the pressure fluctuation being emitted as sound from the exhaust.

1.2.1 Physical Characteristics of Sound

The two basic characteristics of a sound wave are amplitude and frequency. Amplitude is a physical measure related to the subjective loudness of a sound. Frequency corresponds to the subjective pitch of a sound.

1.2.1.1 Frequency Frequency, together with two further characteristics, wavelength and period, can be illustrated by again considering the sound generated by the piston reciprocating in a tube referred to previously. The rate at which the piston reciprocates on its connecting rod determines the frequency of the compressions and hence the frequency of the sound. The frequency, f, is measured as the number of cycles each second and the unit of measurement is named HERTZ (cycles per second), abbreviated Hz. In our simple illustration this will relate to the crank's rotational speed and if this were n revolutions per minute (rpm) then

$$f = \frac{n}{60} \qquad\qquad 1.1$$

where n = revolutions per minute
and f is in Hertz.

Rather than specify the pistons speed as the number of cycles in a second (f Hz), the time taken to execute one complete cycle — there and back — can be defined. This is named the PERIOD, T, and is measured in seconds.

$$T = \frac{1}{f} \text{ seconds} \qquad\qquad 1.2$$

If sound travels at velocity c then the distance between successive compressions, known as the WAVELENGTH, λ, is related to the period and frequency by:

$$f = \frac{1}{T} = \frac{c}{\lambda} \qquad\qquad 1.3$$

as shown in Figure 1.3.

The waveform shown is sinusoidal and is a representation of the change of excess pressure — compression or rarefaction — with either time or distance along the tube.

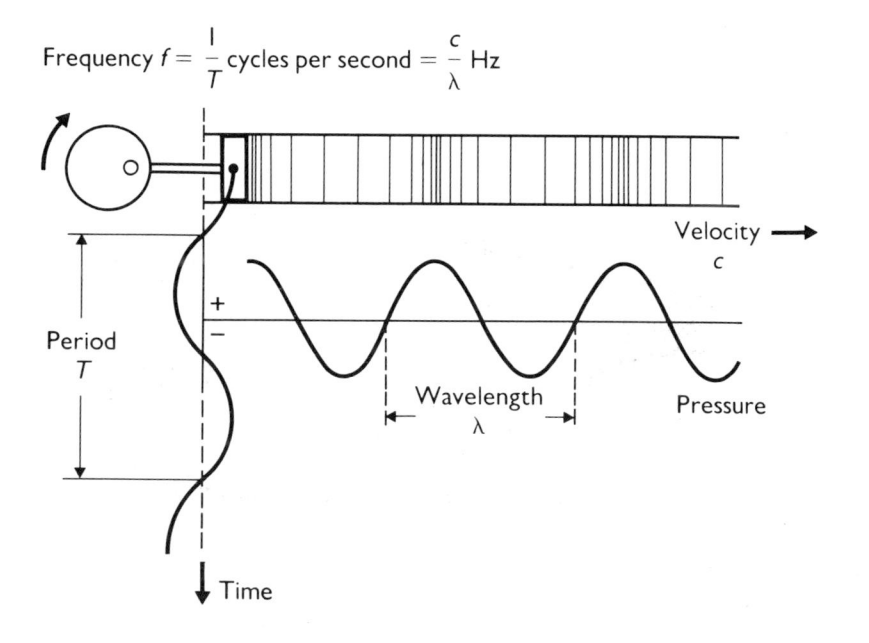

Figure 1.3   Relationship between Frequency, Wavelength and Period

1.2.1.2 Pure Tones A sound such as shown in Figure 1.3, with pressure changes at one frequency, is a pure tone and has the waveform and frequency spectrum shown in Figure 1.4. where frequency or pitch is represented in Hz along the horizontal axis and amplitude along the vertical axis. A frequency spectrum is a representation of the distribution of energy in sound among various frequencies. All the energy in a pure tone is at one frequency and its spectrum is known as a line spectrum.

Figure 1.5 illustrates the frequencies and wavelength of some sounds with strong single frequency characteristics. Humans have a hearing range from 20 Hz up to 20 kHz. (1 kHz = 1000 Hz). This upper limit decreases with age once adulthood is reached and old people would not hear sound above 10 kHz. In air at room temperature and pressure, sound travels at 343 m/sec and so at 20 Hz has a wavelength of 17,000 mm and at 16 kHz a wavelength of 20 mm approximately.

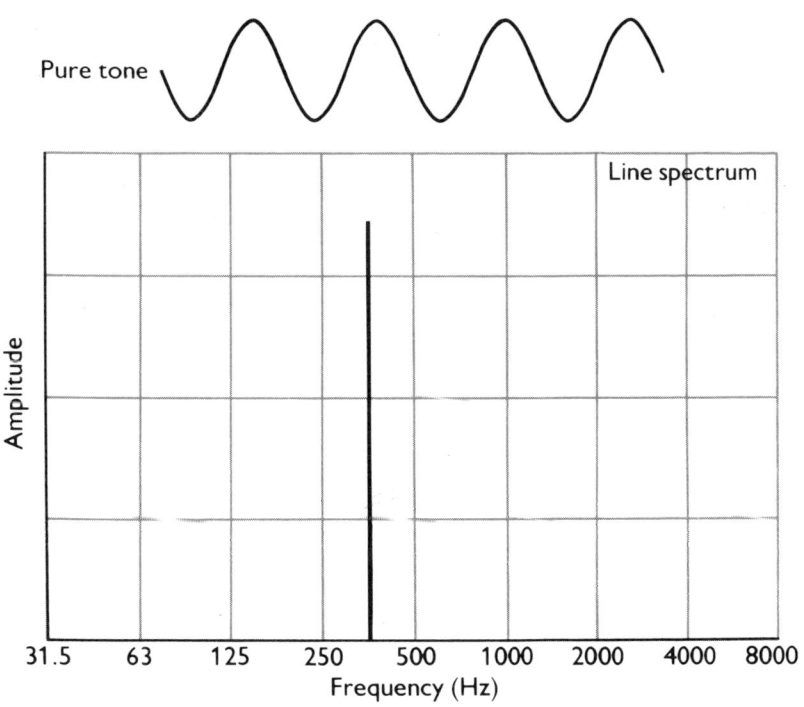

Figure 1.4 Wave Form and Spectrum of a Pure Tone

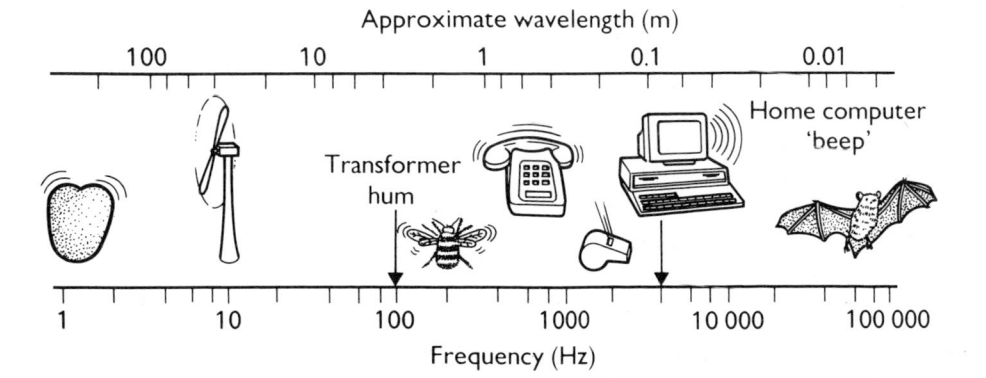

Figure 1.5 Frequency and Wavelength of Everyday Sound

1.2.1.3 Complex Tones Most sounds are not pure tones. If the waveform repeats itself, as in the square wave in Figure 1.6, it is said to be periodic and can be constructed from a number of pure tones of various frequencies and amplitudes. This is reflected in its line spectrum; energy existing at a number of discrete frequencies. Musical notes have periodic waveforms and therefore have line spectra comprised of a number of tones.

There is often a simple mathematical relationship between the frequencies of these different tones such as twice, three times, four times etc, which are then known as harmonics.

The base frequency is the fundamental and it is this frequency which is ascribed to the musical pitch. The notes of a piano keyboard are also shown on Figure 1.5 drawn against the frequency scale. Each piano note has many harmonics with a mathematical relationship between them.

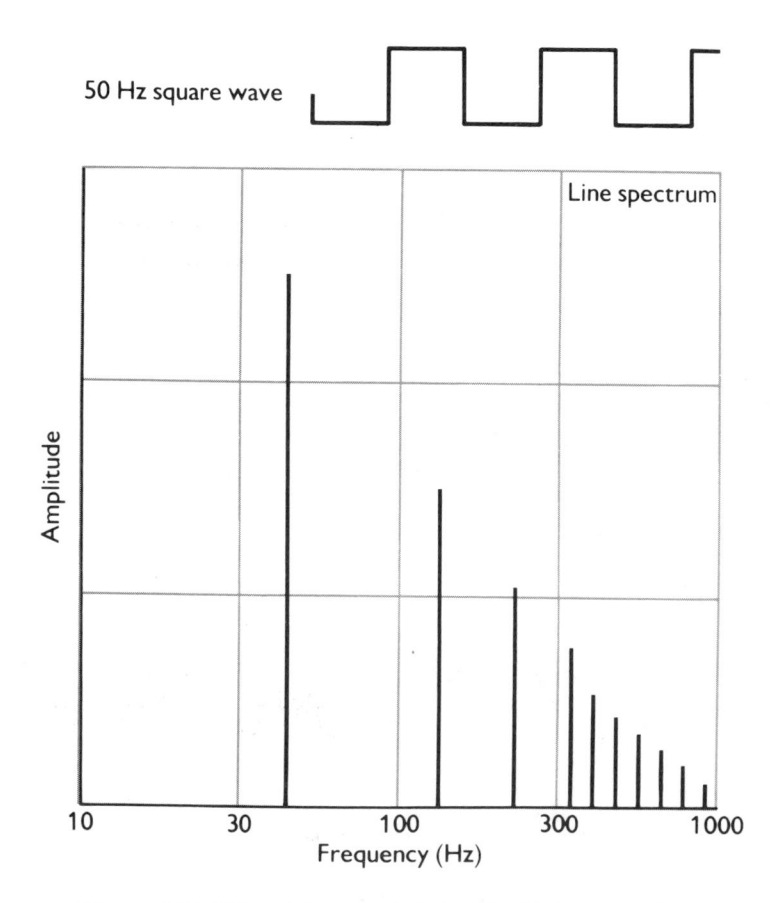

Figure 1.6   Waveform and Spectrum of a Square Wave

Noises are not usually pure tones or periodic but have sound energy spread over all audio frequencies. A typical waveform of broad band noise, shown on Figure 1.7 has no pattern to it and the spectrum is said to be continuous. To represent such sound existing at all frequencies it is measured in frequency bands using filters; that is the sum of all the sound energy between a lower and upper frequency is measured, say between 50 Hz and 100 Hz. The difference between the upper and lower frequencies is the bandwidth in Hz. To standardise measurements, international octave and 1/3rd octave frequency bands have been specified at the centre frequencies shown on Figure 1.8. A 1 kHz octave band ranges from 707 Hz to 1414 Hz, the upper frequency being twice the lower. Each octave band can be divided into three 1/3rd octave bands. For example, the octave band centred about frequency f_0 can be divided into three 1/3rd octave bands with upper and lower frequency limits as follows:

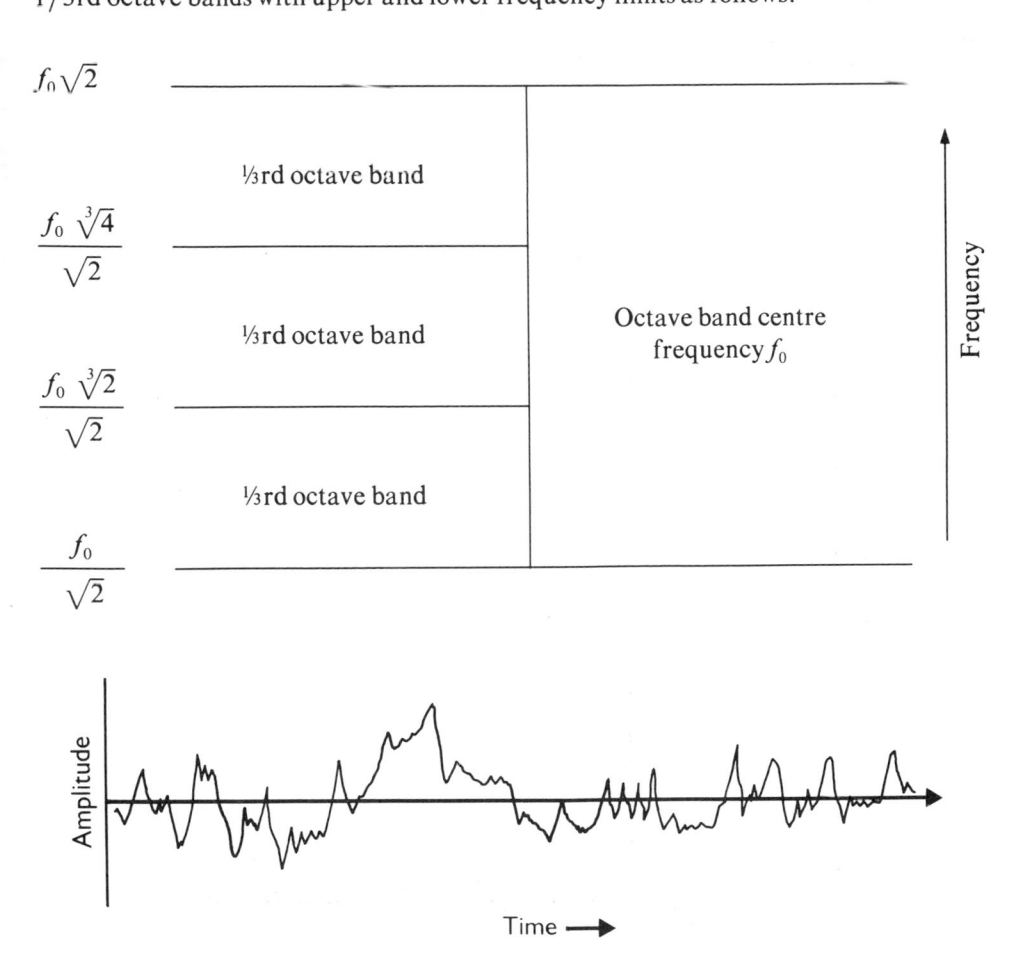

$f_0 \sqrt{2}$

⅓rd octave band

$\dfrac{f_0 \sqrt[3]{4}}{\sqrt{2}}$

⅓rd octave band

Octave band centre frequency f_0

$\dfrac{f_0 \sqrt[3]{2}}{\sqrt{2}}$

⅓rd octave band

$\dfrac{f_0}{\sqrt{2}}$

Frequency

Amplitude

Time ⟶

Figure 1.7   Waveform of Broadband Noise

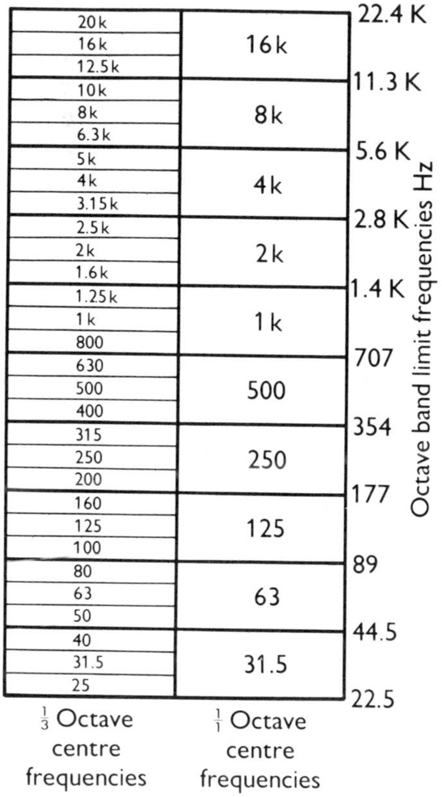

$\frac{1}{3}$ Octave centre frequencies	$\frac{1}{1}$ Octave centre frequencies	Octave band limit frequencies Hz
20k		22.4 K
16k	16k	
12.5k		11.3 K
10k		
8k	8k	
6.3k		5.6 K
5k		
4k	4k	
3.15k		2.8 K
2.5k		
2k	2k	
1.6k		1.4 K
1.25k		
1k	1k	
800		707
630		
500	500	
400		354
315		
250	250	
200		177
160		
125	125	
100		89
80		
63	63	
50		44.5
40		
31.5	31.5	
25		22.5

Figure 1.8   International Preferred Frequencies, Octave Band and 1/3 Octave Band

The measured amplitudes in each frequency band can then be plotted to show the frequency spectrum as in Figure 1.9 which uses octave frequency bands. Strictly a histogram is the correct representation as we do not know the amplitude of different unique frequencies within each band. For instance in Figure 1.9 it is likely from the spectrum shape that there is more sound energy in the upper half of the octave band between 1000 and 1414 Hz than the lower half between 707 and 1000 Hz, but we do not know this and such inferences could be incorrect.

An alternative popular presentation is to draw an implied continuous spectrum as shown in Figure 1.10 by joining the mid band frequency levels with straight lines. Again intermediate discrete frequency levels must not be inferred from this representation. The line is simply to guide the eye from point to point and consequently indicate trends.

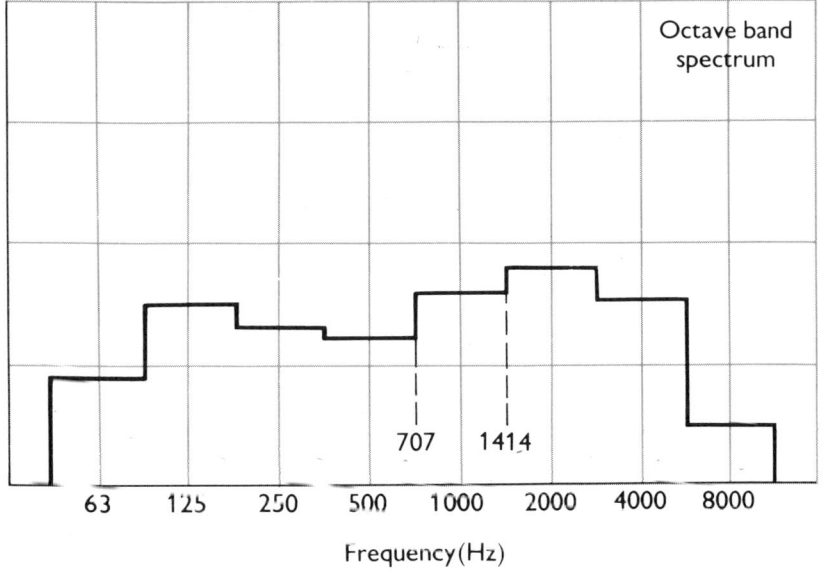

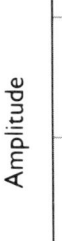

Figure 1.9 Continuous Spectrum — Histogram Representation

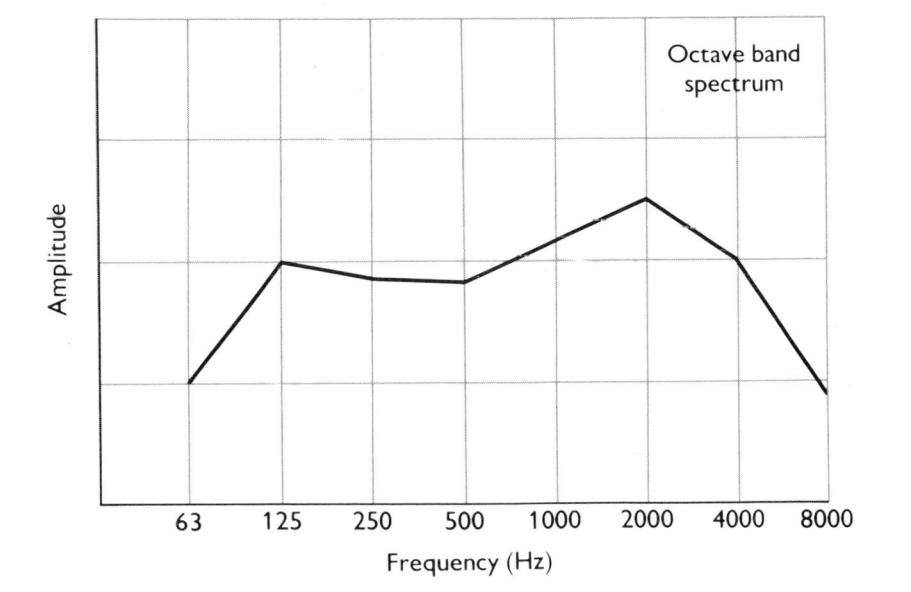

Figure 1.10 Continuous Spectrum — Conventional Representation

If more detailed information about the frequencies and their relative amplitudes is required then a narrow bandwidth is used. 1/3rd octave bands can provide a worthwhile but limited increase in the information but the spectrum shape will look fairly similar to the octave band spectrum. A more radical change is to use what is referred to as narrow band analysis. Narrow band analysis employs smaller frequency bandwidths which are usually of constant bandwidth independent of frequency. (Octave bands have constant percentage bandwidth, the bandwidth being proportional to and increasing with the frequency).

This narrow band analysis reveals substantial variations in amplitude with small changes in frequency. Any pure tones hidden in the continuous spectrum will also be revealed and these frequencies accurately determined. These tonal frequencies may then for instance be compared with the running speed of machinery components to help track the noise sources. Figure 1.11a shows a narrow band spectrum from 0 to 200 Hz with a 0.5 Hz bandwidth showing a tonal component at 120 Hz. When represented in octave bands the tone is combined with all the other frequency components in the 125 Hz octave band and can no longer be distinguished, as shown on Figure 1.11b. This is a histogram representation of the octave band spectrum from 31.5 Hz to 4 kHz centre frequency. The narrow band spectrum was obtained from an analyser where the selection of a narrow bandwidth (0.5 Hz) limited the frequency range (0–200 Hz) so that the higher frequencies are not shown.

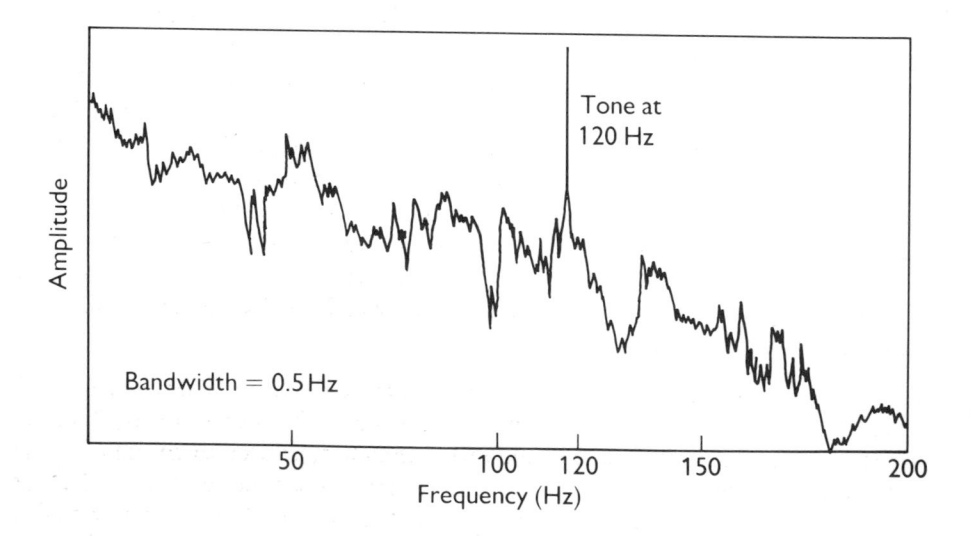

Figure 1.11a   Narrow Band Spectrum (0–200 Hz) of Noise with Tonal Component

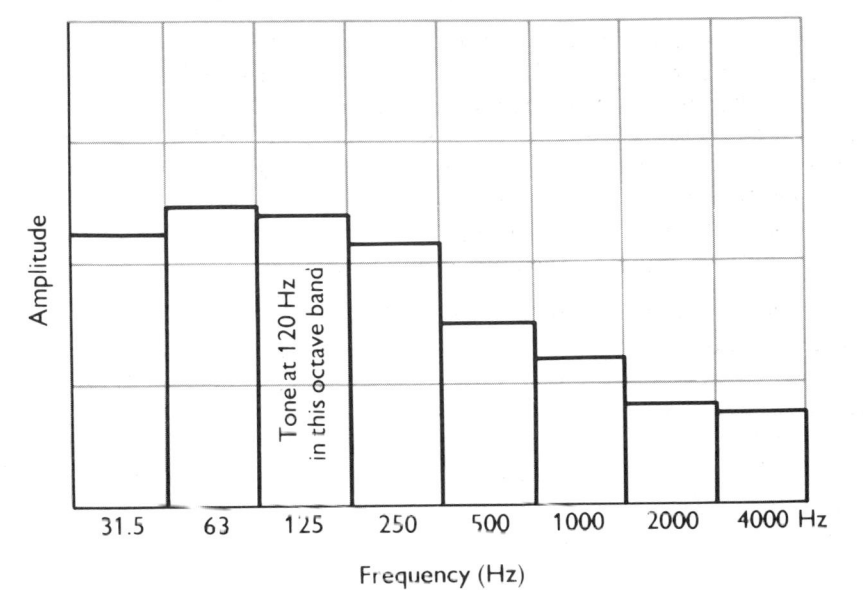

Figure 1.11b   Octave Band Spectrum of Noise with Tonal Component shown in
Figure 1.11a

1.2.1.4 Amplitude The second important characteristic of sound is amplitude. The ear
responds to sound pressure which is measured in units of Newtons/m^2 now known as
Pascals (Pa). Calculations often start with another quantity, sound power, which is
the rate of sound energy output of a source and can be measured in watts. Sound
power is an inherent property of a sound source whilst sound pressure is dependent on
the surroundings. This is illustrated by considering the analogy with an electric fire
shown in Figure 1.12. The heat 'amplitude' of an electric fire is rated in watts, but the
temperature which is experienced also depends on the distance from the fire and the
thermal insulation of the room. Similarly sound power in watts is the output from the
grille, whilst the sound pressure which is heard depends on factors such as distance
and sound absorption of the room.

Figure 1.13 lists the sound power outputs of several sound sources. The range of sound
powers is large from a whisper of 10^{-9} watts to a space age booster rocket at 10^8 watts.
Subjectively people respond to the ratio rather than to the absolute difference if the
sound power of a source changes. This is an example of a general characteristic of
human sensation described by Fechner, that the incremental increase in stimulus
required to produce a given small increase in sensation is proportional to the initial
stimulus. In relation to a sound source of varying power output an increase from 1/10

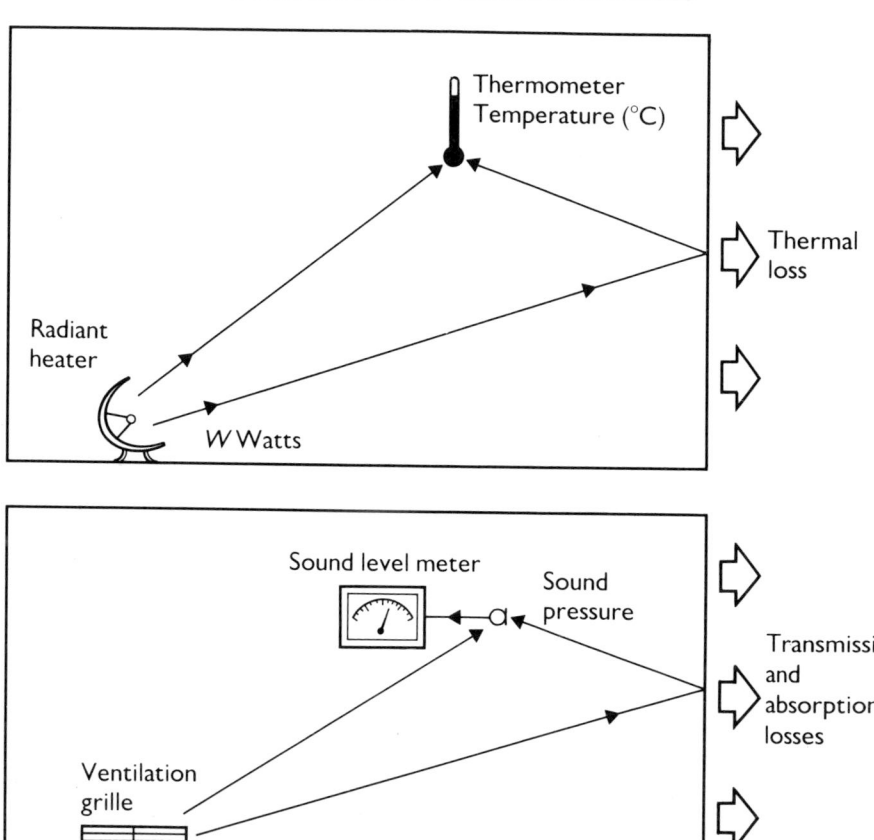

Figure 1.12   Thermal Analogy

to 1 watt sounds the same as from 10 to 100 watts — ten times in both cases. It is therefore convenient to use a logarithmic scale and sound power level (L_W) is defined:

$$L_W = 10 \log_{10} \left(\frac{W}{W_0} \right) \quad \text{decibels (dB)} \qquad\qquad 1.4$$

where W = sound power in watts,
    W_0 = reference sound power = 10^{-12} watts.
    dB = symbol for decibel.

Therefore a sound power of 1 watt has a sound power level of 120 dB. When quoting sound power levels the reference level should be stated, particularly as at one time it

Acoustic Power Watt	Sound Power Level dB re 10^{-12} w	Typical Sources
100 000 000	200	Saturn Booster Rocket
10 000 000		
1 000 000	180	
100 000		
10 000	160	Boeing 707–full power
1000		
100	140	75 Piece Orchestra
10		
1	120	Chain Saw
0.1		
0.01	100	
0.001		Average Motor Car
0.000 1	80	
0.000 01		Normal Voice
0.000 001	60	
0.000 000 1		
0.000 000 01	40	
0.000 000 001		Whisper
0.000 000 000 1	20	
0.000 000 000 01		
0.000 000 000 001	0	

Figure 1.13   Sound Power and Sound Power Level of Typical Sources

was practice in the USA to use 10^{-13} watts as the reference so the numbers are 10 dB higher. Figure 1.13 also lists the sound power levels of the sound sources in dB re 10^{-12} watts.

These are overall sound power levels, that is the total power at all frequencies. It is also possible to distribute this power into different frequency bands and so obtain a sound power level spectrum as shown in Figure 1.14. In this example the total sound power emitted in the 1 kHz octave band, at frequencies between 707 and 1414 Hz, is 67 dB re 10^{-12} W.

Sound power is an inherent property of a sound source but sound pressure is that which is heard and measured. The relationship between sound pressure and sound power in a simple propagating sound wave is analogous to a simple resistive electric circuit. The power dissipated in a resistor is equal to the voltage squared divided by the constant resistance. Similarly in a sound wave the sound power (W_s) is equal to the pressure (p) squared, divided by the acoustic resistance (R_a).

$$W_s = \frac{p^2}{R_a}$$ 1.5

The acoustic resistance is constant so sound power is proportional to pressure squared.

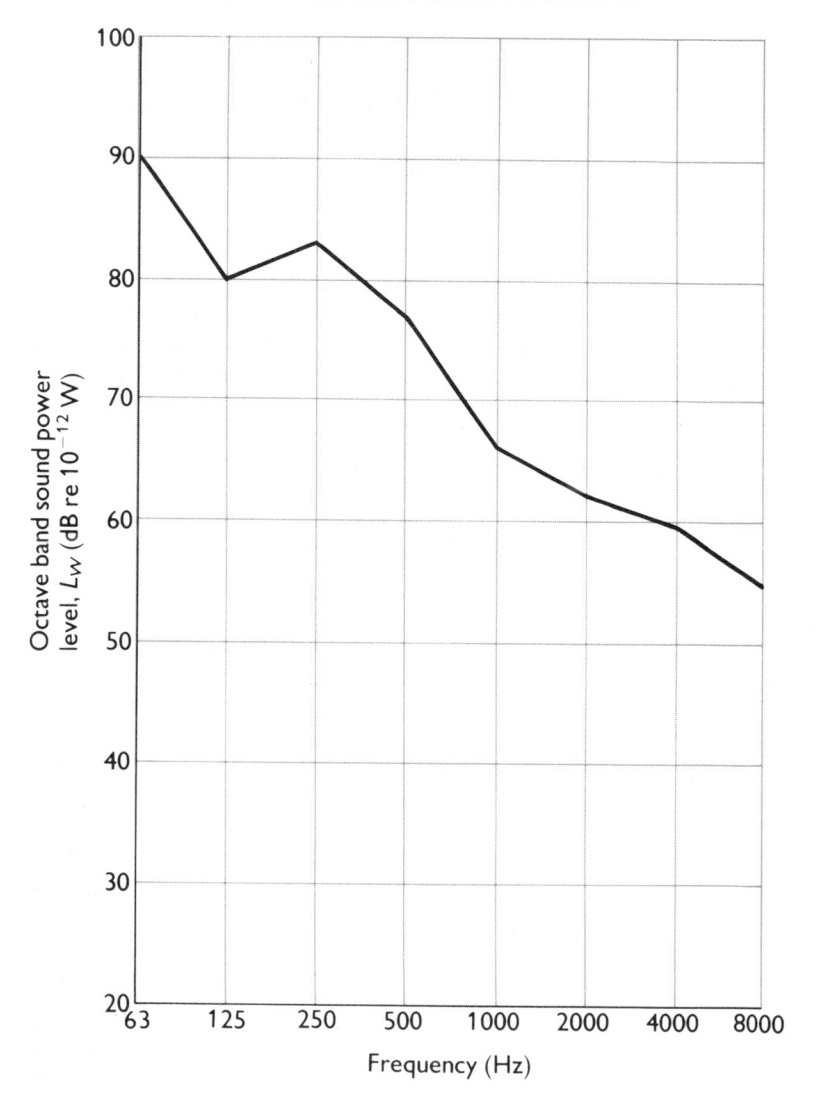

Figure 1.14   Sound Power Level Spectrum

Since human response to sound pressure follows Fechner's Law it is convenient to use a logarithmic unit similar to L_W.

Sound pressure level (L_p) is defined as:

$$L_p = 10 \log_{10} \left(\frac{p^2}{p_0^2} \right) \quad \text{decibels (dB)} \qquad\qquad 1.6$$

or

$$L_p = 20 \log_{10} \left(\frac{p}{p_0} \right) \text{ decibels (dB)} \qquad 1.7$$

where p is the sound pressure in Pa (N/m^2)
and p_0 is the reference pressure $= 2 \times 10^{-5}$Pa.
dB is the symbol for decibel

Because of the way these quantities are defined, although L_w decibels and L_p decibels are different, a given increase in L_w dB produces an equal increase in L_p dB. For instance if you quadruple power, L_w increases by $10 \log_{10} 4 = 6$ dB. As pressure is proportional to the square root of power you have only doubled the pressure so L_p increases by $20 \log_{10} 2 = 6$ dB again.

Figure 1.15 shows examples of typical L_ps of familiar sounds.

As with L_w, L_p can be applied to sound pressure in frequency bands to obtain a sound pressure level spectrum (L_p dB versus frequency).

Acoustic Pressure Pa	Sound Pressure level dB re 2×10^{-5} Pa	Typical examples
200	140	
		Racing Engine Test Cell
20	120	Jet Take Off at 50m
		Chain Saw Operator
2	100	Noisy Factory
		Inside Tube Train
0.2	80	Inside Sports Car
		Raised Voice at 1m
		Department Store
0.02	60	Normal Voice at 1m
		Open Plan Office
0.002	40	Private Office
		Residential Area at Night
		Whisper
0.0002	20	TV Studio
0.00002	0	Threshold of Hearing (Young People – 1 KHz)

Figure 1.15 Typical sound pressure levels

1.2.1.5 Intensity There is a third quantity, sound intensity (I), which is neither an inherent property of a source nor heard or measured, but is an important link between sound power and sound pressure. Sound intensity is the sound power passing normally through a unit area of space.

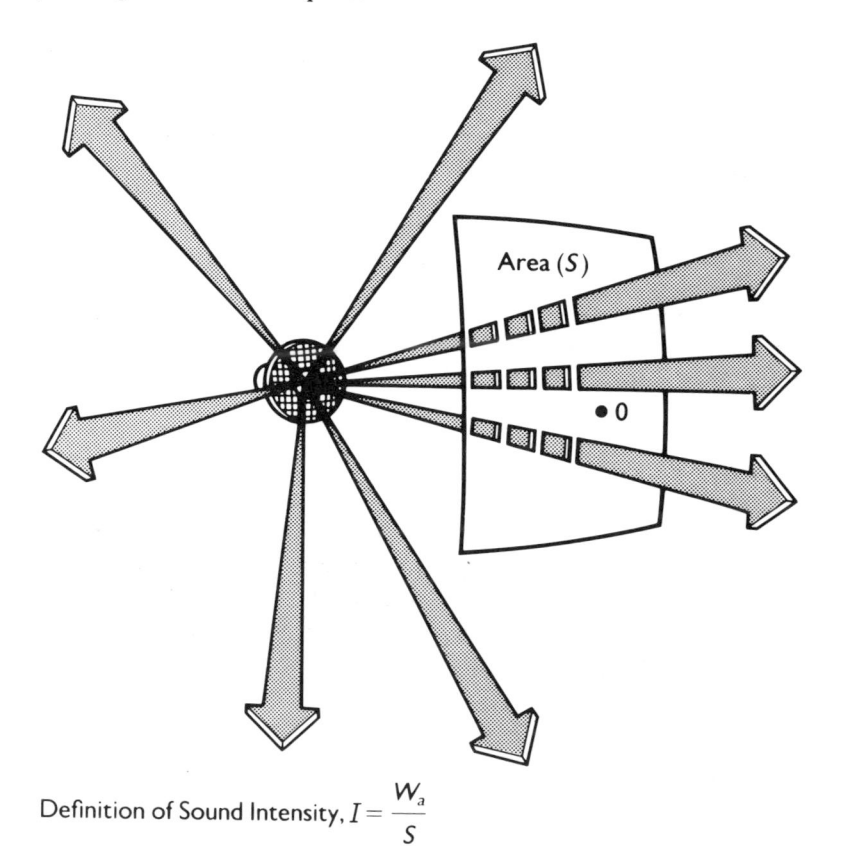

Definition of Sound Intensity, $I = \dfrac{W_a}{S}$

Figure 1.16   Definition of Sound Intensity

The average intensity, I, in the vicinity of the point 0 in Figure 1.16 is the total acoustic power, W_a, passing through the shaded area S divided by S.

$$I = \frac{W_a}{S} \qquad\qquad 1.8$$

Intensity level is defined against a reference as:

$$\text{Intensity level} = 10 \log_{10} \left(\frac{I}{I_0} \right) \text{ dB} \qquad\qquad 1.9$$

where $I_0 = 10^{-12}$ watts/m^2 is the reference sound intensity.

It can be shown that as a result of choosing this reference level the simple relationship between measurable sound pressure level, L_p, and intensity level, I, for simple plane waves becomes:

$$L_p = \text{Intensity Level} \qquad\qquad 1.10$$

so intensity levels dB are numerically equal to sound pressure levels dB that can be measured.

Further, substituting $W_a = IS$ from equation 1.8 into the definition of sound power level, L_w gives:

$$L_w = \text{Intensity Level} + 10 \log_{10} S \qquad\qquad 1.11$$

$$L_w = L_p + 10 \log_{10} S \ (S \text{ in square metres}) \qquad\qquad 1.12$$

This formula will be of value with duct work systems when an induct measurement of sound pressure level will enable the sound power level to be calculated, at least at low frequencies.

1.3 DECIBEL ARITHMETIC

Decibels, being a logarithmic scale, are handled differently from more usual units (e.g. watts, Pa). If you have two identical sources of known sound power level 100 dB, what is the sound power level of both sources together? Using the formula for sound power level "backwards" to convert to watts, these watts are added together and then converted back to sound power level.

Source A   100 dB re 10^{-12} watts = 0.01 watt
Source B   100 dB re 10^{-12} watts = 0.01 watt
A + B = 0.02 watt = 103 dB re 10^{-12} watts.

Given a third source

Source C   100 dB re 10^{-12} watts = 0.01 watt
A + B + C = 0.03 watt = 105 dB re 10^{-12} watts

For a number of sources of equal values of L_w the answer can be obtained by referring to Figure 1.17 which gives the increase in total L_w for a number of equal sources.

If the sources have different values of L_w, for instance two sources of L_w = 100 and 95 dB.

Source D   100 dB re 10^{-12} watts = 0.01 watt
Source E   95 dB re 10^{-12} watts = 0.003 watt
D + E = 0.013 watt = 101 dB re 10^{-12} watts

If an answer is required to only the nearest dB then the following rule of thumb may be used for sources taken two at a time.

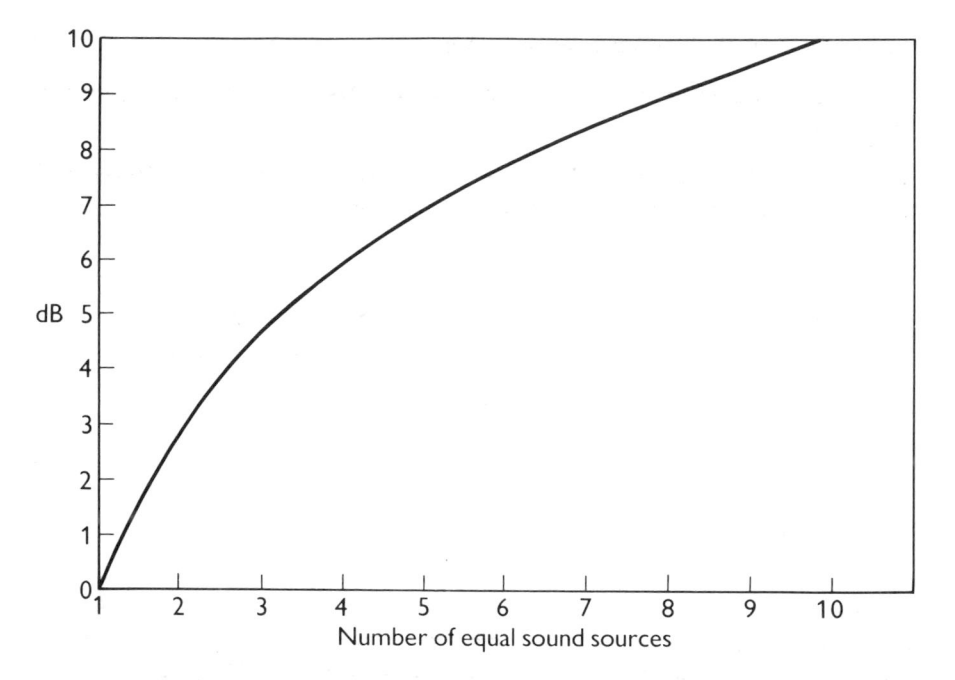

Figure 1.17   Summation of Sound Sources

1.3.1 Decibel Addition

If the levels to be added differ by dB	the total is equal to the larger level plus dB
0 or 1	3
2 or 3	2
4 to 9	1
10 or more	0

For example adding 84 dB and 87 dB the difference is 3 dB so 2 dB is added to the higher (87 + 2) giving 89 dB.

Sound pressure levels expressed in dB may be combined in an identical way. For instance the rule of thumb can be used to combine octave band sound pressure levels to obtain the overall sound level.

Sound pressure levels can also be subtracted. This would be needed where the sound pressure level produced by the operation of a machine was required but "background noise" from other machines which could not be stopped was affecting the measurements. If the machine or process being investigated can be stopped, the following procedure may be adopted:

(i) stop the machine and measure the background level

(ii) start the machine and measure the total sound pressure level of machine plus background levels

The difference between these two figures is used to obtain a factor to subtract from the total sound pressure level following the rule of thumb below.

1.3.2 Decibel Subtraction

If total and background levels differ by dB	the L_p of interest is equal to the total minus dB
10 or more	0
6 to 9	1
4 to 5	2
3	3
2	(4 or 5)

For example if the background level is 62 dB and when the machine is switched on it rises to 69 dB then the difference of 7 dB gives a correction of 1 dB. The sound pressure level due to the machine alone would be 69–1 = 68 dB. If the sound pressure level rises by 10 dB or more when the machine is switched on then the background noise is not affecting the reading whilst if the increase is less than 3 dB the background noise is too high for accurate (± 1 dB) measurement.

1.3.3 Decibels and Decibels — The Future

As you have now read and may have digested, the term "decibel" is employed in acoustics alone with two distinct meanings.

Firstly, it is employed for the measurement of sound pressure level where the reference level of two micropascals is employed and is, of course, related to a pressure unit.

Secondly, the self-same term, decibel, is used in connection with sound power level, but this time the reference level is one micro microwatt and is, of course, related to the power. The reader would be perfectly justified to conclude that this was somewhat clumsy and remains decidedly confusing, for whilst during acoustic training insistence is stressed that the reference level and, hence, reference type — pressure or power — is always stated, mistakes still occur. One solution currently being proposed is that realising the need for a logarithmic scale the BEL is employed solely for the measurement of sound power level. Hence, 72 decibels of power would become 7.2 BELS of power. The decimal will always be quoted, for example, 7.0 BELS, whereupon confusion is a great deal less likely, as the magnitude of the numbers will tend to be incomparable. Even 12.0 BELS of power will be quite rare.

An alternative proposal of several years back now, from America, was that we should do away with the logarithmic scale and quote the fluctuating pressure levels directly in pascals or micropascals, as the ability to print and absorb power factors of ten is now well accepted. Similarly, sound power levels would again be quoted directly in watts. The decibel arithmetic would then, of course, become unnecessary.

The author is slightly concerned about the return to direct units, but very much supports the idea of restricting the use of BELS for sound power levels.

1.4 MEASUREMENT OF SOUND PRESSURE LEVEL

1.4.1 Sound Level Meter

Sound pressure level is generally measured using a sound level meter, SLM, which essentially comprises a microphone, input and output amplifiers and a meter. The SLM shown in Plate 1 also has weighting networks, including A-weighting which is explained later in this chapter and a set of switchable octave band filters. Figure 1.18 is a block diagram of such a meter.

The microphone will be a condenser, electret or piezoelectric type, each of which converts the fluctuating sound pressure on the diaphragm to a varying electric charge. A windshield is used outdoors or where airflow might generate turbulence at the microphone surface causing spurious readings. A high impedance amplifier converts this to a voltage signal suitable for processing by electronics within the sound level meter. The output amplifier increases the signal level so that it is adequate to drive the meter needle. In some modern instruments the needle and scale have been replaced by a digital read out or LCD bar chart as shown on the meter in the Plate.

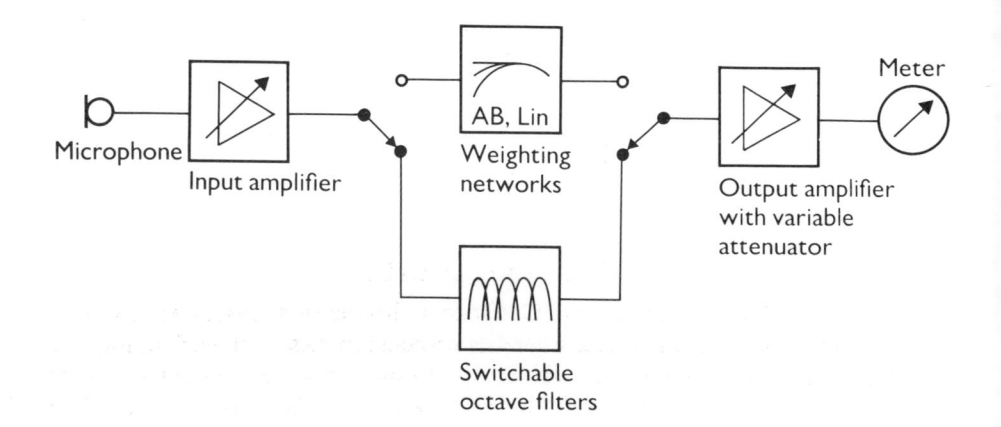

Figure 1.18   Sound Level Meter, Block Diagram

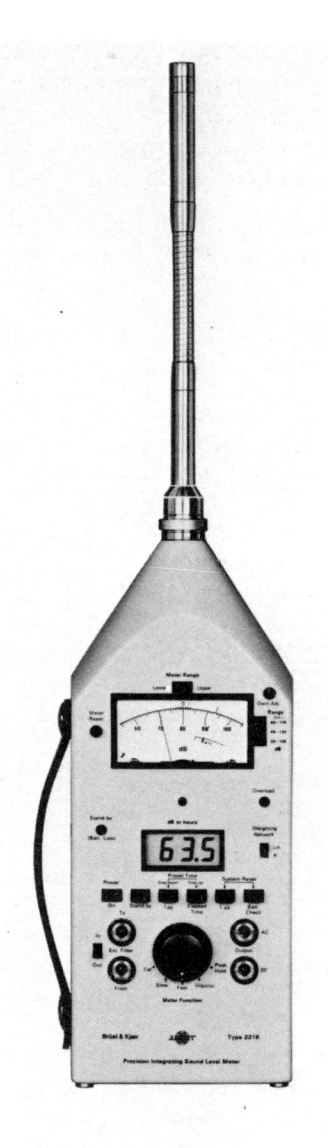

SOUND LEVEL METER

Here a basic sound level meter is illustrated with both genuine analogue and a digital readout device. The analogue meter with four response modes — slow, fast, impulse and maximum hold — allows the trained eye to interpret much about the noise sources, especially when assessing minimum representative background levels. The digital meter used in its time averaging mode yields a good averaging meter justifying its decimal accuracy. Filter sets may be added on the base of the meter.
Courtesy: Bruel & Kjaer

1.4.1.1 Range Setting We may wish to measure the ventilation noise in an empty recording studio or diesel engine noise in a generator room. A sound level meter must therefore be able to register levels below 20 dB whilst not being overloaded at over 100 dB. To achieve this a variable attenuator is built into the output amplifier and sometimes another into the input amplifier. These attenuators are variable in steps of 10 dB; sometimes multiples of 10 dB. Adjusting the range setting on the meter steps the attenuator(s). The meter needle then reads the difference in dB between the sound pressure level and the attenuator setting, usually displayed in a window on the meter. Figure 1.19 shows an example of 55 dB sound pressure level displayed on a meter with the attenuator set to 40 dB.

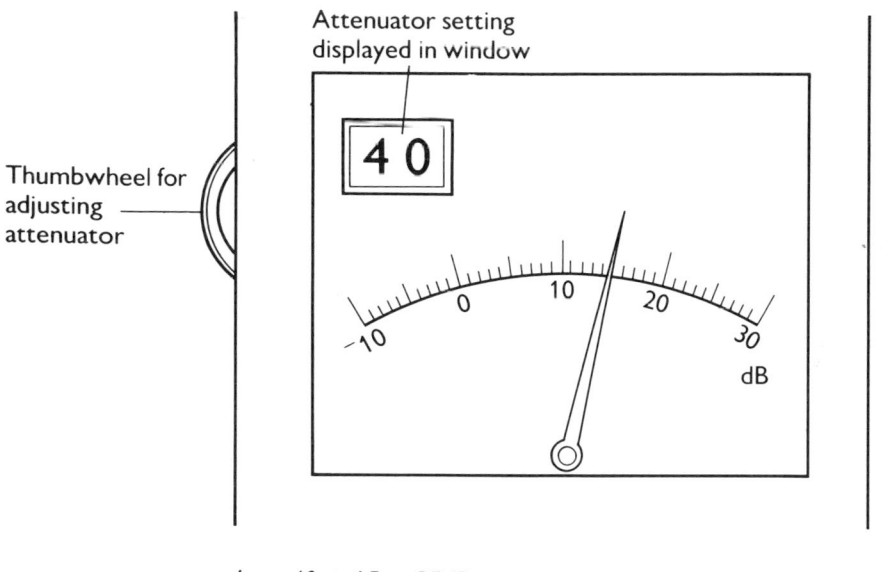

$$L_p = 40 + 15 = 55\text{dB}$$

Figure 1.19   Example of Sound Pressure Level Reading on Sound Level Meter

1.4.1.2 Detector Response Sound pressure oscillates between positive and negative pressure. The corresponding signal from the output amplifier fluctuates the same way; it is an AC voltage. We wish to average the signal without negative values cancelling positive ones. Therefore, a root-mean-square (RMS) detector is located between the output amplifier and meter. This is a rectifier which gives a DC output voltage proportional to the root-mean-square of the AC input voltage. The DC voltage drives the meter needle or other display. How quickly the meter responds to changes in sound pressure level depends on the damping of the needle and the averaging time of the RMS detector. Two standardised response rates, "fast" and "slow", are built into many meters. Either can be selected, "slow" response smoothing out small erratic fluctuations which would otherwise make the meter difficult to read. If impulsive noise, such as from hammer blows, are measured on "slow" response the meter will not respond quickly to the short duration peaks whilst on "fast" response it is difficult to read. A third response time known as "impulse" is available on some meters. This is intended for measurement of impulse sound, the meter rising more rapidly even than "fast" response but falling back slowly to facilitate reading the maximum level.

More comprehensive sound level meters may also have a peak level detector, so that the absolute peak value of the unrectified signal can be measured. As peak measurements are usually made of very short duration impulsive sounds, a peak level hold circuit is built into the detector.

1.4.1.3 Filters and Weighting To enable frequency analysis to be carried out, some sound level meters are provided with octave or 1/3rd octave band filters, either built into them or in separate detachable boxes. The filters are switched manually so that successive readings are taken at each frequency band. Also located between the input and output amplifiers may be weighting networks. These are wide band electronic filters which can be selected to attenuate some frequencies relative to others giving a "weighted sound pressure level". Most commonly used is "A" weighting, which is designed to produce the relative response of the human ear to sound at different frequencies. The "A" weighted sound level is therefore a measure of the subjective loudness of sound rather than physical amplitude (see chapter 2 for further discussion).

1.4.2 Sound Level Meter Grades

The tolerances allowed in measurements by different grades of sound level meter are defined in BS5969:1981. This has replaced BS4197 and BS3489 which defined precision and industrial grade meters. Grades are now defined from the most accurate type 0 laboratory meter to the type 3 "industrial" meter. In practice these are used as follows:

Type 0 — confined to specialist measurements in a laboratory.
Type 1 — slightly more accurate than old precision grade: used when accuracy is required such as during commissioning noise survey, certification of plant noise levels or investigation of noise nuisance complaints.
Type 2 — slightly less accurate than old precision grade: used for routine investigation of noise levels within buildings or from plant.
Type 3 — similar to old industrial grade: used for preliminary checks, as a teaching aid and comparative measurements in quality control. Meters conforming only to this Type 3 standard should not generally be used for measurements in the building services industry.

CHAPTER 2

NOISE AND PEOPLE

2.1 INTRODUCTION

Noise affects different people in different ways. Apart from the generally accepted pleasurable sounds, of which music is perhaps the most obvious, a lot of other sound, including that which is caused by mechanical services, is classed as noise. For this reason, noise is frequently described as "un-wanted sound".

Unwanted sound, however, covers a very wide range of noise, from sounds which are only vaguely disturbing to those which are dangerous to health. Whilst each individual can make a subjective judgement about the various noises which impinge upon him, the acoustician needs a method for making an objective measurement of these subjective effects.

Various indices have been developed to relate the subjective to the physically measurable, some of which are described in this chapter. First the ear, which is the mechanism between sound pressure and the sensation of hearing, will be considered.

2.2 THE EAR

An ear is divided anatomically into three parts (Figure 2.1) the outer ear, middle ear and the inner ear. The function of the outer and middle ear together is to collect sound waves and transform the fluctuating sound pressure into mechanical movement. The inner ear transduces the mechanical movement into electrical nerve impulses, which travel to the brain and register as the sensation of sound.

Sound waves collected by the outer ear travel up the ear canal and vibrate the eardrum at the end of the canal. Tiny bones known as "ossicles" connect the eardrum to the inner ear as a system of levers. Eardrum vibration is transmitted through the ossicles and into the cochlea of the inner ear. The cochlea is shaped like a snail shell and contains fluid around a spiral membrane (Basilar membrane) covered with hair cells. The vibrations transmit through the fluid and indirectly excite the hair cells, which then generate electrical impulses which are sent to the brain. Signals from hair cells along different parts of the membrane are sensed as sounds of different frequencies and different intensities.

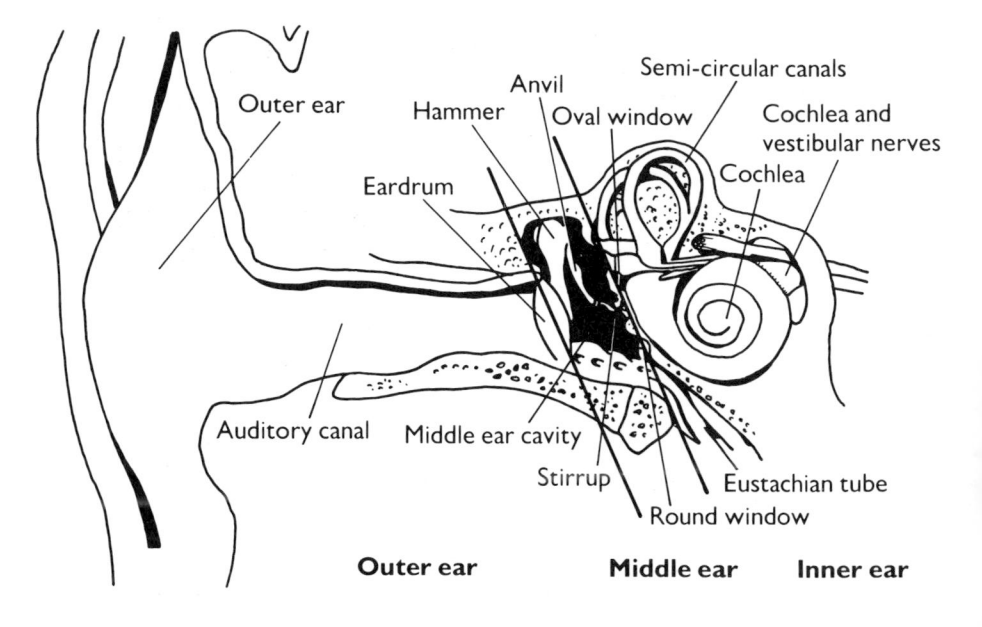

Figure 2.1   Anatomy of The Ear after Bilsom International Ltd.

The ear has an enormous range of sensitivity to sound pressure and can discern sound of 0 dB (2×10^{-5} Pa or 2×10^{-10} bar), whilst the threshold of pain is 120 dB (20 Pa or 2×10^{-4} bar).

The ear is not equally sensitive to all frequencies, 50 dB at 100 Hz does not sound as loud as 50 dB at 500 Hz. Overall sound pressure level in dB is therefore not a good measure of subjective loudness. Figure 2.2 shows equal loudness contours, which are a graphical representation of the relative sensitivity of the ear to tones at different frequencies. The unit of loudness is the phon, for example:

The 60 phon contour indicates the sound pressure levels in dB at each frequency which sound equally loud as 60 dB at 1 kHz. Below approximately 500 Hz the contours rise with decreasing frequency — that is the ear is progressively less sensitive with decreasing frequency. At lower amplitudes this rise is more steep. Above 1 kHz there is a dip down to 4 kHz, which is the frequency at which the ear is most sensitive. To simulate the ear's response, 3 weighting networks, "A", "B" and "C" (Figure 2.3) have been defined, which can be built into a sound level meter as electronic filters to "copy" the ear's frequency response. Originally it was intended that the A-weighting should be used for sound pressure levels up to 55 dB, B-weighting from 55–85 dB, and C-weighting at higher amplitudes.

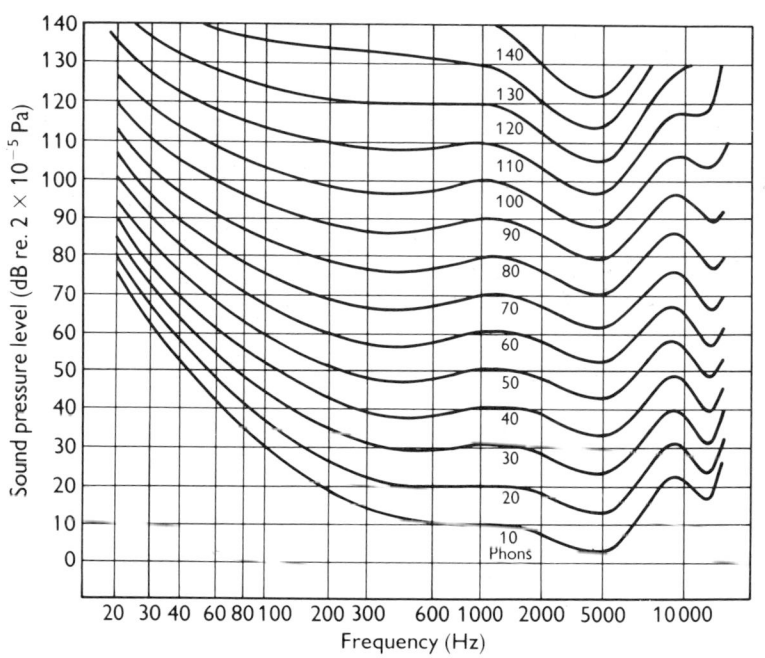

Figure 2.2   Equal Loudness Contours

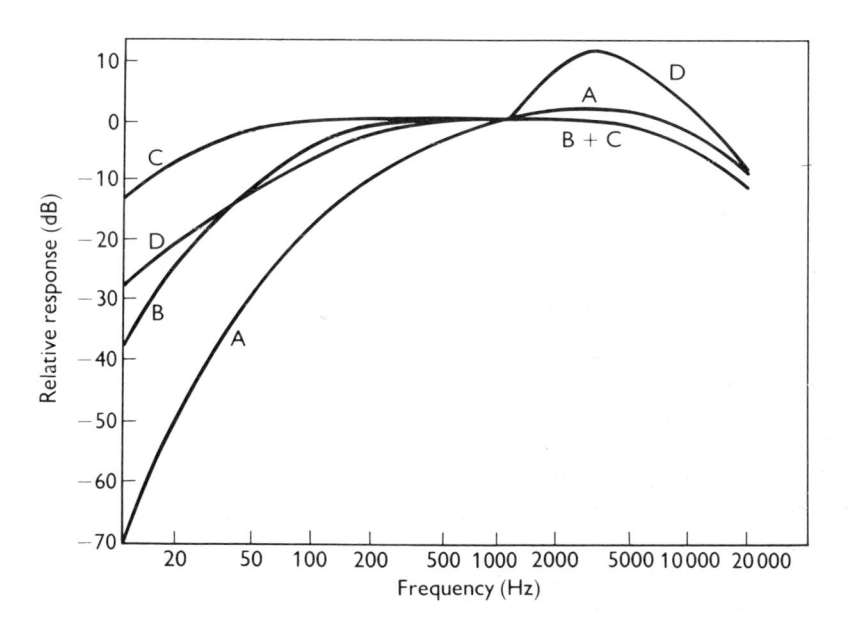

Figure 2.3   Weighting Networks

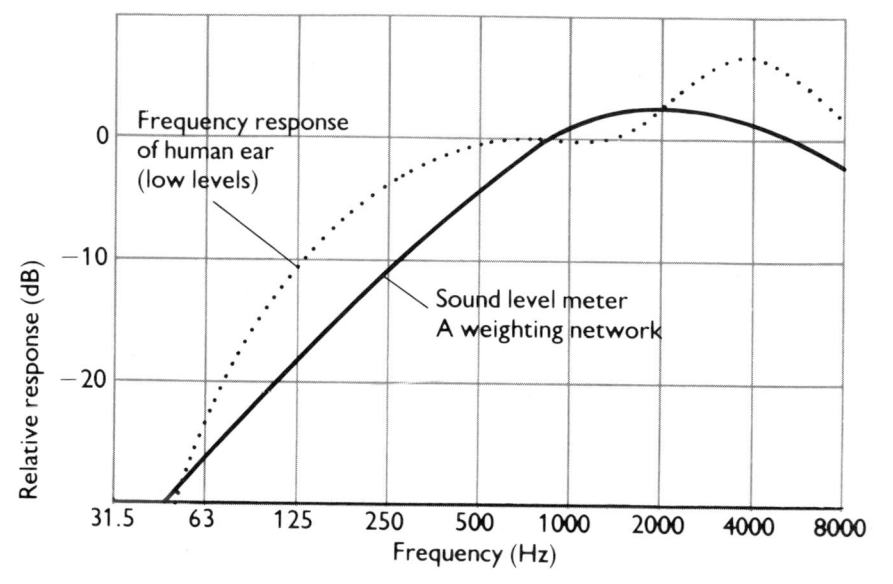

Figure 2.4   A-Weighting Network versus Inverted 40 phon Contour

D-weighting is a further network designed for aircraft noise measurements for which the extra sensitivity of the ear around 4 kHz is very important because of the predominant jet engine compressor whine. A-weighting is an electronic approximation to the inverse 40 phon contour (Figure 2.4) and in practice, it has become generally accepted for measurement of loudness independent of amplitude. B- and C-weighting have fallen into disuse. There are, however, instances where sounds are dominated by low frequency, for which A-weighted sound pressure levels in dBA underestimate the loudness and subjective annoyance.

2.3 FREQUENCY ANALYSIS

Whilst the dBA level is a convenient single number rating of loudness, for design of remedial work information on frequency content is required which is obtained by using filters such as the octave band filters mentioned earlier.

Figure 2.5 shows an example of octave band levels measured 1 m from the open intake of a fan, being combined to obtain the A-weighted level, that is, how loud the fan sounds at a distance of 1 m.

First, A-weighting correction factors (Figure 2.5) are subtracted at each frequency, (i.e. the values of the A-weighting curve at each frequency) to give A-weighted octave band sound pressure levels. These are then added in pairs using the rule of thumb (Chapter 1, page 18) to obtain the overall A-weighted level, in this case, 81 dBA.

	Octave band centre frequency						
	63	125	250	500	1000	2000	4000
Fan octave band L_p	85	86	85	80	73	70	66
A-weighting	−25	−16	−9	−3	0	+1	+1
A-weighted octave band L_p	60	70	76	77	73	71	67
	70		80		75		67
	80				76		
A-weighted L_p	81						

Figure 2.5   Calculation of A-weighted Sound Level

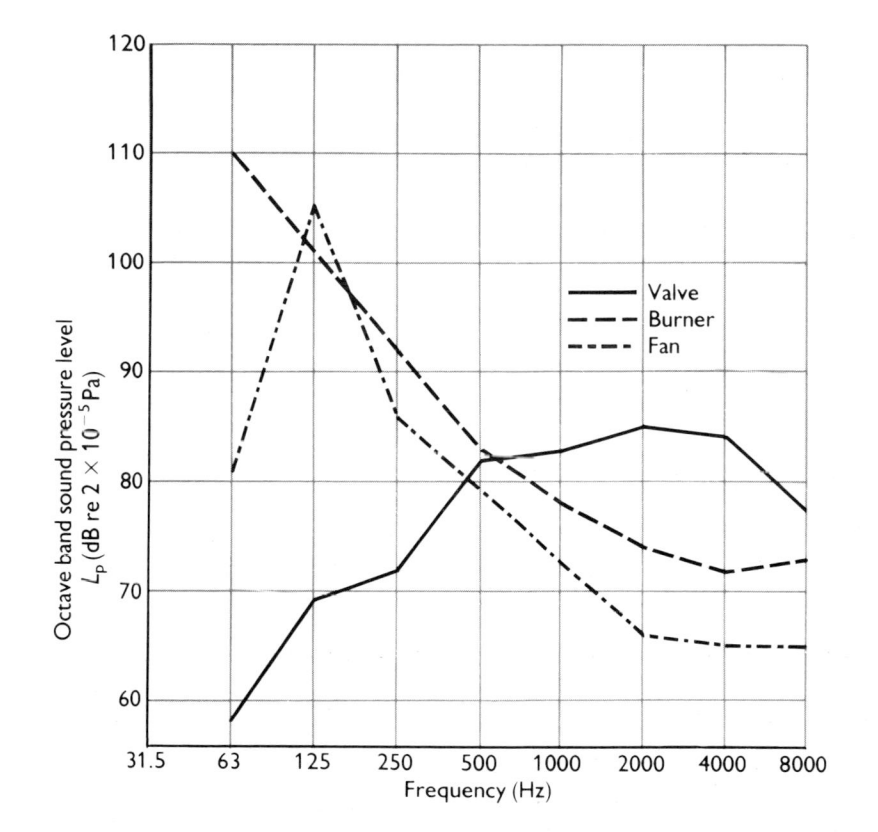

Figure 2.6   Spectra of 3 Noise Sources

However, consider the octave band spectra of the steam valve, furnace burner and paddle bladed fan shown in Figure 2.6. Valve noise is mainly high frequencies with a spectrum rising with frequency, whilst burner noise is dominated by low frequencies. The paddle bladed fan displays a peak in the 125 Hz octave band (not necessarily at 125 Hz exactly) due to a pure tone at the blade passing frequency. Which of these is the louder? This is assessed by calculating the overall A-weighted spectra, as shown in Figure 2.7. These are summed together to obtain 90 dBA for all three sources. Hence, they are said to be equally "loud", although the fan will probably be judged the most annoying because of the pure tone.

		Octave band centre frequency (Hz)							
	dBA	63	125	250	500	1000	2000	4000	8000
A-Weighting		−25	−16	−9	−3	0	+1	+1	−1
Burner dB		110	101	92	83	78	74	72	73
dBA	90	85	85	83	80	78	75	73	72
Valve dB		58	69	72	81	82	85	84	78
dBA	90	33	53	63	78	82	86	85	77
Fan dB		80	105	86	79	72	66	65	65
dBA	90	55	89	77	76	72	67	66	64

Figure 2.7 A-Weighted Level Calculation for the 3 Noise Sources in Figure 2.6

2.4 ACCEPTABILITY

Octave band sound pressure levels can be compared with a set of rating curves to assess the acceptability of noise. Figure 2.8 shows two octave band spectra plotted on a set of Noise Criteria (NC) curves. These curves were originally developed in the USA by L Beranek for use by the Heating and Ventilating industry. To obtain the NC value, the lowest value NC curve is found which is not exceeded by the spectrum. In this case the spectra can be said to meet NC55 and NC40. The NC value can be expressed more precisely in the second case, taking the greatest number of dB by which the next lower curve (NC35) is exceeded, in this case 4 dB in the 125 Hz band, and adding this. Hence the value becomes NC35 + 4.

Similar curves were adopted in Europe originally to assess community noise complaints. These Noise Rating (NR) curves, shown in Figure 2.9, rise a little more steeply towards low frequency and so are more tolerant of low frequency noise, and

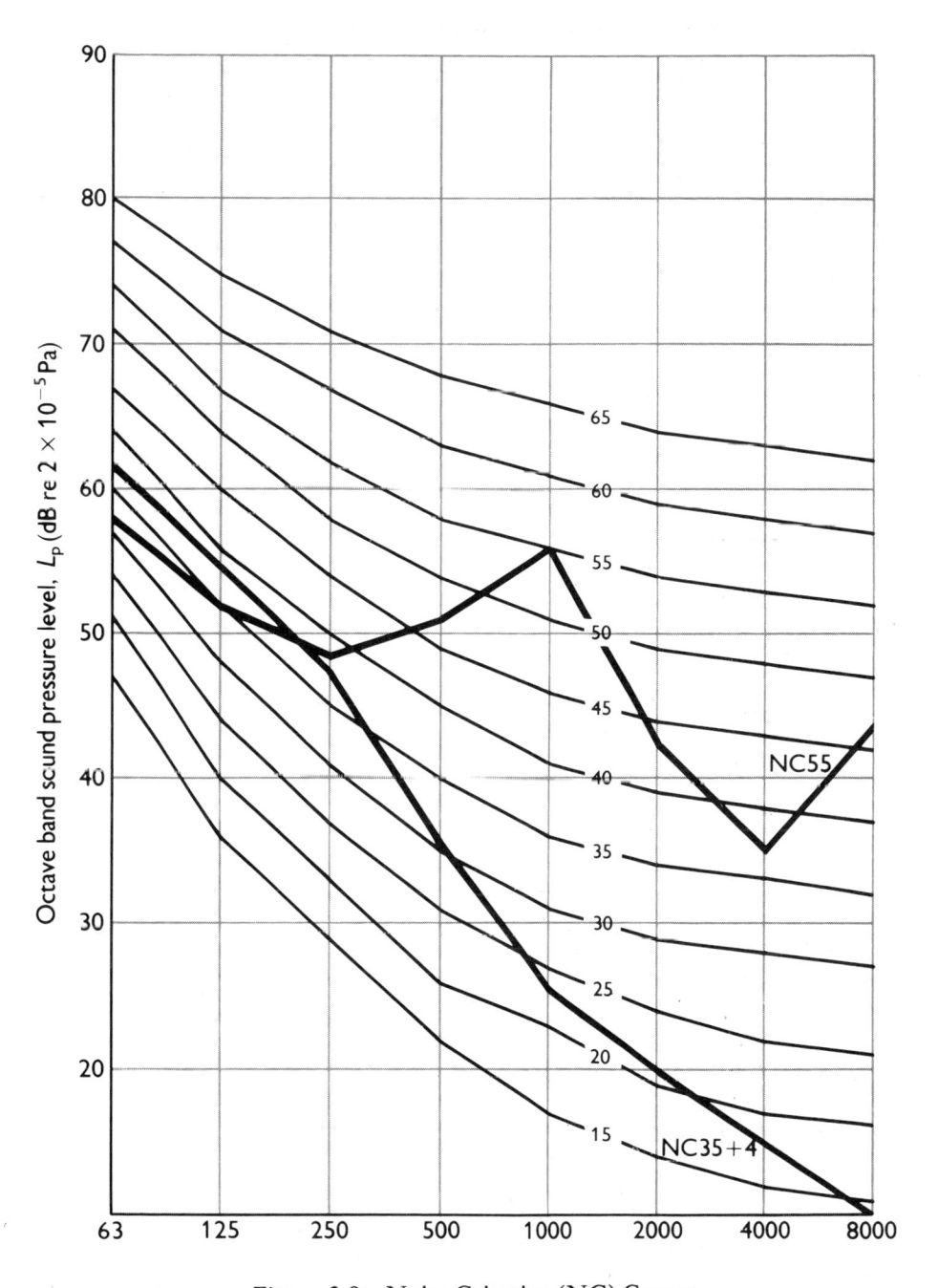

Figure 2.8   Noise Criterion (NC) Curves

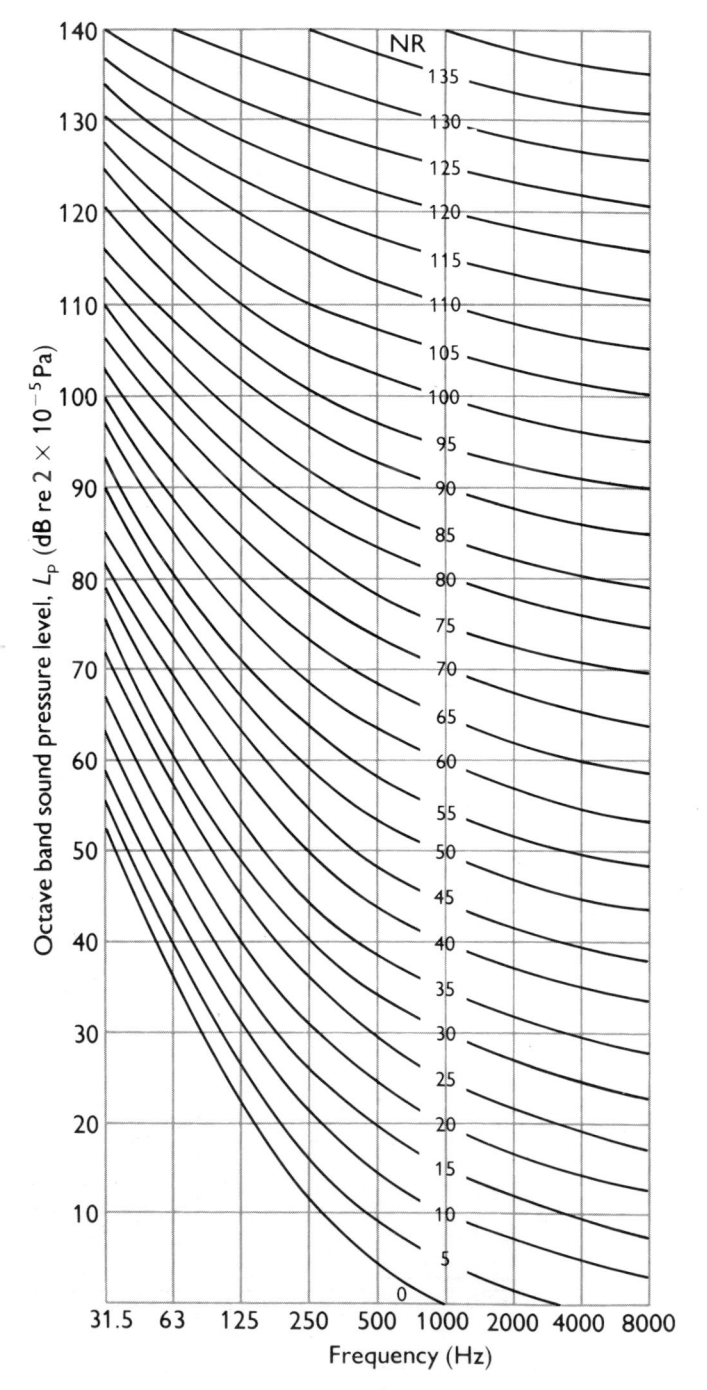

Figure 2.9   Noise Rating (NR) Curves

are slightly stricter at the higher frequencies. They are generated by a formula

$$L = a + bn \qquad 2.1$$

where L is the octave band sound level for NR level n,
and the constants a, b are frequency band dependent and given by

	31.5	63	125	250	500	1k	2k	4k	8k
			Octave Band Centre Frequency (Hz)						
a	55.4	35.4	22.0	12.0	4.8	0	−3.5	−6.1	−8.0
b	0.681	0.790	0.870	0.930	0.930	1	1.015	1.025	1.030

It should be noted that the frequency range of these NR curves is 31.5 to 8 kHz compared with 63 to 8 kHz for NC curves.

In response to criticism in the USA that in offices designed to NC curves the air-conditioning noise was too "rumbly" and "hissy", Beranek produced a modified set of Preferred Noise Criteria (PNC curves) Figure 2.10.

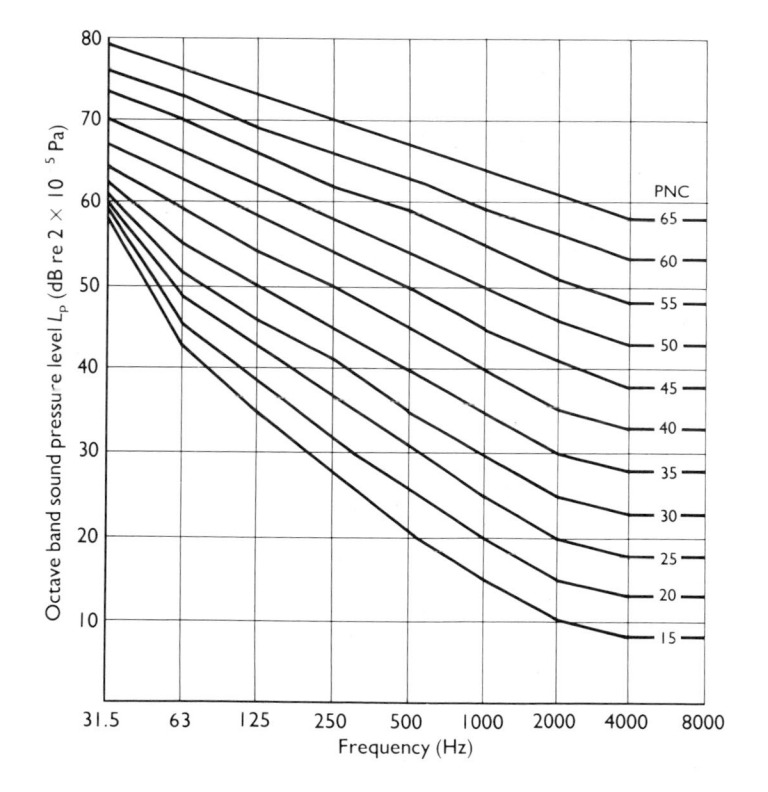

Figure 2.10   PNC Curves

These curves are effectively less tolerant of high and low frequency components when compared with equivalent NC curves. The frequency range of PNC curves was also extended down to 31.5 Hz.

In the recent ASHRAE publication (1980 Systems Volume) a new set of Room Criterion (RC) curves were published, as shown in Figure 2.11.

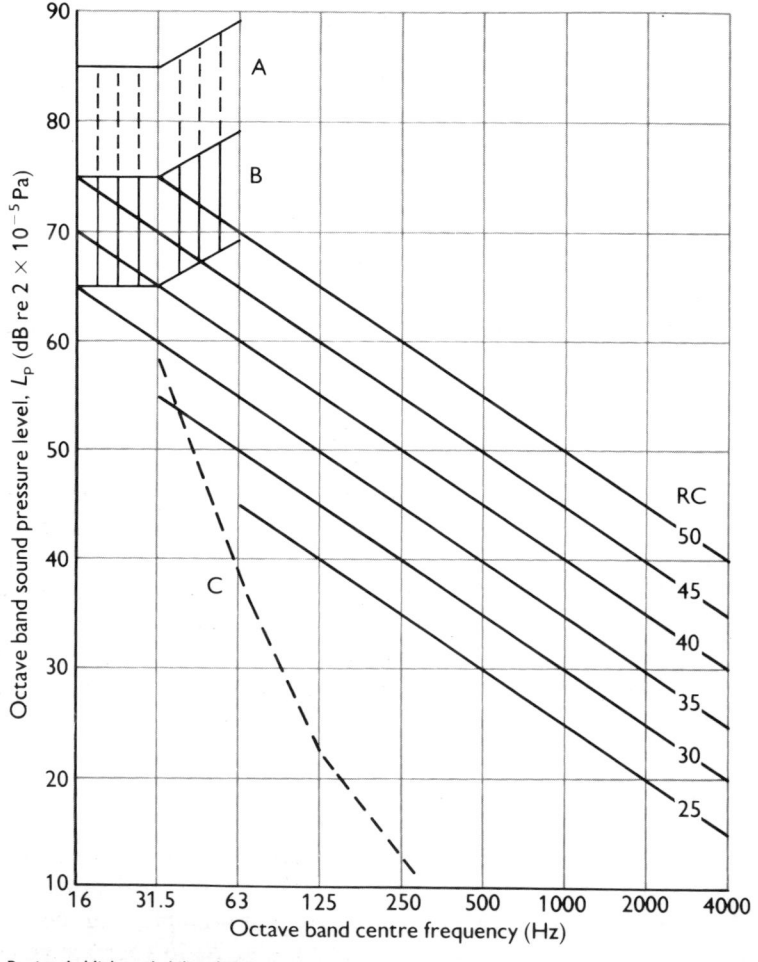

Region A: High probability that noise-induced vibration levels in lightweight wall and ceiling constructions will be clearly feelable; anticipate audible rattles in light fixtures, doors, windows, etc.
Region B: Noise-induced vibration levels in lightweight wall and ceiling constructions may be moderately feelable; slight possibility of rattles in light fixtures, doors, windows, etc.
Region C: Below threshold of hearing for continuous noise.

Figure 2.11   RC (Room Criterion) Curves for Specifying Design Level in Terms of a Balanced Spectrum Shape

As with PNC curves, their development was prompted by complaints that designing to NC curves often did not produce a well-balanced spectrum. Compared with an equivalent NC curve, an RC curve is more restrictive of the sound pressure level at low and high frequencies. This is to reduce the likelihood of rumbly or hissy systems. The frequency range of RC curves has been conditionally extended at low frequency range to 16 Hz, whilst at high frequency the limit reduced from 8 to 4 kHz. As a 16 Hz octave band filter is not generally available on sound level meters, it is recommended that an approximate level can be judged by measuring the overall (linear) sound pressure level on a Type 1 sound level meter and subtracting (using decibel arithmetic as Chapter 1, page 19) the 31.5 Hz octave band level. This would only work for a spectrum with dominant low frequency content, but in other cases the 16 Hz level would not be relevant.

There is no direct conversion from NC or NR (which measure acceptability) to dBA (which measures loudness). However, a rough rule of thumb is

$$dBA = NC/NR + 5 \qquad\qquad 2.3$$

but this varies considerably, depending on spectrum shape. The constant term could lie between 0 and +11.

2.4.1 Indices of Annoyance

Is the annoyance caused by fluctuating noise from road traffic, construction works and factories, or is the general noise climate adequately described by a single dBA, NC, NR or RC level? The answer is "no", mainly because not only do such noises vary in character, but vary considerably in level from minute to minute or hour to hour, unlike the steady building services noises so far covered. Indices have been developed, usually based on dBA levels, which also take into account these variations with time. Some important indices generally accepted for different noises are described below.

2.4.2 Road Traffic Noise

Annoyance caused by noise from road traffic has been shown to correlate best with the L_{A10} noise level. L_{A10} (previously annotated L_{10} dBA) is a statistical sound level, being the dBA level exceeded for 10% of a given time. For example, if the hourly L_{A10} is 70, then during that hour the noise level was greater than 70 dBA for 6 minutes and less than or equal to 70 dBA for the remaining 54 minutes. This is shown on Figure 2.12, which is a time history of varying traffic noise. L_{A10} corresponds with the higher levels, that is, the peaks in the traffic noise.

Similarly, other statistical levels can be defined: L_{A50}, the level exceeded for 50% of the time, which is an "average" level; L_{A90}, the level exceeded for 90% of the time, which corresponds to the "quieter" periods.

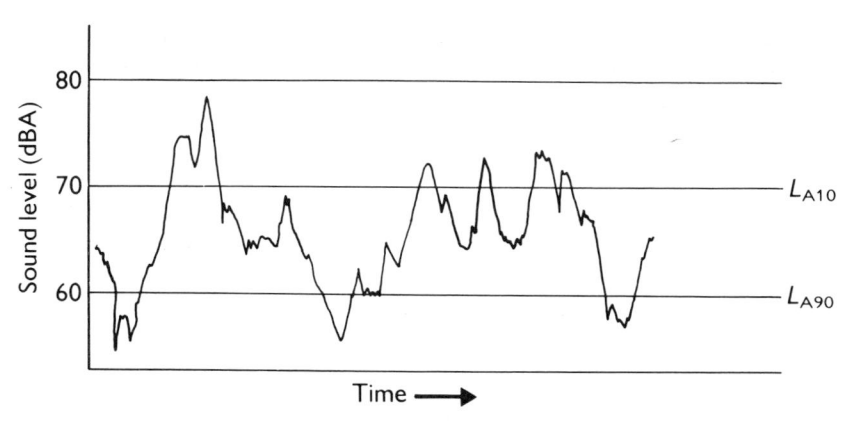

Figure 2.12   Statistical Sound Level

Methods for calculating L_{A10} levels from road traffic based on factors such as traffic flow, speed and distance from the carriageway are set out in the DoE/Welsh Office publication "Calculation of Road Traffic Noise".

2.4.3 Aircraft Noise

Much effort has been spent in developing indices for describing the annoyance due to aircraft. In the UK a complex unit called Perceived Noise (PN dB) is used, which is based on the D-weighted level mentioned earlier in the Chapter. For a jet aircraft, PNdB is approximately equal to dBA + 13. PN dB gives a measure of annoyance for a single aircraft, but close to airports the numbers and mix of aircraft must be taken into account. Noise and Number Index (NNI) was the peak PNdB from various overflights averaged to give average peak $\overline{PN\,dB}$. Then,

$$NNI = \overline{PN\,dB} + 15\log N - 80 \qquad\qquad 2.3$$

where N is the daily number of flights.

Around a busy airport NNI might be expected to vary from 35 a few miles out to 50 close to the boundary.

2.4.4 Rail Traffic

The intermittency of noise from trains is taken into account by using the equivalent continuous A-weighted sound level, L_{Aeq}. This is a form of energy averaging which determines the steady sound level over a period of time, which would have contained the same total sound energy as the noise under consideration for that same period of time. It is essential that the period considered is known and quoted. When a train passes the average will increase, and while no trains pass the average slowly falls. For example, if in a 10 minutes period two trains pass creating an L_{Aeq} 10 minute of 60, then if no trains pass for a further 10 minute, the L_{Aeq} 20 minute will have fallen to 57. "A Guide to Measurement and Prediction of the Equivalent Continuous Sound Level L_{eq}" by the Noise Advisory Council is published by HMSO.

2.4.5 Construction Noise

The variations of noise from construction operations are also taken into account using L_{Aeq}. Generally, the L_{Aeq} is taken over a 12-hour working day, but other times could be taken for night-time or weekend working. Data for calculating L_{Aeq} 12-hour for construction operations are given in BS5228:1985.

Chapter 12 has been dedicated to construction site noise and vibration.

2.4.6 Industrial/Mechanical Services Noise — Noise Nuisance

The method for assessing the annoyance likely from mechanical plant and industrial operations is set out in BS4142:1967. Further explanation of the method is given in Chapter 4. The sound level (dBA) caused by the industrial premises at the nearest dwelling is corrected for its character, intermittency and duration to give the Corrected Noise Level (CNL). This CNL is then compared with the background noise level existing without noise from the premises. If the CNL is 10 dBA or more above the background, then the standard says that complaints are likely.

2.4.7 Background Noise Level — Noise Climate

Background noise level as defined by BS4142 is the statistical sound level L_{A90}, the level exceeded for 90% of the time. It should be noted that the British Standard approach is at variance with the ISO Standard 1996, which uses L_{Aeq} rather than CNL and L_{A95} rather than L_{A90}. Further, the proposed revised ISO Standard would replace statistical levels as a measure of the noise climate or background. The L_{Aeq} with the noise under investigation present, would be compared with the "residual" L_{Aeq} when that noise was not present. It is likely that a revised standard will soon replace BS4142 in 1988 and this will recommend the comparison of disturbing noise in terms of L_{Aeq} with background noise in terms of L_{A90}.

2.5 NOISE INDUCED HEARING DAMAGE

The quietest sound that the ear can detect is named the "Threshold of Hearing". When the ear is exposed for a length of time to noise levels above 70–75 dBA, the sense of hearing becomes less acute and this threshold is temporarily raised. This loss of hearing sensitivity is regained after a period in a quiet environment. It is therefore referred to as temporary threshold shift (TTS). Figure 2.13 shows TTS at one frequency after exposure to sound of varying amplitude and duration.

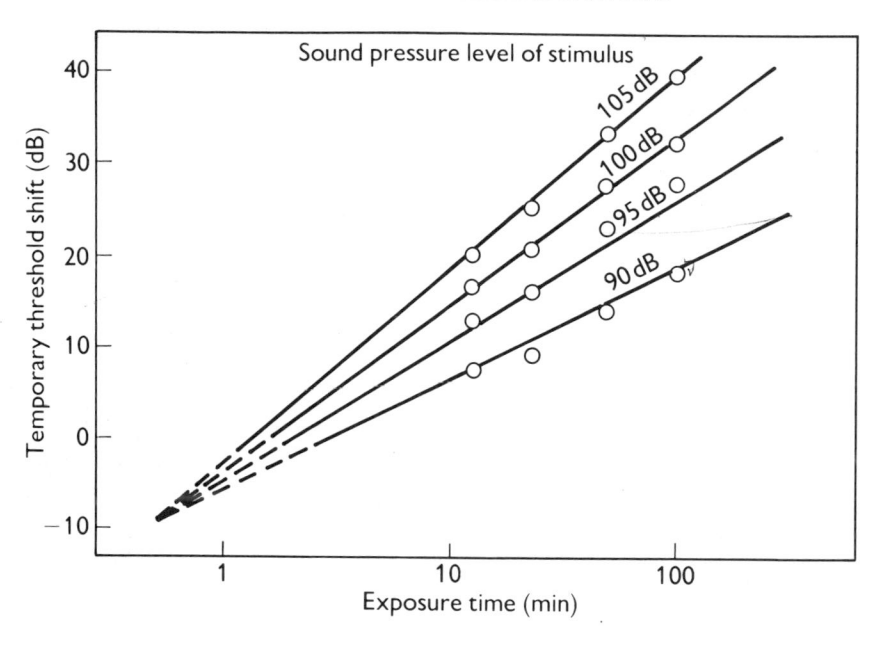

Figure 2.13   Temporary threshold shift (TTS) at 4000 Hz 2 minutes after the end of exposure to a band of noise of 1200–2400 Hz at the sound pressure levels and durations indicated. Points are observed data lines from the equation developed by Ward (From Ward, Glorig & Sklar 1959a).

Figure 2.14 shows the time taken to recover from Temporary Threshold Shift (TTS) for 5 different people.

For moderate levels of TTS less than 20 dB, recovery time is relatively short, minutes rather than hours. As the level of TTS increases, the recovery time increases very rapidly. If TTS is repeatedly suffered, for instance each day at work, and particularly if recovery is not complete, then the ear's sensitivity is never fully recovered. In fact, hair cells in the inner ear die and the threshold shift is permanent — PTS, Permanent Threshold Shift. PTS generally starts at frequencies around 4 kHz. With years of exposure the loss at this frequency steadily increases and the loss spreads to other frequencies. This is shown in Figure 2.15 showing PTS as a function of frequency for increasing years of noise exposure and represents partial deafness.

The index used in the UK for assessing the risk of hearing damage is L_{Aeq} 8 hour, that is the L_{Aeq} averaged over an 8 hour working day.

At the time of writing there is no legal limit on the noise exposure which employees in industry in general are allowed to suffer. There is a duty under common law to provide a safe workplace. The Health and Safety at Work Etc Act 1974 places duty on an employer to "ensure as far as is reasonably practicable the health, safety and

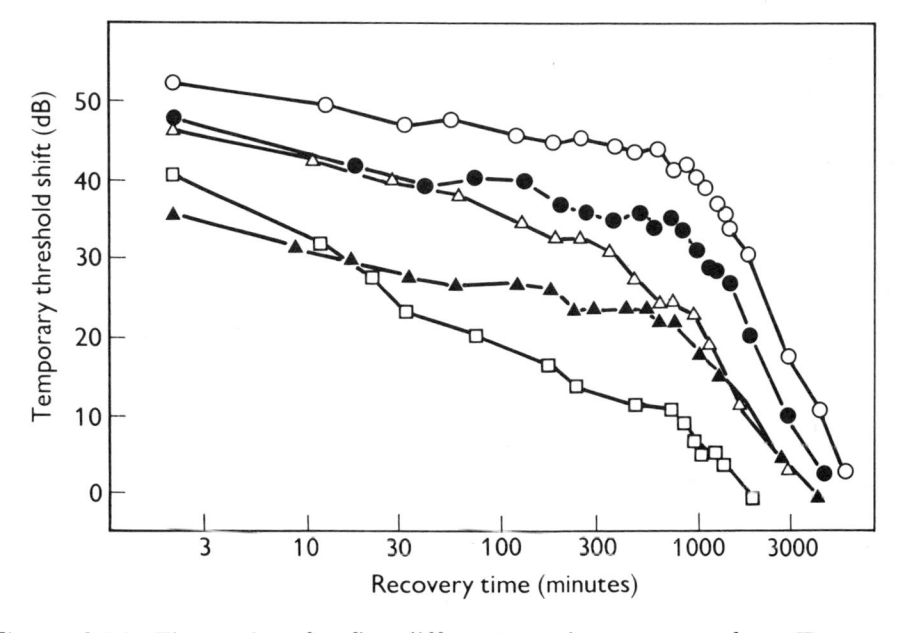

Figure 2.14 Time taken for five different people to recover from Temporary Threshold Shift (TTS).

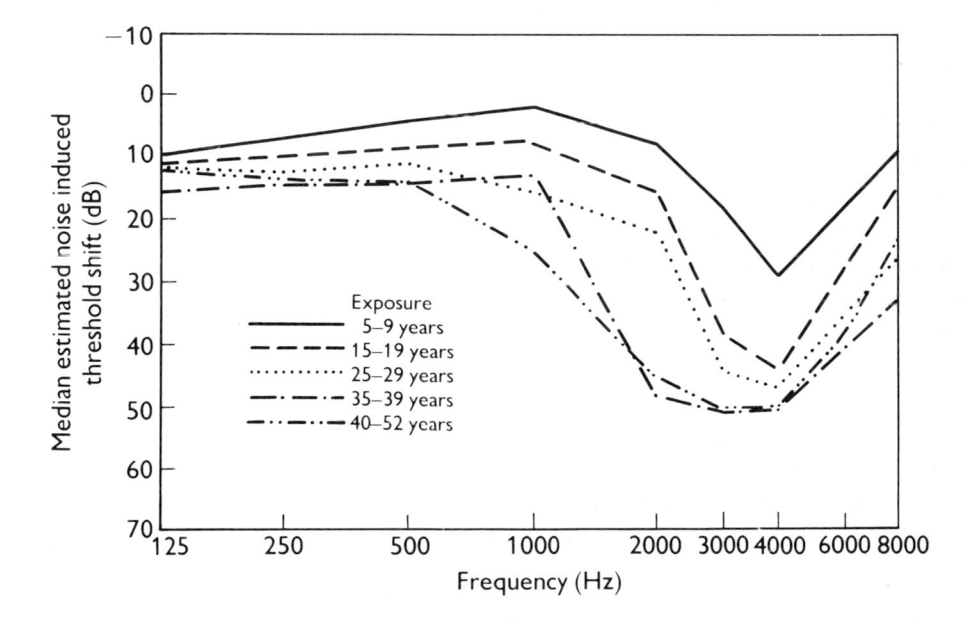

Figure 2.15 Permanent Threshold Shift (PTS) with Increasing Years of Exposure. After Burns.

welfare at work of all his employees" although noise is not mentioned specifically. It is intended that Regulations relating to noise be made under the Act, but apart from Regulations for the Woodworking Industry, these are not yet in force. With respect to mechanical services noise, such Regulations will be relevant to people working in plantrooms or in some industrial work areas where there may be, for example, very noisy air moving systems for drying or moving material.

At present, claims for compensation for hearing damage and the decisions of the Factory Inspectorate are based on the existing Code of Practice (CoP) published in 1972 by the Department of Trade and Industry — "Code of Practice for Reducing the Exposure of Employed Persons to Noise". This recommends a maximum exposure of the unprotected ear of 90 L_{Aeq} for an 8 hour day. Exposure to higher sound levels are allowable for shorter times.

For example, 90 L_{Aeq} 8 hour is equivalent to:

90 L_{Aeq} for 8 hours
93 L_{Aeq} for 4 hours
96 L_{Aeq} for 2 hours
99 L_{Aeq} for 1 hour

The CoP gives a calculation method for determining the L_{Aeq} 8 hour level for a combination of different noise levels and exposure times. This calculation can be done by adding the factor $10 \log_{10} T/8$, where T is the duration in hours of exposure to each noise exposure level. The results are 8 hour L_{Aeq} levels, which can then be combined using decibel arithmetic (Chapter 1, page 17). For example, a machine operator is exposed to

96 L_{Aeq} for 1/2 hour
93 L_{Aeq} for 2 hours
88 L_{Aeq} for 5 hours

His overall exposure is
$$96 + 10 \log_{10} 1/16 = 96 - 12 = 84$$
$$93 + 10 \log_{10} 1/4 \ = 93 - 6 \ = 87$$
$$88 + 10 \log_{10} 5/8 \ = 88 - 2 \ = 86$$
$\left. \right\}$ 91 L_{Aeq} 8 hour

Work by the Health and Safety Commission on proposed noise Regulations under the Health and Safety At Work, etc. Act was set aside, pending outcome of discussions on a Proposal of the European Commission. A draft EEC directive has been published and amended after representations from member states, particularly the UK. The amended proposal directive is likely to form the basis of future Regulations in the UK. This will set a limit insofar as is reasonably practicable of 90 L_{Aeq} 8 hour, with an aim to reduce this to 85 L_{Aeq} 8 hour at a future date when economic circumstances are more favourable. It will be allowable to average the daily noise exposure over a working week. The proposal also calls for regular audiometric screening. Where it is not practicable to reduce employees noise exposure to within the limit, the employees must wear hearing protection.

CHAPTER 3

SOUND IN ROOMS

3.1 INTRODUCTION

To enable calculations of the noise control required in mechanical services to be made, it is necessary to understand the behaviour of "sound in rooms". The main aim of applying noise control is to meet predetermined acoustic standards, the acoustic design criteria, in the various rooms throughout the building. To do this it is necessary to take account of the effect of the rooms on the sound power levels produced and the listening conditions of the occupants. If the same sound power is fed into two different rooms of equal volume, one having a high degree of acoustic absorption, the other being less absorptive, the resulting mean sound pressure levels in these rooms will differ. Furthermore the distribution of sound around the rooms and the hearing conditions within the rooms will be different. In most problems of noise transfer it is important to know the effects of room acoustics so that calculations of the sound pressure level in the room containing a noise source e.g. a plant room or an office, can be made in order that the sound energy that will pass to other areas as air-borne, structure-borne or duct-borne noise can be determined. This chapter aims to provide an understanding of the behaviour of sound in rooms, the terms involved, and the principles behind the calculation procedures used to determine sound levels in rooms.

3.2 SOUND PROPAGATION

3.2.1 Inverse Square Law

The most simple sound source is a point source, where the total sound energy is conveniently considered as concentrated at a mathematical point. If a point source is suspended in mid-air and radiates sound energy in all directions equally, the sound will spread out spherically and the intensity at any location will depend on the surface area of the sphere at that location. This is shown in Figure 3.1. As simple geometry states that the surface area of a sphere equals $4\pi r^2$, where r is the radius of the sphere, it can be seen that the sound intensity (energy per unit area) received at a certain location will be inversely proportional to the square of the distance, from the source.

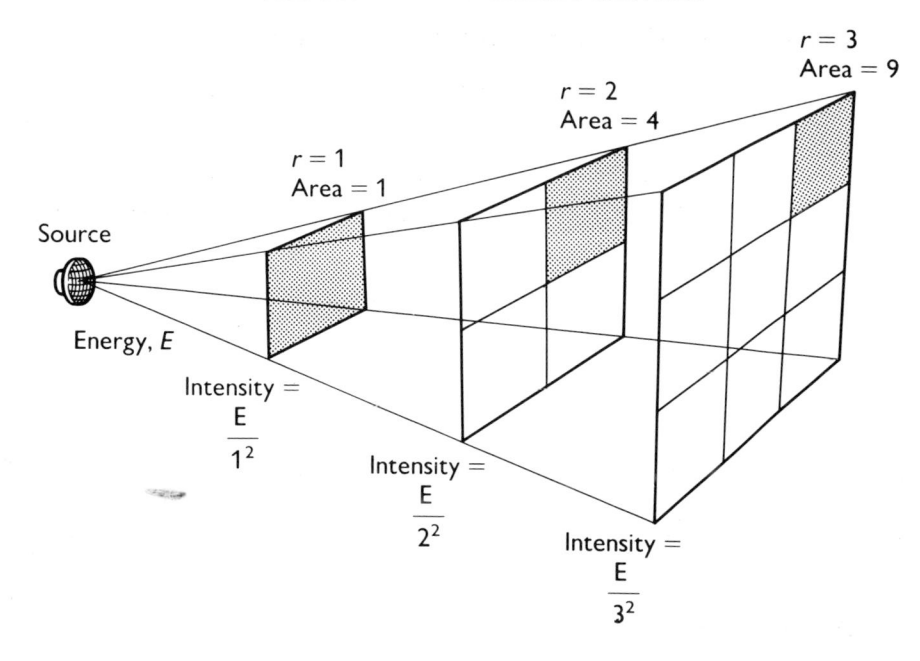

Figure 3.1 Spherical Radiation from a Point Source

This relationship is known as the Inverse Square Law and to consider this in terms of decibels it is necessary to use logarithms. Thus:

$$\text{Intensity } I_1 \text{ at } r_1 \propto \frac{1}{4\pi r_1^{\,2}} \qquad\qquad 3.1$$

$$\text{Intensity } I_2 \text{ at } r_2 \propto \frac{1}{4\pi r_2^{\,2}} \qquad\qquad 3.2$$

Thus:

$$I_1 = \frac{K}{r_1^{\,2}} \qquad\qquad 3.3$$

$$I_2 = \frac{K}{r_2^{\,2}} \qquad\qquad 3.4$$

Sound intensity is related to sound pressure in Chapter 1 equation 1.10:

$$P_1 = cI_1 \qquad\qquad 3.5$$

$$P_2 = cI_2 \qquad\qquad 3.6$$

where c is the speed of sound.

Taking logarithms to base 10 to obtain sound pressure levels, the difference in sound pressure level between r_1 and r_2 can be written as:

$$\Delta L = 10\log_{10}P_1 - 10\log_{10}P_2 \qquad\qquad 3.7$$

$$= 10\log_{10}\left(\frac{P_1}{P_2}\right) \qquad\qquad 3.8$$

$$= 10\log_{10}\left(\frac{I_1}{I_2}\right) \qquad\qquad 3.9$$

$$\text{Sound Pressure Level decrease} = 10\log_{10}\left(\frac{r_1^2}{r_2^2}\right) \qquad\qquad 3.10$$

Thus for a doubling of distance:

$$r_2 = 2r_1$$

Therefore

$$\text{Sound Pressure Level decrease} = 10\log_{10}\left(\frac{2^2}{1^2}\right)$$

$$= 10\log_{10}(4)$$

$$= 6\,\text{dB} \qquad\qquad 3.11$$

Thus if the sound pressure level from a point source radiating omnidirectionally at 5 metres is 85 dB, at 10 metres the sound pressure level will be 79 dB, at 20 metres it will be 73 dB, at 40 metres 67 dB, and so on. Theoretically this applies to all the octave bands, linear sound levels and dBA. However as sound is absorbed in air, especially at higher frequencies the true sound reduction can exceed this.

In practice few noise sources are so small. The complex movement of the surface of larger sources produces a zone close to the source known as the "Near Field" in which the variation in level can be complex. Fortunately this only occurs for a comparatively short distance. Once the source size becomes small compared with distance, the source can increasingly be likened to a point and the Inverse Square Law again becomes applicable.

Noise sources are often divided into three convenient and explicit categories — point, line and plane. A point source can be likened to a small piece of equipment such as a pump, a grille or a fire bell viewed from a relatively short distance. A duct or pipe passing through a room, or a linear diffuser are typical line sources whilst a plantroom wall, a door or the side of a large air handling unit are plane sources. Figure 3.2 shows the reduction in sound level from these three types of source. In each case the sound power of the source is the same. As can be seen the sound pressure at a large distance at all locations is the same irrespective of the source. Variations, however, occur near to the source. Whereas the sound pressure level decreases at 6 dB per doubling of distance at all locations relative to a point source, the sound level decreases by only 3 dB per doubling of distance away from a line source, for a distance equal to the length of the line source divided by π, that is a/π. Close to a plane source the sound level remains constant for a distance equal to the shorter dimension of the plane source divided by π, that is b/π. After that for a distance up to the longer dimension of the source divided by π, that is c/π, the sound level falls off by 3 dB per doubling of distance. In this region the plane source is acting like a line source. Beyond this distance the sound level again falls off by 6 dB per doubling of distance and is, therefore, similar to a point source.

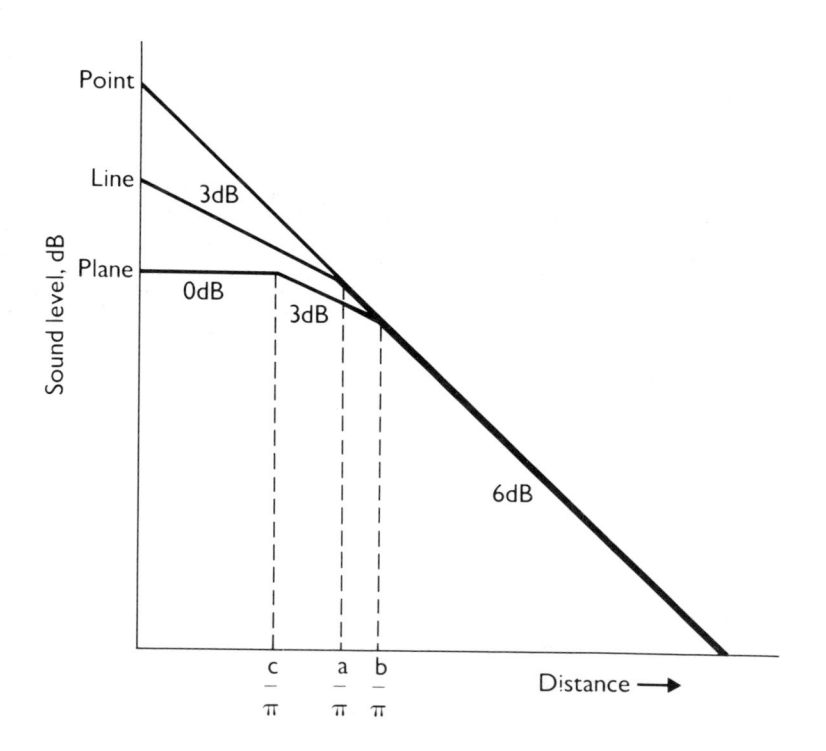

Figure 3.2   Attenuation with Distance from Various Sources
The slope represents 0, 3 and 6 dB per doubling of distance.

POINT SOURCE
A striking example provided by an alarm bell.

PLANE SOURCE

Large plantroom louvres acting as a source of substantial cross-section.

3.2.2 Reflection of Sound

This description of sound propagation assumes that the sound source is located in free field conditions and that there are no surfaces obstructing the free transmission of the sound waves. Once a sound source is located in a room, the energy leaving a sound source travels only a short distance before it strikes a room boundary. When sound energy strikes a surface some is reflected back into the room, some is absorbed within the material and some is transmitted to the other side. This is shown in Figure 3.3.

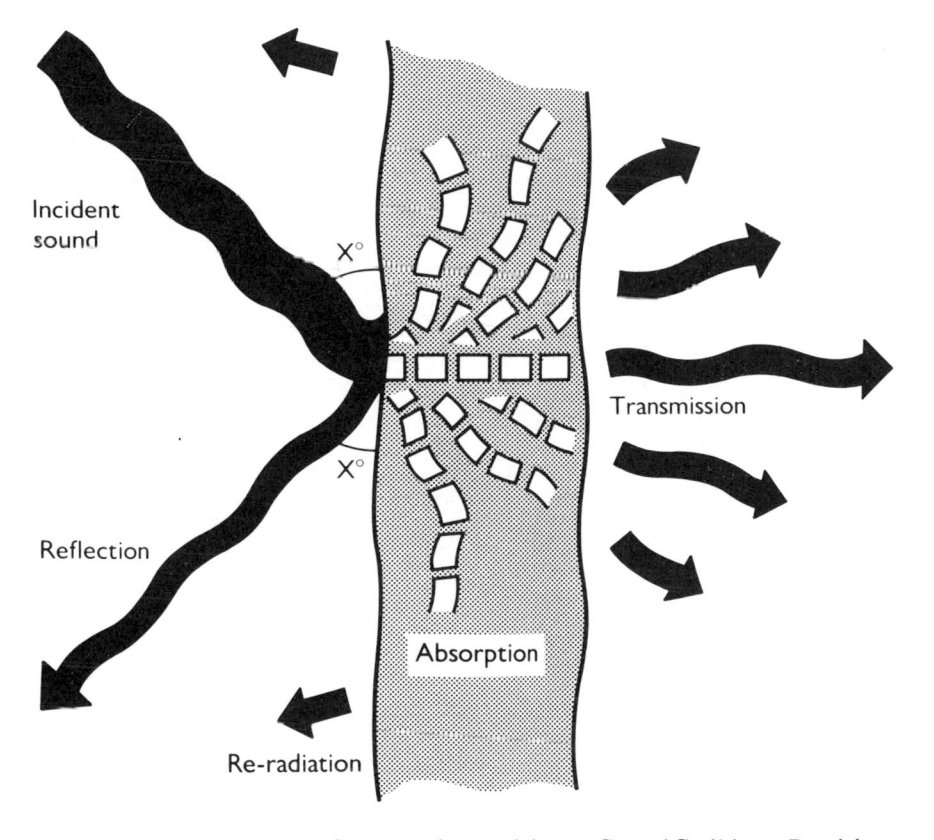

Figure 3.3 Distribution of Energy from Airborne Sound Striking a Partition

Generally the reflection of sound at a surface follows the laws of optics. However it should be noted that the wavelength of sound, especially at low frequencies, can call upon diffraction analogies, and that treating sound waves as "rays" can greatly oversimplify the situation with resultant inaccuracies. However treating sound waves as "rays" can be a useful technique, especially when trying to provide a simple explanation of the behaviour of sound in rooms, and produces reasonable accuracy providing the following conditions apply:

1. The wavelength of the sound being considered is considerably smaller than the dimensions of the reflecting surface — if not, scattering and distortion effects will occur.

2. Diffraction close to absorbent materials and at the edge of any obstacles is not evident. Under these conditions the "rays" no longer travel in straight lines and undergo a change on reflection, which alters the "apparent" plane of reflection.

Fortunately these conditions are generally achieved in normal sized rooms at medium and high frequencies providing none of the surfaces are highly absorbent.

3.3 SOUND FIELDS IN ROOMS

Acoustic "rays" are used in Figure 3.4 to illustrate the build-up of the sound field in a room. The sound energy at any point can be seen to be the summation of a complex series of single and multiple reflections. Because the reflection of sound from the room boundaries affects the sound pressure level within the room it is normal to use the engineering concept of two sound fields within the room, these being:

1. The direct sound — the sound "ray" which transmits directly from the source, S, to the receiver, L, as shown by the line marked 1 on Figure 3.4.

2. The reflected sound — the remaining sound "rays", all of which are reflected at least once from the room boundaries.

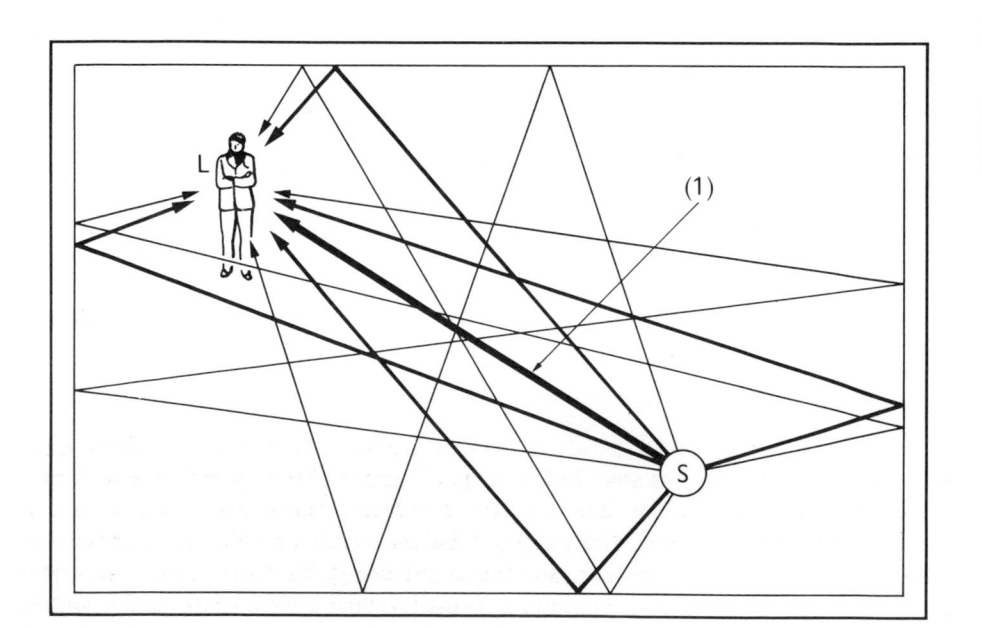

Figure 3.4   Sound Rays in a Room

Figure 3.4 shows only a few of the sound rays in a room. In practice there is an infinite number but only some of the "rays" provide a significant contribution to the overall sound level at the receiver, L. This is because a sound "ray" that has undergone many reflections at the room surfaces will have lost part of its energy at each reflection, and will also have travelled a long distance from the source.

A useful way to appreciate the effect of the various reflected rays is by reference to an echo. A discrete echo occurs when a reflected sound ray which has travelled a long distance before reaching the receiver arrives with sufficient time delay and intensity relative to the direct sound that it is perceived as a separate signal by the ear. If several echoes are superimposed upon each other the ear becomes less capable of distinguishing between each individual reflection. Large churches are full of echoes. Because of the number of echoes, however, the resulting impression is of a "reverberation" or "hanging on" of the sound energy, which arrives at the ear from all directions. Figure 3.5 shows how the sound rays are perceived and form the impression of the total sound field.

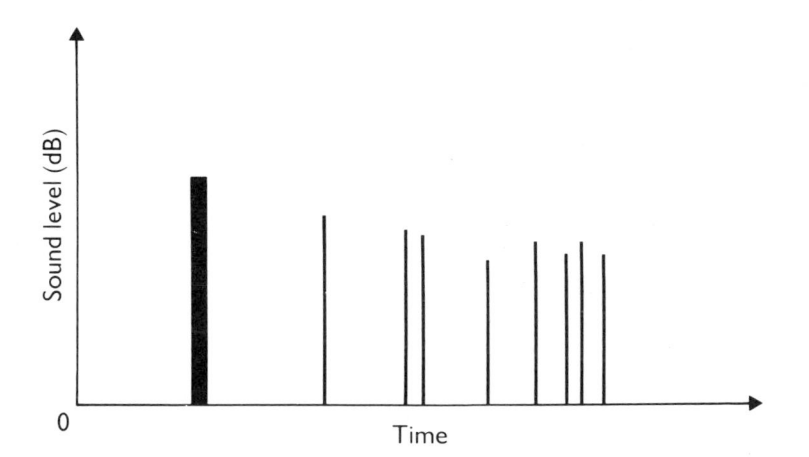

Figure 3.5   Sound Rays Arriving at the Receiver Immediately after Impulse Sound Wave

When a sound source is switched on in a room, the sound level at any location changes as shown in Figure 3.6. Initially the direct ray (1) arrives. Subsequently the reflected sound field increases as the various reflections take a short time to arrive at the receiver. As with other forms of energy, a balance must be obtained between the sound energy supplied to the room and the sound energy lost by the room. After the first few moments of sound emission (2), during which time there is this initial increase in sound level, this balance is obtained, and the sound pressure level due to the reflected energy stabilises (3). Providing the following assumptions are achieved

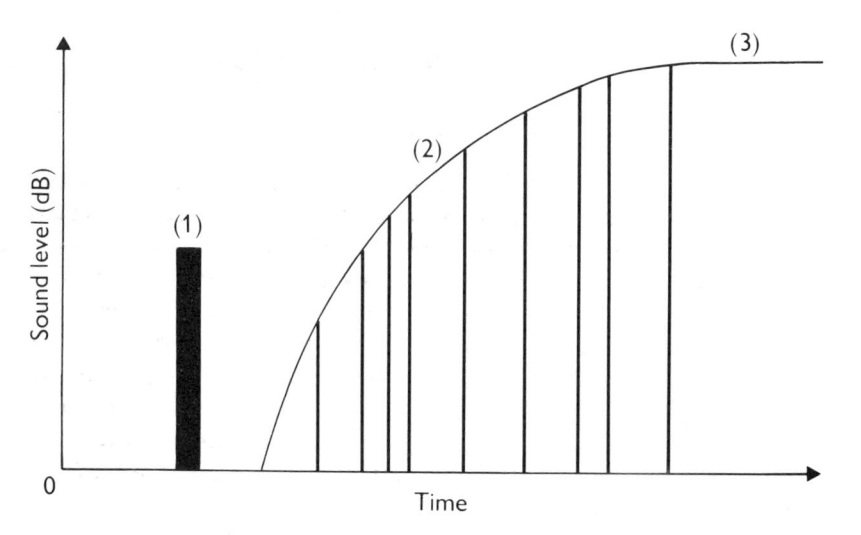

Figure 3.6   Build-up of Sound in a Room over a Longer Period after a Continuous Sound Source is Switched On

a truly diffuse sound field is obtained and the sound pressure level due to the reflected component tends to be independent of position within the room:

(a) The room is of "normal" size and shape ie the room is not too large and one dimension is not significantly greater than the other two.
(b) There are no surfaces of high sound absorption.

If the assumptions are achieved the reflected sound field is normally referred to as the "reverberant field".

The sound levels of the two fields, direct and reflected, can be calculated as follows.

3.3.1 Direct Field

By definition the direct sound field is the sound that transmits from the source to the receiver and, as such, undergoes no reflections at any boundaries. The sound pressure level of the direct field is therefore totally independent of the room and is directly related to the Inverse Square Law:

Sound Pressure Level, L_p(direct) = Sound Power Level of

$$\text{Source, } L_w + 10 \log_{10} \left(\frac{Q}{4\pi r^2} \right) \text{dB} \qquad\qquad 3.12$$

$$L_p(\text{direct}) = L_w + 10 \log_{10} \left(\frac{Q}{4\pi r^2} \right) \text{dB} \qquad\qquad 3.12$$

where r is the distance to the receiver in metres.

Q is called the "surface directivity" factor and is dependent upon the position of the source in relation to the room boundaries. If an omnidirectional source is positioned in mid-air it radiates spherically and the intensity at a point depends on the surface area of a sphere at this point i.e., $4\pi r^2$. In this case $Q = 1$. If the source is positioned against one boundary i.e. in the middle of the floor, it radiates only hemispherically and the surface area is $2\pi r^2$. In this case $Q = 2$. If the source is moved to the junction of the two boundaries, i.e., the floor and one wall — a corner, $Q = 4$, while at the junction of three boundaries i.e., the floor and 2 walls — a double corner or corner/corner, $Q = 8$. In terms of decibels the factor $Q = 2$ means that the direct sound pressure level is 3 dB greater for a sound source located against one boundary than if the sound source is located in mid-air. If positioned against 2 boundaries and $Q = 4$ the direct sound level is 6 dB higher, and against 3 boundaries it is 9 dB higher for $Q = 8$. This is illustrated in Figure 3.7.

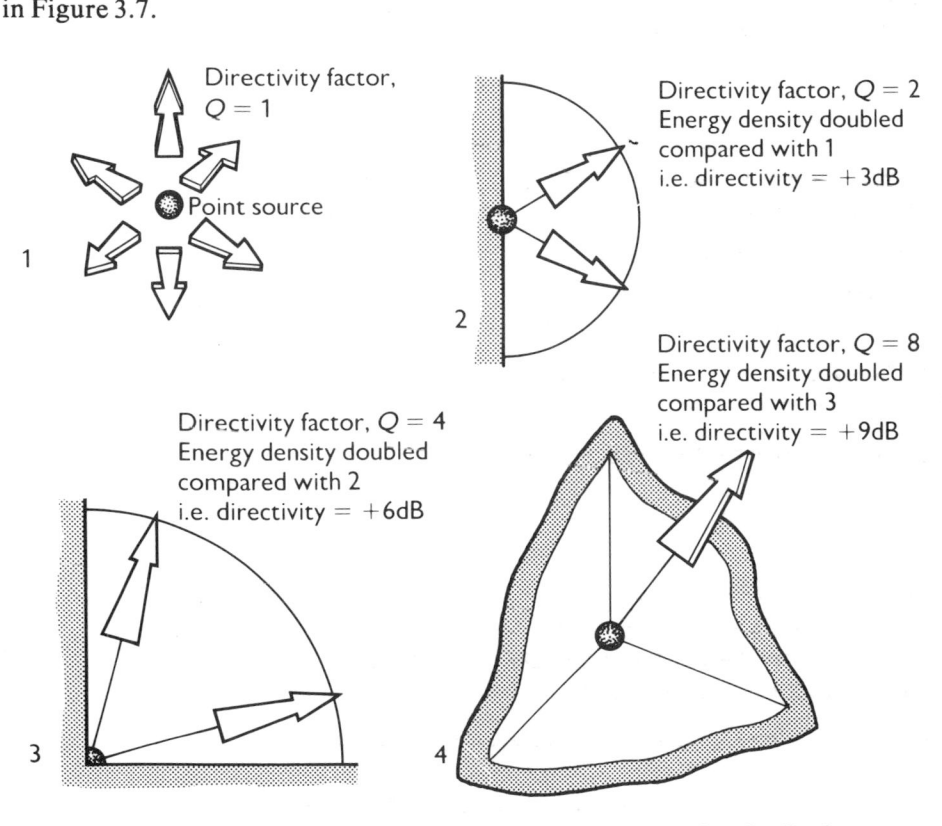

Figure 3.7 Geometric Surface Directivity Patterns of Noise Surfaces

Many items of mechanical services equipment already emit sound in some directions more than in others. This is referred to as the "source directivity" and should, also, where the information is available be taken into account. It is sometimes the case that noise control requirements can be reduced if an item of equipment is located such that the side which emits the least sound energy is faced towards noise sensitive areas. This source directivity does interact with the surface directivity and a complex compound system results. An estimate of the combination of surface and source directivity is shown in Figure 3.8.

	Source location			
Natural source directivity	Free space	Wall	Corner (Wall/floor)	Corner/corner (wall/wall/floor)
0	0	+3	+6	+9
3	+3	+4	+7	+11
6	+6	+7	+8	+11
10	+10	+11	+11	+11
20	+20	+20	+20	+20

Figure 3.8 Combination of Source and Surface Directivity

3.3.2 Reverberant Field

As all the reflected sound rays have struck at least one boundary, the sound pressure level of the reverberant field depends on the effect of the boundary materials. The expression for the reverberant sound field, therefore, takes the effect of the boundaries into account:

Sound Pressure Level, L_p(reflected) =

$$\text{Sound Power Level of Source, } (L_w) + 10 \log_{10} \left(\frac{4}{R} \right) \text{dB}$$

$$L_p(\text{reflected}) = L_w + 10 \log_{10} \left(\frac{4}{R} \right) \text{dB} \qquad 3.13$$

where R is known as the room constant.

$$R = \frac{S\overline{\alpha}}{1 - \overline{\alpha}} \qquad 3.14$$

where S is the total surface area of the room (m²)
and $\overline{\alpha}$ is the mean absorption coefficient of the room boundaries.

The symbol α represents the random incidence absorption coefficient of a material. This is defined as the proportion of incident sound energy arriving from all directions

FREE FIELD
Fan in an absorbent acoustic enclosure.

REVERBERANT PLANTROOM

Diesel generator set located in a hard walled plantroom.

that is not reflected back into the room. Thus $\alpha = 1.0$ represents total absorption and $\alpha = 0.0$ total reflection. The mean absorption coefficient $\bar{\alpha}$ is the average value of the absorption coefficients of all the room boundaries and is expressed mathematically as:

$$\bar{\alpha} = \frac{S_1\alpha_1 + S_2\alpha_2 + S_3\alpha_3 + ... + S_n\alpha_n}{S_{total}} \qquad 3.15$$

where $S_n\alpha_n$ represents the area of a particular surface multiplied by its absorption coefficient and, as such, represents the totally absorbent area provided by the material.

S_{total} is the total geometric area of all the surfaces in the room. The measurement of α is described in Chapter 13.

It should be noted that all surfaces i.e., furniture, screens, people, etc provide equivalent absorbent areas in a room and where known should be taken into account, together with all the room boundaries. All materials have absorption coefficients that are different at different frequencies. It is standard practice to specify sound absorption coefficients in manufacturers literature in the octave bands 125, 250, 500, 1000, 2000 and 4000 Hz. Typical absorption coefficients for common materials are given in Figure 3.9.

It can be seen from the formula for calculating the reverberant sound field, that for any given frequency, the terms involved are all constant ie surface area and absorption coefficient. Thus the formula predicts that the reverberant sound field will be unaltered by changing the location of the receiver or the source. Both source and surface directivity do not effect this ideal reverberant field.

3.3.3 Total Sound Field

The total objective sound field is the logarithmic addition of the direct and reverberant fields:

Sound Pressure Level $L_p(\text{total}) = L_p(\text{direct}) + L_p(\text{reflected})$

$$= L_w(\text{source}) + 10\log_{10}\left(\frac{Q}{4\pi r^2} + \frac{4}{R}\right) dB$$

$$L_p = L_w + 10\log_{10}\left(\frac{Q}{4\pi r^2} + \frac{4}{R}\right) dB \qquad 3.16$$

The resultant relative influence of the two sound fields is shown in Figure 3.10. This shows typical variations in sound pressure level with distance doubling from the source. A "live" room refers to a room in which there is little sound absorption. In this case the reflected sound field becomes significant at a short distance from the source, and the sound level is even throughout the majority of the room. By contrast a

ABSORPTION COEFFICIENTS OF COMMON MATERIALS
– compiled from Sound Research Laboratories tests and other published data
N.B.   The following data should be considered only as a guide since many factors will influence the actual absorption (e.g. method of mounting, decorative treatment, thickness, properties of surrounding structure)

Material	125	250	500	1k	2k	4k Hz
Sprayed Acoustic Plaster	0.30	0.35	0.5	0.7	0.7	0.7
Board on Joist Floor	0.15	0.2	0.1	0.1	0.1	0.1
Breeze Block	0.2	0.3	0.6	0.6	0.5	0.5
Brickwork-plain/painted	0.05	0.04	0.02	0.04	0.05	0.05
Concrete, Tooled Stone, Granolithic	0.02	0.02	0.02	0.04	0.05	0.05
Cork, 25 mm Solid Backing	0.05	0.1	0.2	0.55	0.6	0.55
Fibreboard – solid backing	0.05	0.1	0.15	0.25	0.3	0.3
– 25 mm air space	0.3	0.3	0.3	0.3	0.3	0.3
Glass, 3–4 mm	0.2	0.15	0.1	0.07	0.05	0.05
>4 mm	0.1	0.07	0.04	0.03	0.02	0.02
Plaster, lime or gypsum – solid backing	0.03	0.03	0.02	0.03	0.04	0.05
– on lath/studs, air space	0.3	0.15	0.1	0.05	0.04	0.05
Plywood/Hardboard, air space	0.32	0.43	0.12	0.07	0.07	0.11
Wood Blocks/Lino/Rubber Flooring	0.02	0.04	0.05	0.05	0.1	0.05
Wood Panelling, 12 mm on 25 mm battens	0.31	0.33	0.14	0.1	0.1	0.12
Carpet, Haircord on felt	0.1	0.15	0.25	0.3	0.3	0.3
Pile and thick felt	0.07	0.25	0.5	0.5	0.6	0.65
Acoustic 'blocks'	0.38	0.8	0.43	0.4	0.42	0.50
Acoustic timber wall panelling	0.18	0.34	0.42	0.59	0.83	0.68
Air Absorption (Relative Humidity = 50%) – (×) (per m^3)	0.0	0.0	0.0	0.003	0.007	0.02
Audience per person	0.33	0.40	0.44	0.45	0.45	0.45
Upholstered seat	0.45	0.60	0.73	0.80	0.75	0.64
Proprietary Ceiling Tile (Typical values) A. Fixed to Solid Backing Mineral Wool/Fibre	0.10	0.25	0.70	0.85	0.70	0.60
Perforated Metal, 31 mm thick absorbent infill	0.10	0.30	0.65	0.75	0.65	0.45
B. On Battens 25 mm to 50 mm thick Mineral Wool/Fibre	0.15	0.35	0.65	0.80	0.75	0.70
Perforated Metal, 31 mm thick (absorbent infill)	0.2	0.55	0.80	0.80	0.80	0.75
C. Suspended Mineral Wool/Fibre	0.50	0.60	0.65	0.75	0.80	0.75
Perforated Metal, 31 mm thick (absorbent infill)	0.25	0.55	0.85	0.85	0.75	0.75

Figure 3.9   Absorption Coefficients of Common Materials

"dead" room contains a large amount of absorbent area, with the result that the reverberant sound field is weak and only becomes significant at a relatively long distance from the source. For the majority of the room the sound level is dependent upon the direct sound field and falls off at a rate of 6 dB per doubling of the distance from the source and will be influenced by source and surface directivity.

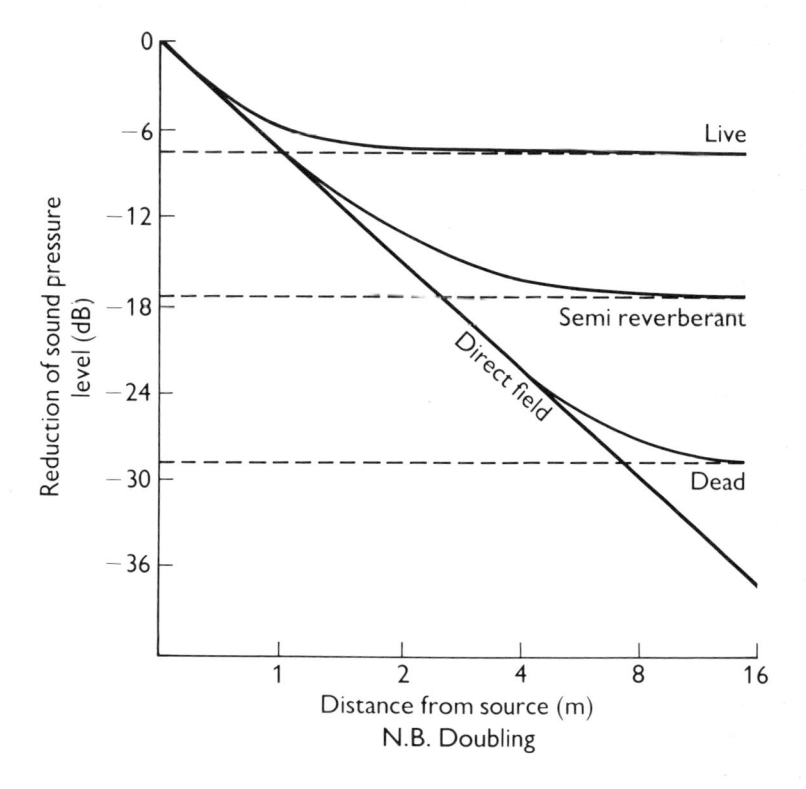

Figure 3.10   Components of Sound Fields

The use of the total sound field formula (Equation 3.16) to calculate the sound pressure level in a room together with the calculation of mean absorption coefficient and Room Constant are shown in the following examples.

Example A. A chiller is to be located in a plantroom which is 10 metres long x 6 metres wide x 4 metres high. The plantroom has a concrete floor, painted breeze block walls and a plasterboard ceiling with an air space behind. The following sound power levels are stated for the chiller:

Octave Band Centre Frequency

	63	125	250	500	1k	2k	4k	Hz
L_w	85	87	88	90	89	85	81	dB

The chiller will be located 1.5 metres from the wall of the plantroom. What will the sound pressure level, L, be at the plantroom wall

The mean absorption coefficient can be calculated from:

$$\overline{\alpha} = \frac{\dot{S}_1\alpha_1 + S_2\alpha_2 + S_3\alpha_3}{S_{total}}$$

			Octave Band Centre Frequency					
	63	125	250	500	1k	2k	4k	Hz
S_1 = wall area = 128m²								
$\alpha_1 = \alpha$ wall	0.09	0.10	0.12	0.12	0.16	0.20	0.20	
$S_1\alpha_1$	11.5	12.8	15.4	15.4	20.5	25.6	25.6	m²
S_2 = floor area = 60m²								
$\alpha_2 = \alpha$ floor	0.01	0.01	0.01	0.02	0.02	0.02	0.03	
$S_2\alpha_2$	0.6	0.6	0.6	1.2	1.2	1.2	1.8	m²
S_3 = ceiling area = 60m²								
$\alpha_3 = \alpha$ ceiling	0.10	0.20	0.12	0.10	0.10	0.08	0.08	
$S_3\alpha_3$	6.0	12.0	7.2	6.0	6.0	4.8	4.8	m²
$S\overline{\alpha} = S_1\alpha_1 + S_2\alpha_2 + S_3\alpha_3$	18.1	25.4	23.2	22.6	27.7	31.6	32.2	m²
S total = 248m²								
mean $\overline{\alpha}$	0.07	0.10	0.09	0.09	0.11	0.13	0.13	

$S_1\alpha_1 + S_2\alpha_2 + S_3\alpha_3 = S\alpha$. This equals the total absorbent area in the room.

From equation 3.14, the Room Constant $R = S\overline{\alpha}/(1 - \overline{\alpha})$. This can, therefore, be calculated.

			Octave Band Centre Frequency					
	63	125	250	500	1k	2k	4k	Hz
$S\overline{\alpha}$	18.1	25.4	23.2	22.6	27.7	31.6	32.2	m²
$1 - \overline{\alpha}$	0.93	0.9	0.91	0.91	0.89	0.87	0.87	
R	19.5	28.2	25.5	24.8	31.1	36.3	37.0	

From equation 3.16 the total sound pressure level

$$L_p = L_w + 10 \log_{10} \left(\frac{Q}{4\pi r^2} + \frac{4}{R} \right) dB$$

As the chiller is to be located away from the wall, but on the floor $Q = 2$ (Figure 3.7). The distance to the wall, $r = 1.5$m. This sound pressure level can be calculated:

	Octave Band Centre Frequency							
	63	125	250	500	1k	2k	4k	Hz
$\dfrac{Q}{4\pi r^2}$	0.071	0.071	0.071	0.071	0.071	0.071	0.071	
$\dfrac{4}{R}$	0.205	0.142	0.157	0.161	0.129	0.110	0.108	
$\dfrac{Q}{4\pi r^2}+\dfrac{4}{R}$	0.276	0.213	0.228	0.232	0.200	0.181	0.179	
$10\log_{10}\left(\dfrac{Q}{4\pi r^2}+\dfrac{4}{R}\right)$	−5.6	−6.7	−6.4	−6.3	−7.0	−7.4	−7.5	dB
L_w	85	87	88	90	89	85	81	dB
L_p	79.4	80.3	81.6	83.7	82.0	77.6	73.5	dB

It would normally be standard to round these figures to the nearest dB:

	63	125	250	500	1k	2k	4k	Hz
L_p	79	80	82	84	82	78	74	dB

3.3.4 Reverberation Time

The room constant R is a term that describes a room's capacity to absorb sound. It is however, a fairly cumbersome term both to calculate or to measure. A more convenient measure of a room's absorption is that of "Reverberation Time".

Reverberation Time is defined as the time taken for the sound energy produced by a source to decay by 60 dB after the sound source has been switched off. A typical decay of sound level is shown in Figure 3.11. As can be seen the sound pressure level decreases linearly with time. When it is not possible to measure a decrease of 60 dB, which is often the case, the Reverberation Time can be obtained by measuring, for instance, the time it takes for the sound level to decrease by only 30 dB and multiplying this result by two. A relationship between the Reverberation Time, the volume and total sound absorbent area of a room was developed by W C Sabine, Sabine's formula is:

$$\text{Reverberation Time (R.T.)} = \frac{0.163V}{S\bar{\alpha}} \text{ seconds} \qquad 3.17$$

where V = room volume (m³),
S = room surface area (m²)
$\bar{\alpha}$ = mean absorption coefficient and is the formula employed for the measurement of Chapter 13.

This relationship is reasonably accurate providing:

1. The sound field is truly diffuse
2. The room does not contain a large area of acoustic absorption and this usually means that the mid-frequency reverberation time should not be less than 0.5s.

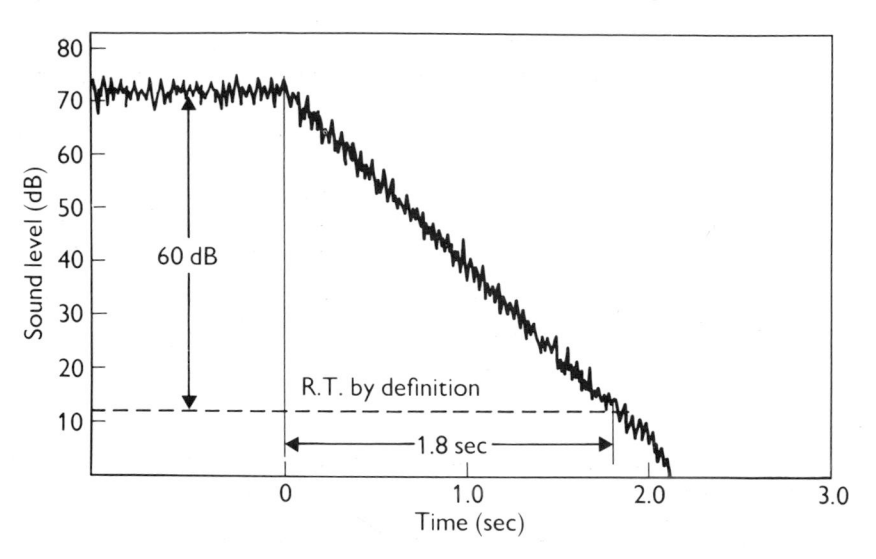

Figure 3.11 Reverberent Sound Decay

A "live" room has a long reverberation time, while a "dead" room has a short reverberation time.

As the absorbent area in a room is increased, the accuracy of Sabine's formula decreases. As can be easily seen, when the room is totally absorbent, that is $\bar{\alpha} = 1.0$, there is still a finite value of the reverberation time predicted by the formula, whereas for total absorption the reverberation time should be zero. For rooms with large areas of absorbent material the following formula devised by Eyring, should be used:

$$\text{R.T.} = \frac{0.163V}{-S\log_e(1-\bar{\alpha})} \text{ seconds} \qquad 3.18$$

As can be seen from the Eyring formula it is more complex than Sabine's and requires more calculation. It is, therefore, often the case that Sabine's formula is incorrectly used for "dead" rooms. However the value of $\bar{\alpha}$ to be employed in Eyring's formula is that from laboratory tests using Sabine's formula. The effect of reverberation time and optimium values for various types of rooms are discussed in Chapter 4 — Design Targets. With regard to sound levels in rooms, Reverberation Time is important as it is directly related to the total absorptive capacity of a room. At the design stage the reverberation time can be estimated by substituting the absorption coefficients of the different room surfaces and contents in the Sabine's formula. At the remedial stage measurement of Reverberation Time allows the amount of excess or insufficient absorbent area to be established.

Example B. The Reverberation Time of the plantroom discussed in Example A can be calculated. The absorbent areas provided by the various surfaces are, as calculated, in Example A.

	63	125	250	500	1k	2k	4k	Hz
			Octave Band Centre Frequency					
$S_1\alpha_1$ (wall)	11.5	12.8	15.4	15.4	20.5	25.6	25.6	m^2
$S_2\alpha_2$ (floor)	0.6	0.6	0.6	1.2	1.2	1.2	1.8	m^2
$S_3\alpha_3$ (ceiling)	6.0	12.0	7.2	6.0	6.0	4.8	4.8	m^2
$S\bar{\alpha} = S_1\alpha_1 + S_2\alpha_2 + S_3\alpha_3$	18.1	25.4	23.2	22.6	27.7	31.6	32.2	m^2
Volume = V	240	240	240	240	240	240	240	m^3
$T = 0.163V/S\bar{\alpha}$	2.1	1.5	1.7	1.7	1.4	1.2	1.2	s

The effect of adding acoustic absorption to the plantroom in the form of wall lining can also be calculated.

The addition of the wall lining alters the absorbent area of the walls:

	63	125	250	500	1k	2k	4k	Hz
			Octave Band Centre Frequency					
α wall lining	0.1	0.2	0.35	0.6	0.8	0.9	0.85	m^2
$S_1\alpha_1$ (wall) ($S = 128m^2$)	12.8	25.6	44.8	76.8	102.4	115.2	108.8	m^2
$S_2\alpha_2$ (floor)	0.6	0.6	0.6	1.2	1.2	1.2	1.8	m^2
$S_3\alpha_3$ (ceiling)	6.0	12.0	7.2	6.0	6.0	4.8	4.8	m^2
$S\bar{\alpha}$	19.4	38.2	52.6	84.0	109.6	121.2	115.4	m^2
V	240	240	240	240	240	240	240	m^3
$T = 0.163V/S\bar{\alpha}$	2.0	1.0	0.7	0.5	0.4	0.3	0.3	s

As can be seen, the wall lining has a large effect at the middle and higher frequencies, reducing the R.T. by a factor of 4 — equivalent to 6 dB reduction on the reverberant or reflected sound pressure level. At low frequencies the effect is less.

3.4 SOUND ABSORBERS

There are three main types of sound absorbers:

1. the porous or dissipative absorber
2. the membrane absorber
3. the cavity absorber.

Each absorbs sound energy by converting it to heat, although the method is different in each case. Also the frequency response of the these types of absorber is different, typical performances being shown in Figure 3.12.

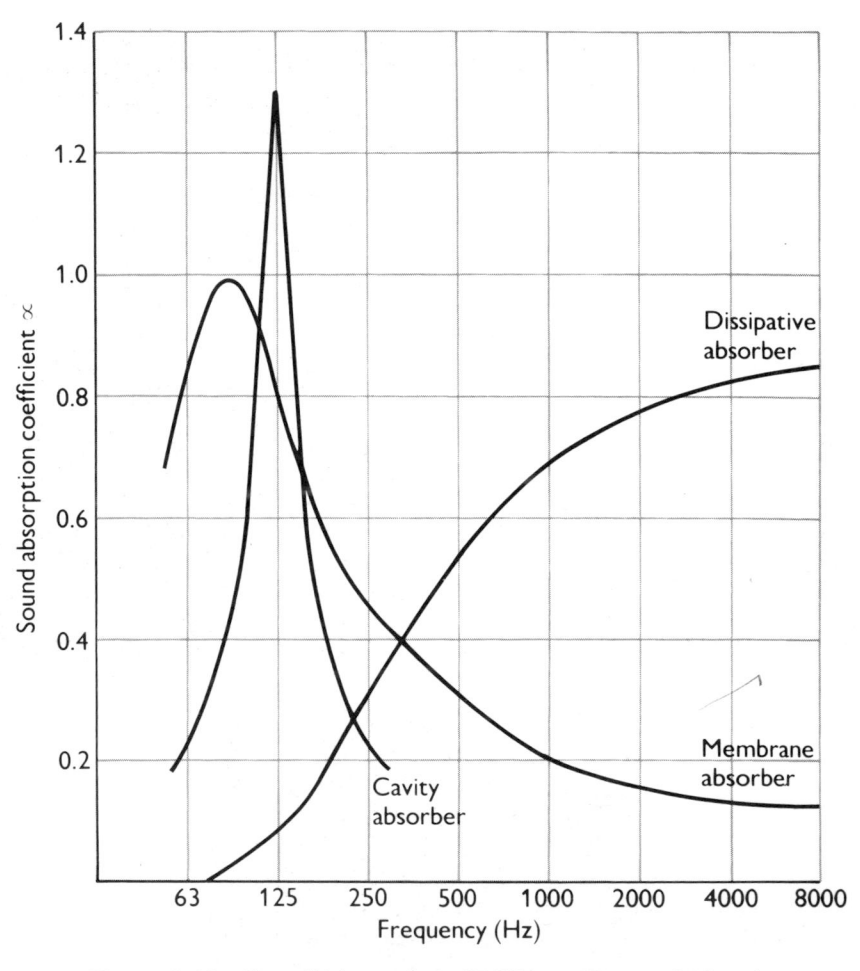

Figure 3.12   Sound Absorption of Different Types of Absorbers

The porous or dissipative absorber works by allowing the air movement or fluctuations into the fabric of the material. Friction between air particles and the materials narrow airways causes sound energy to be dissipated in the form of heat. To enable this mechanism to work it is imperative that the pores of the material are interconnected. For example when a foam is used it must be an open pore material. It is, therefore, normally possible to blow through it. Typical porous absorbers are glass fibre, rock wool, open foam, mineral plaster and hair mat. These are shown in Figure 3.13. There is often confusion between materials which are good for thermal insulation and those for sound absorption. Expanded polystyrene and foam-glass are good for thermal insulation because they contain a large number of unconnected air pockets. They are, however, poor sound absorbers because of the structure of the material. These materials are shown in Figure 3.14.

Glass fibre · Rock wool

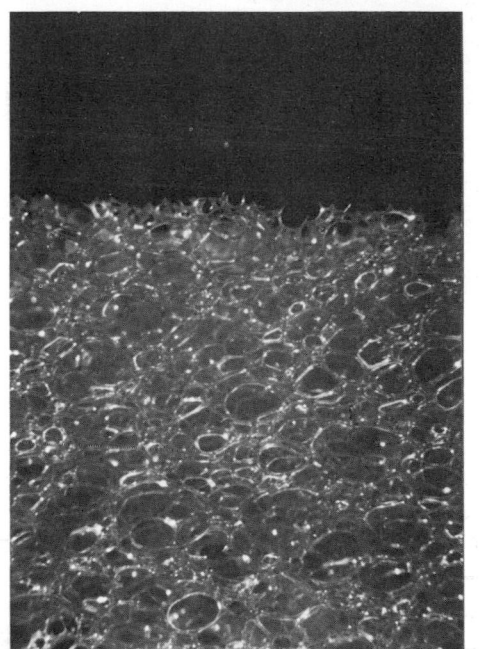

Open foam

Figure 3.13 Porous Absorbers

Mineral plaster

Hairmat

Figure 3.13   (continued)

It is sometimes the case that porous absorbers are contained in a material that cannot be blown through. For example glass fibre and rock wool is often located behind plastic film or in plastic bags in order to stop any fabric erosion. This "facing" does not stop the absorptive effect of the material. It does, however, reduce the performance at higher frequencies. The absorptive capacity reduces with increase in thickness of the facing.

As can be seen from Figure 3.12 porous absorbers are most efficient at high frequencies, but are comparatively poor at low frequencies. The frequency at which useful absorption begins depends on the thickness of the material and is based on the wavelength of sound at that frequency compared with the thickness of material. As

Foam glass

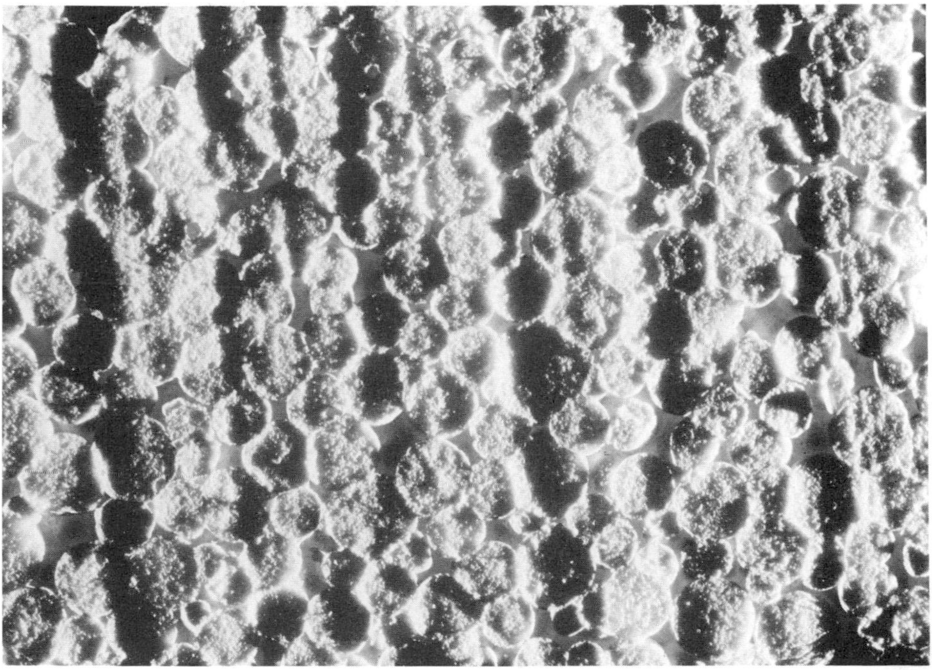

Expanded polystyrene

Figure 3.14 Non-absorbing Foams

the thickness increases so the lowest frequency of useful absorption decreases. Porous absorbers can give useful absorption at low frequencies. This however requires very thick material which, in many cases, is not always practical. The absorption coefficient at any frequency for a given thickness of material will also depend on the density of the material. An increase in density increases the absorption coefficient. The variation of absorption of mineral wool with thickness is shown in Figure 3.15.

Membrane or panel absorbers work by transferring sound energy firstly into vibrational energy in the panel facing. Maximum absorption occurs at the resonance frequency of the panel corresponding to maximum movement. The mode of vibration of most importance is that in which the panel vibrates as a diaphragm over the cushion of air which is behind it. The resonance frequency depends on the mass per unit area of the panel and the stiffness of the airspace, although for small panels it also

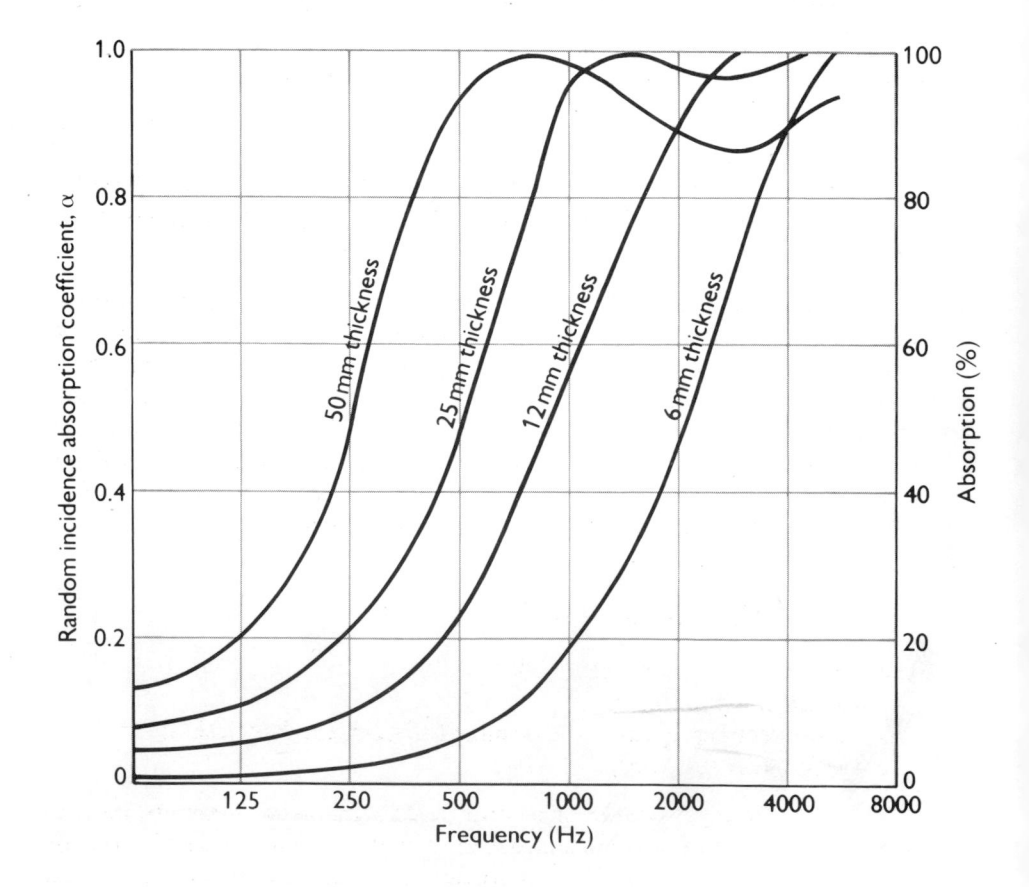

Figure 3.15   Porous Absorber — Effect of Thickness

depends on the stiffness of the panel. Where the panel stiffness can be neglected the resonance frequency can be approximately calculated from:

$$f = \frac{60}{\sqrt{md}}$$ 3.19

where f = resonance frequency (Hz),
m = mass per unit (kg/m^2) and
d = depth of air space (m).

As can be seen from this relationship, as the mass of the panel or the depth of the air space is increased, so the resonance frequency is decreased. Although this formula enables the resonance frequency to be calculated, it gives no indication of the quantity of absorption provided. In order to determine the absorption coefficient it is necessary either to test the panel or refer to previous test results on similar panels.

The most common membrane absorber is the suspended ceiling. Walls, especially studwork, windows and raised floors are, also, common membrane absorbers.

As can be seen from Figure 3.12 the absorption provided by a membrane absorber occurs at low frequencies. This would be typical of a suspended ceiling consisting of a "hard" material such as plasterboard. If, however, the "hard" material is replaced by a porous absorber such as a mineral fibre tile, it is possible to obtain a broadband absorber. The absorption of such a ceiling is shown in Figure 3.16.

The most common cavity absorber is a Helmholtz resonator. As shown in Figure 3.12 these provide narrow band absorption and, therefore, have limited application. Because of the limited bandwidth of their absorption they require careful tuning and are only used in cases using specialist acoustics such as studios, auditoria and for pure tone noise attenuation. Cavity absorbers take the form of an enclosed volume of air connected to the outside world by means of a small neck, for example a medicine bottle. Sound energy is transferred into heat by the piston action of the air contained in the neck of the cavity. The resonant frequency can be calculated from:

$$f_r = \left(\frac{c}{2\pi} \right) \left(\frac{sl}{v} \right)^{1/2}$$ 3.20

where f_r = resonant frequency (Hz),
c = speed of sound (m/s),
s = cross-sectional area of neck opening (m^2), l = length of neck (m) and
v = volume of cavity (m^3).

Panelling with a low percentage open area perforated facing material, such as pegboard, can produce the effect of a large number of cavity absorbers. Each cavity has an imaginary boundary where the air pressure fluctuations meet those in the next cavity. If the perforation pattern is carefully designed a broad band absorber can be

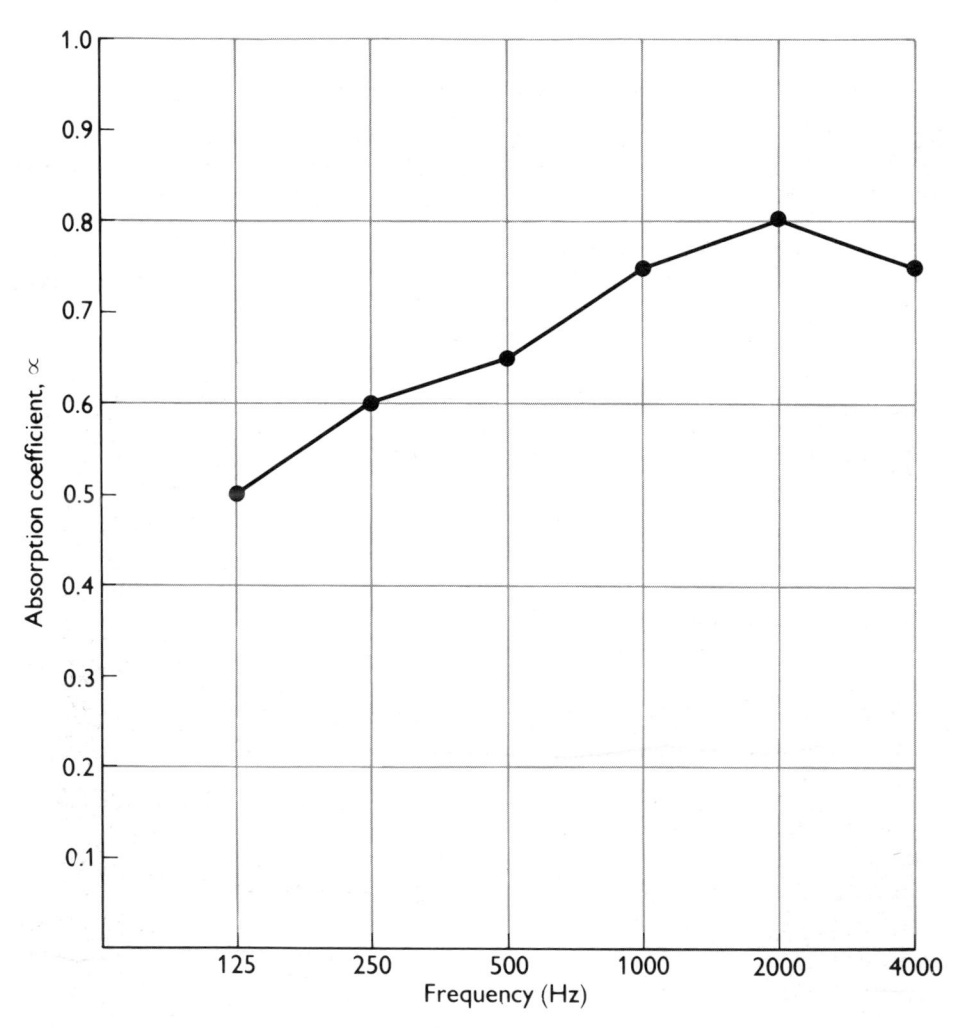

Figure 3.16   Absorption Coefficient of Suspended Mineral Fibre Tile Ceiling

produced with good low frequency absorption. Cavity absorbers can, also, either be made from material that provides porous or membrane absorption or combined with such material to produce a broadband absorber. Figure 3.17 shows examples of such materials. Figure 3.17a is a timber wall panel which is made from chipboard. The chipboard provides some mid and high frequency absorption to combine with the cavity absorption. Furthermore if the panel is constructed in front of a cavity, low frequency membrane absorption can be added to the overall effect. Figure 3.17b shows a brick which has a slit leading to a central cavity. By placing mineral wool in the cavity the brick is both a cavity and porous absorber.

Figure 3.17a Absorbent Timber Wall Panel and Detail Courtesy of Sound Acoustics Ltd.

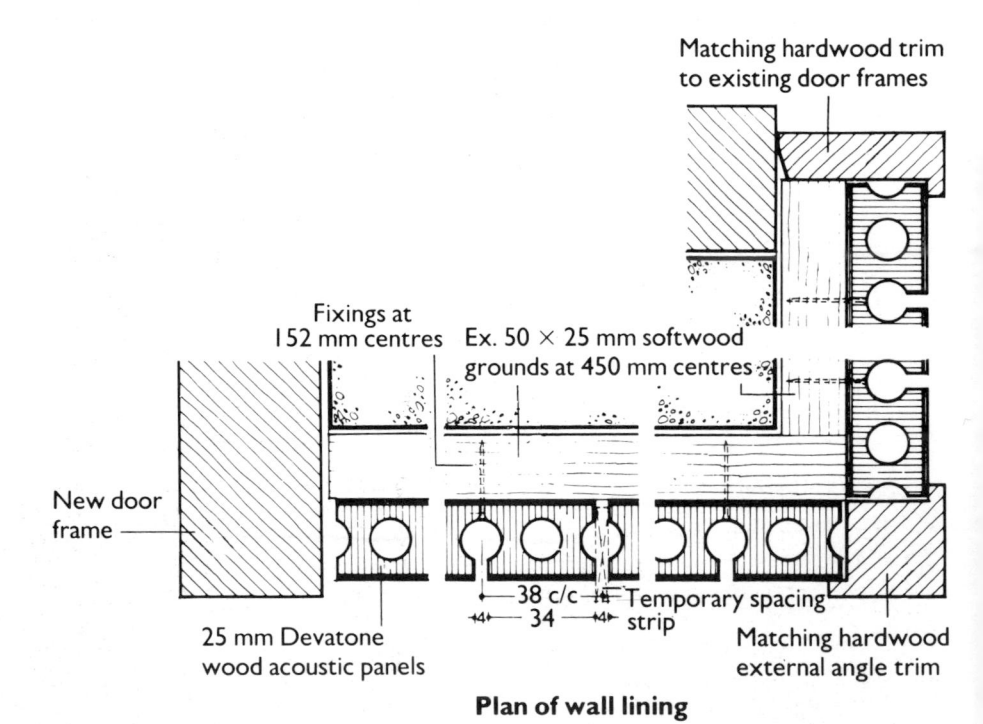

Plan of wall lining

Figure 3.17a   (continued)

Figure 3.17b Acoustic Block Wall with Component Tuned Brick Courtesy of Edenhall

3.5 ALTERNATIVE CALCULATION METHOD

The formula previously shown for calculating the sound pressure level in a room is quite cumbersome to use. A simplified formula, however, can be used providing the room is not too "dead". In this formula the Room Constant is replaced as follows:

$$\text{Room Constant } R = \frac{S\bar{\alpha}}{1-\bar{\alpha}} \qquad 3.14$$

Providing $\bar{\alpha}$ is small, which is the case in most rooms, this can be approximated to:

$$R = S\bar{\alpha} \qquad 3.21$$

This latter term is defined from Sabine's formula, equation 3.17 as

$$S\bar{\alpha} = \frac{0.163V}{T} \qquad 3.13$$

Thus the room constant in the formula for the sound pressure level of the reflected sound field, can be replaced by this term to yield:

$$L_p(\text{reflected}) = L_w + 10\log_{10}T - 10\log_{10}V + 14\,\text{dB} \qquad 3.22$$

Thus by knowing the volume of the room in which a sound source is to be located and by either measuring or calculating the Reverberation Time the sound pressure level of reflected sound field can be calculated. This is added logarithmically to the sound pressure level of the direct sound field:

$$L_p(\text{direct}) = L_w + 10\log_{10}\left(\frac{Q}{4\pi r^2}\right)$$

This can be simplified by using the nomograms shown in Figure 3.18. It can be seen from this simplified formula that as the room volume increases so the sound pressure level decreases. Providing the reverberation time remains the same, a doubling of the volume results in a decrease in sound pressure level of 3 dB. It can also be seen that if the reverberation time decreases the sound pressure level decreases. If the reverberation time is halved, which would require a doubling of the total absorbent area, so the sound pressure level will decrease by 3 dB.

Example C. The simplified formula in conjunction with the nomograms can be used to calculate the sound pressure level due to the chiller in the plantroom previously discussed.

To use the nomograms it is necessary first to calculate the direct field, then the reflected field and then logarithmically add the two fields, an engineering concept. The volume of the plantroom is 240 m³ and the reverberation times, R.T., are as calculated in Example B (no wall lining).

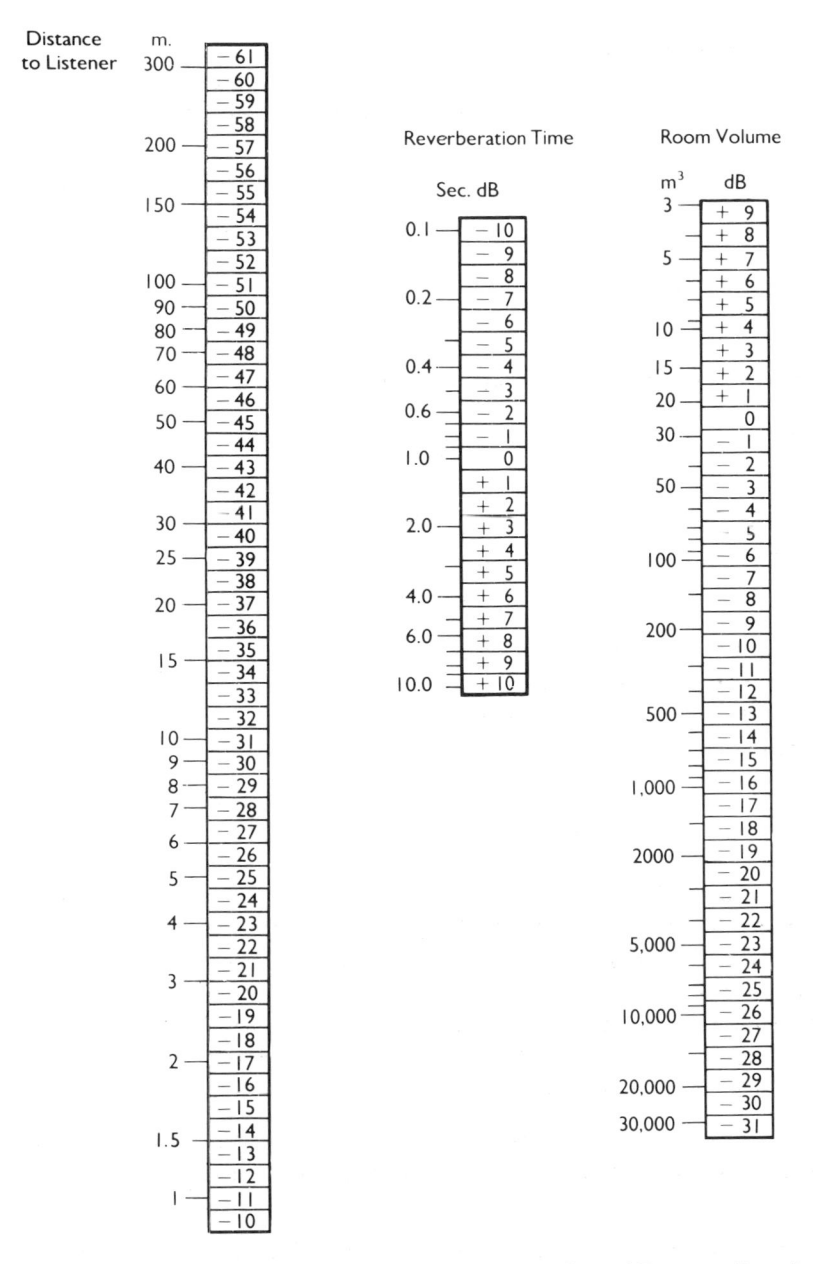

Figure 3.18   Nomograms for Establishing Room Sound Pressure Levels

	Octave Band Centre Frequency							
	63	125	250	500	1k	2k	4k	Hz
L_w	85	87	88	90	89	85	81	dB
Correction for 1.5m ($Q = 2$)	−11	−11	−11	−11	−11	−11	−11	dB
Direct L_p	74	76	77	79	78	74	70	dB
RT	2.1	1.5	1.7	1.7	1.4	1.2	1.2	s
L_w	85	87	88	90	89	85	81	dB
Correction for RT	+3	+2	+2	+2	+2	+1	+1	dB
Correction for Volume	−10	−10	−10	−10	−10	−10	−10	dB
Reflected L_p	78	79	80	82	81	76	72	dB
Total L_p	79	81	82	84	83	78	74	dB

The effect of adding the wall lining to the plantroom can also be calculated by using the revised reverberation times.

	Octave Band Centre Frequency							
	63	125	250	500	1k	2k	4k	Hz
L_w	85	87	88	90	89	85	81	dB
Correction for 1.5m ($Q = 2$)	−11	−11	−11	−11	−11	−11	−11	dB
Direct L_p	74	76	77	79	78	74	70	dB
RT	2.0	1.0	0.7	0.5	0.4	0.3	0.3	s
L_w	85	87	88	90	89	85	81	dB
Correction for RT	+3	0	−2	−3	−4	−5	−5	dB
Correction for Volume	−10	−10	−10	−10	−10	−10	−10	dB
Reflected L_p	78	77	76	77	75	70	66	dB
Total L_p	79	80	80	81	80	75	71	dB

Looking only at the reverberant sound pressure level it can be seen that the effect of the wall lining makes no difference at 63 Hz, but reduces the sound pressure level by 6 dB at the higher frequencies. Thus, providing the noise source provides the medium/high frequency sound the wall lining provides some benefit regarding the reflected sound pressure level. If the source produced low frequency sound the effect is negligible.

Looking at the total sound pressure level, the effect of wall lining provides a maximum reduction in any octave band at the plantroom wall of 3 dB. This would not normally be considered significant. The wall lining only produces a small reduction in sound pressure level at the plantroom wall because of the short distance between the chiller and the plantroom wall. This shows the importance of calculating both the direct and reverberant (or reflected) sound pressure level and not just considering the latter.

3.6 CALCULATION PRINCIPLES

Although the calculations already shown provide examples of how sound pressure levels in rooms can be calculated, it is worthwhile describing the basic principles that should be observed when carrying out any of these calculations.

Initially it is necessary to know the sound power output of all noise sources in the room, the frequency content and where possible any directional effects of the sources. This information is then combined with the room dimensions, the location of the sources and the capacity of the room surfaces and contents to absorb sound in calculating the resultant overall sound pressure level. The basic principles of the calculations can be shown in the following two examples.

3.6.1 Example 1

The sound pressure level in a plant room containing a single noisy item has already been calculated using three methods. In each method the first step was to obtain the sound power level of the source in octave bands. Because the source was considered to be small and to be omnidirectional no corrections were made for source directionality. Thus the sound pressure level of the direct field was calculated at the required location using the inverse square law. It may, however, be necessary to estimate the source directionality for large sources, in which case reference can be made to the values shown in Figure 3.19.

The contribution of the reflected energy depends on the total amount of sound absorption within the room. Although the sound pressure level of the reverberant field can be calculated using the room constant, it is more convenient to consider the room volume and reverberation time. Typical reverberation times of plant rooms are shown in Figure 3.20. The total sound pressure level is obtained by logarithmically adding the direct and reverberant components.

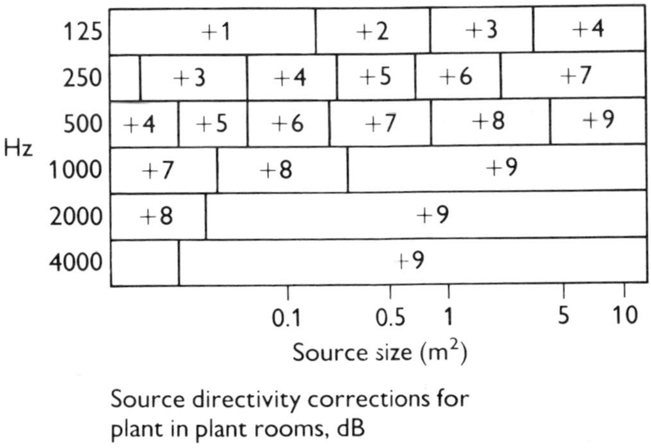

Source directivity corrections for
plant in plant rooms, dB

Figure 3.19 Directivity Coefficients for Plant in Plant Rooms, in dB

In many situations the procedure is complicated by several variables. For instance there are often many sound sources in a plantroom, each one contributing an individual direct field at a specific location. Each contribution must, therefore, be calculated. The reverberant component is more simple as the sound sources can be logarithmically added prior to allowing for the room conditions. The effect of the addition of acoustic lining can be calculated as previously, remembering that it only affects this reflected sound pressure level component. Reference to the different types of absorbers allows the correct selection of material. The effect of different absorbers and different surface areas on the total sound pressure level can then be assessed. Normally plantrooms tend to be "live", although the reverberation time can spread over a fairly wide range, depending on the degree of ductwork and lagging material which affects the total absorptive capacity of the room. The resultant sound pressure levels can also be used to calculate the sound transfer to adjacent spaces. This is described in later chapters.

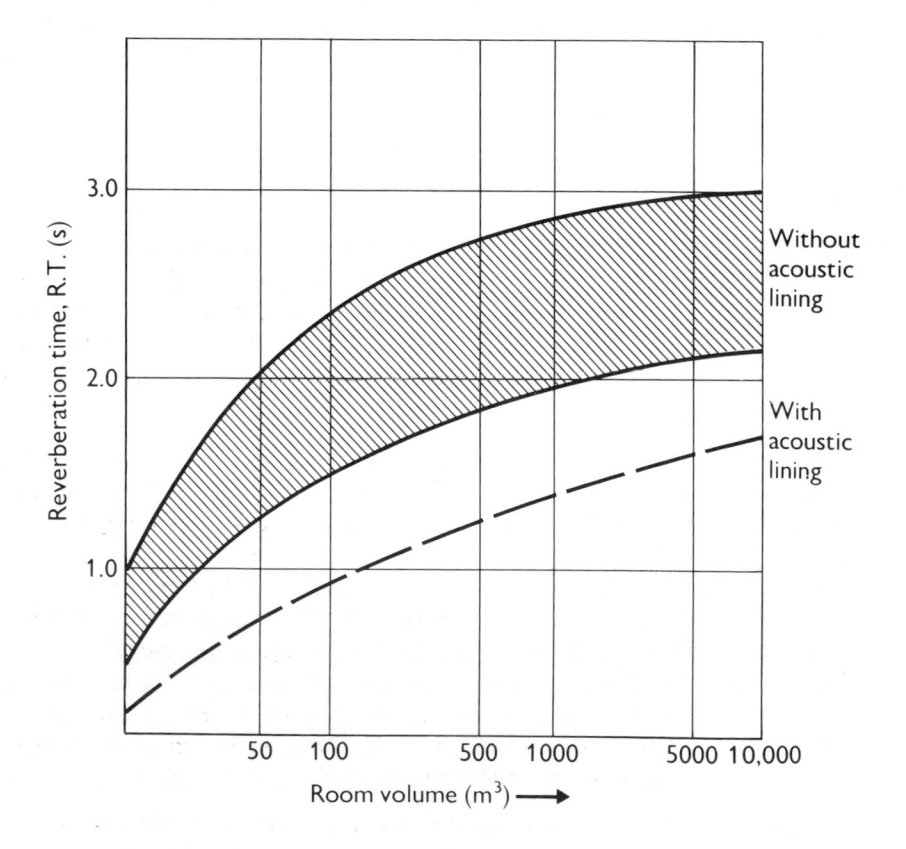

Figure 3.20   Typical Plant Room Reverberation Times

3.6.2 Example 2

A similar case might be the total sound pressure level in a "critical" space such as a conference room. The sound power output from a grille can be obtained from the ductborne noise calculation described in Chapter 7. The size of the grille affects the directional characteristics, whilst the location of the grille adjacent to other surfaces, such as in the corner of a room, will modify the directional characteristics due to local reflection. Thus the direct field can be calculated by reference to the inverse square law and the reflected field by reference to the room conditions. This is an extremely important procedure where ventilation systems are being designed to meet a specific criterion. A variation of only a few decibels, especially at low frequencies will result in variations in lengths of attenuators of 300 mm or more. Care must be taken to allow for all noise sources. For instance in the conference room there may be several supply diffusers and several extract grilles. Noise will be ductborne from the fan, but will also be generated at dampers and grilles. Any units such as variable air volume dampers, fan coil units or induction units will also add to the total sound field. Any long lengths of ductwork may also contribute due to noise breakout, especially at low frequencies. Ignoring any of these potential noise services can lead to the design criteria being exceeded.

3.7 ROOM FAULTS

Although room faults will, generally, be of more interest to the architect than the mechanical services designer, it is useful for everybody involved in the design team to understand these particular acoustic problems. The geometry and boundary conditions of a room can have an effect on the resultant aural conditions and it should be appreciated that the sound field in a room is not always simple and that "faults" can occur.

Focussing, as shown in Figure 3.21, is similar to that which occurs in light reflecting from a concave mirror. The sound reflecting off a hard concave surface is concentrated to a "focal" point. This results in large variations in sound level over quite small distances. The "acoustic focal point" can be predicted using sound rays and assuming that the angle of reflection at the hard surface equals the angle of incidence. This technique can be used to ensure that the focal point does not occur close to an area requiring good listening conditions. In Figure 3.21 the hall should be designed so that the focal point occurs as close to the ceiling and as far above the audiences' heads as possible. Alternatively the effect of the focussing can be greatly reduced by finishing the concave surface with highly absorbent material. This will reduce the effect of the focussing, although it will not be totally negated.

Echoes (Figure 3.22) occur when the reflected sound arrives at the listener with sufficient level and time delay compared with the direct sound so that it sounds distinct. This normally involves a time delay of greater than 40–50 milliseconds, and

Figure 3.21   Focussing of Sound

is associated with large rooms or halls. The simplest method of avoiding echoes is to acoustically line any walls that will produce reflections with long time delays.

Flutter echoes (Figure 3.23) occur when sound is reflected backwards and forwards between two parallel reflecting surfaces, thus taking a long time to die away. Flutter echoes are normally heard as a "buzzing" after an impulse such as a handclap.

Although much reference has been made to sound rays in this chapter, it should always be remembered that sound is transmitted as waves not rays.

As described in Chapter 1 every frequency of sound has a corresponding wavelength. When sound is produced at a frequency the half wavelength of which matches a dimension of the room a standing wave or room mode is set up (see Figure 3.24). When this occurs the sound pressure level at certain locations is reinforced whilst at other locations it is decreased. Also the sound decays more slowly when the sound source is switched off. Standing waves that are set up between an opposite pair of boundaries are referred to as axial modes. Standing waves set up between non-opposite boundaries are either tangential or oblique waves.

At low frequencies the room modes are widely spread and, if excited, can be easily heard by the variation in sound levels across the room. At higher frequencies the standing waves are more numerous and occur closer together. They are, therefore, less noticeable.

Standing waves are, normally, most noticeable in small rooms with hard surfaces which contain low frequency noise sources. Plantrooms containing air handling units are such rooms and standing waves are often heard when moving around the plantroom.

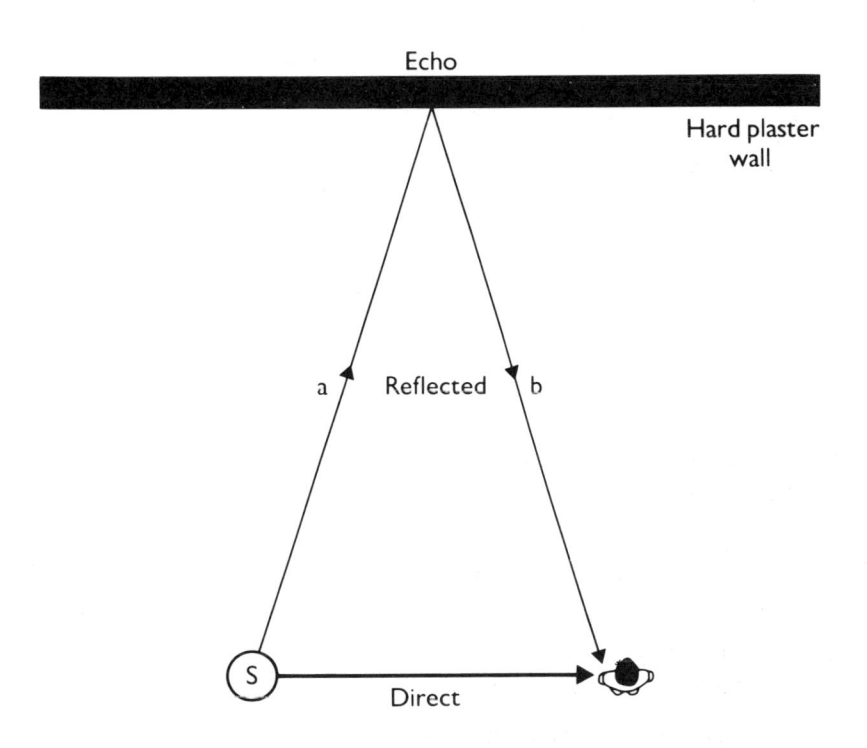

If the reflected path (a+b) is greater than the direct path by more than 15 m, an echo may be heard

Figure 3.22 Echo

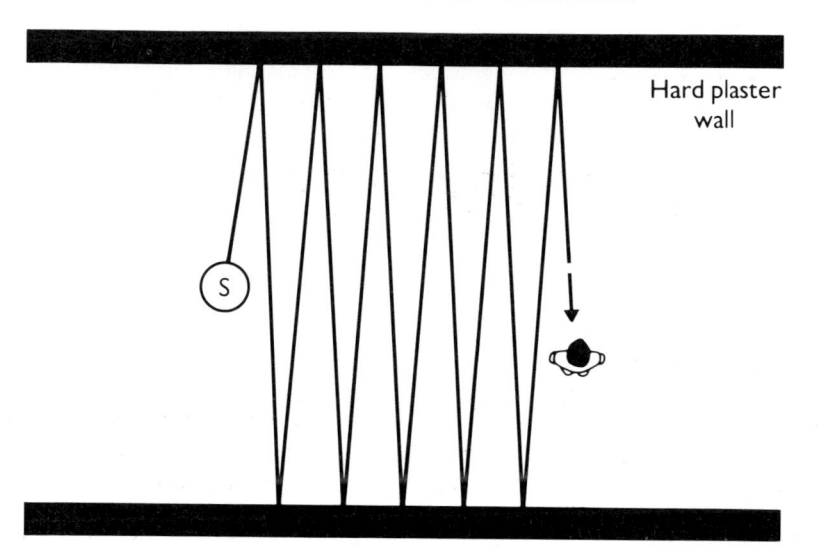

Figure 3.23   Flutter Echo

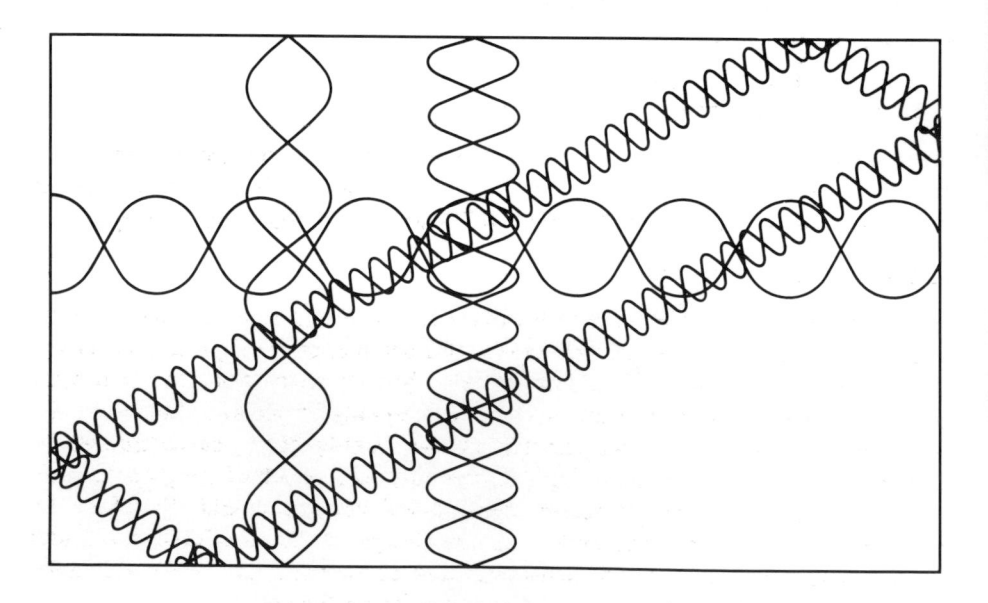

Figure 3.24   Some Room Resonant Modes

CHAPTER 4

DESIGN TARGETS

4.1 INTRODUCTION

At the preliminary planning stages of any building project where consideration is to be given to the internal or external environments certain design targets are formulated in order that the desired comfort conditions are attained.

In nearly all buildings these design targets will include temperature, number of air changes, lighting level etc. In many buildings there will also be acoustic design targets which are often referred to as the Acoustic Design Criteria. These design criteria vary according to the type of usage of the building and refer broadly to either noise control or to room acoustics. Although many of the acoustic requirements of a building are of a qualitative nature so that all members of the design team have acoustic maxims to refer to, certain fundamental aspects of the acoustic environment can be defined by quantitative design targets. These are:

1. Permissible background noise levels
2. Optimium reverberation times
3. Permissible levels of structural vibration

It may be that only one or two of these design targets will be defined. Providing the permissible background noise levels are defined it is possible, if the likely noise levels associated with the usage of the building are known, to set a fourth design target:

4. Minimum acoustic insulation.

As their general classification suggests each criteria can be defined in absolute terms and checked by standard survey techniques at the completion of the project. Often the absolute terms so specified tend to dominate the ultimate acoustic success of a project rather than the qualitative maxims which are equally important. Achieving the design criteria may not always mean that a project is subjectively successful, whilst some projects that exceed the design criteria may still be subjectively acceptable. Much will depend on the ultimate users. These criteria should, therefore, be considered as invaluable guidelines for the design of the mechanical services installations, selection of room furnishings and equipment, specification of anti-vibration treatment and the specification of partition performance.

4.2 BACKGROUND NOISE LEVELS

The background noise level is the minimum sound level that exists in a particular location. Thus speech and other short duration noise sources will cause the background noise level to increase. The sound level will, however, return to the background noise level in the absence of these short duration sources. The background noise level that is encountered both inside and outside buildings varies enormously. External noise levels will be discussed later in the chapter. Inside a building the background noise level will depend on its usage. They will vary between the very high noise levels, which can be hazardous, encountered in industrial buildings, through the audible but not obtrusive levels often referred to as the comfort levels, which occur in standard office, residential and recreational buildings, down to the very quiet levels, bordering on the inaudible, which are required in acoustically critical rooms such as theatres, recording studios, audiology suites and acoustic laboratories. Even within these groupings the correct background sound level can vary quite significantly depending on the actual usage of the building/room.

The groupings will be discussed in turn.

4.2.1 Industrial Noise Levels

Although industrial noise levels are generally high, they must, wherever possible, be controlled so that they do not become dangerous. It is normally accepted, at present, that workers should not be exposed to an equivalent noise level of 90 dBA or more over an eight hour day, although this may be reduced to 85 dBA in the next few years. Furthermore workers should not be exposed to impulse levels of greater than 135 dBA, without ear defenders, or 150 dB with ear defenders. These are advisory limits which do not guarantee against noise induced hearing loss, especially amongst people with sensitive hearing.

They are, however, generally, used in industrial design work and when calculating noise control requirements. There are other factors that should be considered.

Care should be taken that workers are not exposed to very high noise levels for just a short time during the day or to high impulsive noise levels as these can lead to temporary threshold shift. If background noise levels are too high it can be difficult for workers to perceive hazard warnings such as moving vehicles or even fire alarms. Both fire alarms and public address systems in areas of high noise levels require careful design in order that they can be heard and understood in all locations. It should also be remembered that in nearly all areas of work it is necessary to be able to understand speech at close quarters.

4.2.2 Comfort Levels

The majority of mechanical services designs which consider the resultant sound level as an important aspect are associated with "comfort levels". The sound levels are much below the dangerous levels, but must be controlled in order to provide an acceptable and, wherever possible, pleasing level for the occupants. At the higher end of the range the basic aim will be to provide a comfortable noise environment which avoids noise-induced fatigue. For most sedatory workers the aim will be to provide noise levels that do not interfere with speech or telephone communication, but, at the same time, are not so low that other noises cause distraction to the work or even startle. Also within the "comfort" range are the background noise levels associated with most, but not all, recreation activities, relaxation, domestic and sleeping accommodation, whether it be at home or an hotel.

4.2.3 Critical Standards

Although speech communication is important for many sedatory workers, the distance between speaker and listener is, usually, fairly small. Where speech has to be audible over long distances, and has to clearly understood it is essential that the background noise level is low enough that it causes neither distraction nor annoyance, as this would be detrimental to the use of the area. For example speech has to travel long distances from the stage in a theatre. At the same time it is often necessary for the speech to be of a low level e.g., whispering, asides etc. If the background sound level is too high it will be possible neither to hear in all parts of the theatre nor provide the correct range to cover all levels of speech.

Similar critical listening conditions are necessary where music is to be performed. Low background noise levels are even more critical in areas such as broadcasting or recording studios as excess sound levels will render the area unusable for its designed purpose. Similarly anything higher than the absolute minimum background noise level will mean that audiometric suites or acoustic laboratories cannot be properly utilised.

In an acoustic laboratory it may be necessary to either test the noise produced by a quiet piece of equipment or the sound transmitted through a partition providing high sound reduction. If the background sound level is too high the microphone will record the background sound level and not that of the equipment or sound transmitted. As an audiometric suite is used for testing hearing, it is essential that the listener is hearing the test sound and that the background sound level does not interfere.

4.3 NOISE RATING SYSTEMS

Prior to the development of noise rating systems specifically aimed at representing the internal noise level by a single index, acousticians widely used the A-weighted sound level meter response already introduced in Chapter 2 and listed in equation 2.1. It was found, however, that although the A-weighted sound level was useful during acceptance tests at the completion of buildings this advantage was more than offset by its deficiency as a design criteria. Among the shortcomings of the A-weighted sound level with regard to internal noise levels was that it tends to reject the low frequency components of the noise and, also, fails to discriminate against high levels of narrow band noise. Furthermore the dBA level provides no specified information that can be used as an aid to pinpointing any levels that are measured in excess of the design criteria. dBA has, therefore, become only infrequently used in the "comfort" range, although it is extensively used in industrial environments where hazardous levels are encountered.

4.3.1 NC and NR Curves

In order to provide more suitable systems, prototype rating curves were developed in the early 1950s. These were based on the equal loudness contours of the human ear and the sound levels were defined in each of the octave bands. Today the most commonly encountered noise rating spectra are the NC curves developed by Beranek and the NR curves developed by Kosten and Van Os. As has been mentioned in Chapter 2, the NC system was evolved from empirical considerations of the equal loudness contours whilst the NR system is generated by the equation 2.2.

$$L = a + bN \qquad\qquad 4.1$$

where L is the octave band sound level for NR level N,
and a and b are constants given in Chapter 2.

The whole value octave band sound pressure levels associated with NC curves and the NR curves in steps of 5 are given again in Figures 4.1 and 4.2. Following complaints that the NC and NR systems produced sound spectra that were "rumbly" at the low frequencies and "hissy" at the high Beranek developed a further system known as Preferred Noise Criteria (PNC) during the 1970s which provides a subjectively more acceptable spectra by reducing the sound level at the lower and higher frequencies. PNC curves are shown in Figure 4.3.

Both NC and NR systems have found wide acceptance among building services engineers in the design of air conditioning systems, although the PNC system has not been widely adopted. NC, NR and PNC curves are interchangeable, having the same index for similar spectra. A comparison of NC, NR and PNC curves is shown in Figure 4.4.

In setting out to lay down design targets it is worth reviewing the significant differences between NC and NR curves already mentioned in Chapter 2:

(a) at low frequencies NR sound levels are higher than the equivalent NC level
(b) at high frequencies NR sound levels are lower than the equivalent NC level
(c) NC levels are defined between 63 Hz and 8 kHz whereas NR levels are defined between 31.5 Hz and 8 kHz
(d) NR levels are produced from a formula. It is therefore possible to represent any sound level by an NR value. However NC curves are defined in steps of 5 (i.e. NC30, NC35, NC40). As the difference between the curves is not 5 dB throughout the

OFFICE ENVIRONMENT

Only appropriate background noise levels can complement a cordial office atmosphere.

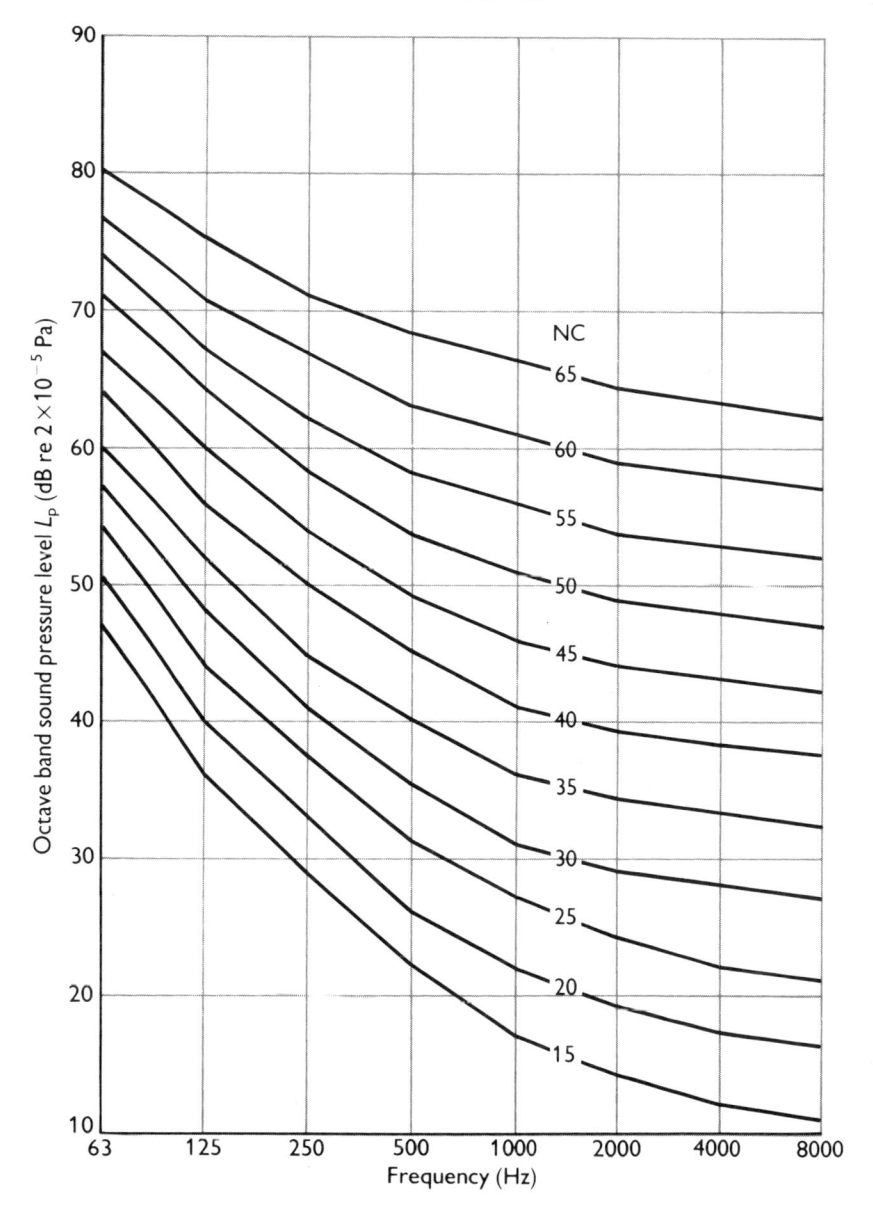

Frequency	63	125	250	500	1000	2000	4000	8000	Hz
NC65	80	75	71	68	66	64	63	62	dB
NC60	77	71	67	63	61	59	58	57	dB
NC55	74	67	62	58	56	54	53	52	dB
NC50	71	64	58	54	51	49	48	47	dB
NC45	67	60	54	49	46	44	43	42	dB
NC40	64	57	50	45	41	39	38	37	dB
NC35	60	52	45	40	36	34	33	32	dB
NC30	57	48	41	35	31	29	28	27	dB
NC25	54	44	37	31	27	24	22	21	dB
NC20	51	40	33	26	22	19	17	16	dB
NC15	47	36	29	22	17	14	12	11	dB

Figure 4.1   NC Curves and Table

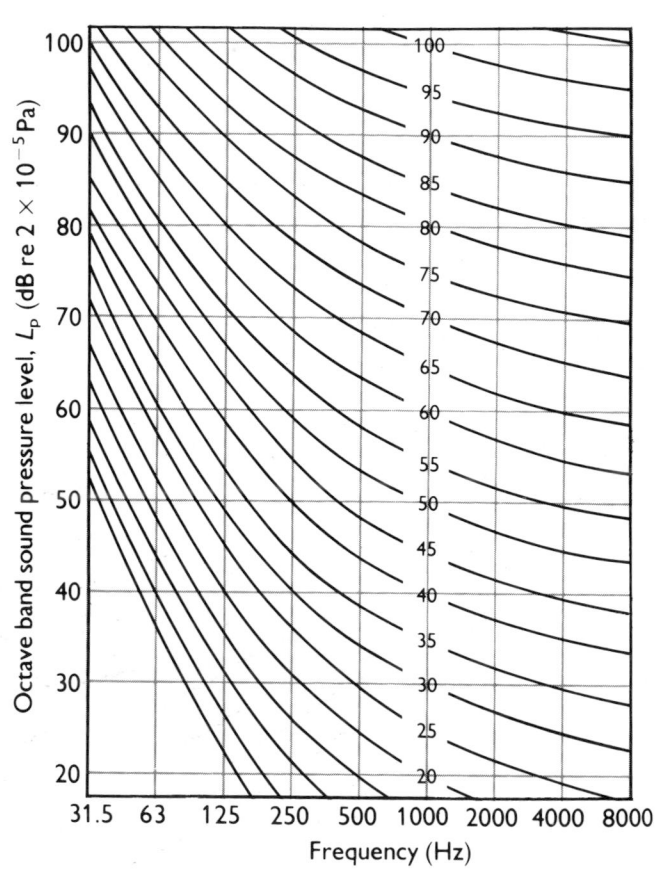

Frequency	31.5	63	125	250	500	1000	2000	4000	8000 Hz
NR100	123	115	109	105	102	100	98	96	95 dB
NR95	120	111	105	100	97	95	93	91	90 dB
NR90	117	107	100	96	92	90	88	86	85 dB
NR85	113	103	96	91	87	85	83	81	80 dB
NR80	110	99	91	86	82	80	78	76	74 dB
NR75	106	95	87	82	78	75	73	71	69 dB
NR70	103	91	83	77	73	70	68	66	64 dB
NR65	100	87	78	72	68	65	62	61	59 dB
NR60	96	83	74	68	63	60	57	55	54 dB
NR55	93	79	70	63	58	55	52	50	49 dB
NR50	89	75	65	59	53	50	47	45	43 dB
NR45	86	71	61	54	48	45	42	40	38 dB
NR40	83	67	57	49	44	40	37	35	33 dB
NR35	79	63	52	45	39	35	32	30	28 dB
NR30	76	59	48	40	34	30	27	25	23 dB
NR25	72	55	44	35	29	25	22	20	18 dB
NR20	69	51	39	31	24	20	17	14	13 dB
NR15	66	47	35	26	19	15	12	9	7 dB
NR10	62	43	31	21	15	10	7	4	2 dB
NR5	59	39	26	17	10	5	2	−1	−3 dB
NR0	55	35	22	12	5	0	−4	−6	−8 dB

Figure 4.2   NR Curves and Table

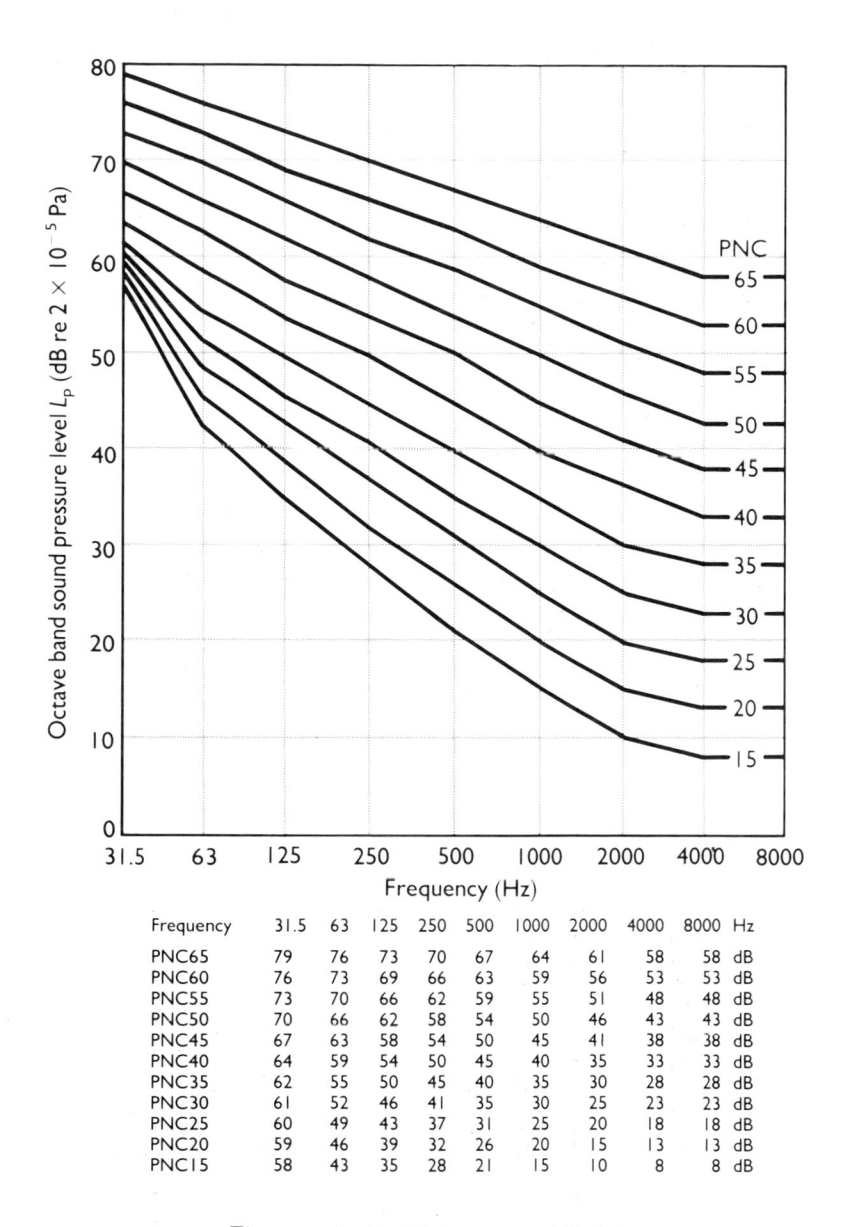

Frequency	31.5	63	125	250	500	1000	2000	4000	8000	Hz
PNC65	79	76	73	70	67	64	61	58	58	dB
PNC60	76	73	69	66	63	59	56	53	53	dB
PNC55	73	70	66	62	59	55	51	48	48	dB
PNC50	70	66	62	58	54	50	46	43	43	dB
PNC45	67	63	58	54	50	45	41	38	38	dB
PNC40	64	59	54	50	45	40	35	33	33	dB
PNC35	62	55	50	45	40	35	30	28	28	dB
PNC30	61	52	46	41	35	30	25	23	23	dB
PNC25	60	49	43	37	31	25	20	18	18	dB
PNC20	59	46	39	32	26	20	15	13	13	dB
PNC15	58	43	35	28	21	15	10	8	8	dB

Figure 4.3   PNC Curves and Table

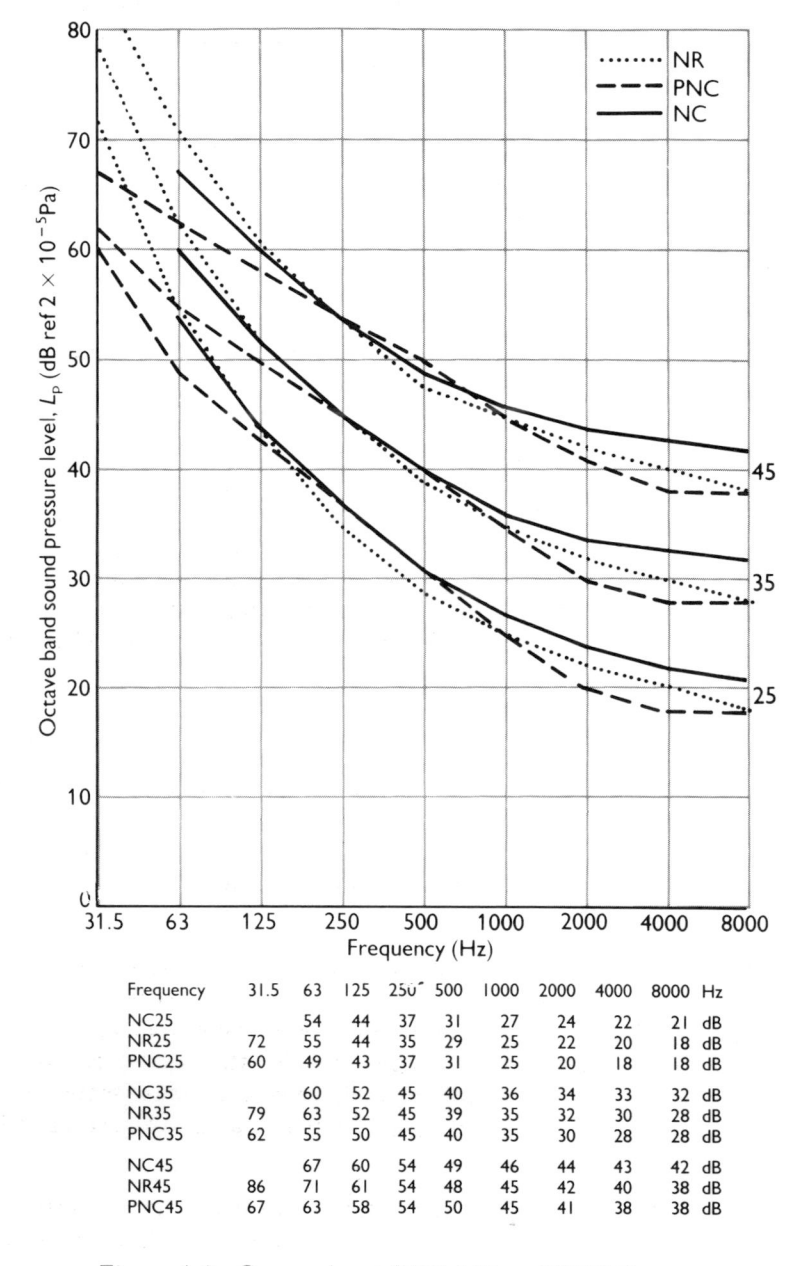

Frequency	31.5	63	125	250	500	1000	2000	4000	8000	Hz
NC25		54	44	37	31	27	24	22	21	dB
NR25	72	55	44	35	29	25	22	20	18	dB
PNC25	60	49	43	37	31	25	20	18	18	dB
NC35		60	52	45	40	36	34	33	32	dB
NR35	79	63	52	45	39	35	32	30	28	dB
PNC35	62	55	50	45	40	35	30	28	28	dB
NC45		67	60	54	49	46	44	43	42	dB
NR45	86	71	61	54	48	45	42	40	38	dB
PNC45	67	63	58	54	50	45	41	38	38	dB

Figure 4.4   Comparison of NC, NR and PNC Curves

spectrum (i.e., at 63 Hz NC35 is 60 dB whereas NC30 is 57 dB) it is not possible to define intermediate levels by an NC value. Intermediate levels should be defined as being the level in excess of one of the defined curves ie NC35 + 3 dB. An intermediate value such as NC38 is technically not definable.

The fact that NR levels are higher at low frequencies means that if the design target is set in terms of NR rather than NC it is normally possible to use shorter attenuators as these are often sized on the low frequency requirement. The resultant spectrum, however, is more likely to sound "rumbly".

4.3.2 ASHRAE

Introduced in Chapter 2 and available as a design target are the Room Criteria (RC) curves. ASHRAE recommends these curves for air conditioning noise on the basis that they produce a well balanced, bland-sounding spectrum. RC Curves are shown in Figure 4.5. It is suggested that if the sound spectrum approximates the slope of the curve to within ±2 dB over the entire frequency range there will be an optimum balance in the sound quantity. If, however, the sound level exceeds the curve by 5 dB in the lower frequencies (up to 250 Hz) the sound will be rumbly whilst a similar excess at high frequencies (above 2 kHz) sounds hissy. Although following a set curve to ±2 dB may appear good principle it is, in fact, extremely difficult to achieve in practice which may account for the lack of use of the RC curves and current non-acceptance as design targets.

4.3.3 Guidance

For the vast majority of occupied areas, advisory limits can be placed on the maximum permissible background noise level generated by the mechanical services installations. These are based on practical experience in the use of NC and NR criteria since their development. Figure 4.6 shows a summary of the major sub-divisions of background noise criteria.

Within the broad outlines indicated in Figure 4.6, individual variations are permissible dependent upon the volume of the rooms, acoustic treatment, their precise usage and their location in relation to all other sources of noise and vibration. There is an unfortunate tendency for designers to select the lower value of the range of criterion curves on the basis that the quieter the environment the more subjectively acceptable it will be. However, selecting the upper value can have the following benefits:

1. The background sound level produces masking to any additional intrusive noise such as speech or noise from office equipment. In relation to speech this can reduce the requirement of any partitions in relation to adequate privacy.
2. There is an economic incentive to design to higher sound levels as the amount of noise control will be reduced accordingly.

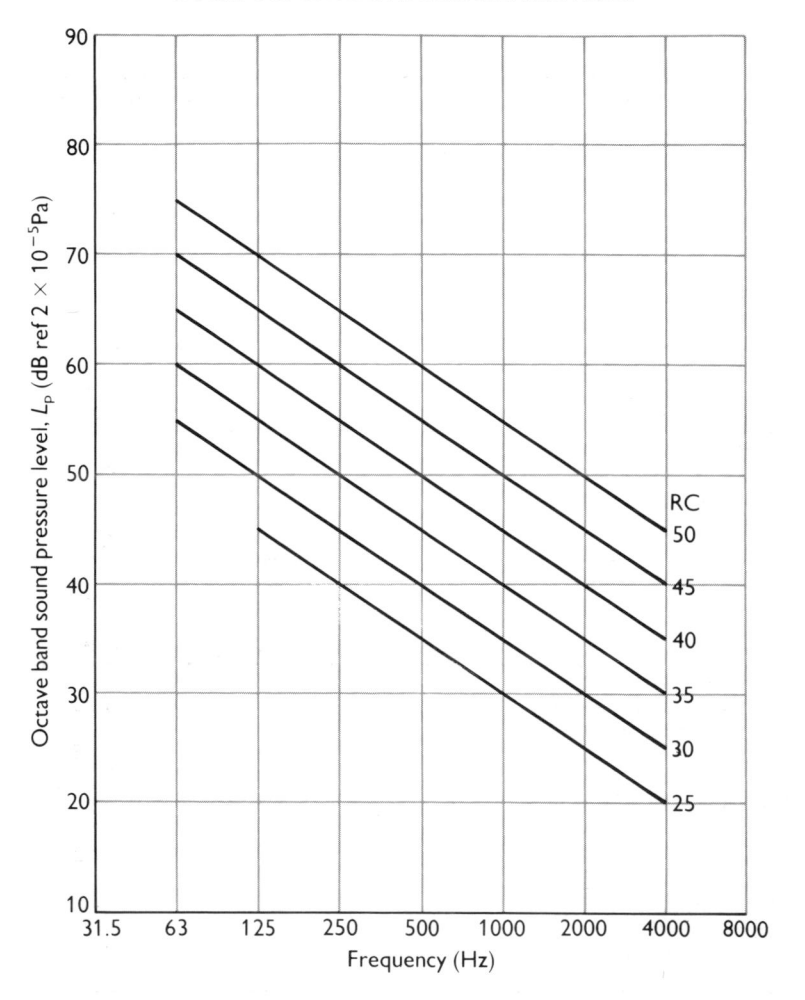

Figure 4.5   RC Curves Represented only over the 63 to 4 kHz Range

4.3.4 Other Criteria

Some organisations have their own criteria based on their experience of levels acceptable to their particular situation. For example the specialist nature of broadcasting and recording studios has enabled such organisations as the BBC and IBA to establish their own background noise criteria. The BBC, in fact, differentiate between the sources of background sound and have laid down three criteria for mechanical ventilation/air conditioning, as shown in Figure 4.7 and separately three criteria, as shown in Figure 4.8, for other extraneous noise levels either originating from within, for example, clocks, camera trolleys etc or outside for example, traffic, aircraft, structure borne noise etc.

Section 1 – Studios and Auditoria	NC/NR Index
Sound Broadcasting (drama)	15
Sound Broadcasting (general), TV (general), Recording Studio	20
TV (audience studio)	25
Concert Hall, Theatre	20–25
Lecture Theatre, Cinema	25–30

Section 2 – Hospitals	
Audiometric Room	20–25
Operating Theatre, Single Bed Ward	30–35
Multi-bed Ward, Waiting room	35
Corridor, Laboratory	35–40
Wash Room, Toilet, Kitchen	35–45
Staff Room, Recreation Room	30–40

Section 3 – Hotels	
Individual Room, Suite	20–30
Ballroom, Banquet Room	30–35
Corridor, Lobby	35–40
Kitchen, Laundry	40–45

Section 4 – Restaurants, Shops and Stores	
Restaurant, Department Store (upper floor)	35–40
Night Club, Public House, Cafeteria, Canteen, Retail Store (main floor)	40–45

Section 5 – Offices	
Boardroom, Large Conference Room	25–30
Small Conference Room, Executive Office, Reception Room	30–35
Open Plan (Bürolandschaft) Office	35
Drawing Office, Computer Suite	35–45

Section 6 – Public Buildings	
Court Room	25–30
Assembly Hall	25–35
Library, Bank, Museum	30–35
Washroom, Toilet	35–45
Swimming Pool, Sports Arena	40–50
Garage, Car Park	55

Section 7 – Ecclesiastical and Academic Buildings	
Church	25–30
Classroom, Lecture Theatre	25–35
Laboratory, Workshop	35–40
Corridor, Gymnasium	35–45

Section 8 – Industrial	
Warehouse, Garage	45–50
Workshop (light engineering)	45–55
Workshop (heavy engineering)	50–65

Section 9 – Private Dwelling (Urban)	
Bedroom	25
Living Room	30

Figure 4.6   Subdivision of Background Noise Criteria

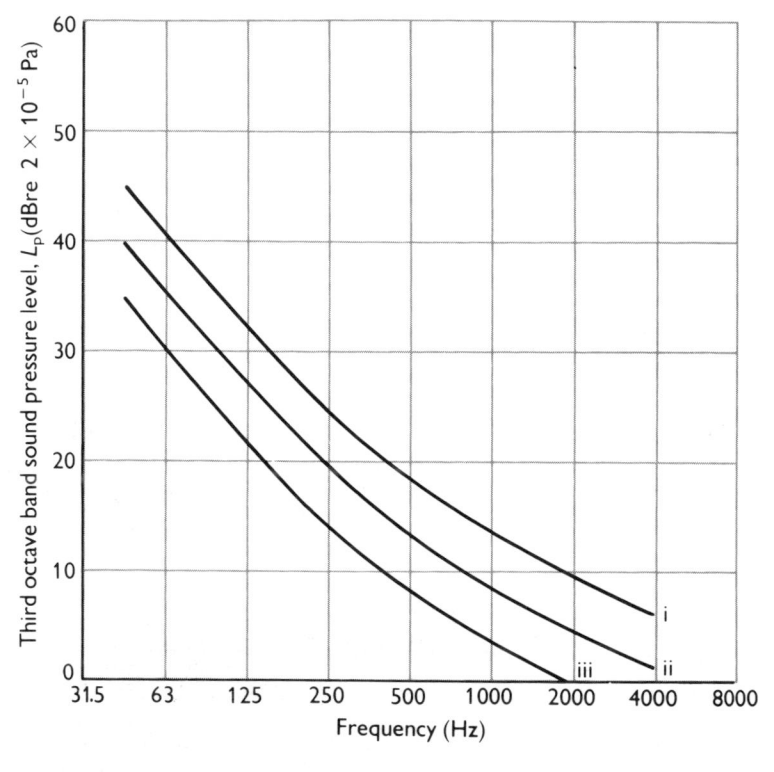

Figure 4.7 BBC Criteria for Maximum Tolerable Background Noise in Studios from Ventilation Noise Sources: (i) radio, other than classes (ii) and (iii); (ii) radio, talks, continuity and recital, television — all categories; (iii) radio — drama

These criteria are presented in 1/3 octave bandwidths to enable tonal characteristics to be detected and highlighted. For convenience, however, octave band equivalents have been evolved and are shown in Figure 4.9. These are only employed for convenience and are not used when testing for compliance with specifications, which are stated in 1/3 octave. The IBA has developed its own background sound level requirements based on the traditional NC curves specified in octave bands. The IBA requirements for local radio stations are shown in Figure 4.10.

4.3.5 Interpretation

Traditionally background noise criteria have been construed as meaning "less than", but for studios a tolerance band with maximum deviations above and below may be introduced, especially for ventilation noise. This is because if the background sound level is too low this leads to problems of noise transfer both from adjoining areas and from within. The achievement of sound levels between maximum and minimum

STUDIO ACOUSTICS
Low noise levels and optimum reverberation times are required for studios. Modular
studio tuning absorbers are clearly visible.

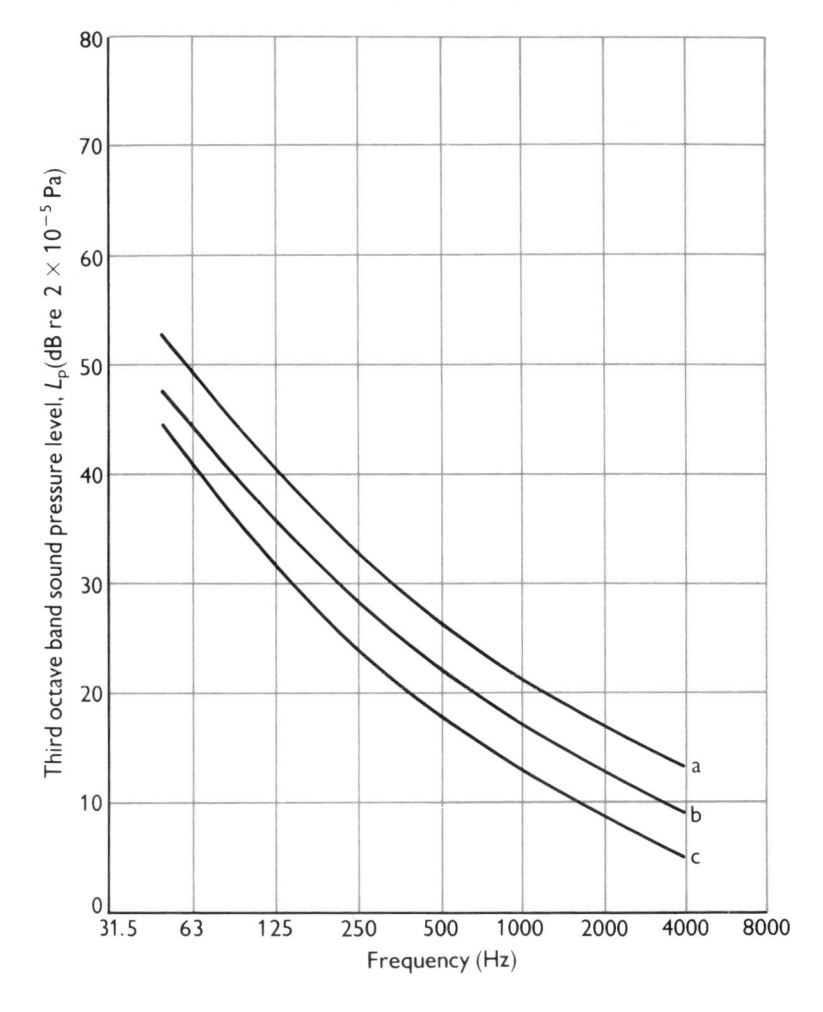

Figure 4.8   BBC Permissible Background Noise Levels in Studios from all Sources: (a) sound studios for light entertainment; (b) sound studios (except drama), all television studios; (c) sound drama studios—(discontinued in 1982)

limits may well prove difficult to achieve and contrived additional background noise levels such as flow noise from perforated plates, grilles or secondary attenuators, as well as electronically generated noise, are being considered.

Variable Air Volume Systems which incorporate terminal units in the conditioned space which vary in noise level based on the volume delivered provide a slightly different requirement because of these fluctuating levels.

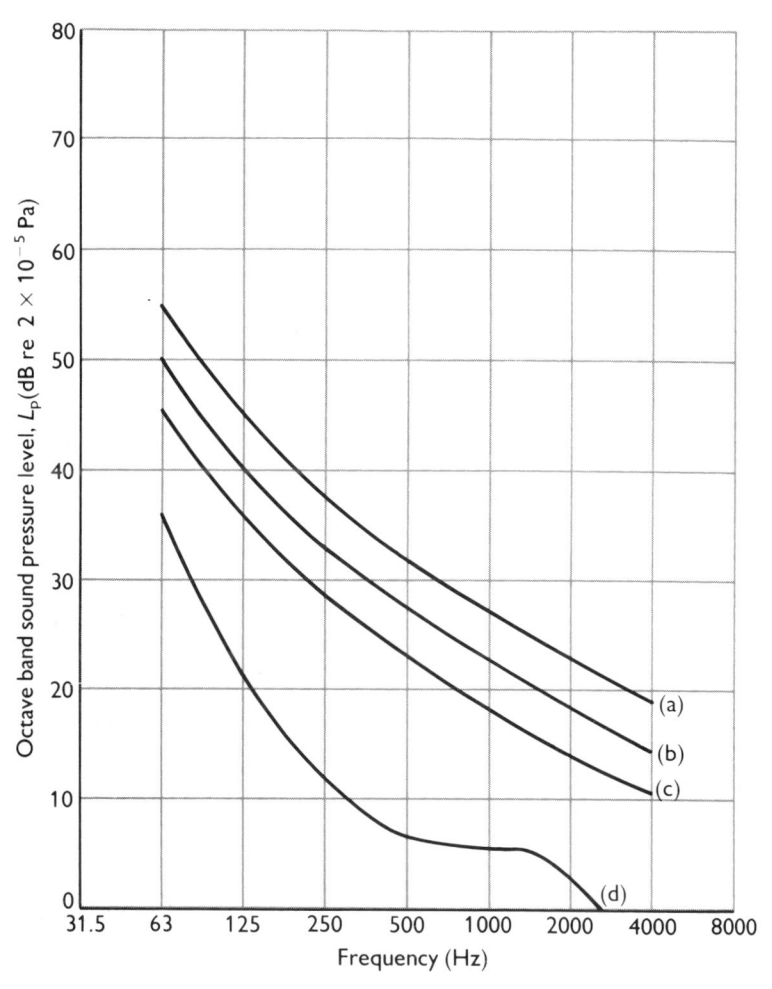

Figure 4.9   BBC Permissible Background Noise Levels in Studios from all Sources — Octave Bands: (a) sound studios for light entertainment; (b) sound studios (except drama), all television studios; (c) sound drama studios; (d) threshold of hearing for continuous spectrum noise (Robinson and Whittle 1964)—(discontinued in 1982)

It is found that it is normally acceptable to allow the background sound level to increase during periods of maximum flow rate as during standard occupancy hours this only occurs for a short period of time during the year. Maximum flow rate may also occur during initial warm-up. This, however, should be outside normal occupancy hours. The design criteria may often be specified by means of a level to be achieved under normal operating conditions (that is, 60–70% volume) and another higher level not to be exceeded at maximum volume. For standard offices this may be NC35 under normal conditions, NC40 at maximum.

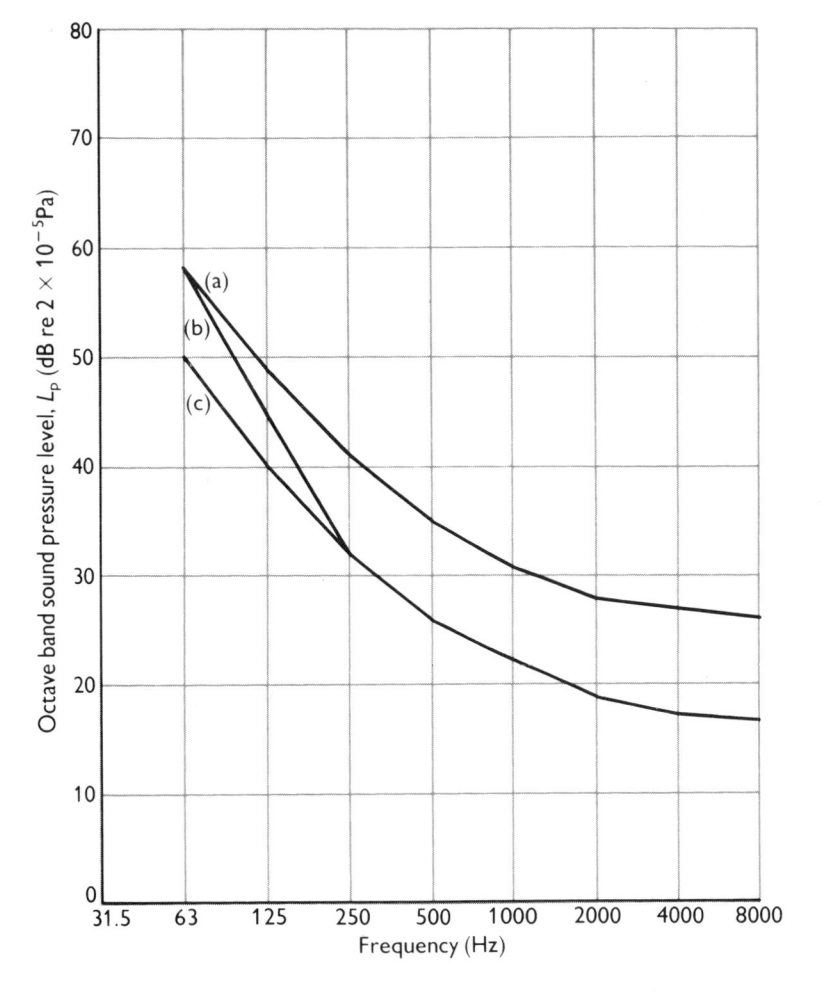

Figure 4.10   IBA Requirements for Local Radio Stations: (a) studios on-air control rooms (category a), (b) studios/on-air control rooms (category b), (c) control rooms for monitors and booths

4.4 NOISE FROM OUTSIDE

Although the mechanical services will often produce the background sound level in a building, noise break-in from outside, particularly traffic noise can affect the overall sound level and in some cases can dictate the resultant sound level. It is, therefore, important that design criteria be set for traffic noise break-in and steps taken to ensure compliance. As has been stated in Chapter 2, traffic noise is normally measured in terms of L_{A10} levels, the level exceeded for 10% of the time. In setting internal design criteria it should be remembered that the L_{A10} also means that the level will be lower for 90% of the time. Because of this fluctuation it is normally acceptable to design the traffic noise break-in to a level just above the background design criteria set for the mechanical services. Figure 4.11 shows the L_{A10} traffic noise break-in criteria that SRL has found, over several years experience, to be acceptable in association with a background design criteria of NC35. There are some situations, where traffic noise must be controlled to the background level or below, so that it does not interfere with the activity in the room. No traffic noise break-in will be acceptable for a recording studio, whilst Auditoria and Conference Halls will only tolerate minimum traffic noise break-in.

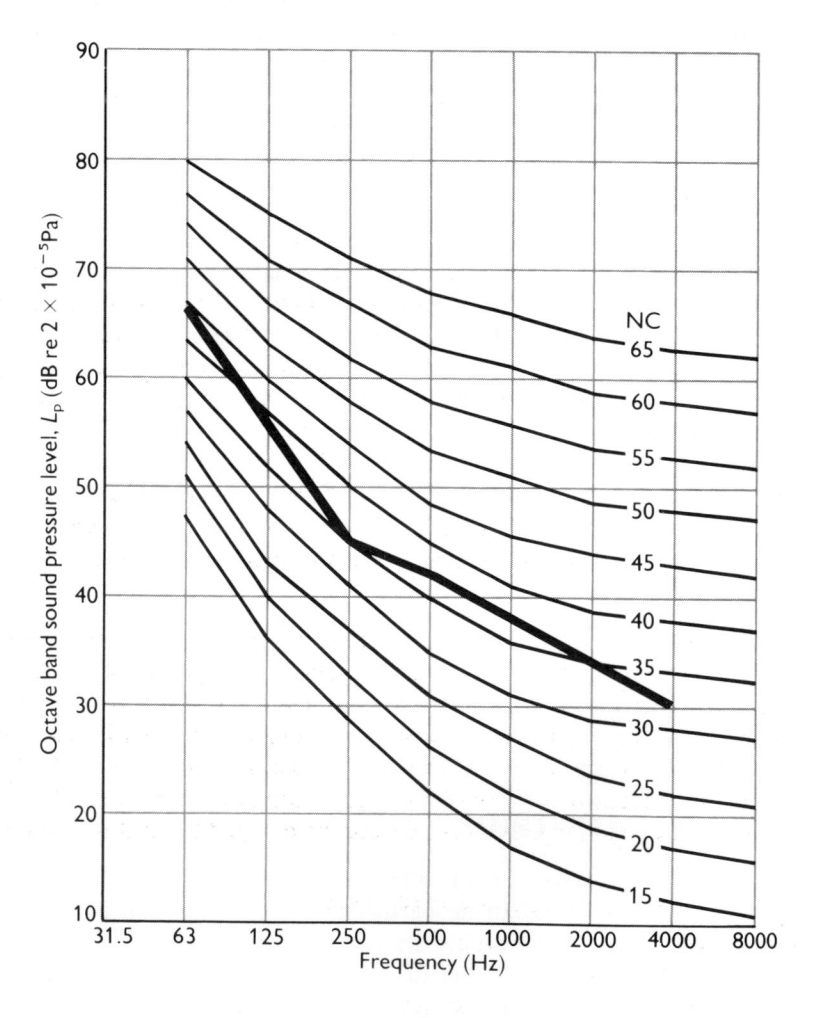

Figure 4.11   Traffic Noise Break-in

4.4.1 External Noise Limits

Although the mechanical services engineer is normally concerned with the internal environment of a building he should also be aware of the need to control the noise emitted from the services equipment to outside down to an acceptable level. There are no legal requirements as to acceptable external noise levels and every location has its own acceptable level based on the environmental noise level and the position of nearby buildings.

Although there are no acceptable noise levels defined by law, if a noise nuisance is considered to exist, the local authorities, who are responsible for maintaining control over noisy premises, will normally refer to BS4142:1967 — Method of Rating Industrial Noise Affecting Mixed Residential and Industrial Areas. The standard provides a method of assessing the potential annoyance caused by noise and, although developed to provide guidance with regard to industrial premises, is often used by local authorities with noise originating from commercial and residential premises. The standard is essentially a noise rating system and not a noise control guide. The assessment is based on measurements in dBA.

BS4142 can be used to provide guidelines for the determination of acceptable environmental design levels, or where a nuisance is alleged to have occurred, the determination of the annoyance. The basic guide provided by BS4142 is that if the noise produced is more than 10 dBA above the background level (L_{A90}) then complaints may be expected. Differences of 5 dBA are considered of marginal significance, whilst if the noise is more than 10 dBA below the measured background level this is a positive indication that complaints should not be expected. Thus if the noise from the building causes the background level to rise by less than 5 dBA, complaints may not be expected. It should be noted that if the noise from the building equals the existing background noise level, this will cause an increase in the measurement of 3 dBA due to the addition of the two sources (background and building).

This rating is normally acceptable for broadband noise operating during the day, which is the case for many items of mechanical services equipment. When the noise has discrete frequency components, such as can be associated with induced draught fans on boiler stacks, cooling towers, and fan noise radiated from ventilation louvres, the noise is considered more annoying and corrections have to be made for this factor. Further corrections are also necessary for any intermittency in the noise, whilst noise at night is normally less acceptable than during the day and lower design levels are necessary.

The following corrections are provided in BS4142:

Corrections to Measured Noise Level
Pure tone etc Add + 5 dBA

Impulsive characteristic + 5 dBA

Intermittency and duration see BS4142

Where it is not possible to measure the background sound level BS4142 provides a method of predicting a notional acceptable background level based on a basic criterion of 50 dBA together with the following adjustments for type of building, location and time of day:

New or modified factory or process	+ 0 dBA
Established factory in non-industrial area	+ 5 dBA
Established factory in industrial area	+ 10 dBA
Rural districts	− 5 dBA
Suburban	0 dBA
Urban	+ 5 dBA
Urban with light industry	+ 10 dBA
Intermediate	+ 15 dBA
Predominantly industrial	+ 20 dBA
Weekdays only 8am to 6pm	+ 5 dBA
Night time 10pm to 7am	− 5 dBA
All other times	0 dBA

Care, however, should be taken when notional levels are used as large variations compared with actual levels have been found. Wherever possible site measurements should be used.

If measurements are made in dBA, it is often necessary to "convert" to octave band levels in order to provide a workable design criteria. A rule of thumb, also mentioned in Chapter 2 equation 2.3, shows that most environmental noise levels expressed in dBA are between 5 and 10 dBA higher than the equivalent NC or NR index. For broad-band noise:

$$\text{NC/NR rating} = \text{dBA} - 5 \text{ (approximately)} \qquad 4.2$$

It should be remembered that this is only an approximate conversion and should not be taken as exact.

The most dependable method of ensuring that complaints from neighbouring properties do not occur is to relate the design to the existing background noise level, which should be measured in each of the octave bands. If the equipment is to operate only during the day it is normally acceptable to limit the noise emitted to atmosphere so that it does not exceed the existing background sound level at the neighbouring property. By this method the maximum increase in background noise level is 3 dBA, which most local authorities consider acceptable. If the equipment is to run throughout the night, it may be necessary to ensure that the existing background sound level is not caused to increase by the new installation. Under these circumstances it is necessary to reduce the sound level emitted to atmosphere to 10 dBA below the existing background sound level. Some local authorities may accept an increase of 1–2 dBA during the night. As any tonal or impulsive characteristics of the sound will have a subjectively bad effect on any source containing such characteristics, they should be designed to a lower level than normally considered acceptable. A reduction of 5 dBA should be made to compensate for such characteristics.

4.5 REVERBERATION TIME

As noted in the chapter "Sound in Rooms", every room or enclosure has a reverberation time which varies with frequency and, to a lesser extent, with the position in the room. In commercial and residential buildings the reverberation time is of importance as a method of noise control. As previously shown the reverberant sound level is dependent on the reverberation time, such that a halving of the reverberation time reduces the reverberant sound level by 3 dB. Thus useful reduction in noise control in the form of shorter attenuators on fans, less sound reduction required through partitions etc can be obtained by reducing the reverberation time. In open plan offices a short reverberation time will help to reduce the sound transmission between workstations as a low reverberation time indicates a high degree of absorption of the floor and ceiling. In most standard offices containing an acoustic ceiling and some form of carpeting on the floor the reverberation time will be of the order of 0.5 s. This figure can usually be used in noise control calculations in the absence of actual selections of room finishes, providing it is known that a sound absorbent ceiling is to be used.

Although reverberation times are usually not too critical in most commercial and residential buildings, there are some types of rooms and buildings that require optimum reverberation times based on their usage. These critical situations normally involve rooms where the hearing conditions are important, such as rooms used for speech and music.

In rooms used for speech the reverberation time must not be too long otherwise the sound level of one syllable will not have decayed sufficiently before the sound of the next syllable arrives at the listener and clarity will be confused. A common example of long reverberation time being detrimental to speech is that of a large church, where the reverberation time can be in excess of 4 s. In such churches it is normally necessary to speak slowly, thus increasing the gap between syllables, so as to counteract the effect of the reverberation time. Unfortunately although a short reverberation time can increase the clarity of speech, it also reduces the reverberant sound level. Also if sound absorbing material is located between the speaker and the listener it reduces the transmission of sound.

A high reverberant sound level together with selected hard reflective surfaces between the speaker and the listener are, therefore, also required in rooms used for speech. In order to obtain clarity together with good speech transmission rooms for speech should not be "dead" and should have sound absorbing materials correctly located.

In auditoria, where both speech and music are performed there are different and potentially conflicting requirements. The decay of the sound level of one musical note must be such that it is sufficient not to interfere with the next note, but at the same time must not decay too quickly otherwise the auditoria will sound "dry". The requirement for music is further complicated by the need to provide different conditions for different types of music.

The reverberation time for a concert hall can, therefore, require to vary between 1.5 s and 2.0 or more seconds depending on the music to be performed. The effect of playing music in the wrong acoustic conditions is most dramatically exemplified when church organ music, which is composed based on a long reverberation time, is played in a standard room.

An auditorium used for both speech and music can require a reverberation time of the order of one second for the former and 2 seconds for the latter. To overcome this problem some auditoria have been designed with revolving wall/ceiling panels which provide an acoustically hard surface on one side and an absorbent surface on the other and by revolving the panels the reverberation time can be altered. Reverberation time has, also been increased electronically by feeding a delayed signal through a series of loudspeakers located in the walls and ceiling at the correct volume in relation to the natural sound. One such system is known as "assisted resonance".

The reverberation times shown in Figure 4.12 relate to the mid-frequency region of 500 Hz, as these values usually provide a relative indication of the listening conditions in most rooms. In critical situations, it is necessary to consider the reverberation times in all the octave bands. In this case it is normal to retain the same reverberation time through the bands above 500 Hz. For music an increase of the lower frequencies rising to about one and a half times the 500 Hz value at 125 Hz is normally acceptable, but for speech the reverberation time should remain the same down to 100 Hz.

A deviation of 5–10% from the selected criteria is generally considered acceptable, especially in auditoria with a high degree of diffusion.

Multi-purpose halls are used to accommodate many activities which require different acoustic conditions. For instance a classical music concert normally requires a much longer reverberation time than a conference. Unless costly methods of altering the reverberation time are used, it is necessary to design multi-purpose halls to have a compromise reverberation time which does not provide ideal conditions for any one activity but does provide acceptable conditions for all activities.

LECTURE THEATRE ACOUSTICS

Reflections and absorptions need careful juxtaposition to create good lecture room intelligibility.

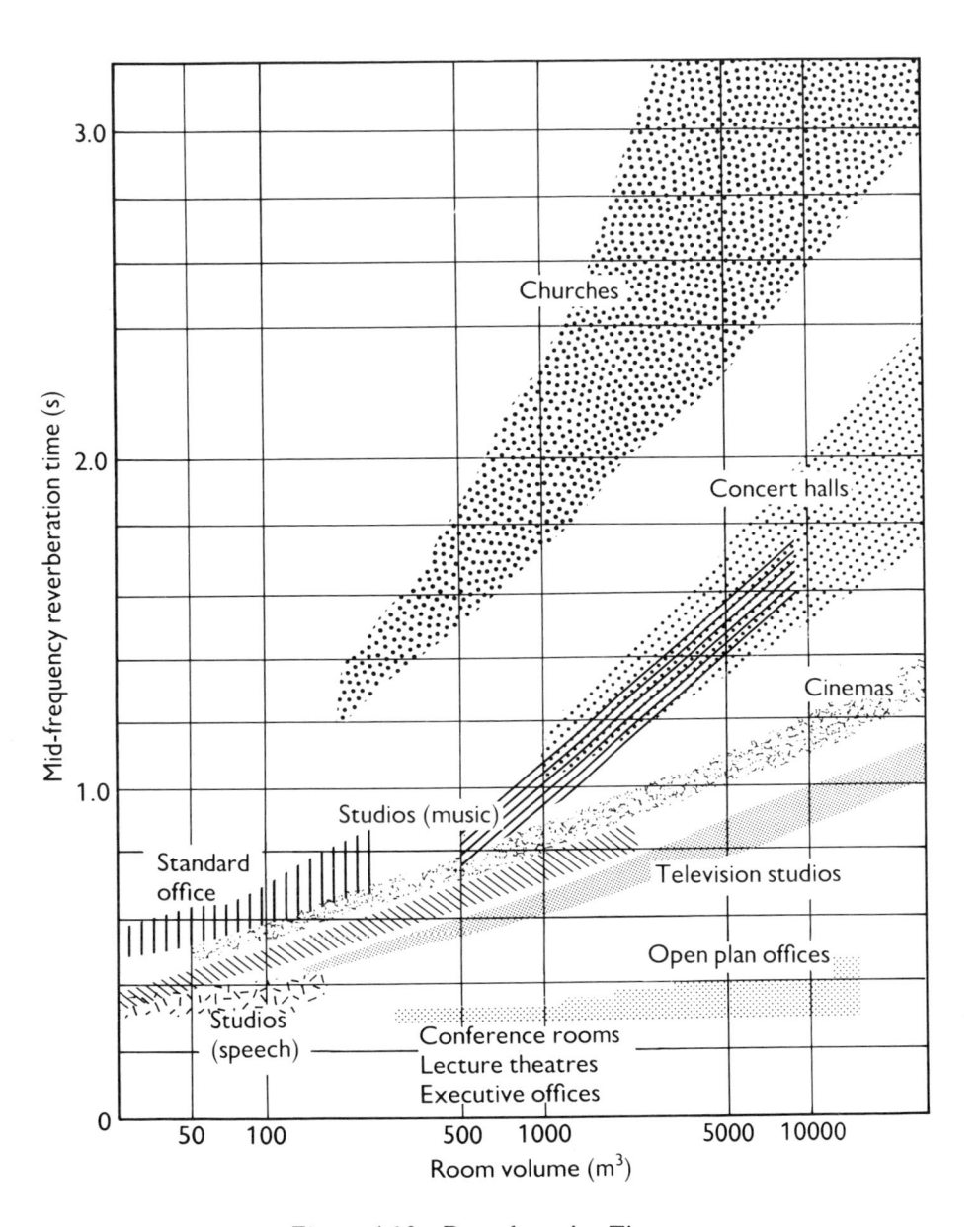

Figure 4.12 Reverberation Times

4.6 VIBRATION

The major source of vibration inside a building is normally the mechanical services. As mechanical services installations have become larger and more complex, with a subsequent increase in the amount of larger reciprocating or rotating equipment, so the potential sources of vibration have increased.

Furthermore buildings have become more lightweight, thus increasing the resultant level of vibration. Buildings can, also, suffer from vibration which originates outside.

Traffic, especially HGVs and rail traffic, can produce quite high levels of vibration in buildings. There is, therefore, the possibility that vibration may be objectionable to the people in the building or in some instances may interfere with the work and processes conducted in the building.

The frequency of vibrations in buildings lies mostly within the range 5–50 Hz. When the frequency exceeds about 20–30 Hz it passes into the audible range. Even if the vibration frequency is sub-audio, a partition or slab vibrating with high amplitude at a low inaudible frequency can cause fixtures and fittings such as ducts, pipework or suspended ceilings to rattle and re-radiate audible noise components based on the resonant frequencies of the attached materials. Vibration isolation must, therefore, certainly not be solely aimed at the audio frequencies.

The human body is a very sensitive detector of vibration. A standing person can, under certain conditions, detect amplitudes as small as one micron (10^{-6}m) whilst amplitudes of the order of 0.05 micron can be detected under the fingertips. Vibration levels broadly describing the sensitivity of a human to vibration were developed by Reiher-Meister. These are shown in Figure 4.13 but it should be noted that these are only typical values as the actual sensitivity will depend on the individual.

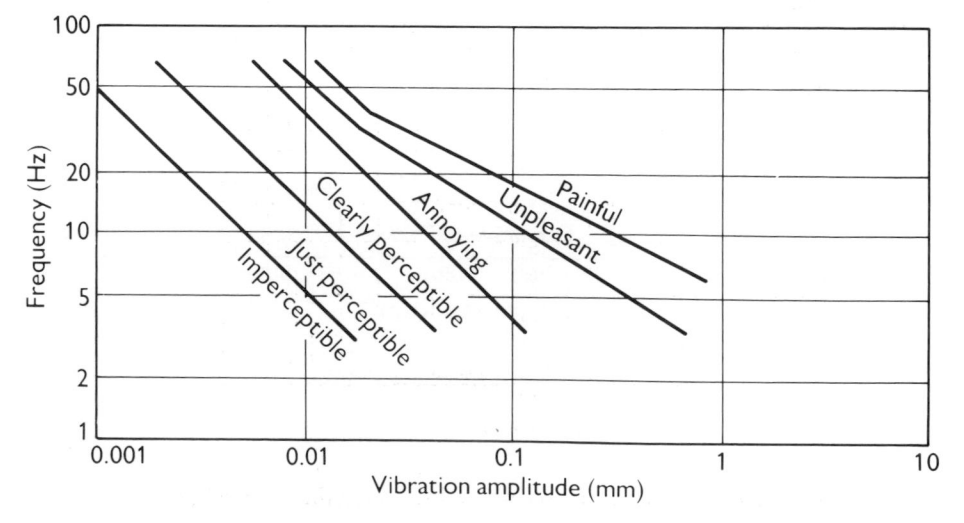

Figure 4.13 Vibration Levels Describing Human Sensitivity

A recent guide to human sensitivity and response to building vibration is provided by British Standard BS6472:1984 — Evaluation of Human Exposure to Vibration in Buildings. This standard provides guidance to acceptable levels of vibration in buildings dependent upon circumstances such as the type of building (workshop, office, residential or critical area) and to the time of day. Suggested levels of maximum vibrations are given for three orthogonal directions, vertical and two horizontal. Figure 4.14 shows the suggested continuous levels (over a 16-hour day) in the vertical direction. The suggested intermittent levels vary according to the number of events and their duration. Intermittent levels may be obtained by multiplying the

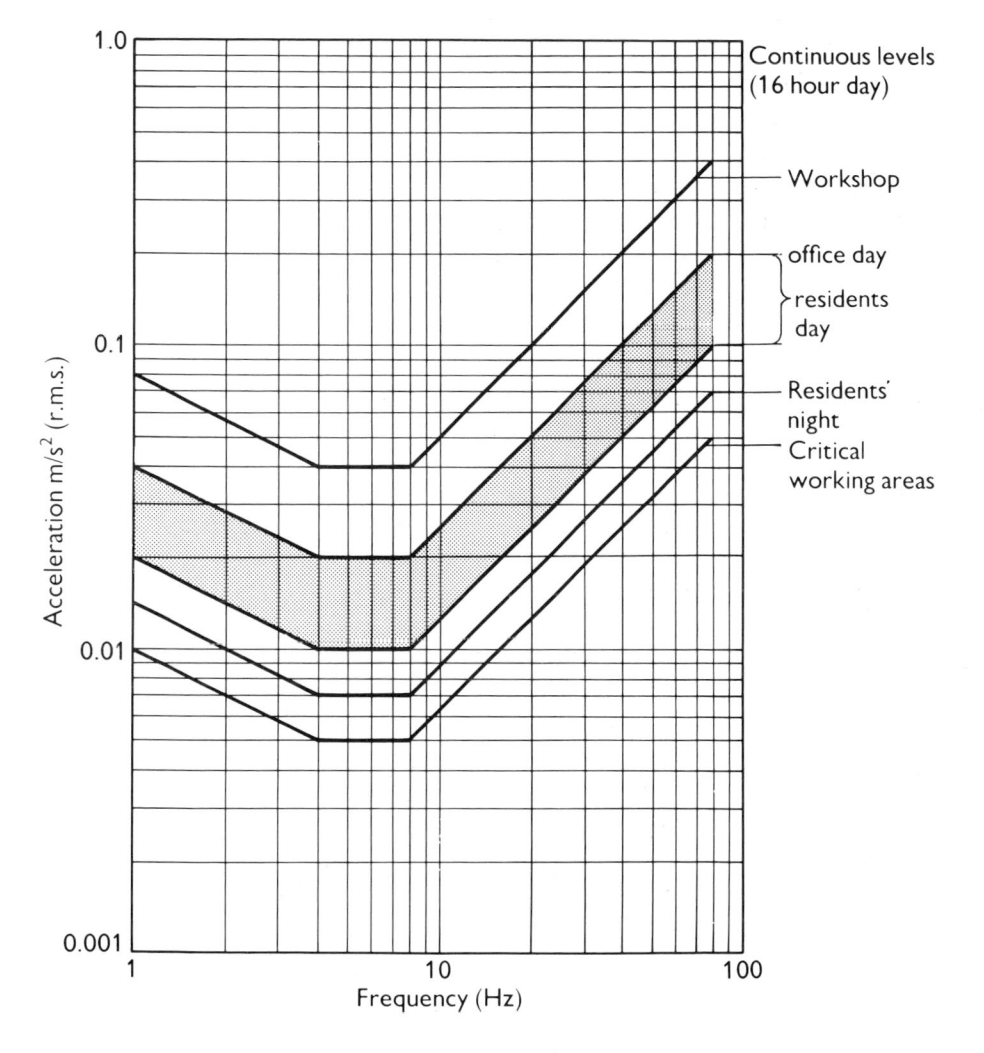

Figure 4.14 Continuous Levels of Vibration in the Vertical Direction

continuous value by the appropriate intermittent factor (Figure 4.15). Where vibration criteria are required for a project reference can be made to this standard.

The most critical levels shown refer to hospital operating theatres and precision laboratories. Lower levels are sometimes required. The most critical vibration requirements currently encountered often occur in the micro-technology industry. Because of the large amount of circuitry inscribed on very small chips any slight movement during manufacturer can ruin the complete process. In these situations suitable vibration criteria must be defined at the earliest stages of design and extensive steps taken to ensure that the criteria is achieved.

Intermittent = Continuous × Factor (below)
N = Number of events per 16 hour day

Duration time (s)

N	5	10	15	20	25	30
5	6.9	5.8	5.2	4.9	4.6	4.4
10	5.8	4.9	4.4	4.1	3.9	3.7
15	5.3	4.4	4.0	3.7	3.5	3.4
20	4.9	4.1	3.7	3.5	3.3	3.1
25	4.6	3.9	3.5	3.3	3.1	3.0
30	4.4	3.7	3.4	3.1	3.0	2.8

Figure 4.15   Intermittency Factors

In less sensitive cases it is still necessary to avoid vibration-induced damage occurring to the building. There is little consistent information because the levels of vibration that can damage buildings depend on factors such as the size and type of building, construction method, the fatigue properties of the materials and the possibility of resonance and have to be taken into account. The levels for the frequency range 5 to 50 Hz shown in Figure 4.16, are the most reliable current information relating to the possibility of damage to buildings. When assessing the likelihood of damage being caused by vibration in a building it is normally found that vibration was unpleasant and painful to the occupier long before damage occurred.

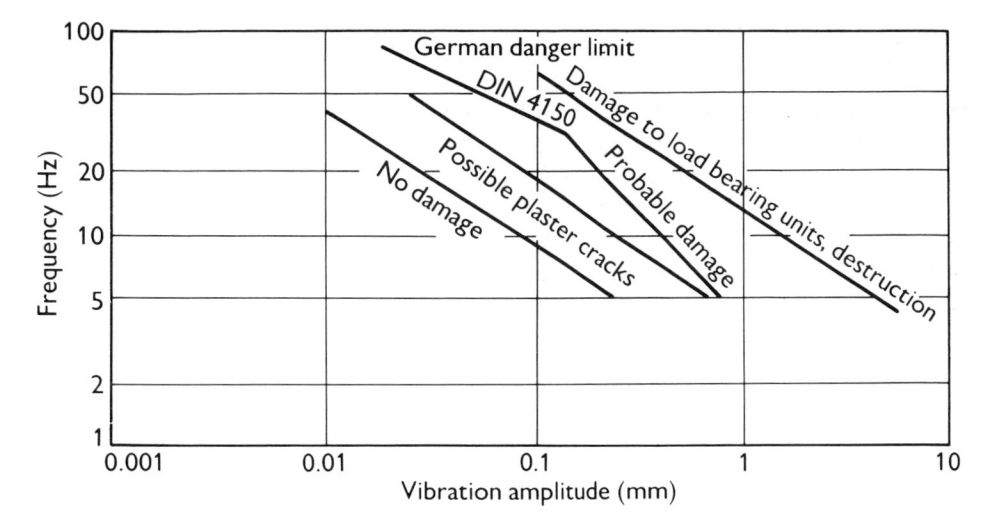

Figure 4.16   Vibration Damage Levels for Buildings

CHAPTER 5

SOUND INSULATION

5.1 INTRODUCTION

When airborne noise is generated in a room it can be transferred to adjacent rooms via a number of transmission paths, for example through walls, floors, building framework and interconnecting ductwork. The nett reduction of airborne sound energy caused by noise transmission from one room to another via all of these paths is termed the airborne sound insulation.

Other energy may be transferred from one area to another as structure-borne noise. Structure-borne noise is caused by vibration from, for example, footsteps, slamming doors or rotating machines being directly transferred into the structure. These structural vibrations are then transferred along various elements of the structure to be re-radiated as noise from the wall, ceiling or floor surfaces of adjacent areas. In order to control this source of noise, vibration isolation, normally in the form of resilient material between the source of vibration and the supporting structure, is introduced. This is discussed in Chapter 6.

5.2 REDUCTION OF AIRBORNE SOUND

Consideration of the requirements for the reduction of airborne sound between areas is of great importance in designing buildings, whether they be offices, schools, auditoria, studios or any other building requiring a reasonable degree of acoustic comfort and privacy. Failure to consider these aspects at a design stage can ruin an otherwise well-planned facility.

Sound can transfer from one area to another via many different paths, as illustrated in Figure 5.1. Airborne sound waves may be incident on the party wall between the two areas, either through travelling directly from the source or after reflection. These then excite "bending waves" in the wall, which re-radiate airborne sound waves in the receiving room. Airborne sound waves may be incident on other surfaces of the source room, again exciting bending waves which will transfer to other building elements before being re-radiated in the receiver room as airborne sound waves. Finally, the

101

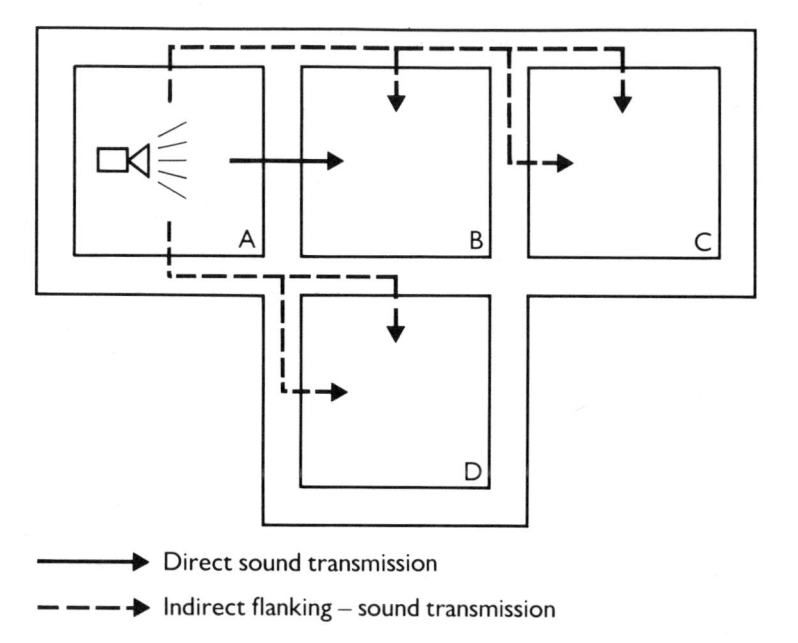

──────▶ Direct sound transmission

─ ─ ─▶ Indirect flanking – sound transmission

Figure 5.1   Sound Transmission by Structural Flanking

mechanical vibration of the sound source may induce bending waves in the floor or walls, which again transfer structurally to other building elements to also produce airborne sound waves in the receiver room.

The intensity of the transmitted sound depends on a number of factors, including:

the frequency of the sound source
the intensity of the sound source
the angle of incidence of the incident sound waves on walls, floors and ceilings
the dimensions of the dividing wall
the mass of the dividing wall
the manner in which the various elements of the structures are linked together
the amount of damping present
other flanking paths such as air-conditioning ductwork
any cracks around pipework or ductwork which penetrate the party wall.

The basic measure of the sound insulation of a partition is termed the Sound Reduction Index. The Sound Reduction Index is equal to the number of decibels by which sound power, which is incident on a partition, is reduced by transmission through it.

This can be expressed mathematically as

$$\text{Sound Reduction Index, SRI} = \log_{10} \frac{W_i}{W_t} \text{ dB} \qquad 5.1$$

where W_i = sound power incident on one side of a partition
and W_t = the sound power transmitted into the air on the other side of the partition.

It should be noted that the Sound Reduction Index is also referred to as the Sound Transmission Loss. Therefore, if a wall has a high Sound Reduction Index, then the sound power transmitted by the wall is very small; for example the sound power transmitted through a wall having a sound reduction index of 60 dB is only one millionth of the incident sound power.

As mentioned above, the sound insulation characteristics of a partition depend on the frequency of the incident sound, being relatively poor at low frequencies and relatively good at high frequencies (Figure 5.2). To more fully describe the sound insulation characteristics of a partition, it is necessary to evaluate and express the Sound Reduction Index at various frequencies, normally in terms of octave or 1/3 octave bands of noise.

5.2.1 Average Sound Reduction Index

When discussing the performance of typical partitions in a general manner, it is common to use the averaged Sound Reduction Index over the 100 Hz to 3150 Hz 1/3 octave bands. This will nearly always be found to be very similar to the single value measured at 500 Hz. There are, however, a number of other single figure sound insulation ratings for airborne sound insulation of partitions, the most common of which are discussed below.

5.2.2 Sound Transmission Class (STC)

This is based on the 1/3 octave band Sound Reduction Indices in the frequency range 125 Hz to 4000 Hz, and originated in the USA. These values for a given partition are compared with a set of transmission class contours until certain conditions are fulfilled. The Sound Reduction Index curve should be matched to the highest STC contour such that the total adverse difference between the contour and the values of 1/3 octave band Sound Reduction Indices (over all sixteen bands) falling below the contour does not exceed 32 dB, and no individual Sound Reduction Index is more than 8 dB below the contour. The STC rating is then the value at 500 Hz for the matched contour. An example of the STC rating system is shown in Figure 5.3, where the dip at 3150 Hz determines STC44 through the 8 dB rule.

5.2.3 Weighted Sound Reduction Index, R_w

This rating is, in most instances, numerically equal to the STC rating and was originated in Europe and is covered by BS5821:1984. The difference between the two

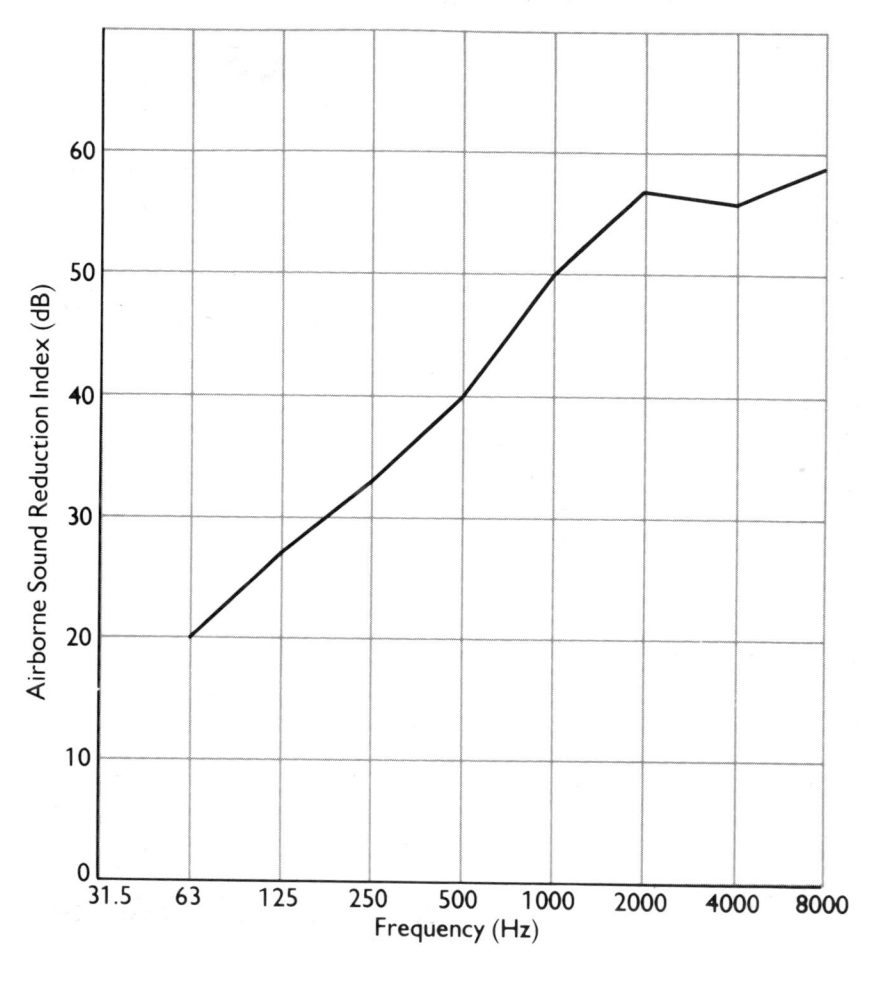

Figure 5.2 Airborne Sound Reduction Index vs. Frequency (125 mm thick solid clinker blocks plastered both sides weight 145 kg/m²)

systems is that the weighted Sound Reduction Index contours are between 100 Hz and 3150 Hz, while the STC contours cover the range between 125 Hz and 4000 Hz.

All of these rating methods depend on accurate measurements of the Sound Reduction Index. Such measurements need to be undertaken in a purpose built laboratory conforming to the requirements of BS2750:1980, or ISO R140–1978. Such facilities give measurements which are representative of the partition under test and minimise all sound flanking paths. A typical test layout is shown in Figure 5.4.

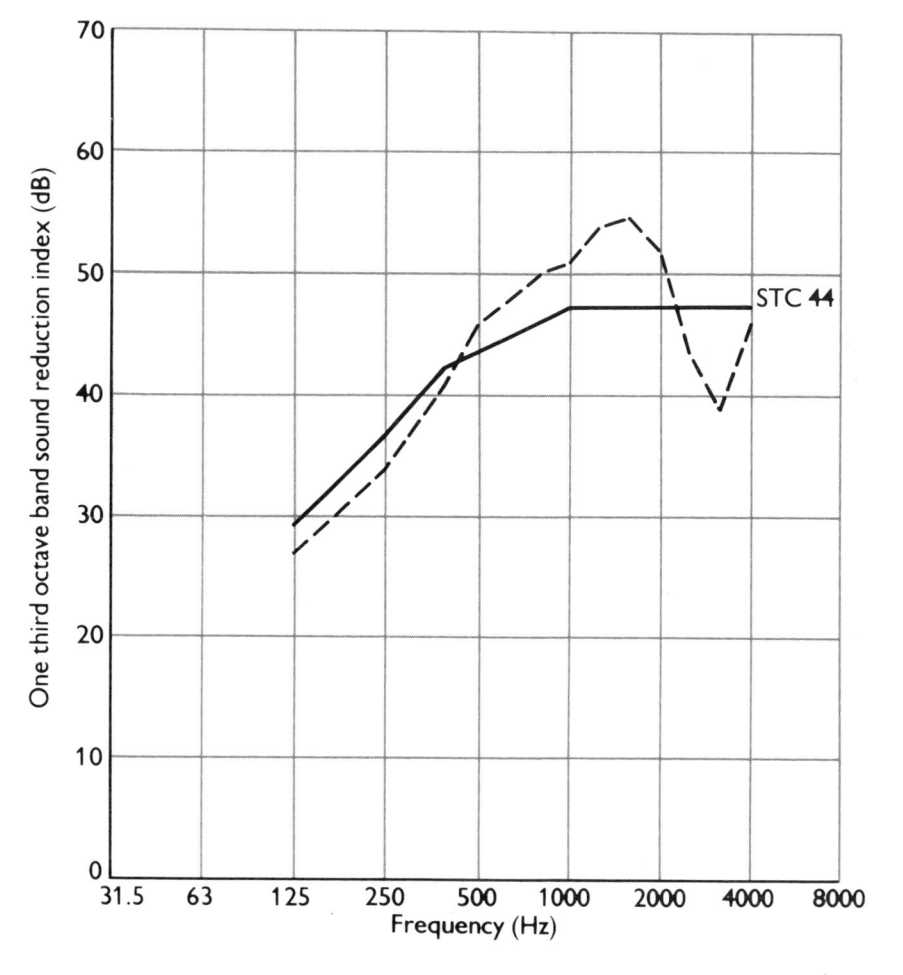

Figure 5.3 Example of STC rating

The source room contains a diffuse sound field of level L_1 dB and is separated from a receiving room of volume V m^3 by a partition of area S m^2 and a compound Sound Reduction Index, R. If the sound level in the receiving room is L_2 and the receiving room reverberation time is T seconds then:

$$R = L_1 - L_2 + 10\log_{10} S + 10\log_{10} \frac{T}{0.163V}$$
5.2

from which R may be computed.

This measurement and calculation should be carried out in 1/3 octave frequency bands.

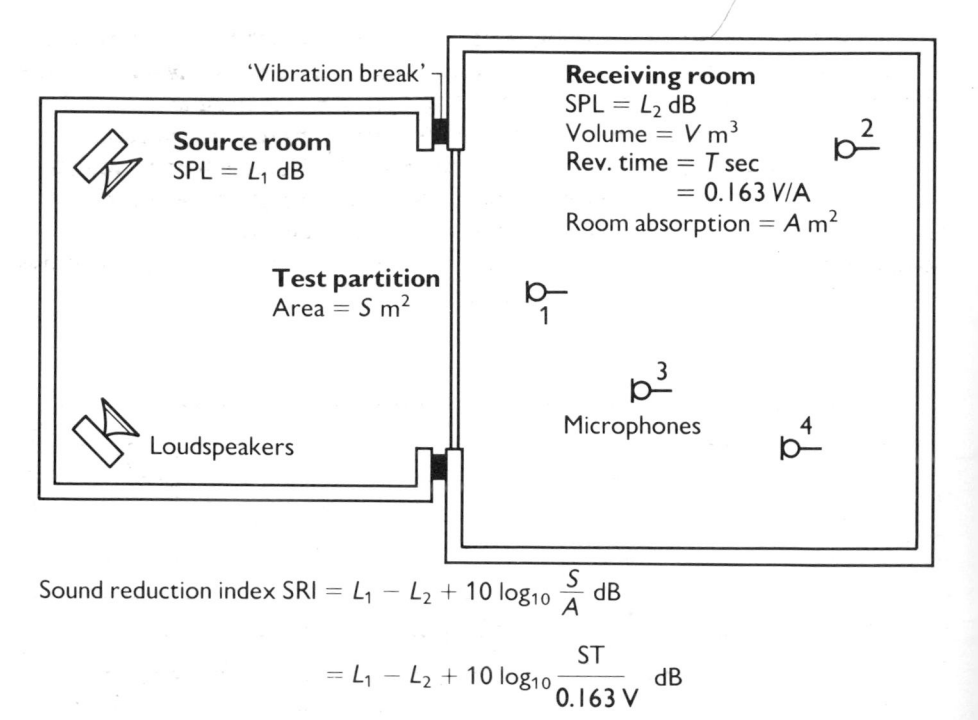

Sound reduction index SRI $= L_1 - L_2 + 10 \log_{10} \dfrac{S}{A}$ dB

$$= L_1 - L_2 + 10 \log_{10} \dfrac{ST}{0.163\, V} \text{ dB}$$

Figure 5.4   Measurement of Sound Reduction Index, R, in the Laboratory

5.3 SOUND INSULATION OF A SINGLE LEAF PARTITION

A single leaf partition is one having both exposed surfaces rigidly connected. Partitions falling into this classification include homogeneous panels of brickwork, timber, plasterboard, concrete blockwork and solid or hollow core concrete. Sandwich constructions may also act as single leaf partitions if the core forms a rigid connection between the two surfaces.

The Sound Reduction Index of this type of partition depends on three main factors, namely:

mass
stiffness
internal damping
mounting arrangement

The transmission of sound results from the partition being forced into vibration by pressure variations caused by incident sound waves. The vibrating structure, mainly bending waves, then radiates acoustic energy into the space on the opposite side of the partition. It would seem reasonable to expect that under these circumstances the more massive the wall, the more difficult it is to excite into bending motion. This dependence of the average Sound Reduction Index on the weight per unit area of the partition is termed the Mass Law and is shown in Figure 5.5. It will be noted that in general there is about a 5 dB increase in Sound Reduction Index for each doubling of superficial weight of the partitions. It is important, however, to note that the insulation provided by a partition also varies with frequency, normally increasing also by about 5 dB for each doubling of frequency.

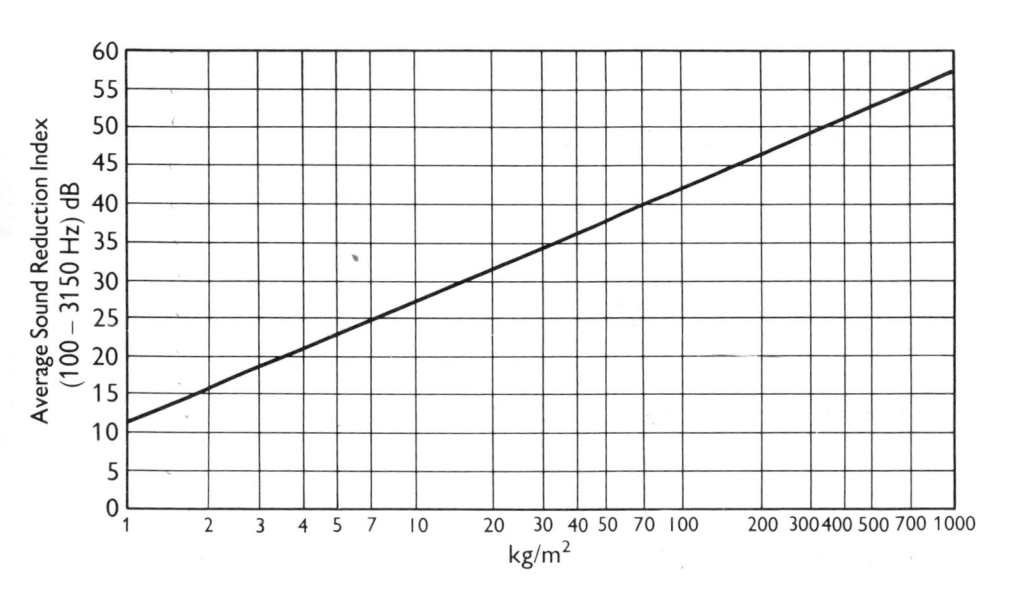

Figure 5.5   Mass Law

At very low and very high frequencies the transmission loss characteristics of a single leaf partition depart from the mass law, due to low frequency resonances within the partition, the stiffness of the partition and a phenomenon known as coincidence. Figure 5.6 represents a generalised frequency response of a partition and is divided into three regions:

Region 1 — Stiffness control with resonances
Region 2 — Mass control
Region 3 — Wave coincidence – damping control

The stiffness of the partition affects its transmission loss performance in both Regions 1 and 3. In Region 1 at low frequencies the partition tends to move as a

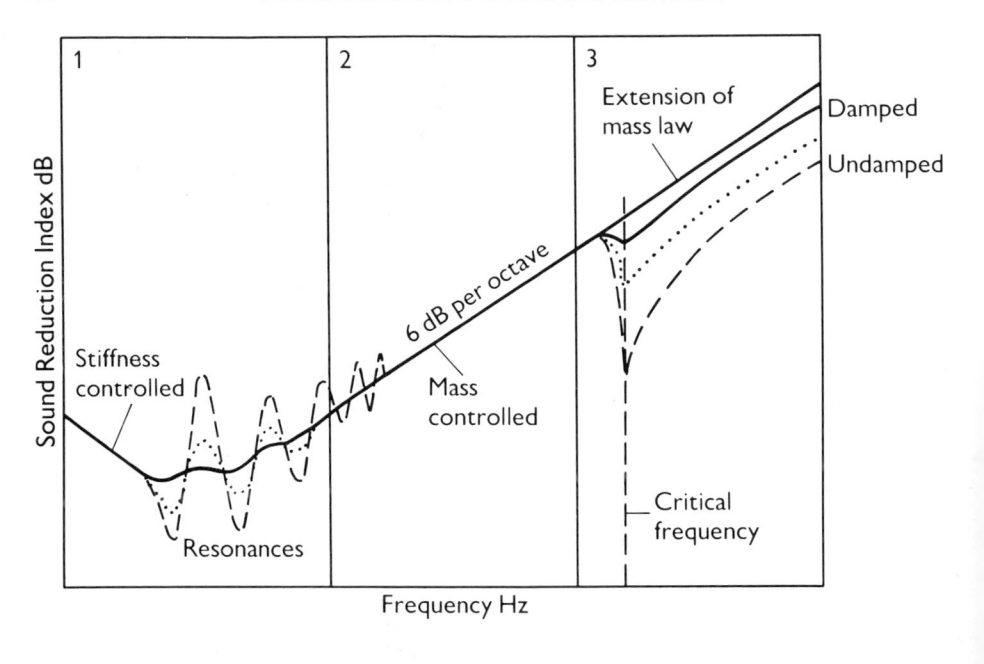

Figure 5.6   Sound Transmission Characteristics of Single Partitions

membrane exhibiting bending motions. The more resistance there is to this bending motion, that is, the greater the panel stiffness, the higher will be the low frequency transmission loss which is obtained. For this reason, lightweight partitions having rigid foam, plastic or honeycomb stiffening cores often exhibit very good low frequency sound transmission characteristics. Such materials, however, also exhibit a complex series of resonances which tend to limit their performance at mid frequencies.

A further effect in Region 3 of the frequency response curve is known as the coincidence dip. This gives a reduction in sound transmission performance above a critical frequency, associated with the partition where the wavelengths of the free bending waves in the material coincide with the tangential components of the incident wave velocity, as indicated in Figure 5.7, together with the trace velocity equalling the bending wave velocity.

The sharpness of the coincidence dip depends on the homogeneity of the material and its internal damping. Also the mounting size can assist the formation of standing bending waves and affect the transmission. In the mass law region, Region 2, the Sound Reduction Index is given approximately by the formula:

$$\text{SRI} = 20 \log_{10} fm - 48 \text{ dB} \qquad 5.3$$

where m = the mass per unit area, kg/m^2
and f = frequency in Hz.

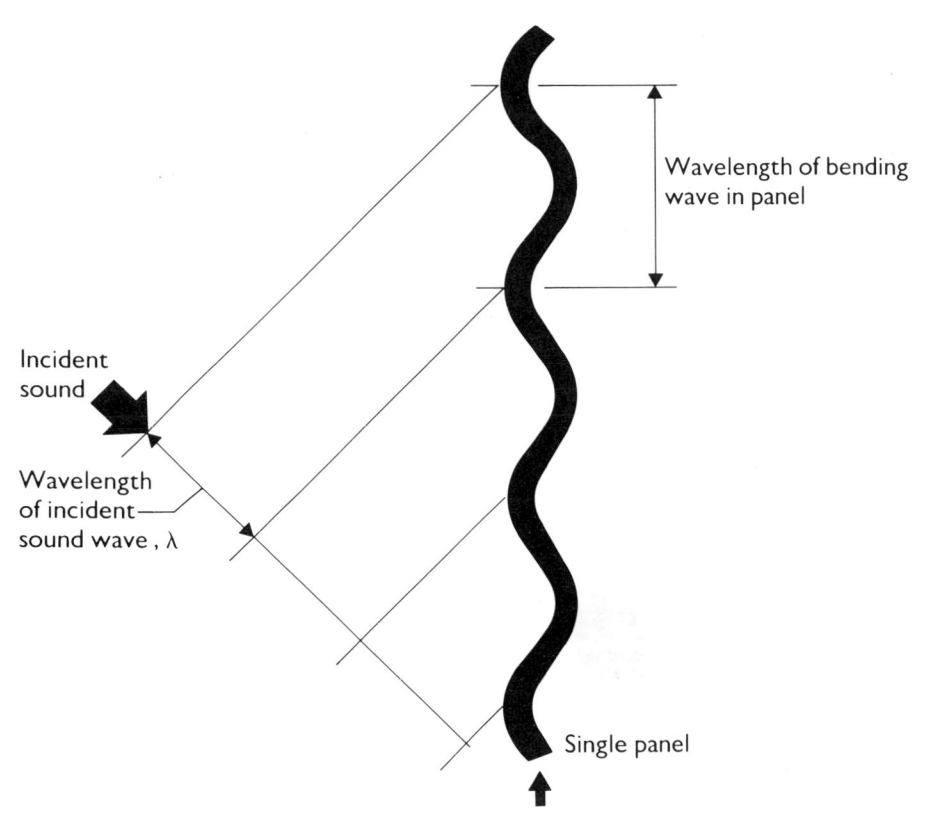

Wavelength of bending wave in panel

Incident sound

Wavelength of incident sound wave, λ

Single panel

Figure 5.7   Wave Coincidence Effect

Hence, the Sound Reduction Index of a single plasterboard sheet partition of mass per unit area 7kg/m² at a frequency of 500 Hz would be approximately 23 dB.

Figure 5.8 shows the octave band Sound Reduction Indices of a variety of common building materials. It will be noted that many of the materials exhibit values of Sound Reduction Index below the mass law prediction. This is mainly attributable to the effect of coincidence dip and to the porosity of the material. Masonry walls are particularly vulnerable to reductions in sound transmission characteristics, due to sound leakage through small unsealed gaps in the construction, or transmission through the porous material of the block itself. Even walls made of solid, dense blockwork or brickwork often exhibit cracks in the mortar joints, which can seriously limit the sound transmission performance. For this reason it is normal practice to seal one side of masonry walls with 12 mm thick plaster.

	Superficial Thickness mm	weight kgm²	63	125	250	500	1k	2k	4k	8k
Panels of sandwich construction										
1.5 mm lead between two sheets of 5 mm plywood	11.5	25	19	26	30	34	38	42	44	47
9 mm asbestos board between two sheets of 18 g steel	12	37	16	22	27	31	27	37	44	48
"Stramit" compressed straw between										
two sheets of 3 mm hardboard	56	25	15	22	23	27	27	35	35	38
Single masonry walls										
Single leaf brick, plastered both sides	125	240	30	36	37	40	46	54	57	59
	255	480	34	41	45	48	56	65	69	72
	360	720	36	44	43	49	57	66	70	72
Solid breeze or clinker blocks, plastered										
(12 mm both sides)	125	145	20	27	33	40	50	57	556	59
Solid breeze or clinker blocks, unplastered	75	85	12	17	18	20	24	30	38	43
Hollow cinder concrete blocks, painted										
(concrete base paint)	100	75	22	30	34	40	50	50	52	53
Hollow cinder concrete blocks, unpainted	100	75	22	27	32	37	40	41	45	48
"Thermalite" blocks	100	125	20	27	31	39	45	53	38	62
Glass bricks	200	510	25	30	35	40	49	49	43	45
Double masonry walls										
280 mm brick, 56 mm cavity, strip ties,										
outer faces plastered, 12 mm	300	380	28	34	34	40	56	73	76	78
280 mm brick, 56 mm cavity, expanded metal ties,										
outer faces plastered, 12 mm	300	380	27	27	43	55	66	77	85	85
Stud partitions										
50 mm × 100 mm studs, 12 mm insulating board										
both sides	125	19	12	16	22	28	38	50	52	55
50 mm × 100 mm studs, 9 mm plaster board and										
12 mm plaster coat both sides	142	60	20	25	28	34	47	39	50	56
Single glazed windows										
Single glass in heavy frame	6	15	17	11	24	28	32	27	35	39
	8	20	17	18	25	31	32	28	36	39
	9	22.5	18	22	26	31	30	32	39	43
	16	40	20	25	28	33	30	38	45	48
	25	62.5	25	27	31	30	33	43	48	53
Double glazed windows										
2.44 mm panes, 7 mm cavity	12	15	15	22	16	20	29	31	27	30
9 mm glass panes in separate frames, 50 mm cavity	62	34	18	25	29	34	41	45	53	50
65 mm glass panes in separate frames, 100 mm cavity	112	34	20	28	30	38	45	45	53	50
6 mm glass panes in separate frames, 188 mm cavity	200	34	25	30	35	41	48	50	56	56
6 mm glass panes in separate frames, 188 mm cavity										
with absorbent blanket in reveals	200	34	26	33	39	42	48	50	57	60
Doors										
Flush panel, hollow core, normal cracks										
as usually hung	43	9	9	12	13	14	16	18	24	26
Solid hardwood, normal cracks as usually hung	43	28	13	17	21	26	29	31	34	32
Typical proprietary "acoustic" door, double										
heavy sheet steel skin, absorbent in airspace,										
special furniture and seals in heavy steel frame	100	–	37	36	39	44	49	54	57	60
Floors										
T & G boards, joints sealed	21	13	17	21	18	22	24	30	33	63
T & G boards, 12 mm plasterboard ceiling under,										
with 3 mm plaster skim coat	235	31	15	18	25	37	39	45	45	48
As above with boards "floating" on glass wool mat	240	35	20	25	33	38	45	56	61	64
Concrete, reinforced	100	250	32	37	36	45	52	59	62	63
	200	460	36	42	41	50	57	60	65	70
	300	690	37	40	45	52	59	63	67	72
126 mm reinforced concrete with "floating" screed	190	420	35	38	43	48	54	61	63	67
6 mm and 9 mm panes in separate frames, 200 mm cavity,										
absorbent blanket in reveals	215	42	27	36	45	58	59	55	66	70

Figure 5.8   Representative Values of Airborne Sound Reduction Index for Some

5.4 SOUND INSULATION OF DOUBLE LEAF PARTITIONS

High values of sound insulation can be achieved through the use of cavity constructions. To achieve equivalent sound insulation using single leaf partitions would require very massive structures because the sound insulation only increases by 5 dB per doubling of mass as discussed above (Figure 5.5).

The cavity construction would typically comprise of two impervious layers spaced apart with a minimum of structural connections between them. Mineral wool or glass fibre mats should be hung in the cavity. With lightweight partitions cavities up to 50 mm wide will improve the insulation at mid and high frequencies, but will give little benefit at low frequencies. To achieve significant low frequency benefits requires the cavity width to be increased to 150 mm or above.

Where very high values of sound transmission loss are required from lightweight partitions, the two leaves should be supported off independent frames. Cavity brick walls should utilise lightweight wire "butterfly" ties rather than heavy metal strips, whilst in critical applications, such as studio wall constructions, flexible wall ties should be employed. In all cases, a great deal of care should be taken to prevent any builders' rubble or mortar from falling into the cavity, which could bridge the two leaves and act as a "short circuit".

When considering the sound transmission characteristics of a double leaf partition, one might expect the resulting performance to be the sum of the performance of the individual leaves. In practice, this is not the case, since with reasonable spacings the two leaves will be acoustically coupled by the springiness of the enclosed air.

Examples of two leaf partitions include plasterboard, metal and wood stud partitions, cavity masonry walls, double glazed windows and timber joist flooring systems. The sound reduction indices of various double leaf systems are also shown in Figure 5.8.

5.5 SOUND TRANSMISSION THROUGH NON-HOMOGENEOUS PARTITIONS

The sound transmission characteristics of a given partition will be affected by the introduction of, for instance, a door or window which has sound transmission characteristics differing from those of the main partition. The overall Sound Reduction Index of such a compound system can be computed through the use of the chart shown in Figure 5.9. In order to carry out the computation, it is necessary to know the relative areas of the various elements forming the wall and their respective Sound Reduction Indices.

5.5.1 Example

A plantroom wall which must contain a single leaf access door requires an average composite sound transmission loss of 45 dB to reduce noise levels in an adjacent area to acceptable standards.

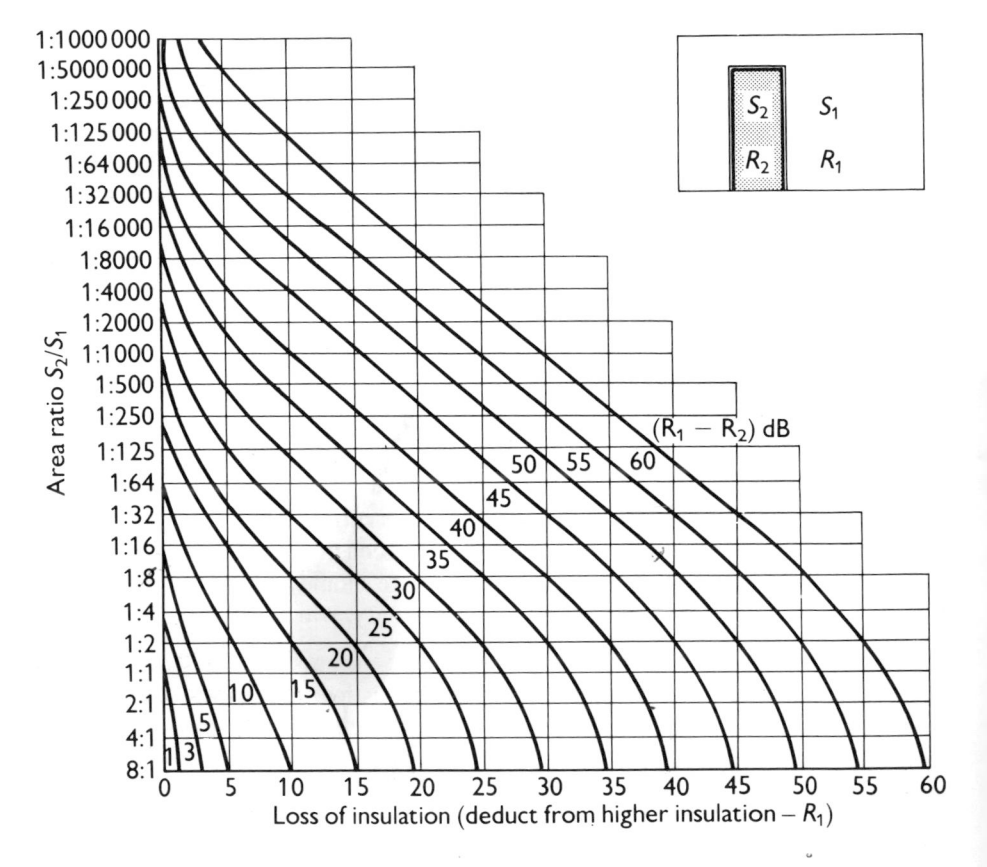

Figure 5.9   Computation of Octave Band Sound Reduction Index of Non-uniform Partitions

If the wall structure itself exhibits an average Sound Reduction Index of 45 dB, then reference to Figure 5.9 shows that the door would need to exhibit the same characteristics to achieve the desired result as any deviation below 45 dB will negate any chance of achieving 45 dB for the compound whole. A heavy metal, single leaf door would therefore, be required, which could be very costly.

Alternatively, if the wall is constructed of a material which gives an average Sound Reduction Index of 50 dB and the wall has an area ten times that of the access door (Figure 5.10) reference to Figure 5.9 indicates that the door would need to give an average Sound Reduction Index of approximately only 37 dB to achieve the overall 45 dB partition performance. This is derived by noting that the 50 dB value of wall, the higher insulation, R, allows a 5 dB loss to fall to 45 dB for the combination wall plus door. Looking up to 5 dB loss line on Figure 5.9 until reaching the 1:10 area ratio indicates that the other component, the door, R_2, can be down on the higher insulation

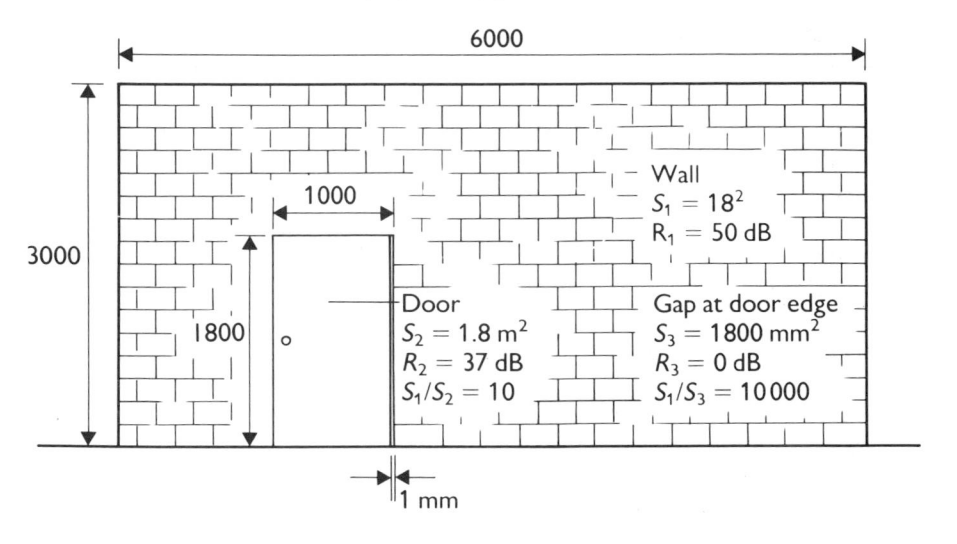

Figure 5.10 Partition SRI Loss Example — Plant Room Wall

by 13 dB, that is, $R_1 - R_2 = 13$ dB. Hence the door sound reduction index of $50 - 13 =$ 37 dB. Such doors are standard items in the catalogue of most acoustic door manufacturers.

As mentioned earlier in this Chapter, the sound transmission characteristics of a partition construction can be seriously affected by sound leakage through small gaps or cracks.

5.5.2 Example

From Figure 5.9 or 5.11 it can be deduced that there is a limit on the aperture free area which can be allowed in order to maintain the acoustic integrity of a particular partition construction. For instance, if a partition made from a material which has a nominal average sound reduction index of 60 dB is required to maintain the performance of a wall to, say, an average Sound Reduction Index of 40 dB, then the free area of the aperture must be limited to one ten thousandth of the area of the wall (see Figure 5.9 or 5.11). This is obtained from the chart thus:

The loss of insulation is $60 - 40$ dB $= 20$ dB.
The difference in transmission loss between the partition (60 dB) and the aperture (0 dB) is 60 dB.

Therefore, entering the graph on Figure 5.11 on the horizontal axis at 20 dB, read vertically upwards until the $R_1 - R_2 = 60$ curve is intersected. Read across horizontally until the vertical area ratio axis is intersected. This is then the limitation on area ratio required to achieve the 40 dB performance required, that is, 1:10,000.

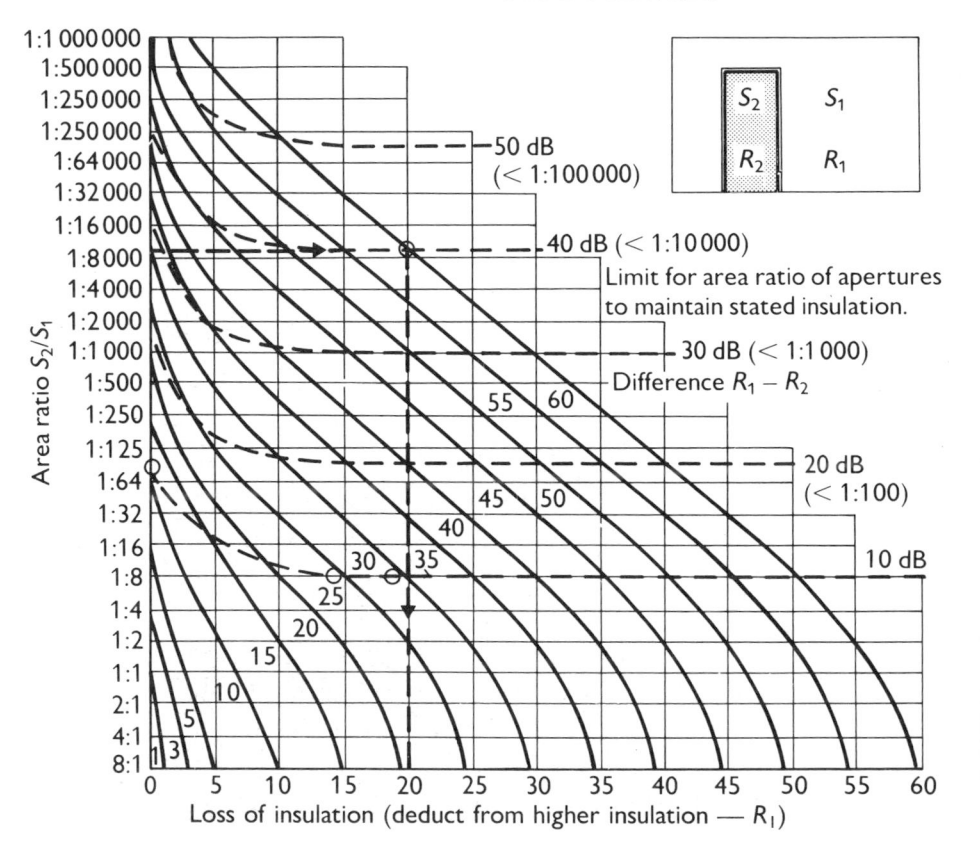

Figure 5.11   Partition Loss of Insulation.

Further consideration of this effect will yield the set of area ratio limitation curves shown on Figure 5.11.

To further illustrate this point, let us return to the plantroom wall with its access door. The provision of a 37 dB door in a 50 dB wall to give an average sound transmission loss of 45 dB assumes that no leakage occurs via the compression seal on the door. Suppose the door has been installed badly and a 1mm gap exists under the compression seal along one edge of the door; from Figure 5.10 it can be seen that the area of the gap would be one ten thousandth of the total partition area. Figure 5.11 shows that the area ratio intercepts the 45 dB insulation difference curve at a point giving a loss of insulation of 6 dB. Therefore, the overall performance of the 50 dB plantroom wall with a 37 dB poorly sealed door would be 39 dB, as opposed to the 45 dB achievable with good door sealing arrangements. Further reference to Figure 5.11 indicates that however high the performance of the plantroom wall, the overall structure performance would be limited to 40 dB due to the poor door sealing detail.

COMPLEX WALL

Only a detailed calculation will yield the effective overall performance of this plant-room wall with its acoustic doors and acoustic louvre in a larger area of higher Sound Reduction Index wall.

These examples serve to demonstrate the importance of sealing structures to maintain their sound insulation. It is common to find that a particular construction fails to achieve the required performance due to small holes through the structure itself, or gaps left between adjacent panels or around the edges of the partition where it abuts the flanking structures. These are often left untreated due to the common misapprehension that such small air paths will not significantly affect the acoustic performance of the wall. This is not so. In a sound insulating partition, any hole or gap whatever its size must be sealed. Non-hardening mastic materials can be successfully used for such applications.

There are other air paths which can exist between the two sides of a partition other than those discussed above. These are illustrated in Figure 5.12 and are termed flanking transmission paths. Any of the indicated transmission paths can reduce the effective acoustic transmission performance of a partition and make the resulting condition unacceptable in many situations. Such flanking transmission paths can be

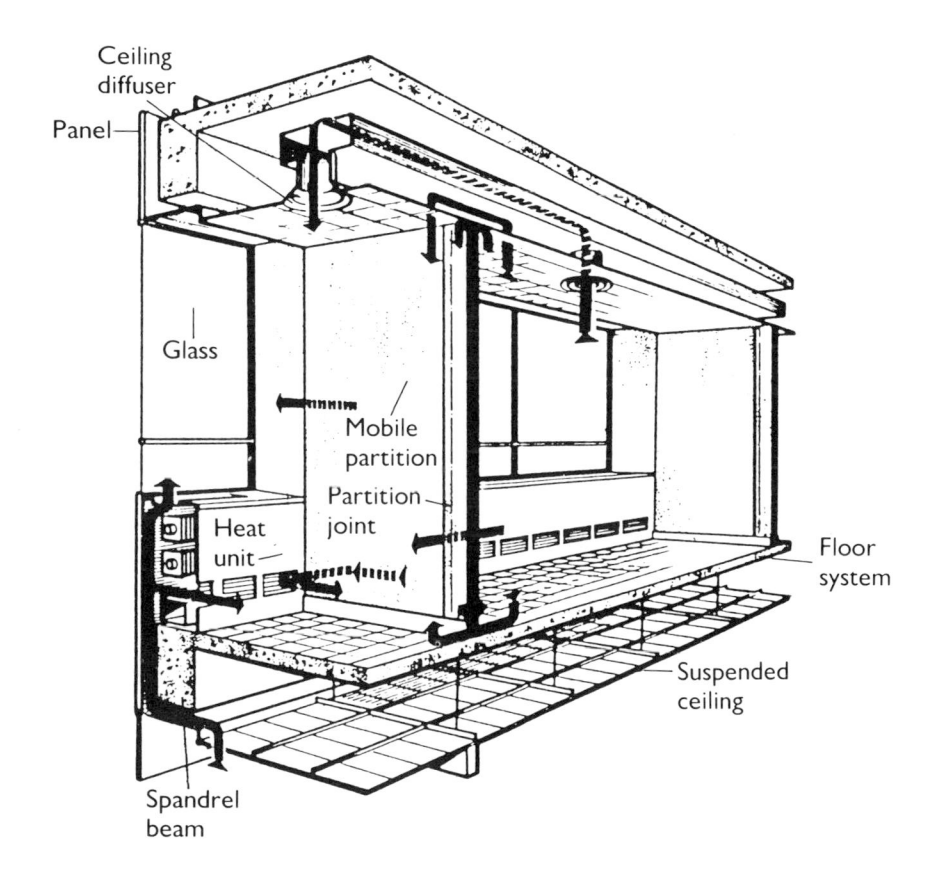

Figure 5.12   Flanking Transmission Between Rooms

controlled by careful acoustic design details. For example, for the situation shown in Figure 5.12, the following control measures should be adopted:

compression seals should be included for the movable partition at all edges and joints;

a septum partition extending up to the structural soffit should be included above the line of the partition to prevent transmission through the ceiling void;

crosstalk attenuators should be included in the air-conditioning ductwork;

sound insulating baffles included to control sound transmission via the heat unit past the spandrel beam. Alternatively, attenuation should be introduced at the supply and discharge louvres.

5.6 SOUND INSULATION REQUIREMENTS

When considering the sound insulation requirements of a partition between two areas, several factors should be taken into account. Firstly, the required function of the two rooms should be considered, that is, do we require conversation in one room to be inaudible in the next room as we would require in a studio situation, or can we accept that some speech will be audible, although not intelligible, as we would require for a routine office? The other factor that must be considered is the background sound level which exists in the critical area. The intelligibility of speech and the perception of extraneous noises is controlled by the masking created by the background sound level present, due to various sources such as air-conditioning systems and traffic. The higher the background sound level, the more effective it will be in masking unwanted sounds. However, the background noise must not become intrusive in itself, and, hence, a balance between the background sound level and the partition sound reduction performance must be achieved.

Figure 5.13 below gives a guide to the selection of the sound insulation performance of a partition between two areas under various background sound level conditions and for the various requirements stated. This guidance is appropriate to sound transmission between hotel bedrooms, studios and offices.

Sound as heard by occupant	Average Sound Insulation Plus Ambient Noise	
	dBA	NR
Intelligible	70	65
Ranging between intelligible and unintelligible	75–80	65–75
Audible but not obtrusive (unintelligible)	80–90	75–85
Inaudible	90	85

Figure 5.13 Conversation Noise Insulation Guidance

5.6.1 Example

We need to specify the sound insulation requirements of a partition between two offices, one occupied by a person carrying out confidential conversations, and the other by his secretary. Suppose the background sound level in both offices is NR30 — this falls into the category in Figure 5.13 of "audible but not obtrusive". The sum of the average Sound Reduction Index of the wall and the background NR will, therefore, require to be about 80. Therefore, the average Sound Reduction Index of the wall will need to be 50 dB to achieve the required privacy conditions. From Figure 5.8 it can be seen that this requirement could be met by single leaf plastered brickwork, or a double leaf plasterboard construction with mineral wool in the cavity.

It should be emphasised that having a very low background sound level in a room can reduce the effectiveness of the room structure in controlling the perception of extraneous noise in the room. For example, in studios the background sound level design requirements can be typically NR20. The major source of such noise is normally the air-conditioning system serving the studio. Thus, from Figure 5.13 it can be seen that in order to render sound such as speech in adjacent areas inaudible in the studio, the structure surrounding the studio will need to have an average Sound Reduction Index of approximately 65 dB. This is only achievable by the use of specially designed wall and floor systems. However, if when the studio was built, it was found that the background level was only NR5, then the sum of the average sound reduction index and the background NR would be only 70, which from Figure 5.13 would mean that speech in adjacent areas would not only be audible in the studio, but on occasions also intelligible. This is clearly unacceptable for this situation.

Such problems do occur in practice, since the insertion loss requirements for crosstalk attenuators for flanking control in the air-conditioning systems can often reduce the mid and high frequency noise levels produced by the air-conditioning systems to very low levels. The background sound levels in the studio must then be increased artificially. This may be achievable through the use of perforated plates in air-conditioning ductwork close to the outlet grilles to produce airflow generated noise, or possibly through the introduction of electronically generated wide–band noise through strategically placed loudspeakers.

The latter solution is termed "sound conditioning", and although it has not to date been used in studio applications, it has been successful in open plan office environments. The idea here is to introduce a non-intrusive optimised wide band noise into the open plan office space to improve aural comfort and mutual privacy for the office occupants. This is achieved through arrays of loudspeakers located in the void above the suspended ceiling — a typical system is shown schematically in Figure 5.14. Normally it is found that best results are achieved by generating the octave band sound spectrum levels shown in Figure 5.15. This provides a high degree of speech masking, whilst the sound in itself is not intrusive. It should be noted that in

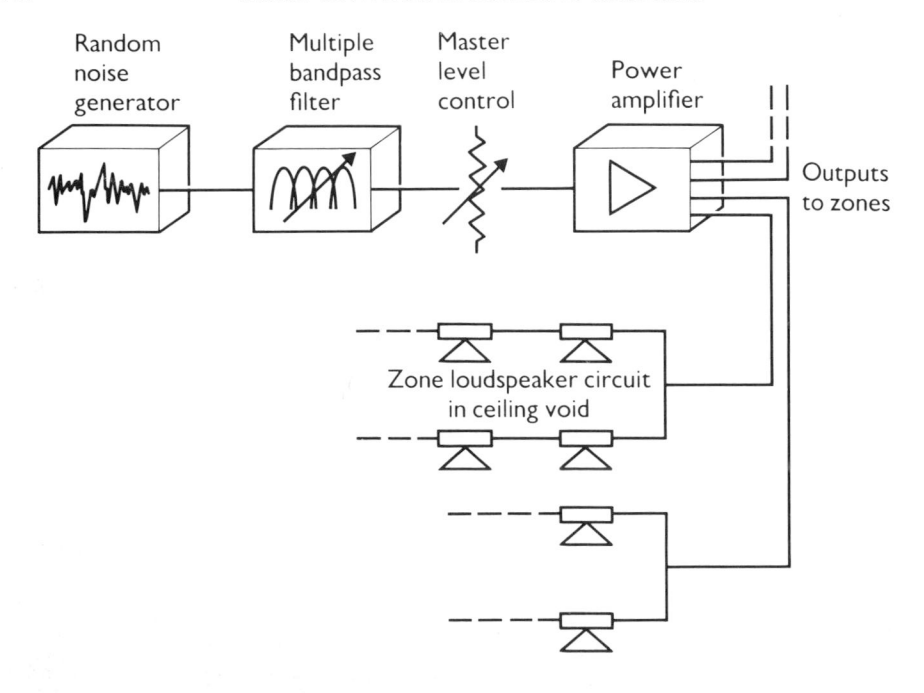

Figure 5.14 Diagrammatic Representation of a Typical Sound Conditioning System

order to achieve the most effective result using such a system, the sound absorption characteristics of the suspended ceiling and partition screens should be high with typical mid-frequency (500 Hz) absorption coefficients between 0.6 and 0.7. Further, certain areas of wall may require treating with acoustically absorptive panels. With careful design, it is possible to achieve a high degree of privacy between adjacent work stations separated by acoustically absorbent screens only 2 m high.

5.7 PLANTROOM BREAKOUT

One of the important factors which must be considered when controlling noise from mechanical services serving a building is that of noise breakout from the plantroom structure, both into other noise sensitive areas of the building and into the surrounding environment.

Plantrooms used to be traditionally located at basement level, or on the roof of multi-storey buildings. With the introduction of high rise buildings, plantrooms at intermediate levels are now commonly found with supplementary plant at various locations throughout the building. These plantrooms are obviously a potential source

SOUND CONDITIONING

Whilst the use of extra sound to create an acceptable noise environment may seem unusual, here ceiling-mounted loudspeakers are generating a constant background level in a library application.

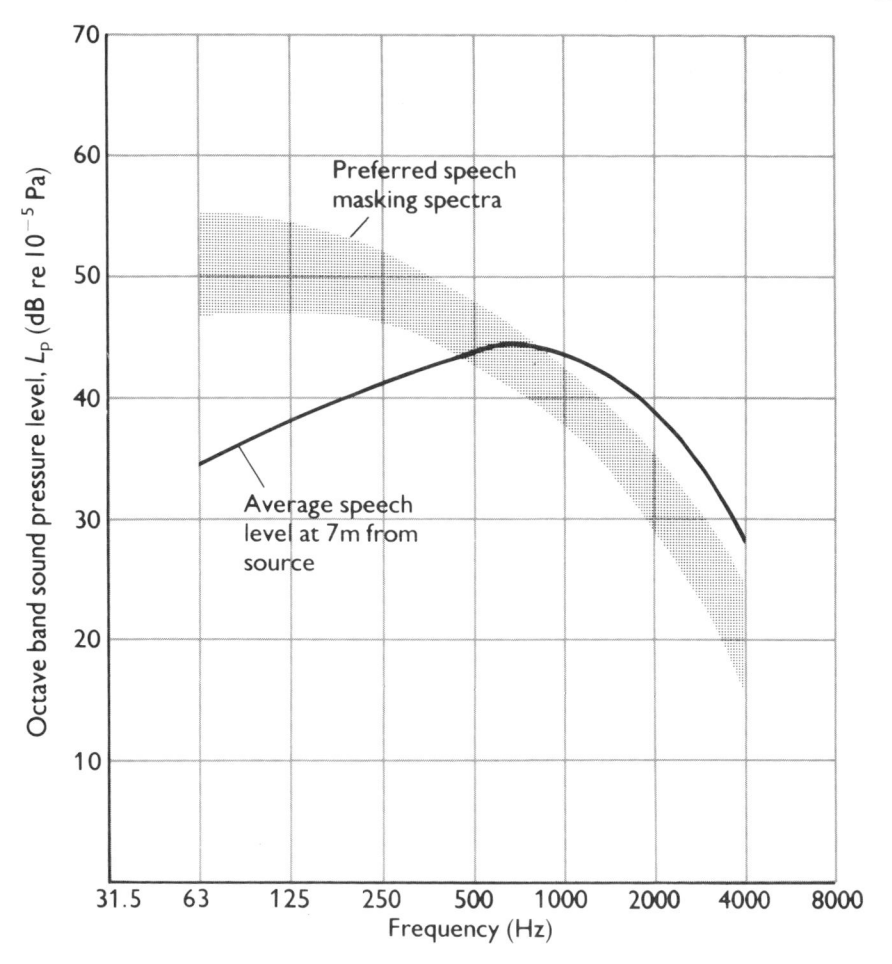

Figure 5.15   Speech-masking Spectrum

of high levels of airborne and structure-borne noise. Structure-borne noise is discussed in the Chapter on vibration isolation (Chapter 6). It is therefore essential to consider in detail the required sound transmission characteristics of the plantroom structure at the early design stage of the building, as this can have a considerable bearing on the overall weight and costs of the building. It will also avoid the high costs of remedial work that can be incurred by failure to meet the required noise standards.

In calculating the required plantroom structure sound insulation characteristics, it is first essential to have a good idea of the total octave band reverberant sound pressure levels in the plantroom due to the various noise sources present. Often it is possible to obtain sound power level data from plant manufacturers at the plant duties envisaged. From this data the reverberant sound pressure level within the plantroom can be calculated, as shown in a later section of this Chapter. If such information is not available, then it may be necessary to measure plantroom reverberant sound pressure levels in another plantroom housing similar plant, or use empirical data to estimate the required information. For guidance, typical plantroom sound pressure levels are shown in Figure 5.16. Most acoustic consultants hold data banks of sound power and/or sound pressure level data relevant to most situations which are encountered.

Equipment	63	125	250	500	1k	2k	4k	8k
Chillers (100 tons +)								
Package recip.	90	89	92	93	92	90	86	81
Package centrif.	91	93	95	96	99	102	97	92
Package Screw	76	80	92	89	85	80	75	73
Boilers (50hp +)								
Boilers	92	92	89	86	83	80	77	74
Steam Valves	70	70	70	75	80	85	90	95
Fans (up to 75mm static pressure)								
25hp	95	94	91	84	79	74	69	64
40hp	98	97	94	87	82	77	72	67
100hp	101	100	87	90	85	80	85	80
250hp	104	103	100	93	88	83	78	83
Fans (150mm static pressure or over)								
50hp	107	106	103	96	91	86	81	76
100hp	111	110	107	99	94	89	84	79
250hp	113	112	109	102	97	92	87	82
Cooling Towers (intake noise)								
Centrif. blow thru	85	85	83	81	79	76	73	68
Prop induced draft	98	97	94	90	85	80	75	70
Pumps (1400 rpm)	83	86	90	90	86	83	80	75
25hp	86	89	93	93	89	86	83	78
50hp	89	92	96	96	92	89	86	81
100hp	92	95	99	99	95	92	89	84
250hp								
Pumps (2800 rpm)								
25hp	84	88	92	93	93	92	86	79
50hp	88	92	96	97	97	96	90	83
100hp	91	95	99	100	100	99	93	86
Air Compressors (reciprocating and centrifugal)								
1.2hp	83	83	83	86	89	89	89	84
3.10hp	86	86	86	89	92	92	92	87
10.100hp	89	89	89	92	95	95	95	90
Engines (internal combustion or diesel)								
25gp	94	97	96	92	92	91	85	78
50hp	96	99	98	94	94	93	87	80
100hp	100	103	102	96	96	95	91	84
250hp	104	107	106	102	102	101	95	88
500hp	106	109	108	104	104	103	97	90
1000hp	110	113	112	108	108	107	101	94
2500hp	114	117	116	112	112	111	105	98

Figure 5.16   Sound Pressure Level (dB) of Mechanical Equipment at 1 m

5.8 PLANTROOM BREAKOUT TO OTHER AREAS WITHIN THE BUILDING

The following example shows how the noise breakout from a fan plantroom into an adjacent area within the same building may be computed from a knowledge of the fan sound power levels and certain details relating to the room acoustics on either side of the dividing wall.

5.8.1 Example

A committee room measuring 5 m x 7 m x 4 m is adjacent to a 200 m^3 plantroom. The common area of wall between the two areas measures 5 m x 4 m and is constructed of 230 mm brickwork plastered on one side. The plantroom and committee room reverberation time in both the 63 Hz and 125 Hz bands is 2 seconds, and in all higher bands is one second. The fan manufacturers have provided the following unit sound power level (L_w) data, under free air inlet conditions:

Frequency	63	125	250	500	1k	2k	4k	8k	Hz
L_w	96	104	103	98	91	86	84	79	dB

First compute the reverberant sound pressure level (L_p) in the plantroom. This is done using the formula:

$$L_p = L_w + 10\log_{10}T_p - 10\log_{10}V_p + 14 \qquad\qquad 5.4$$

where L_p is the reverberant sound pressure level in the plantroom, dB;
L_w is the sound power level of the fan, dB;
T_p is the plantroom reverberation time, s;
V_p is the plantroom volume, m^3.

It is necessary to calculate L_p in the octave bands between 63 Hz and 8 kHz as follows:

Frequency	63	125	250	500	1k	2k	4k	8k	Hz
Fan sound power level, L_w (Manufacturers data)	96	104	103	98	91	86	84	79	dB
$10\log_{10}T_p$	+3	+3	0	0	0	0	0	.0	dB
$10\log_{10}V_p$ ($V_p = 200\text{m}^3$)	−23	−23	−23	−23	−23	−23	−23	−23	dB
+14	+14	+14	+14	+14	+14	+14	+14	+14	dB
Plantroom sound pressure level, L_p	90	98	94	89	82	77	75	70	dB

The reverberant sound pressure level, L_c, in the committee room is then given by the formula:

$$L_c = L_p - R + 10\log_{10}S + 10\log_{10}\frac{T_c}{0.163V_c}\qquad\qquad 5.5$$

where R is the octave band Sound Reduction Index of the plantroom wall, dB;
S is the common area of plantroom wall between the plantroom and committee room, m²;
T_c is the reverberation time in the committee room, s;
V_c is the volume of the committee room, m³.

Carrying out this calculation in each octave band produces the following results:

Frequency	63	125	250	500	1k	2k	4k	8k	Hz
Plantroom sound pressure level L_p	90	98	94	89	82	77	75	70	dB
R (Sound Reduction Index of plastered 230mm brickwork)	−39	−41	−45	−49	−57	−63	−62	−65	dB
+ 10$\log_{10}S$ (Area = 5 × 4 = 20m²)	+13	+13	+13	+13	+13	+13	+13	+13	dB
+ 10$\log_{10}T_c$	+3	+3	0	0	0	0	0	0	
− 10$\log_{10}$ 0.163V_c (volume = 5 × 7 × 4 = 140m³)	−14	−14	−14	−14	−14	−14	−14	−14	dB
Committee room sound pressure level L_c	53	59	48	39	24	13	12	4	dB
NR40	67	57	49	44	40	37	35	33	dB

The resultant sound level in the committee room would be NR40 + 2 determined by the 125 Hz octave band, which is considerably above the level normally considered acceptable for this type of room, that is to say, NR30. This domination of the resultant noise level by the 125 Hz octave band is typical of problems associated with fan noise breakout from plantrooms.

It is all too common to find this type of situation where noise critical areas are located close to plantrooms. For example, prestige office accommodation for directors and managerial staff is frequently located at the top of buildings close to plantrooms. Further, basement plantrooms are often located below noise critical areas. This situation not only occurs in office accommodation, but also in hospitals with plantrooms close to wards and operating theatres, in teaching establishments with plantrooms close to lecture theatres, and in broadcasting establishments with plantrooms close to studios and control rooms. Such situations occur through poor planning, or lack of awareness on the part of the architect or designer. If such matters are not considered at a very early planning stage, then the cost implications of redesign can be high and once a problem has occurred, then there is no simple solution to resolve the problem. However, the noise control measures established over the last ten to twenty years do now make well understood and engineered solutions available to solve the noise problems of the inevitable clashes of requirements. More especially of note is the lighter weight of these measures.

Where it is anticipated that high background noise levels are going to occur in areas surrounding plantrooms, it is possible at the planning stage of a project to consider a variety of techniques to control the noise. Such techniques include:

Relocate the plantroom
Relocate the noise critical area
Upgrade plantroom insulation. NB Floating Floors and walls
Reselect equipment for quiet running
Provide noise control by absorbent materials
Provide local insulation
Provide local attenuation
Apply local vibration isolation.
Each of these techniques is discussed below.

5.8.2.1 Relocate the Plantroom This can only be considered at the very early planning stage of a project. It may be that the advantages of providing a plantroom away from the noise sensitive areas it is serving might be outweighed by the higher capital and running costs of the system.

5.8.2.2 Relocate the Noise Critical Areas Again, this approach can only be considered at the very early planning stage of a project, and the cost implications and convenience will vary widely from project to project. For example, instead of surrounding the plantroom with noise critical areas, introduce buffer zones comprising storage areas, toilets, service voids, quiet mechanical services areas or areas given over to less critical commercial activities between the plantroom and the critical areas.

5.8.2.3 Upgrade Plantroom Insulation If considered at the early planning stage of a project, it should be feasible to increase the sound transmission characteristics of the plantroom structure without incurring significant cost penalties, unless this also entails a general improvement of the load bearing capabilities of the structure and foundations. Where improvements in acoustic insulation have to be effected after commissioning the plant, the cost implication can be very high since the remedial work may, for example, entail the removal of lightweight partitions and substitution with brick and concrete constructions.

5.8.2.4 Reselect Equipment for Quiet Running Where at the planning stage high noise levels are anticipated from plantrooms, consideration should be given to the reselection of plant. Such reselection can often result in significant reductions in plantroom noise levels, albeit at some capital cost penalty. Such penalties are often off-set by the reduced sound insulation requirements of the plantroom structures and plant running costs. For example, lower pressure fans and larger ductwork. For existing plantrooms, excessive noise can often be associated with faulty or poorly maintained equipment, and also bad running conditions. In such cases, significant improvements can be made by replacing the offending equipment and/or improving the running conditions.

5.8.2.5 Provide Noise Control by the use of Absorbent Materials Plantrooms are generally reverberant in nature, with little acoustic absorption provided by the structure. Some reductions in plantroom reverberant noise levels are therefore possible, which in turn will lead to correspondingly lower noise levels in adjacent areas. It should be appreciated, however, that only a 3 dB reduction in sound level can be expected for each halving of reverberation time. Consequently, as most plantroom breakout problems occur at low frequencies, the inclusion of rockwool or acoustic plasters on plantroom walls and ceilings will generally only lead to a 2–3 dB reduction in sound levels in adjacent areas. This is because such materials do not exhibit good acoustic absorption characteristics at low frequencies, although their mid and high frequency performance is reasonably good. Therefore, reductions above 2–3 dB in plantroom breakout noise using such techniques can only be expected where the reverberant sound levels are dominated by contributions of mid and high frequencies.

5.8.2.6 Provide Local Sound Insulation High levels of noise in plantrooms are often associated with one or two items of plant rather than contributions from the total system. Therefore, effective reductions in plantroom noise levels could be achieved by enclosing the offending plant producing the high noise levels in a purpose built acoustic enclosure. These could take the form of builders work enclosures or, more commonly and conveniently, proprietary modular acoustic enclosures which can be assembled on site around the plant. Such enclosures would typically be comprised of double skinned panels consisting of an outer skin of 1.5 mm (16 g) mild steel, a 50 mm cavity, and a 1 mm (20 g) perforated inner skin. The cavity would normally be filled with 50 mm of mineral wool to provide an absorptive material behind the perforated inner face to control reverberant sound building up within the enclosure. The sound transmission loss characteristics of typical enclosures of this type are shown in Figure 5.17. When using such enclosures it is important, for high performance, that any penetrations for pipework and ductwork through the enclosures are sealed (N.B. Figure 5.11 re gaps) and all flanking paths eliminated, more especially that through a common adjacent floor slab by vibration isolation.

Noise transmission directly through the plantroom floor can also prove to be a problem to areas below. In such cases, it is common to increase the thickness of the floor directly beneath the plant to provide additional local sound insulation. This has limited effect and where very high plantroom insulation characteristics are required, it is necessary to use floating floor systems in conjunction with acoustic enclosures. A typical floating floor installation is shown in Figure 5.18, together with the typical

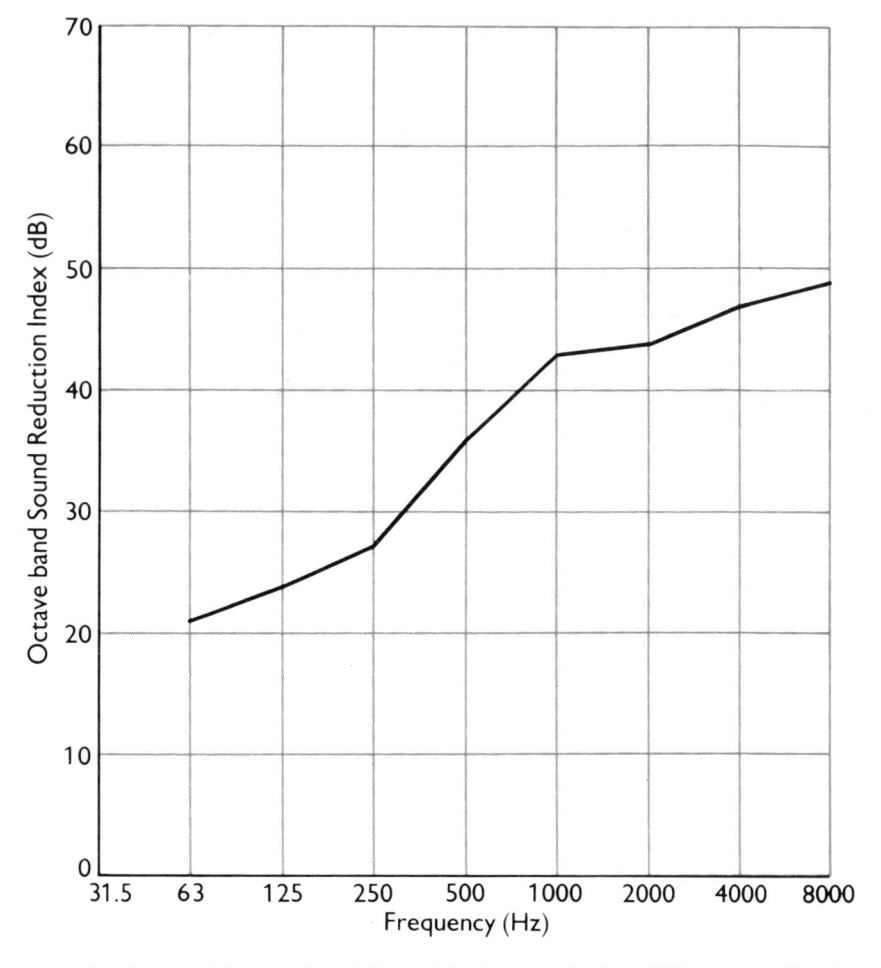

Figure 5.17   Typical Octave Band Sound Reduction Index of Plantroom Enclosure

airborne sound transmission loss characteristics that can be expected from such a system. Such composite constructions achieve a Sound Reduction Index which exceeds the mass law capabilities. This is achieved through effectively isolating the noise source from the main building structure. With careful design and construction, such isolation systems can accomplish a sound insulation which is higher than a conventional structure of three times the mass.

5.8.2.7 Provide Local Attenuation In fan plantrooms excessive noise is often associated with unattenuated open inlets or discharges. Such noise levels may be reduced by installing appropriate attenuating devices.

5.8.2.8 Apply Local Vibration Isolators The above considerations have been mainly limited to the control of airborne noise breakout from plantroom structures. It should, however, be appreciated that structure-borne vibration, so relevant to

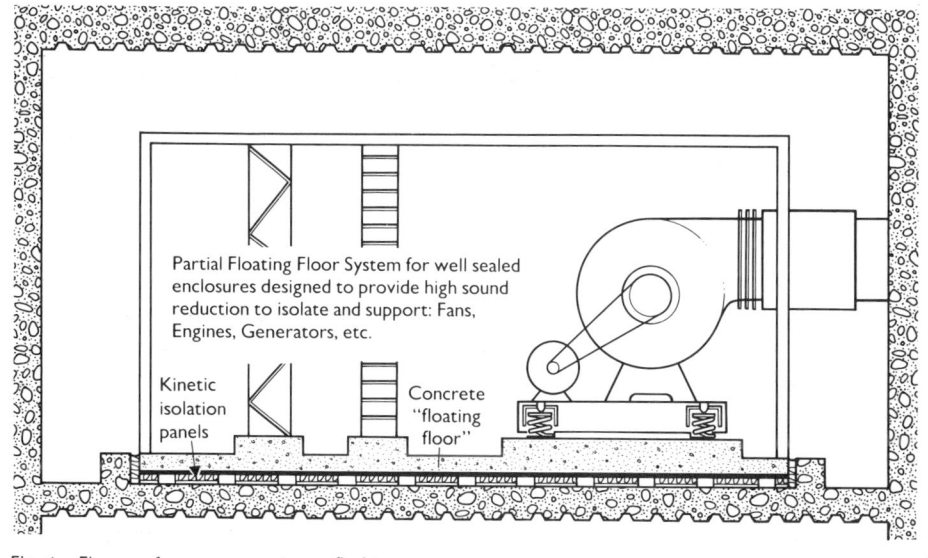

Floating Floor performance assuming no flanking

Frequency	63	125	250	500	1000	2000	4000 Hz
Sound reduction index	52	52	64	78	90	96	100 dB

Figure 5.18   Floating Floor Installation and Performance Details

flanking, is often the primary transmission mechanism of noise originating in the plantroom. Therefore, although the plantroom may be subjectively quiet, the absence of vibration isolators, or the incorrect selection of the isolation required, can lead to structural vibration being radiated as high levels of noise into adjacent and distant noise critical areas. The correct vibration isolators can eliminate such structure-borne noise problems at very low cost.

5.9 PLANTROOM NOISE BREAKOUT TO THE ATMOSPHERE

When designing plantroom structures, due consideration should be given to limiting any noise breakout to the local environment to acceptable standards. The acceptable external noise standards will vary from situation to situation, depending on the pre-existing ambient sound level in the area, the type of locality in which the plantroom is situated, that is whether urban or rural, etc, the proximity of any residential property and whether the plantroom is to operate during the day-time only or also at night-time. Such standards are often set by the the Local Environment Health Department and would typically be in terms of a site boundary dBA or NR level, or similar parameters as measured at the facade of the nearest residential property. For design purposes, the external noise levels at a distance from the plantroom can be calculated from the equation:

$$L_2 = L_1 - R + 10\log_{10}S - 20\log_{10}r - 14\,\text{dB} \qquad\qquad 5.6$$

where L_2 is the sound pressure level in dB at a distance of r metres from the plantroom wall;

L_1 is the sound pressure level in dB inside the plantroom adjacent to the wall where breakout will occur;

S is the area of the wall, m²;

R is the Sound Reduction Index of the plantroom wall, dB.

Calculations are normally carried out for each octave band in the frequency range 63 Hz to 8 kHz. Any air inlet or extract points in the plantroom wall may contain sound attenuating louvres. The acoustic performance of these louvres will be determined by the area ratio of louvre to wall and the external noise criterion to be achieved, as previously discussed above for non-homogeneous partitions (Figure 5.9).

5.9.1 Example

A plantroom faces onto the windows of residential flats some 32 m away. The plantroom wall facing the property is constructed of 112 mm thick brickwork and measures 8 m long by 4 m high. The plantroom requires a fresh air intake louvre, which measures some 3 m x 1 m. This is to be of a proprietary acoustic type, with the Sound Reduction Index stated. If the reverberant sound pressure level in the plantroom, when measured in octave bands, gives the readings shown below, calculate the sound pressure level just outside the flat windows.

Frequency	63	125	250	500	1k	2k	4k	8k	Hz
(i) Sound pressure level in plantroom L_1	105	93	91	88	91	89	83	81	dB
Sound Reduction Index of wall (Figure 5.8)	30	36	37	40	46	54	57	59	dB
Sound Reduction Index of louvre (manufacturers data)	5	5	7	12	18	21	16	16	dB
(ii) Composite Sound Reduction Index of wall and louvre (Figure 5.9) area ratio approximately 1:10)	16	15	17	22	28	31	26	26	dB
(iii) $10\log_{10}S$ ($S = 8 \times$ 4m²)	15	15	15	15	15	15	15	15	dB
(iv) $20\log_{10}r$ ($r = $ 32m)	30	30	30	30	30	30	30	30	dB
Sound pressure level L_2 outside flats = (i) − (ii) + (iii) − (iv) − 14	60	49	45	37	34	29	28	26	dB
NR35	63	52	45	39	35	32	30	28	dB

It will be seen that the resulting sound level at the flats is NR35.

As mentioned previously, the environmental noise criterion appropriate to any situation depends on the prevailing circumstances, but as a general rule one should aim to reduce noise levels from plantrooms to 6 dB below the previous ambient noise level in all octave band frequencies.

5.10 NOISE BREAK-IN THROUGH BUILDING FAÇADES

It is often necessary to calculate the internal sound levels in a building, due to external noise sources. Typical examples would be traffic noise break-in from busy roads affecting the internal noise climate of office accommodation, or roof mounted plant noise break-in into residential accommodation.

For design purposes, the internal noise level due to external noise sources can be calculated from the equation:

$$L_2 = L_1 - R + 10\log_{10}S - 10\log_{10}A + 10\log_{10}4\cos\theta \qquad 5.7$$

where L_2 is the internal noise sound pressure level, dB;
  L_1 is the external noise sound pressure level at the building façade, dB;
  R is the composite sound reduction index of the façade, dB;
  S is the room façade area, m^2;
  A is the total room absorption, $S_t\overline{\alpha}$, m^2;
  $\overline{\alpha}$ is the room average absorption coefficient at each frequency band;
  S_t is the total room surface area, m^2;
  θ is the angle of incidence of sound onto the façade.

5.10.1 Example

Some services plant is located on top of a building. Directly adjacent to this plant some 20 m away is the window of an air-conditioned office measuring 5 m x 4 m x 3 m high, with sealed thermal double-glazing and an interior background sound level of NR35. The window forms 30% of the area of the 112 m thick 12 m^2 brickwork facade of the office, and the office has a surface area of 94 m^2 with an average absorption coefficient as stated below. Maximum sound pressure level which can be tolerated at 20 m from the plant so that the existing internal background level of NR35 is not exceeded is calculated below.

Frequency	63	125	250	500	1k	2k	4k	8k	Hz
NR35	63	52	45	39	35	32	30	28	dB
(i) Maximum tolerable intrusive sound level = NR − 10	53	42	35	29	25	22	20	18	dB
Sound Reduction Index of wall (Figure 5.8)	30	36	37	40	46	54	57	59	dB
Sound Reduction Index of glazing (Figure 5.8)	22	27	29	31	41	37	47	52	dB
(ii) Composite sound reduction index area ratio 1:3 (Figure 5.9)	26	32	33	36	44	42	52	56	dB
(iii) $10\log_{10}S$ ($S = 12\text{m}^2$)	11	11	11	11	11	11	11	11	dB
α	0.1	0.2	0.2	0.3	0.3	0.4	0.4	0.5	dB
(iv) $10\log_{10}S_t\alpha$	10	13	13	14	14	16	16	17	dB
(v) $10\log_{10}4_{\cos}\theta$ ($\theta = 0°$)	6	6	6	6	6	6	6	6	dB
Then maximum permissible level at 20m from plant becomes L_p = (i) + (ii) − (iii) + (iv) − (v)	72	70	64	62	66	63	71	74	dB

NOISE TO ATMOSPHERE

A non-acoustic weatherline louvre acts as a visual but not acoustic barrier for this plantroom inlet.

Courtesy: Sound Attenuators Ltd

CHAPTER 6

VIBRATION ISOLATION

6.1 INTRODUCTION

Just as the acoustics designer plans his strategy for dealing with the effects of machinery noise upon the building containing or adjacent to the equipment, his solutions will invariably be centred upon one, or possibly more, of three main considerations

- the source
- the transmission path
- the receiver

Similarly when we consider the effects of disturbing vibration, there are two scenarios which may apply.

1. To protect the building or a part of its structure from the vibration effects of the machinery operation.
2. To protect equipment or people from the vibration effects of the building.

The interplay between the machinery as a source or receiver and the building structure as the transmission path provides the basic link which results in the designer needing to include a programme of vibration control or isolation as part of the overall acoustic strategy.

When considering transmitted vibration, we are essentially concerned with frequencies at the very low end of the spectrum normally below 200 Hz. Within the frequency band between 0 and 200 Hz there is a further split which takes account of the difference between those vibrations which are felt, on the one hand, and those which are audible on the other. The point in the frequency spectrum at which this difference becomes apparent is between 30 and 35 Hz. The upper band or audible spectrum of vibration, when considered, leads the designer to reviewing methods of acoustic isolation.

Clearly, the perception of vibration in terms of human subjective response is by other body senses and not just the ear.

With the ever increasing use of lighter weight methods of building construction and materials in the last 10 to 15 years, the acoustic designer has seen that instances of inadequate or non-existent vibration isolation provisions result in major acoustic problems. In a large number of cases, these shortcomings have had direct consequences insofar as control design criteria (NC/NR) having not been achieved. Such problems occur despite the fact that air conditioning duct attenuators, acoustic lagging, acoustic enclosures and noise reducing doors have been wisely and extensively used in the overall design of many modern buildings.

The two principal scenarios for which vibration isolation may be considered have been detailed earlier.

The first of these relates to a consideration of fluctuating forces developed by a machine as a result of some rotary or reciprocating out of balance which is transmitted as vibration into the structure upon which the machinery is placed. The introduction of some type of resilient support between the machine and the structure should, if carefully selected, result in a reduction of the transferred force. The relationship between the developed force and the transferred force is defined as the "transmissibility".

In the second case, the considerations relate to the effects of an extraneous vibrating displacement through the structure being transmitted into a piece of equipment giving rise to a transmitted displacement. Again the relationship is defined as the transmissibility and the introduction of a carefully selected resilient support will result in a reduced transmissibility.

Generally in the area of building services design, we are concerned with force transfer from equipment such as fans, pumps, compressors, engine and motor driven systems of various types. However, consideration of the second classic situation has become increasingly important over the last 10 years typically with the installation of highly sensitive optical and electronic equipment as well as computers into modern buildings such as laboratories, hospitals and hi-tech offices. Much more recently with the growth of the micro electronic industry producing micro-electronic circuits on chips, the need to achieve very high levels of vibration isolation has been essential. This has resulted in extensive analysis of both force transfers in mechanical plantrooms as well as displacement transfer from the floor of the micro-electronics production room.

In Figures 6.1 and 6.2, the two situations are illustrated in typically general forms in which they are most usually recognised — that of a compressor when the solution is defined with respect to attenuated force transfer and that of a precision electron microscope where the solution is a function of transmitted displacement.

We are concerned with reducing the transmissibility, or conversely, improving the isolation. Isolation commences when the value of transmissibility is less than unity and it follows that as the ratio decreases, so the level of isolation improves.

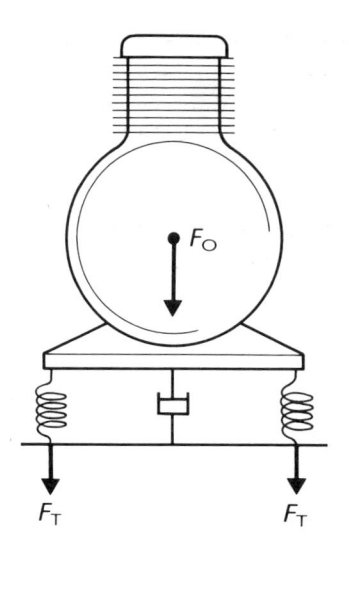

Transmissibility $T_F = \dfrac{F_T}{F_O}$

F_T = Transmitted force

F_O = Source force

Figure 6.1   Compressor on Vibration Isolators

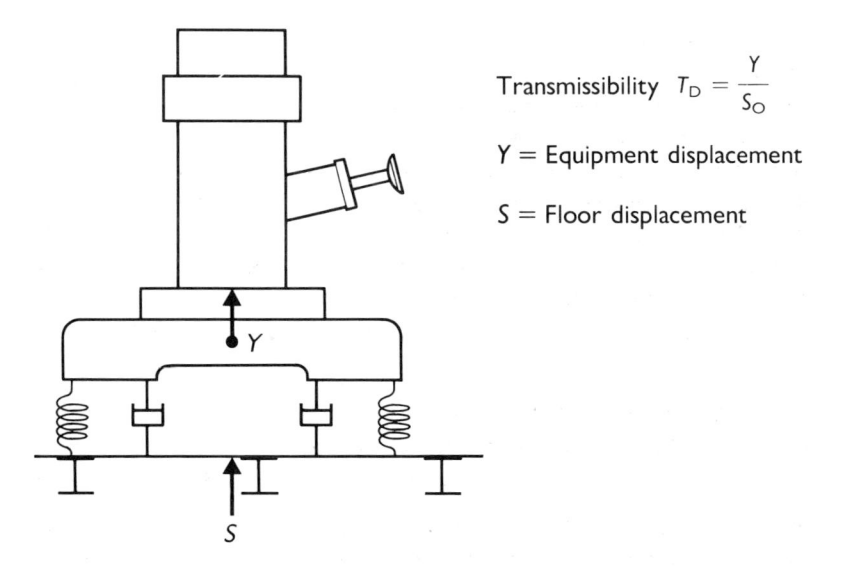

Transmissibility $T_D = \dfrac{Y}{S_O}$

Y = Equipment displacement

S = Floor displacement

Figure 6.2   Electron Microscope on Vibration Isolators

There is a third model condition to be considered and this refers to the need to control the motion of the vibration isolated machine. The assessment requires consideration of both forces and displacements but with the need to attenuate displacement.

In Figure 6.3 the traditional method of depicting the vibration isolated machine is illustrated where the dashpot represents the damping which may be inherent in the isolating material or which may be purpose designed into the system.

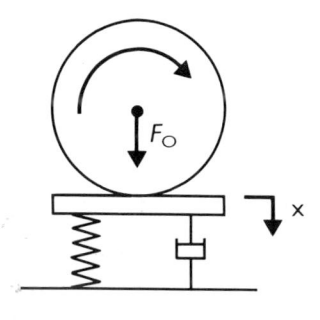

Figure 6.3   The Vibration Isolation Model—Spring and Dashpot Representation

6.2 FUNDAMENTAL THEORY

Where mounted machinery develops vibration forces, as a consequence of its own operation, the motion which results is governed by the effect of that force acting upon the mass of the machine. For the classic case, steady state conditions and periodic sinusoidal force effects and an idealised single degree of freedom system are assumed (Figure 6.4).

Assuming that the mass M is acted upon by a linear vertical force F

$$F = F_0 \sin\omega t \qquad\qquad 6.3$$

where F_0 = maximum force; $\omega = 2\pi$ x forcing frequency, f_d; and t = time, s, then the steady state differential equation of motion is

$$M\ddot{Y} = F_0 \sin\omega t - K_y Y \qquad\qquad 6.4$$

where K_y = axial stiffness – force per unit displacement; Y = axial displacement; and $\ddot{Y}$ = axial acceleration.

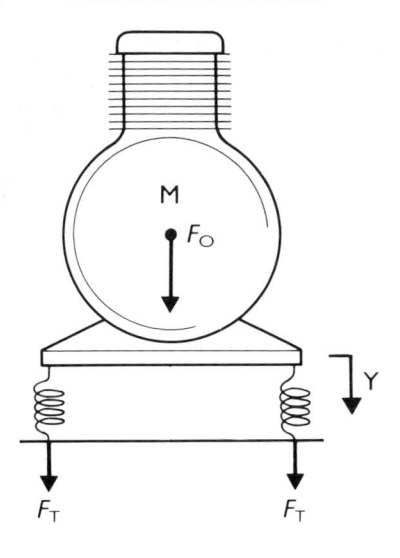

Figure 6.4   Idealised Single Degree of Freedom System

The solution of this equation assuming the transients are ignored; for displacements of the mass M is

$$Y = \frac{F_0 \sin\omega t}{K_y[1 - (\omega/\omega_n)^2]} \qquad\qquad 6.5$$

$\omega_n = 2\pi$ x vertical natural frequency of vibration (f_n)

$$\text{now } \omega_n = (K_y/M)^{1/2} \text{ rads/s} \qquad\qquad 6.6$$

therefore $2\pi f_n = (K_y/M)^{1/2}$ Hz

$$\text{and} f_n = \frac{1}{2\pi}\left(\frac{K_y}{M}\right)^{1/2} = 0.16 \left(\frac{K_y}{M}\right)^{1/2} \text{ Hz} \qquad\qquad 6.7$$

Assuming that the support ground is essentially immovable $(S = 0)$, the force experienced by the support is therefore equal to the product of the stiffness K_y and the displacement of the mass from its neutral position Y. Thus

$$F_T = K_y Y \qquad\qquad 6.8$$

It follows therefore from equation 6.1 that the force transmissibility, T_F, is given by

$$T_F = \frac{F_T}{F_0} = \frac{1}{1 - \left(\dfrac{\omega}{\omega_n}\right)^2}$$

becoming more familiarly

$$T_F = \cfrac{1}{1 - \left(\cfrac{f_d}{f_n}\right)^2} \qquad\qquad 6.9$$

Where in the second classic case (Figure 6.2) the motion of the supported machines is as a result of vibration from the support, S, the steady rate differential equation of motion is given by

$$M\ddot{Y} = K_y (S - Y) \qquad\qquad 6.10$$

where $S = S_0 \sin\omega t$ and S_0 = maximum displacement of the support.

From the above, the response of the supported machine of mass, M is

$$Y_0 = \cfrac{S_0}{1 - \left(\cfrac{\omega}{\omega_n}\right)^2} \qquad\qquad 6.11$$

Allowing that the transmissibility T_0 is a function of the ratio of the maximum displacement of mass M, to the maximum displacement of support, S, then

$$T_D = \cfrac{Y_0}{S_0} = \cfrac{1}{1 - \left(\cfrac{\omega}{\omega_n}\right)^2} \qquad\qquad 6.12$$

becoming more familiarly

$$T_D = \cfrac{1}{1 - \left(\cfrac{f_d}{f_n}\right)^2} \qquad\qquad 6.13$$

Thus it can be seen that the equations for force transmissibility 6.9 and displacement transmissibility 6.13 are identical. Further, the transmissibility of the system can be simply considered as a function of the frequency of the disturbing vibration and the vertical natural frequency of the supported machine.

Figure 6.5 is a plot of transmissibility against f_d/f_n for this simple single degree of freedom system ignoring any damping/friction effects.

Several points are worthy of note.

1. when $f_d > \sqrt{2}f_n$, the transmissibility is less than unity, and useful isolation commences.
2. when $f_d < \sqrt{2}f_n$, amplification of force or displacement occurs.
3. The higher f_d or the lower f_n, then the greater is the degree of isolation.

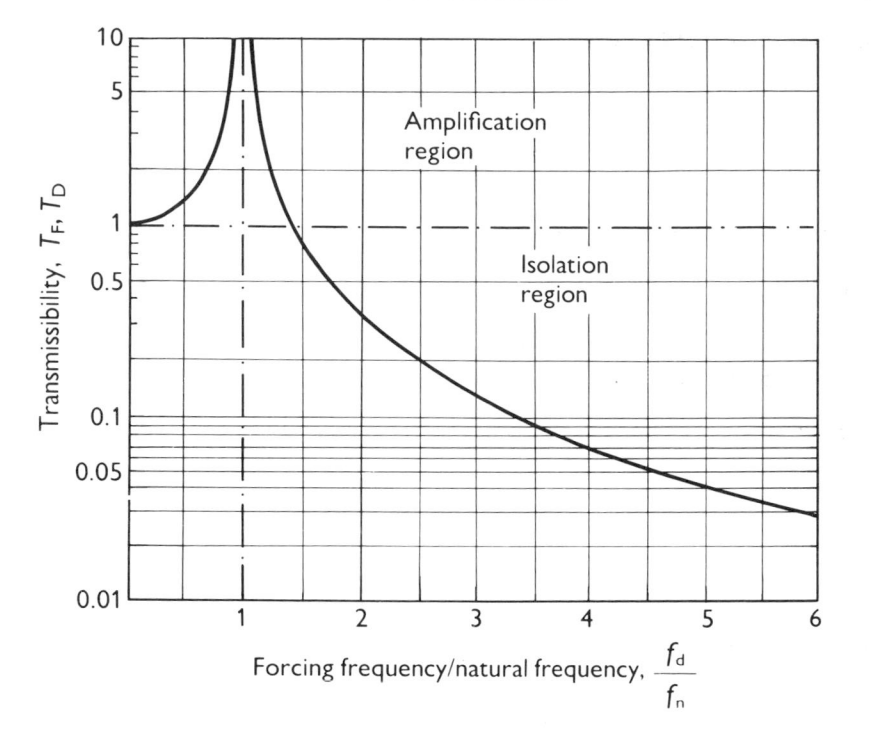

Figure 6.5 Transmissibility Curve for an Undamped System

4. When $f_d = f_n$, then T_F and $T_D \rightarrow \infty$, although in practice this does not occur due either to the presence of damping or mechanical restraint/constraints within the system. This is known as the "resonance condition" — when the disturbing frequency equals the natural frequency.

From equation 6.7 we know that

$$f_n = (1/2\pi)(K_y/M)^{1/2} \qquad 6.14$$

The static deflection, δ, undergone by the spring of axial stiffness K_y is given by

$$\delta = Mg/K_y \qquad 6.15$$

where g = acceleration due to gravity.

By substitution in equation 6.14, then natural frequency is related to static deflection of the spring by

$$f_n = \frac{1}{2\pi}\sqrt{\frac{g}{\delta}} \qquad 6.16$$

By introducing the constant value for *g*, the equation can be reduced to the more useful form

$$f_n = 15.8 \qquad \left(\frac{1}{\delta}\right)^{1/2} \quad Hz \qquad\qquad 6.17$$

where δ = static deflection in mm.

The graphical relationship between static deflection and natural frequency is illustrated in Figure 6.6.

It follows therefore that the general rule for the designer is that the greater the static deflection, then the lower the natural frequency of the system resulting in greater isolation efficiency of the spectrum of disturbing frequencies in which isolation occurs.

Thus having established the conditions under which varying amounts of vibration isolation may be achieved, the equations 6.9 and 6.13 can be modified as follows:

Isolation I = 1 - Transmissibility T_F, T_D

$$I = 1 - \frac{1}{1 - \left(\dfrac{f_d}{f_n}\right)^2} \qquad\qquad 6.18$$

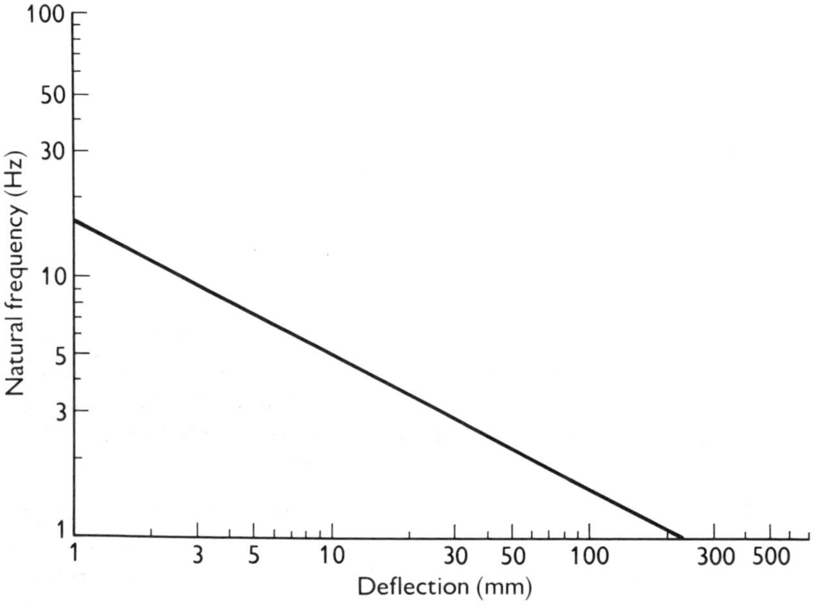

Figure 6.6   Deflection and Natural Frequency Relationship

6.2.1 Damping

So far the effect of damping has been ignored in deriving the simple relationships useful to the designer in establishing the performance of vibration isolators. However damping is present to varying degrees in all linked systems and as a consequence does modify some of the relationships.

The inclusion of damping mechanisms into an isolating system results in a small change of value for the natural frequency f_n of the system. Also there is a change in the overall stiffness of the isolating system since inevitably the damping system will have its own inherent additional stiffness characteristics.

The force transmissibility of a system where damping is considered is given by the expression below:

$$T_F = \sqrt{\frac{1+4\left(\frac{\omega}{\omega_n}\right)^2\left(\frac{C}{C_0}\right)^2}{\left\{1-\left(\frac{\omega}{\omega_n}\right)^2\right\}^2 + 4\left(\frac{\omega}{\omega_n}\right)^2\left(\frac{C}{C_0}\right)^2}} \qquad 6.19$$

where $\omega_n = 2\pi \times$ natural frequency f_n;
$\quad \omega = 2\pi \times$ disturbing frequency f_d;
$\quad C =$ actual damping present;
$\quad C_0 =$ critical damping, this is the value of damping at which oscillating motion just commences, otherwise known as dead beat damping.

It is very rare in vibration isolation systems to require levels of damping to approach critical damping and thus normally the value of C/C_0 is less than unity.

Typical values of C/C_0 are 0.005 for steel and 0.02 for rubber with a Shore hardness of 40 degrees.

The expression given in equation 6.19 takes on more meaning when presented graphically and Figure 6.7 shows the effect of varying C/C_0.

It will be seen from examining the family of curves that increasing the value of damping not only reduces the peak transmissibility at resonance, but also reduces the performance of the vibration isolator in the useful isolating region. In order to achieve maximum isolation then damping should be as low as possible and this is acceptable for a large majority of applications encountered by the building services engineer.

Nevertheless damping may be necessary in certain applications such as reciprocating engines or where as a result of being supported on vibration isolators the machinery motion becomes excessive perhaps leading to instability. Similarly some damping can effectively reduce motion such that flexible connections between the machine and piped services can function effectively.

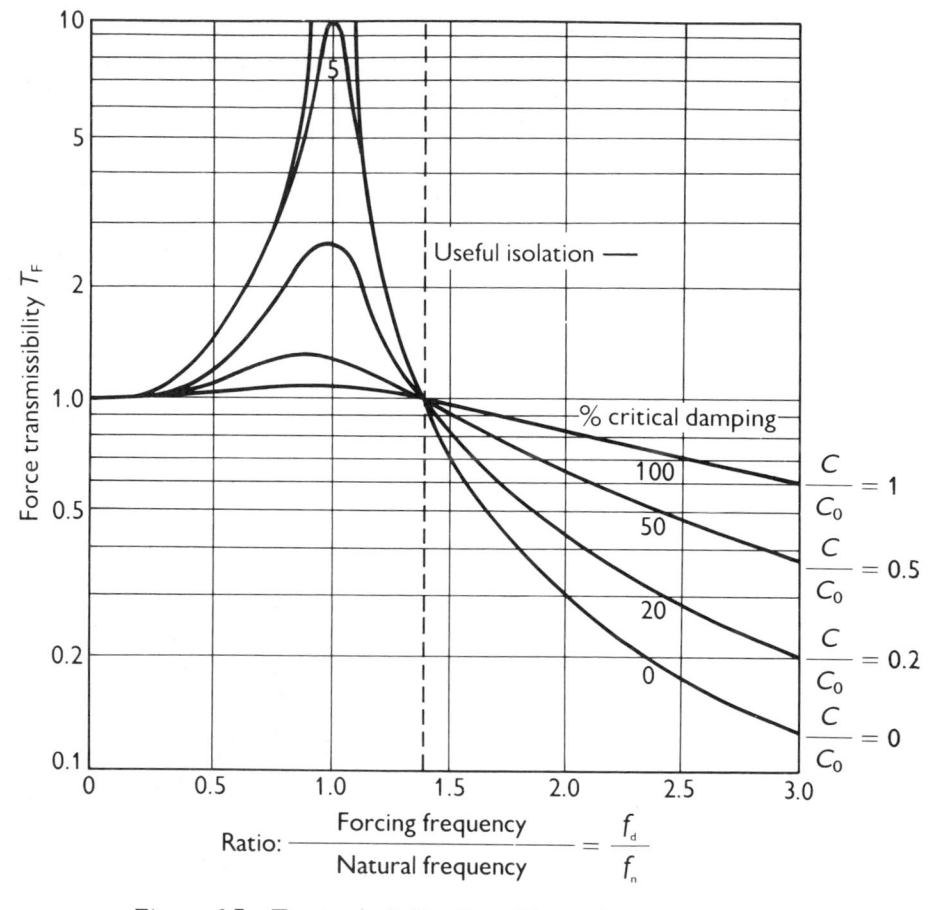

Figure 6.7 Transmissibility for a Viscously Damped System

The damping may take various forms; it may be inherent in the flexible mechanism, viscous effects in a liquid or resulting from friction between sliding surfaces. Some consideration is given to the two most widely applied approaches.

6.2.1.1 Viscous Damping. Considering the first of the two applied methods of damping normally employed, in principle this takes the form of viscous damping by virtue of a moving piston, attached to the mounted unit, being immersed in an oil filled cylinder. The damping force is proportional to the velocity of the system and hence is a function of frequency and amplitude of motion across the damper. The principal problems and disadvantages of viscous damping are viscosity changes due to temperature variations and design difficulties in ensuring freedom of movement in all planes.

6.2.1.2 Friction Damping. In friction damping, the movement of the mass–spring system generates rubbing friction between a stationary element and one attached to the mass. This results in a damping force which is constant and independent of displacement and velocity. When vibration energy levels are low, the initial friction will be high enough to prevent movement and thus no isolation occurs because the damper "sticks". Conversely if the friction force is very low and allows movement at low vibration energy levels, then it will be inadequate to restrict motion when the amplitudes at the resonance condition are high. A disadvantage of rubbing friction systems is that they generate high frequency effects generally in a similar manner to those generated by a bow being drawn across violin strings. The problem of initial friction "sticking" or "breakaway" can be resolved by designing suitable clearances to allow for the amplitude of movement. A further improvement can be achieved by including a spring in series with the damper such, that at higher frequencies all the deflection occurs in the spring with the friction pad "locked" in position.

A section through a vibration isolator with such a damping device is given in Figure 6.8.

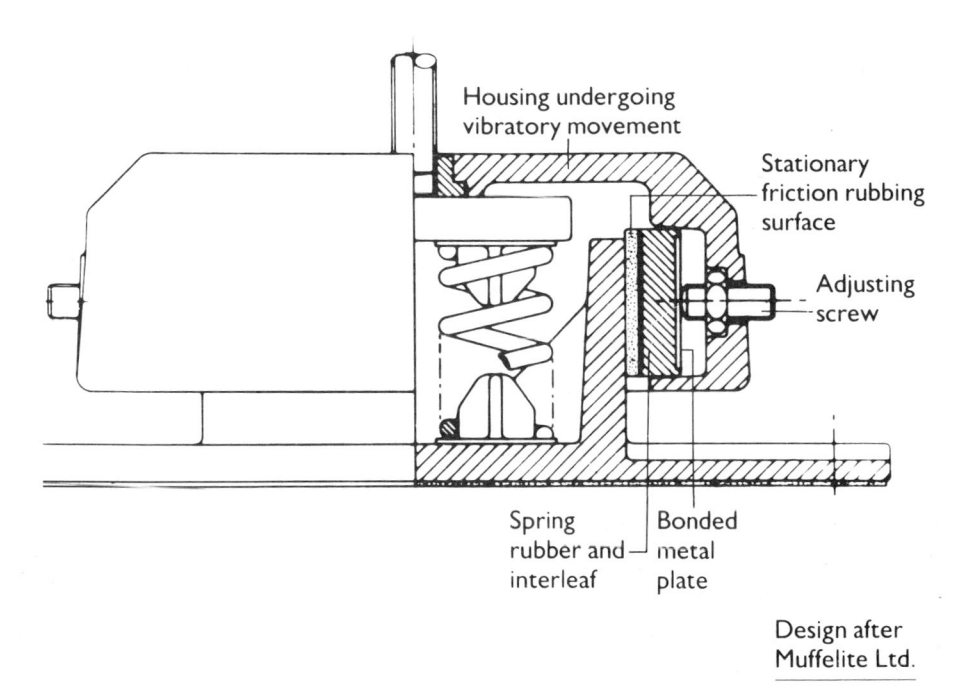

Figure 6.8 Friction—Spring Isolator

One final comment on the methods of achieving useful damping is made with reference to the widely held practice of including "snubbers" in the design of certain vibration isolators. This may take the form of a simple "stop" which limits motion, or by virtue of increasing the stiffness of a rubber pad when being compressed in a non linear manner. In this latter case, the snubbing action is progressive, dependent upon a rapidly increasing force. Whilst being the preferred method, an awareness of the possibility of "rebound" between the two parts of the snubber system could lead to a vibration pulse of large and unacceptable amplitude.

6.3 MOUNTING ARRANGEMENTS

There are four ways of mounting a machine such that the transmission of vibration through the supports may be modified as illustrated in Figure 6.9.

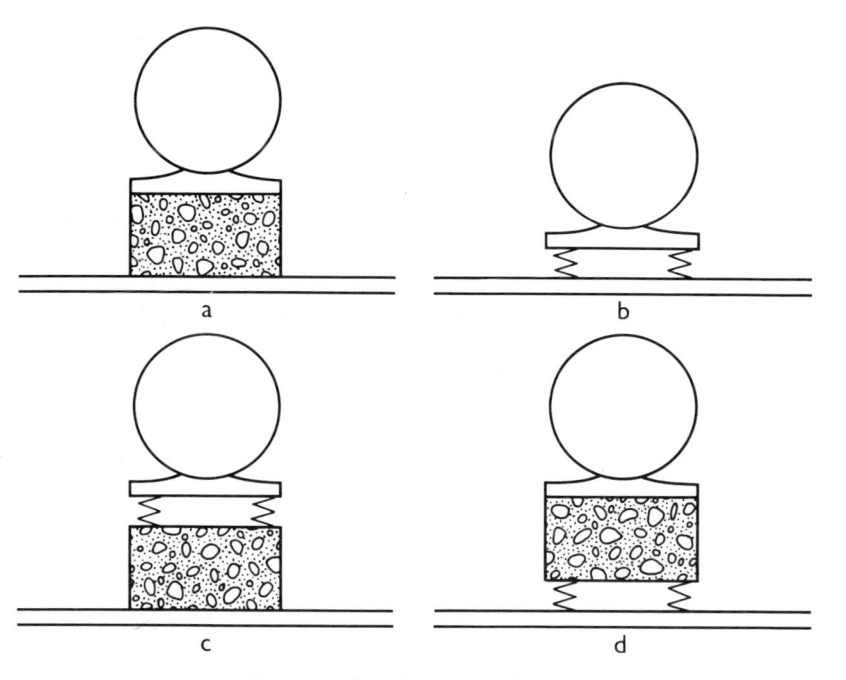

Figure 6.9   Mounting Options

In situation 6.9a where the machine sits upon an inertia block, there is no change to the force which is transferred into the floor, but it is likely that the resultant displacement of the floor will be reduced. Dependent upon the flexibility or conversely stiffness of the floor, then the main structure supporting the floor will, to some degree, be isolated from the machine vibration. In situation 6.9b when the machine sits directly on isolators, the isolation efficiency which is achieved so that the level of transferred force is reduced, depends primarily upon the static deflection which is achieved on the vibration isolators. In situation 6.9c if the floor is flexible (springy) with a low stiffness as for a wooden floor, then the introduction of the inertia block produces a compound spring system. If the natural frequency of response of the floor/inertia block is greater then the disturbing frequency due to the machine, then problems should not occur. More simply if the static deflection of the selected isolator is greater than that of the floor, then it is not necessary to increase the floor's deflection by the addition of the inertia mass. In situation 6.9d a satisfactory degree of vibration isolation will be achieved if the static deflection of the vibration isolators is adequate, whilst the effect of the inertia block will be to reduce the motion of the isolated machine due to the effects of rotary or reciprocating action.

6.4 MODES

So far, we have considered the effects of vibration in the vertical plane only, otherwise known as a single degree of freedom system. However, it is apparent in real life that for even a simple system such as a mass supported on a spring, there is more than one direction in which the mass could move if subjected to a forced displacement from its position of rest.

Allowing that the minimum number of mounting points is three if the mass is to be statically stable, then one can imagine that there is an infinite number of ways in which the body can move when subjected to vibratory forces. In practice a rigid resiliently supported system exhibits six natural modes of vibration each of which may be excited separately or together. These six modes are, in reality, the six classic degrees of freedom in which the system may move if excited.

The six modes are made up of three linear motions and three rotational motions around their respective axes. Figure 6.10 illustrates the six modes as defined. Each of the modes has its own natural frequency.

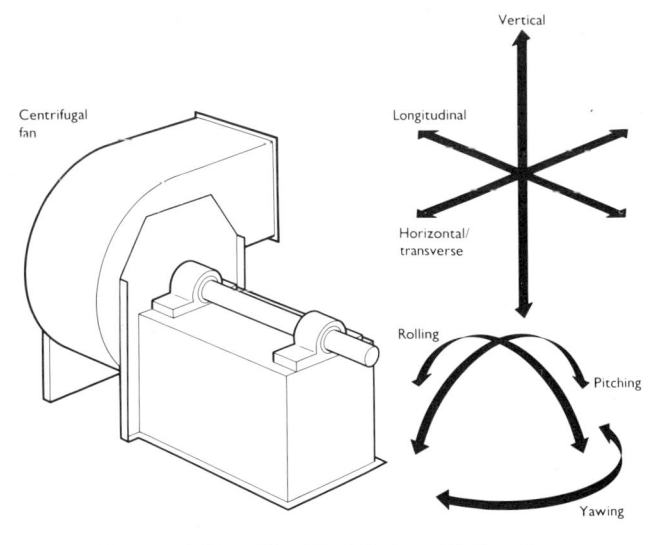

Figure 6.10 The Six Modes of Vibration

If, as a result of the system being subjected to a vibratory force in a particular direction, it moves not only in the direction of the excitation, but also in another direction, then it is said a coupled mode has been excited. If conversely the system moves only in the direction of excitation then it is said that the mode is decoupled.

When coupled, the system will appear to "rock". Generally the lower rocking mode occurs at a frequency below that of the vertical mode, whereas the upper rocking mode occurs at a frequency above that of the vertical mode.

The transmissibility curves for a system where all the component modes are analysed, are illustrated in Figure 6.11.

The six natural frequencies of response are different as suggested previously. If all the other natural frequencies were below the vertical mode then it would be acceptable to maintain the simple approach for analysis of the transmissibility/isolation efficiency adopted earlier in this chapter. However, this is rarely so and if, as often happens, one

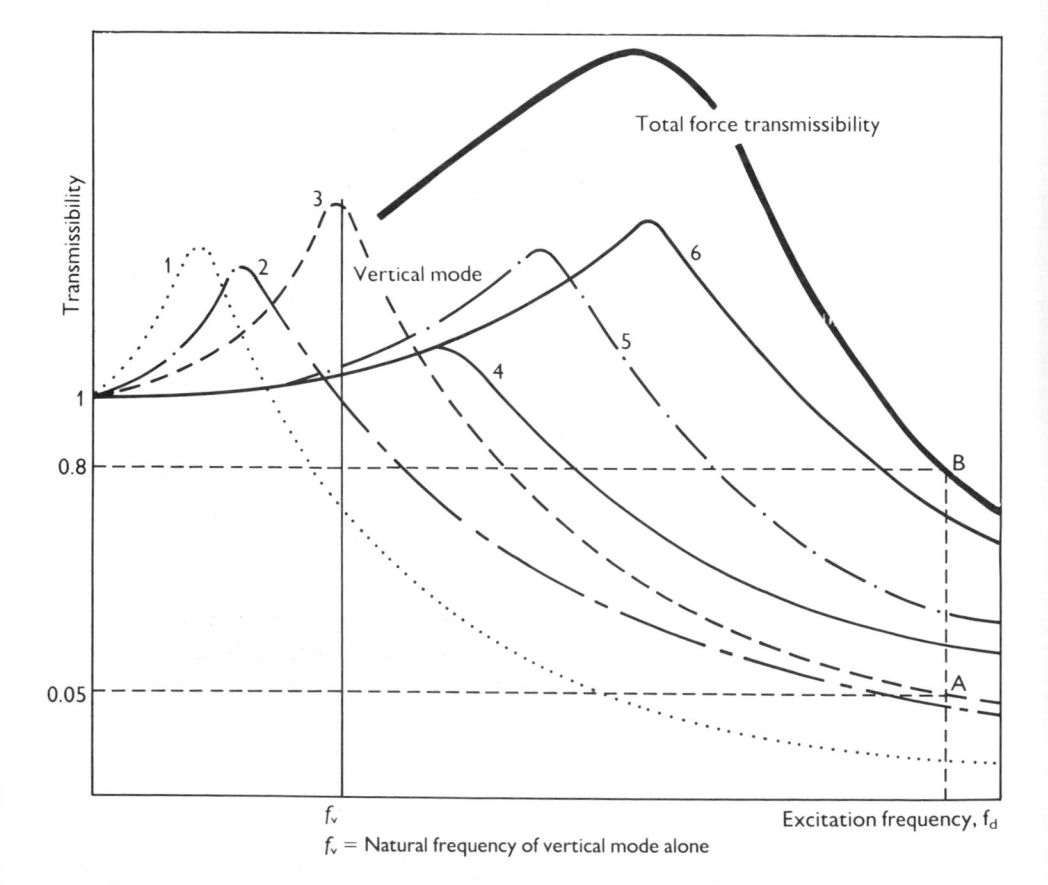

f_v = Natural frequency of vertical mode alone

Figure 6.11   Six Mode Transmissibility including Coupling

of the other modes has a natural frequency higher than the vertical mode, then it may be close to the frequency of the disturbing vibrationary force and thus the transmissibility will be higher than expected.

Figure 6.11 clearly illustrates that rather than achieving the 95% isolation expected from considering the vertical mode only — Position A on Curve 3 — the actual isolation is 20% — Point B — due to this particular combined effect of the coupled modes.

Thus for the correct application of isolators, it is essential that the frequency of the disturbing vibration does not coincide with any of the modal natural frequencies; if resonance is to be avoided. This can be done by the careful selection of the isolating mounts, but in practice it may be easier to arrange for the natural frequencies of the other modes to fall in the region of the vertical natural frequency and not at higher frequencies for three of the modes as Figure 6.11 illustrates. Similarly, to avoid the risk of exciting the coupled modes, the following procedures may be adopted.

1. Ensure that the horizontal stiffness of the isolation support is similar to the vertical stiffness.
2. Ensure that the centre of gravity of the isolated system is as low as possible, preferably in the plane of the isolator supports.
3. Ensure that the axis of the vibration force passes as near as possible through the centre of gravity of the isolated system.

From time to time, the designer may well need to resort to a more elaborate mounting arrangement in order to avoid the problems of coupled modes and such an installation is shown in Figure 6.12, where a vee-twin compressor is mounted on a stepped foundation block. In this case, the centre of gravity of the system is in the plane of the isolation supports and in this way the rocking modes are decoupled and high levels of movement at the pipe connections on the inclined cylinders are avoided.

It is the case that the torsional mode is rarely a problem and the designers main concern will be to estimate the natural frequency for the vertical translation mode — up and down — and then the coupled rocking modes in the two principal directions — fore and aft, left and right.

To assist in the analysis of the coupled modes, the isolated machine is modelled as a uniformly dense cuboid supported at four points. From the analysis it can be shown that rocking mode natural frequency is related to the vertical natural frequency, by the mathematical relationship.

$$\frac{f_c}{f_n} = \frac{1}{\sqrt{2}} \left[\frac{4Kc^2 + K + 3}{c^2 + 1} \pm \left\{ \left(\frac{4Kc^2 + K + 3}{c^2 + 1} \right)^2 - \frac{12K}{c^2 + 1} \right\}^{1/2} \right]^{1/2} \qquad 6.20$$

where f_c = coupled natural frequency;

f_n = vertical natural frequency;

c = ratio of height to width of cuboid;

and K = the ratio of the horizontal stiffness of the isolator to the vertical stiffness

Figure 6.12   Isolated Vee-twin Compressor

A graphical illustration of equation 6.20 is given in Figure 6.13 and shows clearly that providing the vertical natural frequency may be determined, then the rocking modes may be calculated provided that the manufacturer of the vibration isolators is able to supply data on stiffness characteristics for the resilient unit.

Further, by treating the machine as a cuboid, where the base dimensions correspond to that of the centres of support, then the height may be established for the homogenous cuboid by arranging for its mass to be the same as the machine in question.

The designer is thus able to assess the frequency of the coupled mode with relative ease and it may be noted that for large values of c any error due to approximation will have relatively little effect on the value of f_c/f_n since the curves when plotted tend to the horizontal.

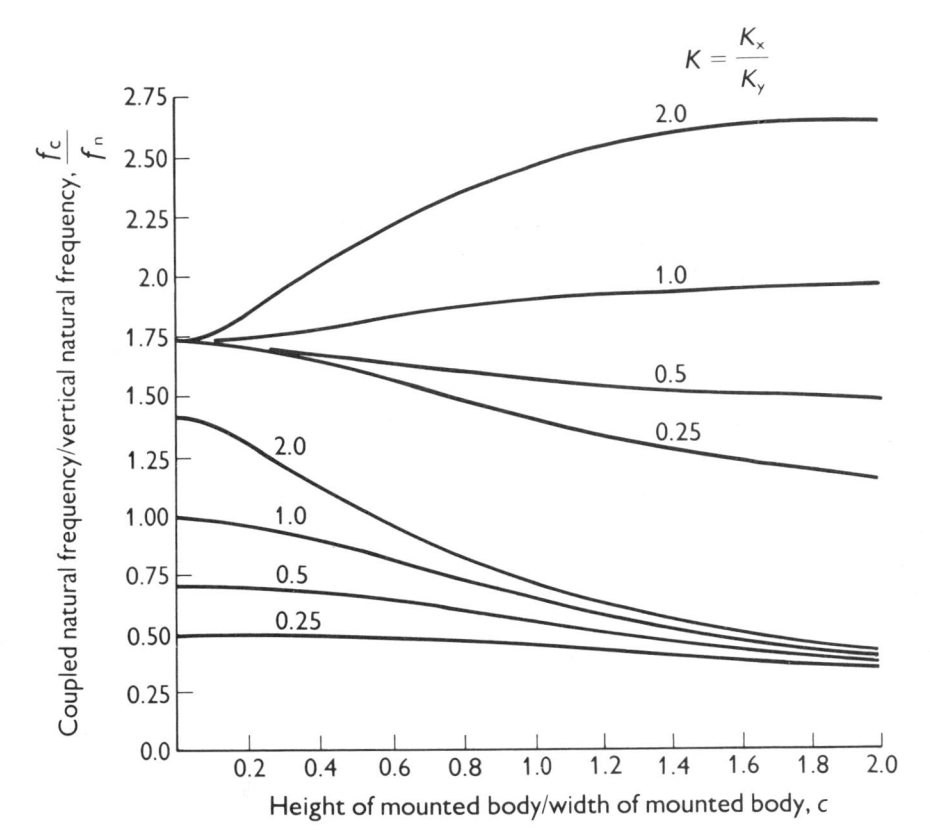

Figure 6.13 Mode Coupling for Cuboid Geometry for Various Isolator Stiffness Ratios

6.5 MOTION

Some consideration of the motion of the isolated system is appropriate since we have so far generally ignored the effect of vibratory force on the isolated mass. Following Figure 6.9d, it was stated that the introduction of an inertia block did not influence the amount of vibratory force transmitted to the supporting structure, since this was governed by the deflection of the isolating support. Rather, the effect of the inertia block was to reduce the level of motion of the isolated system due to the vibrating force.

If from Figure 6.3 we assume that as a consequence of the vibratory force F, the isolated system undergoes a dynamic motion X, then the magnitude of X is determined by:

for low damping
$$X = \frac{F}{K_y} = \frac{1}{1 - \left(\dfrac{f_d}{f_n}\right)^2} \qquad \text{6.21a}$$

for finite damping
$$X = \frac{F}{K_y} = \frac{1}{\left[1 - \left(\dfrac{f_d}{f_n}\right)^2\right]^2 + 4\left(\dfrac{f_d}{f_n}\right)^2\left(\dfrac{C}{C_0}\right)^2} \qquad \text{6.21b}$$

These equations are plotted in Figure 6.14 and show a number of interesting features.

Firstly, increased damping C/C_0 is always advantageous in reducing the dynamic motion X. Secondly, the value of C/C_0 is largely unimportant for ratios of f_d/f_n greater than 3.0. This, of course, is at variance with the trend of the various transmissibility curves illustrated in Figure 6.7.

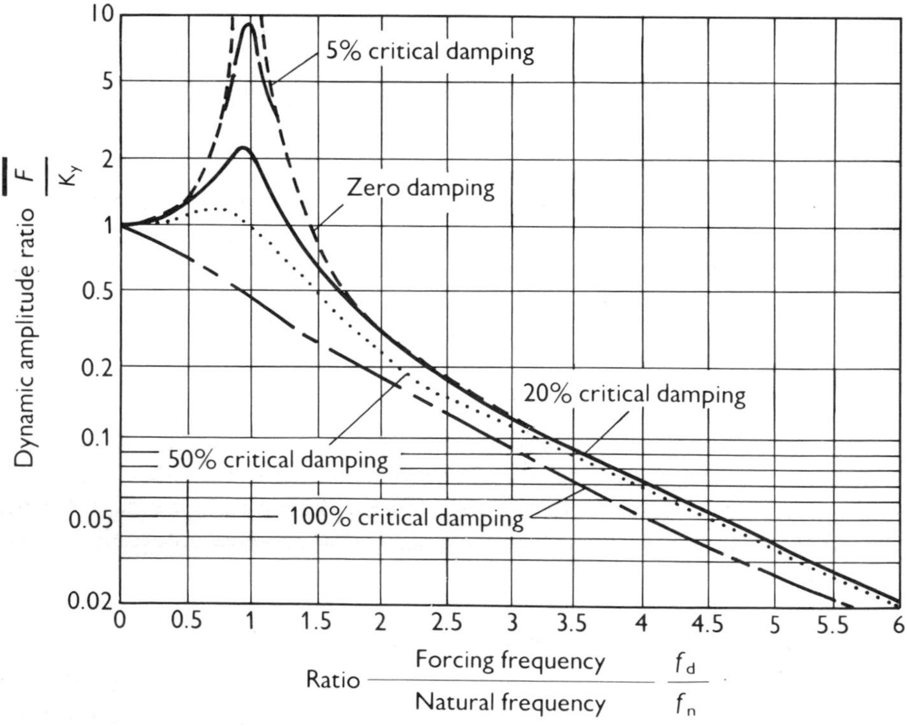

Figure 6.14 Dynamic Amplitude of Vibration Isolated Machine when Subjected to an Oscillating Force, F

Referring again to Figure 6.14, it can be seen that the dynamic motion increases with the increase of the vibrating force and is inversely proportional to the isolation stiffness. However, whilst increasing the isolation stiffness will effectively reduce the dynamic motion, for a given value of f_d/f_n this can only be achieved by the addition of mass, normally in the form of an inertia base, to maintain a chosen value for f_n.

Inertia bases or foundation blocks are usually made of concrete poured into a structural steel frame with reinforcing bars, brackets for the attachment of vibration isolators and mounting bolts for the equipment. Although they are generally referred to as inertia blocks and are normally installed to reduce the movement of the mounted equipment, in many cases they are used for other reasons. These are:

To give more stability to the system, 6.5.1
To lower the centre of gravity of the system, 6.5.2
To give a more even weight distribution, 6.5.3
To minimise the effect of external forces, 6.5.4
To add rigidity to the equipment, 6.5.5
To reduce problems due to coupled modes, 6.5.6
To minimise the effects of errors in the position of the estimated centre of gravity of the equipment, 6.5.7
To act as a local acoustic barrier, 6.5.8

The above eight points are considered in greater detail.

6.5.1 To Increase the Stability of the System

With many machines the mounting locations originally intended for attachment to a rigid concrete slab are too close together to provide adequate stability when the equipment is mounted on vibration isolators. The concrete inertia base provides a means of widening the support and a more stable geometry. This can also be achieved with a steel rail base as illustrated in Figure 6.15.

6.5.2 Lowering the Centre of Gravity

Mounting equipment on a substantial concrete base has the effect of lowering the centre of gravity of the complete assembly. As mentioned this adds to the improvement of the stability provided by extending the width of the base, but also has the effect of reducing the likelihood of rocking motion. A typical section through such a base is illustrated in Figure 6.16.

6.5.3 To Give a More Even Weight Distribution

In many cases, equipment items are very much heavier at one end than the other. This means that, if they are mounted directly on vibration isolators, very different arrangements are needed at opposite ends of the equipment to cope with the uneven weight distribution. If the equipment is mounted on a concrete block, the weight

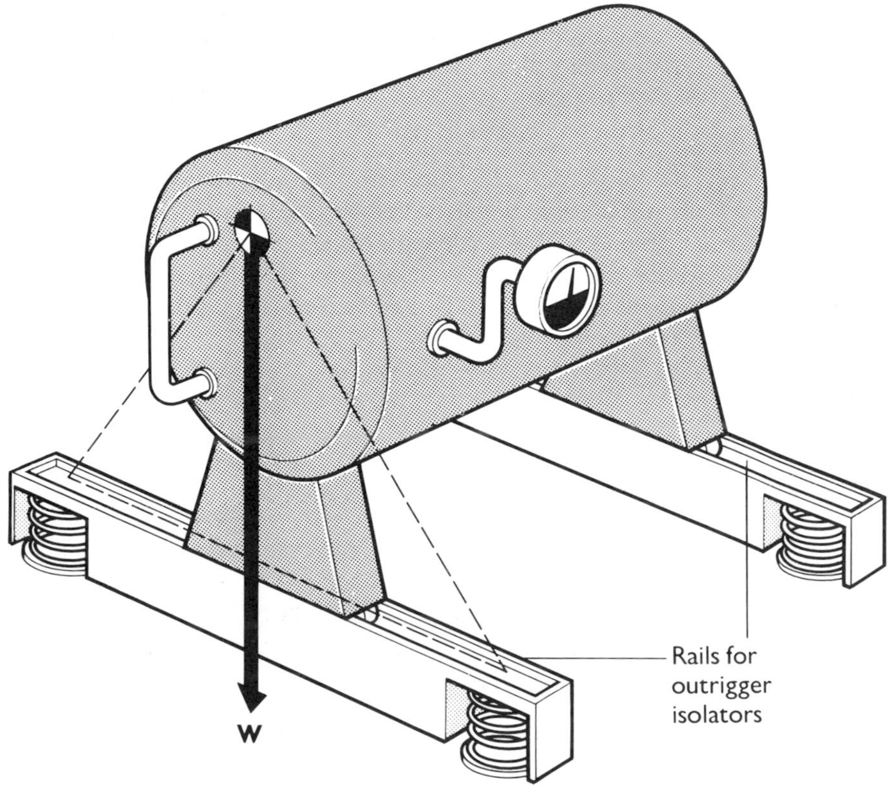

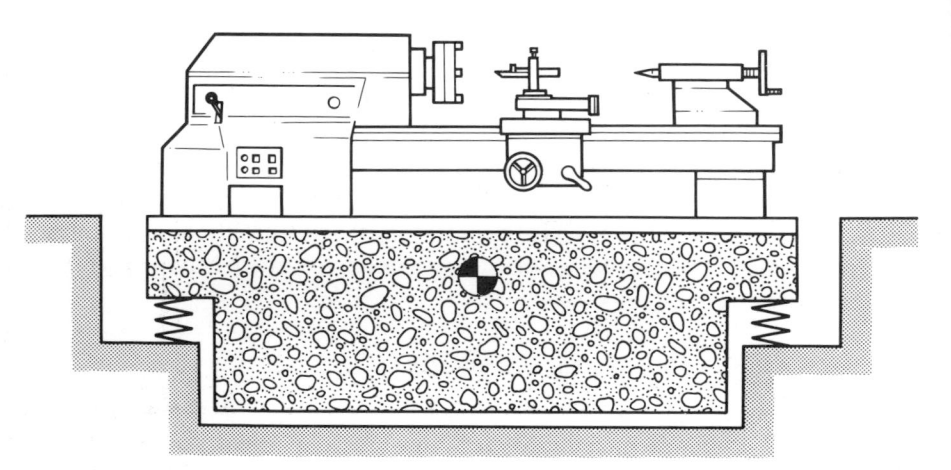

W

Rails for
outrigger
isolators

Figure 6.15   Boiler on Load Spreading Rails for Stability

Figure 6.16   Machine with T-Section Foundation Block to Effect Centre of Gravity
Mounting

distribution will be more even and, providing the block is heavy enough, it may enable a symmetrical mounting arrangement to be used. Figure 6.17 illustrates an example where not only uniform loading of the isolators is achieved, but also a better method of supporting the pipework.

6.5.4 To Minimise the Effect of External Forces

Although the use of an inertia block does not improve the transmissibility for a given static deflection, it does mean that very much stiffer isolators can be used for the same static deflection, that is, if the mass of the equipment is doubled, the stiffness of the isolators necessary to support it is also doubled. This means that the equipment is far less susceptible to the effects of external forces such as fan reaction pressures and transient torques due to changes in speed or load.

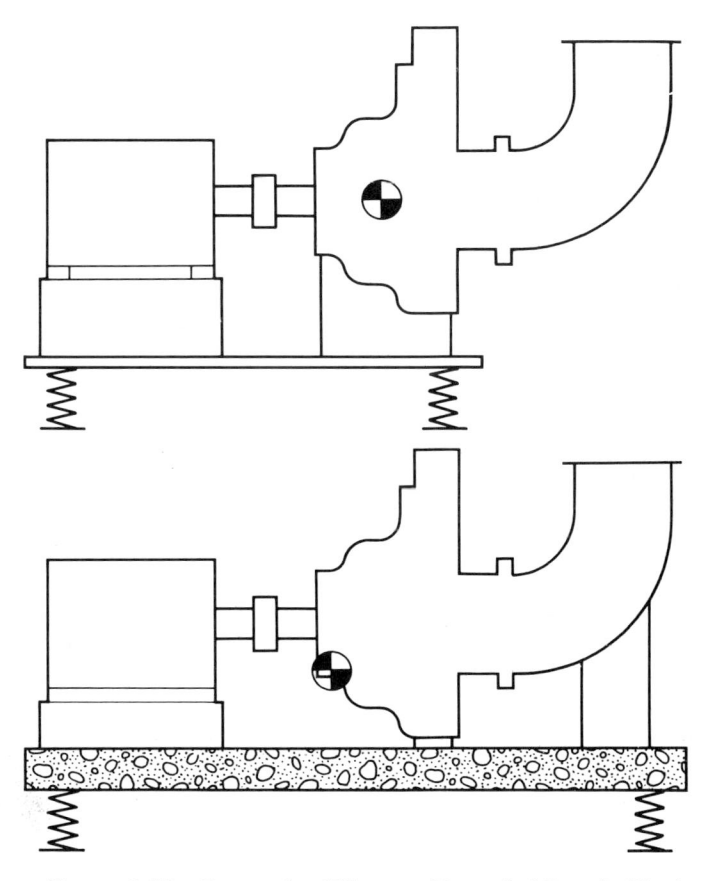

Figure 6.17   Pump plus Elbow on Extended Inertia Block

6.5.5 To Provide or Replace Rigidity

An inertia base can be used to provide rigidity for the mounted equipment in the same way that a steel base is used. This consequently leads to reduced wear.

6.5.6 To Reduce Problems Due to Coupled Modes

The higher of the two rocking sideways coupled movements for a tall item of equipment may occur at two to three times the frequency of the vertical natural frequency. This can lead to resonance problems. Adding an inertia base has the effect of lowering the rocking natural frequency which helps to avoid the problem.

6.5.7 To Minimise the Effects of Errors in Estimated Positions in the Equipment's Centre of Gravity

When vibration isolators are being selected, it is necessary to calculate the total load on each isolator so that the appropriate isolator can be chosen. This normally has to be done before the equipment is available and estimated positions of the centres of gravity of each item have to be used. If this information is inaccurate, the estimated loads may be considerably different from the ones which occur in practice. This may lead to vibration isolators being grossly under or overloaded, or to the equipment sitting at an unacceptable tilt. The latter problem becomes increasingly likely as vibration isolators with high static deflections are used. If a concrete inertia base is used, the centre of gravity of this is normally known accurately and, if the mass of the base is comparable with the mass of the rest of the equipment, it means that, even if the equipment information is not accurate, the possible inaccuracies in the final estimated centre of gravity are small. This reduces the possible errors in isolator loading and reduces the likelihood of a tilted installation. The probability of a tilted installation is also further reduced because of the stiffer springs that will be used to carry the additional weight of the inertia base.

6.5.8 To Act as a Local Acoustic Barrier

When very noisy equipment is mounted directly on the floor of an equipment room, the floor immediately under the equipment may be subject to very high sound pressure levels in the immediate vicinity of the equipment. This local area where the floor is exposed to these high levels may cause problems of noise transmission into the room below. A concrete inertia base can act as an effective barrier, protecting the vulnerable areas of the floor.

Finally, Figure 6.18 illustrates four cases where the effect of changing the mass of the isolated system on the one hand and changing isolation stiffness on the other result in two entirely different outcomes.

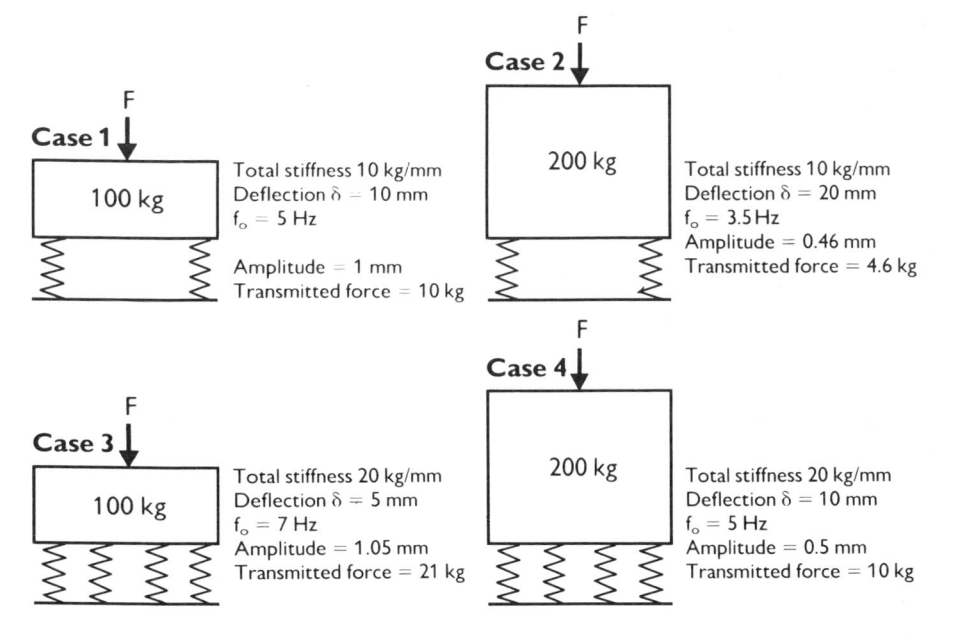

1. Transmitted force depends on static deflection, cases 1 and 4.
2. Increasing mass for the same deflection reduces movement but not transmitted force, cases 1 and 4.
3. Reducing stiffness for the same mass reduces transmitted force and does not increase movement, cases 1 and 2, cases 3 and 4.

Figure 6.18 The Effects of Changing Mass and Stiffness

6.6 VIBRATION ISOLATORS

Most materials exhibit some degrees of resilience and since static deflection under applied load is a prerequisite for a vibration isolator, it is possible to consider many materials for use. However, in normal practice it is the low stiffness materials that are employed since these are capable of greater deflection.

The load bearing capacity of various materials together with the required natural frequency means that more generally the classification of isolators is in terms of stiffness and more particularly of static deflection.

The range of values both in terms of static deflection and natural frequency given below classifies the materials more commonly used in vibration isolator design.

Material	Range of Natural Frequency (Hz)	Range of Static Deflection (mm)
Cork Composites	25 – 45	0.4 – 0.15
Rubbers and Elastomers	7 – 35	5 – 0.5
Engineering glass fibre	7 – 35	5 – 0.5
Metal Springs	2 – 8	75 – 4.5
Air	1.5 – 4	150 – 15

The form in which these materials are engineered into a vibration isolation varies considerably and consequently the performance features and the preferred types of application need to be understood.

Figure 6.19 indicates the broad range of isolator configurations, the type and configuration of materials employed and some, but not all, comments on the performance of the isolator system.

Some further more detailed information is given later in this chapter on the types of vibration solution more generally favoured for use in the field of building services.

6.7 SELECTION

There is a regime of facts and information which is essential for the selection of vibration isolators as well as an additional set of details which would assist in the selection process. The two sets of information required are detailed below.

Essential Facts
1. Type of equipment to be isolated
2. Weight of equipment or machine
3. Estimation of equipment or machine centre of gravity position
4. Number and position of mounting pads
5. Operating speed(s) and nature of operating mechanism (rotary, reciprocating etc)
6. The resulting forcing frequency of vibration

Desirable Facts
1. Fixing details of isolator
2. Operational environmental conditions
3. Type and description of surface upon which the machine is to be mounted
4. The required isolation efficiency
5. The permitted transmissibility at resonance
6. Cost limitations

VIBRATION ISOLATOR ELEMENTS

Type and Configuration	Material	Features & Applications
Simple panel	rubber, rubberised fabric/ cork, felt	simple inexpensive for stationary machines, may/ may not be bonded in place
Area mounts	rubber, rubberised fabric/ cork, cork composites, felt, ridged/constrained rubber mount	usually in the form of profiled carpet mounts laid under machine bases on concrete floors
Helical steel spring	steel	excellent vibration isolation over a wide range of frequencies, also good as a shock absorber, has little damping, useful over a wide frequency range
Damped helical steel spring	steel plus damping material	all the advantages of a spring with adjustable/non adjustable damping by friction or viscous techniques
Rubber mounting in compression	bonded/moulded natural/ neoprene rubber	simple and inexpensive can be captive, has little inherent damping
Rubber mounting in shear	bonded/moulded neoprene rubber	simple, relatively inexpensive capable of higher static deflection – not fail safe
Pneumatic mount	externally pressurised air spring	expensive, high isolation efficiency, self levelling natural frequency down to 1.5 Hz
Woven metal mesh	woven stainless steel, cushion rolled	good damping system, used in conjunction with steel spring for high temperature applications

Figure 6.19   Vibration Isolator Types and Configurations

To assist in the selection of vibration isolators many of the designers and suppliers resort to a simple engineers selection chart of which a typical example is illustrated in Figure 6.20.

To illustrate the primary selection procedure for isolation performance by this limited procedure, the following three examples are worked using the selection chart shown in Figure 6.20.

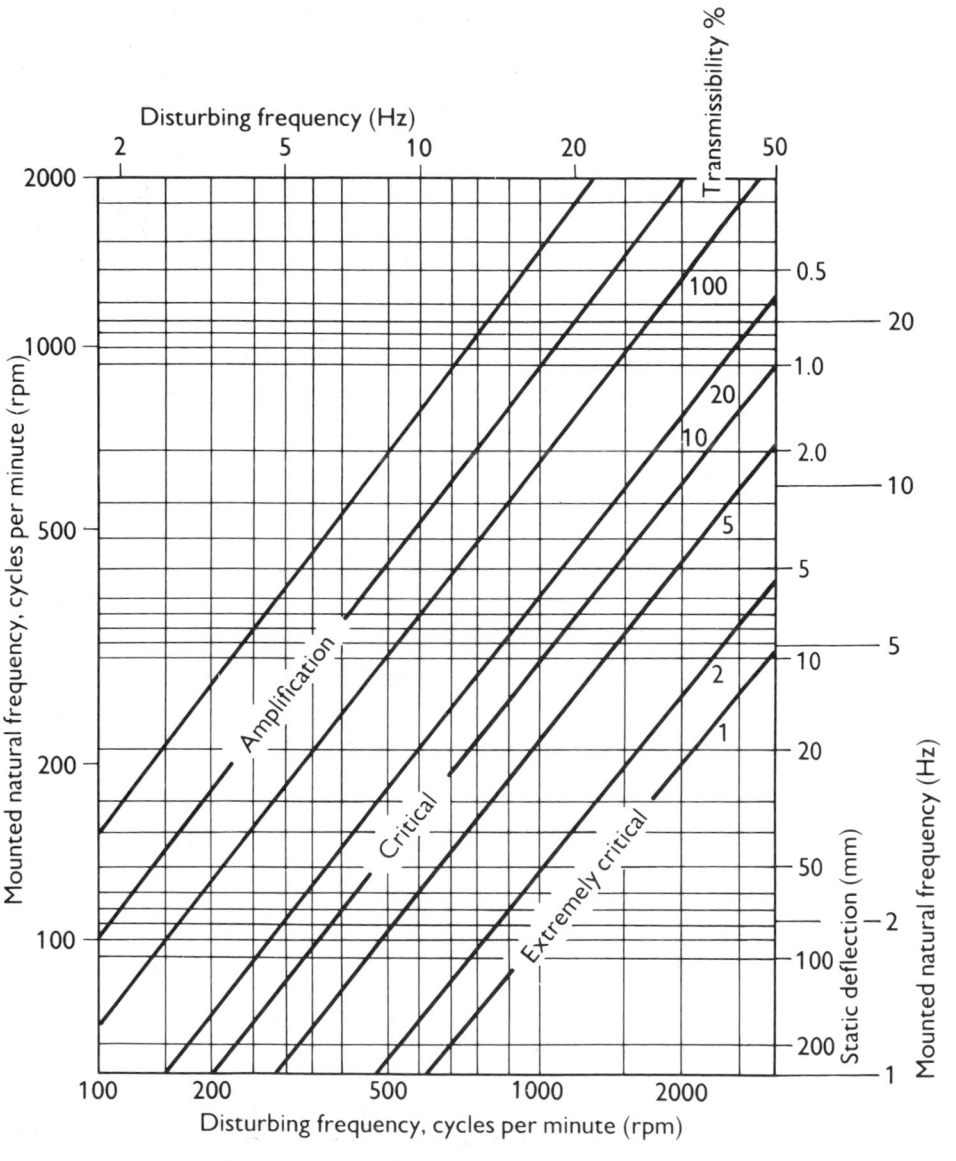

Figure 6.20   Vibration Isolator Selection Chart

Example 1. A fan with a rotational disturbing speed of 1440rpm (24 Hz) is to be located adjacent to a critical area, where an isolation efficiency of 95% is preferred. This requires a vibration isolator with a natural frequency of 5 Hz and a corresponding static deflection of 10mm.

Example 2. A centrifugal fan with a rotational speed of 600rpm (10 Hz) located in a plantroom over a critical area, where an isolation efficiency of at least 95% is preferred. The selection would be of a spring isolator with a natural frequency of 2.3 Hz and a static deflection of approximately 65mm.

Example 3. An axial flow fan with a rotational speed of 960rpm (16 Hz) is located in a non critical area at basement level and an isolation efficiency of 80% is stated. The required natural frequency for the isolator is 6 Hz and a static deflection of 6mm. For this application a rubber in shear isolator could be installed.

6.7.1 Weight Distribution

The aspect of the primary selection procedure which is generally not appreciated but which is essential to the process is the determination of the load at each of the proposed isolation positions.

The simplest situation is encountered when the centre of gravity lies at the geometric centre of fan mounting points and thus each isolator carries 25% of the total weight. However, in the majority of cases, the arrangement of mounting points is not symmetrical and thus the loading is not uniform at the mounting positions. In such cases the loading can be determined by appplying the Principle of Moments and a typical situation is illustrated in Figure 6.21.

Let *W* represent the weight of the object concentrated at the centre of gravity and 1, 2, 3 and 4 the mounts on which the load rests. The position of the centre of gravity with respect to the mounts is given by the dimensions *y*, *Y*, *z* and *Z*.

The results, by taking moments, is given below:

$$\text{Load on mount 1} = W \times \frac{(z-Z)}{z} \times \frac{Y}{y} \qquad \text{6.22a}$$

$$\text{Load on mount 2} = W \times \frac{Z}{z} \times \frac{Y}{y} \qquad \text{6.22b}$$

$$\text{Load on mount 3} = W \times \frac{Z}{z} \times \frac{(y-Y)}{y} \qquad \text{6.22c}$$

$$\text{Load on mount 4} = W \times \frac{(z-Z)}{z} \times \frac{(y-Y)}{y} \qquad \text{6.22d}$$

However this is not a unique answer as in fact only three points of support are really needed — the classic three legged stool illustrates this. In fact the fourth support point is redundant except to assist in a particular load distribution to avoid base strains or add stability. Even Figure 6.21 illustrates a situation in which the centre of gravity lies within the triangle "124" and location "3" could be ignored. For the three legged case the "UNIQUE" solution does exist.

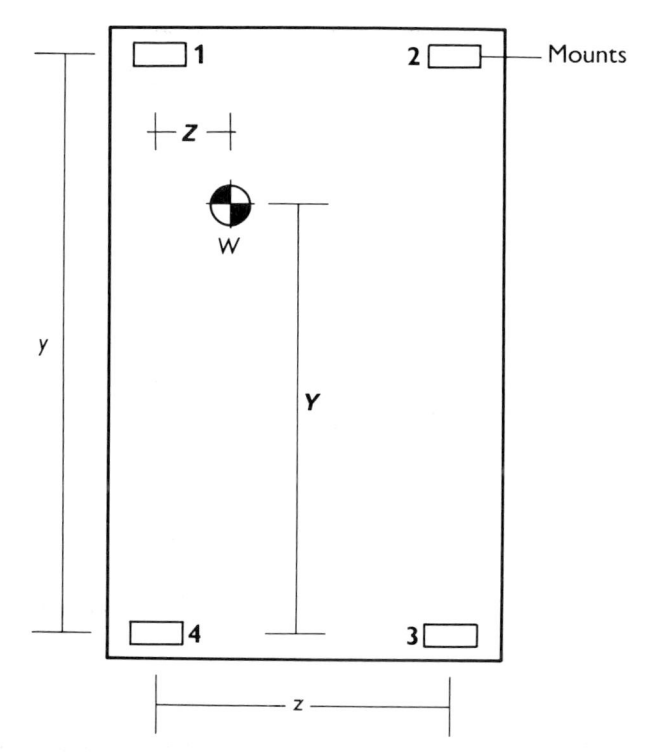

Figure 6.21   Four Point Mounting System with Uneven Weight Distribution

For a more centrally located centre of gravity position mentioned above, an allowable solution is also found by supporting 40% of the load at each of two diagonally opposite corners and 10% at each of the others — illustrating a non-unique situation.

Even loading can be restored to the case illustrated in Figure 6.21 in one direction by providing a pair of intermediate isolation mounts as shown in Figure 6.22. Here, for simplicity, the centre of gravity has been located centrally in one direction while remaining offset in the other. The location of the two additional mounts has been derived by taking moments but on this occasion each of the mountings will have been chosen to support an equal proportion of the total weight. The calculation method is also indicated below Figure 6.22.

If the centre of gravity is located centrally in one direction but offset in the other as shown then two more mounts 2 and 5 may be added at a distance z_1 as shown, such that all six mounts support the same weight — that is $W/6$.

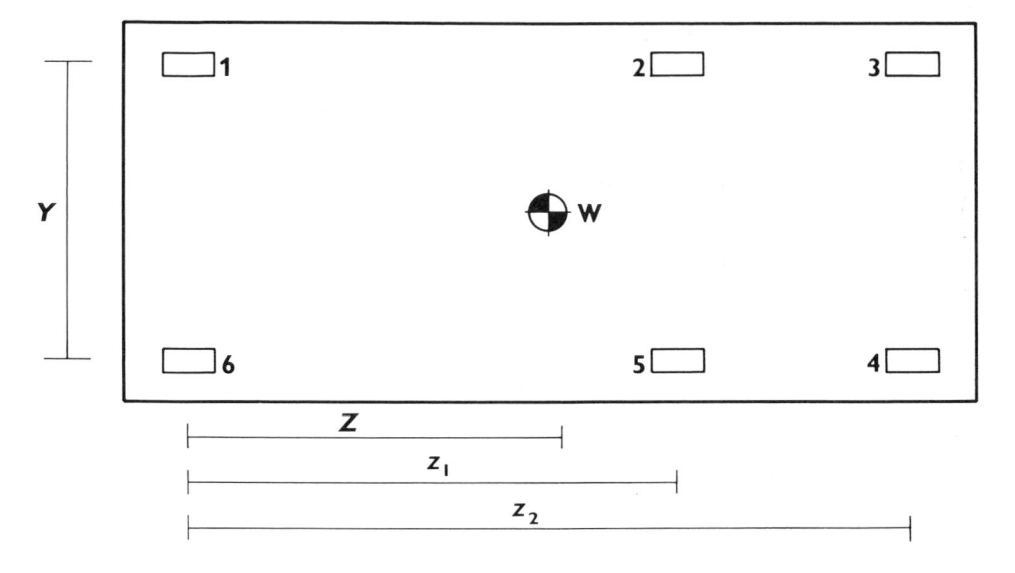

Figure 6.22   Six Point Mounting System with Even Weight Distribution

To calculate z_1 take moments about the line joining mounts 1 and 6. Thus

$$W/3 \times z_1 = W \times Z - W \times z_2/3 \qquad\qquad 6.23$$

So

$$z_1 = 3Z - z_2 \qquad\qquad 6.24$$

6.7.2 Generalized Location

However, the location of mounting positions for vibration isolators may well be dictated by others rather than the acoustic engineer. There is also the consideration of the plan shape and geometry of the machinery and its base, which may not be regular.

Another consideration is that a larger number of support points may be necessary if the stiffness of the machine base is thought to be suspect, particularly if large mounting centre separations might otherwise have been chosen.

Many references and design/selection aids have been promoted in text books on vibration control, but these have been invariably tedious and complex leading to possible error. Alternative "quick" methods have been promoted which work on the principle of weight apportioning across a scale plan shape of the machine base using a square matrix system of coordinates. However, such methods are limited in their application to simple base designs.

The advent of microcomputers enables the designer of vibration isolation systems to easily determine with accuracy the loading at any number of isolator mounting points for machine bases of any shape.

Two methods of calculation are given below for which a given set of mounting locations has been set down by others. The first method assumes that the centre of gravity is centrally located in one direction, but offset in the other direction. The second method covers an assymetric base plan where the centre of gravity is offset in both directions.

Method 1 for centre of gravity position offset in one direction - see Figure 6.23.

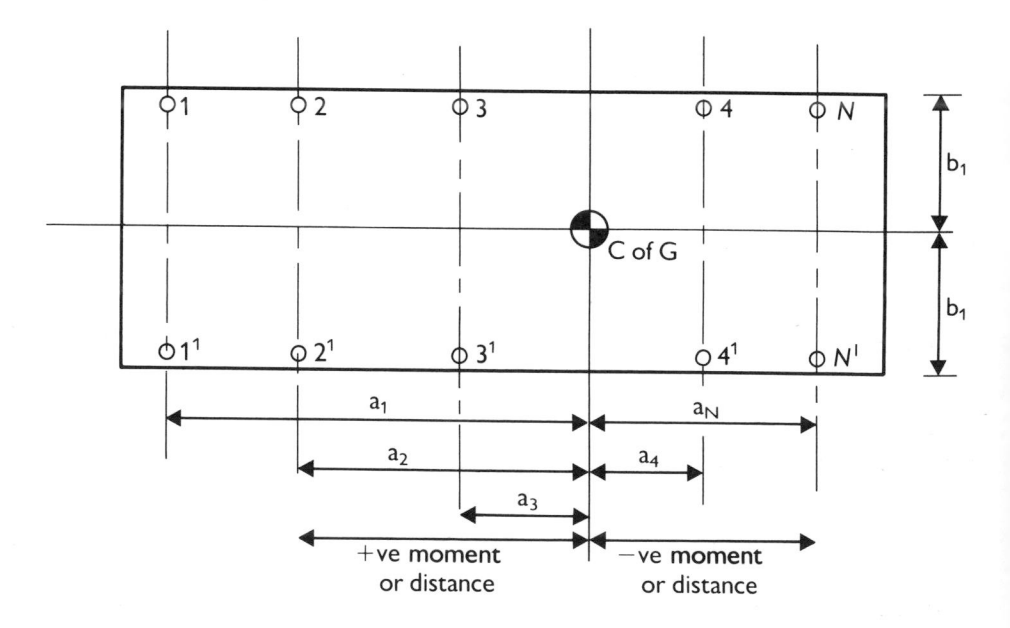

Figure 6.23   The General Case for Uneven Weight Distribution over an Arbitrary Number of Mounting Points where the Centre of Gravity is Offset in ONE Direction.

Load P at any loading point n is given by

$$P_n = \frac{1}{2N} \left[\frac{W\,(\Sigma a^2 - a_n \Sigma a)}{\Sigma a^2 - \dfrac{(\Sigma a)^2}{N}} \right]$$

           6.25

where n = total number of mounting points; W = total weight; a_n = distance of the nth mounting point from the centre of gravity

Calculation Procedure

1. Tabulate all mounting position distances from centre of gravity with the appropriate sign (+/-)

Position	a	a^2
1	$+a_1$	a_1^2
2	$+a_2$	a_2^2
3	$+a_3$	a_3^2
4	$-a_4$	a_4^2
.	.	.
.	.	.
N	$-a_N$	a_N^2
sum	Σa	Σa^2

2. Determine $(\Sigma a)^2 =$

3. Determine $1/N =$

4. Determine $\dfrac{(\Sigma a)^2}{N} =$

5. Determine $\Sigma a^2 - \dfrac{(\Sigma a)^2}{N} =$

6. Determine $\dfrac{W}{\dfrac{\Sigma a^2 - (\Sigma a)^2}{N}} =$

7. Determine $\dfrac{1}{2N} \times \dfrac{W}{\dfrac{\Sigma a^2 - (\Sigma a)^2}{N}}$

8. Load per mounting point
$P_1 =$ item 7 x $[\Sigma a^2 - a_1 \Sigma a]$
$P_2 =$ item 7 x $[\Sigma a^2 - a_2 \Sigma a]$
$P_n =$ item 7 x $[\Sigma a^2 - a_n \Sigma a]$

Method 2 for centre of gravity position offset in two directions — see Figure 6.24.

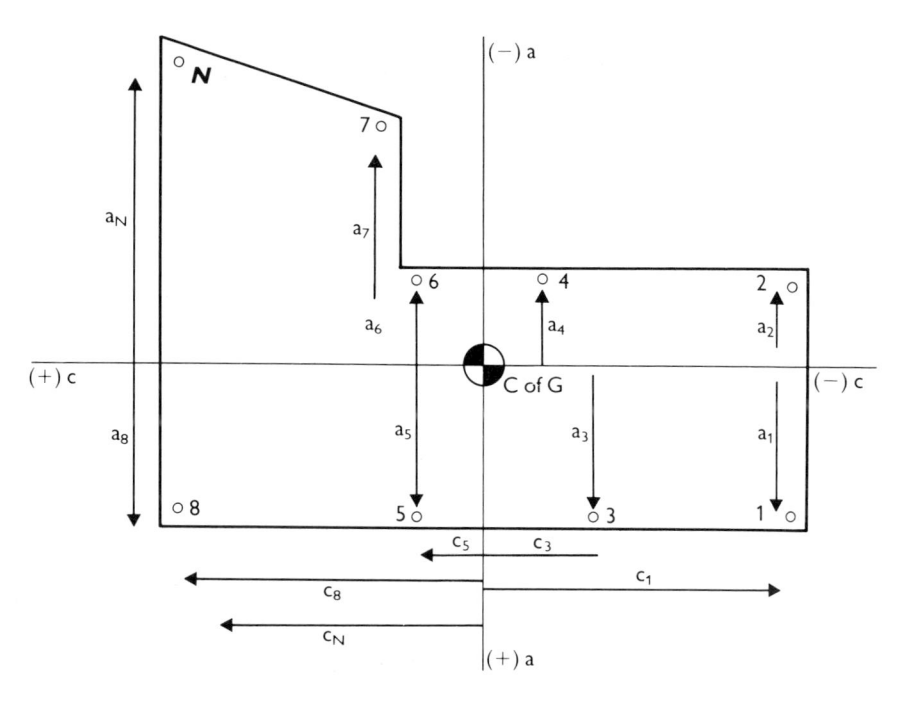

Figure 6.24   The General Case for Uneven Weight Distribution over an Arbitrary Number of Mounting Points where the Centre of Gravity is Offset in TWO directions

Calculation Procedure

1. Tabulate all mounting position distances from centre line in both directions with the appropriate sign $(+/-)$ and determine (ac), a^2, c^2

Position	a	c	ac	a^2	c^2
1	$+a_1$	$-c_1$	$-a_1c_1$	a_1^2	c_1^2
2	$-a_2$	$-c_2$	a_2c_2	a_2^2	c_2^2
3	$+a_3$	$-c_3$	$-a_3c_3$	a_3^2	c_3^2
4	$-a_4$	$-c_4$	$-a_4c_4$	a_4^2	c_4^2
.	.	.	.	.	.
.	.	.	.	.	.
N	$-a_N$	$+c_N$	$-a_Nc_N$	a_N^2	c_N^2
Sum	Σa	Σc	Σac	Σa^2	Σc^2

Square Sum $(\Sigma a)^2$ $(\Sigma c)^2$ $(\Sigma ac)^2$

where N = total number of mounting points; W = total weight; a_n and c_n = distance of the nth mounting point from the centre of gravity in each direction.

2. Using the data determined above determine the following:

$$B = \Sigma a^2 \Sigma c^2 - (\Sigma ac)^2 + (1/N)[2\Sigma a\Sigma c\Sigma ac - \Sigma a^2(\Sigma c)^2 - \Sigma c^2(\Sigma a)^2] \qquad \text{6.26a}$$

$$E = (\Sigma ac)^2 - \Sigma a^2 \Sigma c^2 \qquad \text{6.26b}$$

$$F = \Sigma a\Sigma ac - \Sigma c^2 \Sigma c \qquad \text{6.26c}$$

$$G = \Sigma c^2 \Sigma a - \Sigma c\Sigma c \qquad \text{6.26d}$$

$$D = E/B \qquad \text{6.26e}$$

$$\beta_x = F/B \qquad \text{6.26f}$$

$$\beta_z = G/B \qquad \text{6.26g}$$

$$\text{Load per mounting point} = P_n = (-W/N)[D - a_n\beta_z - c_n\beta_x] \qquad \text{6.26h}$$

6.8 TOWARDS THE FINAL SELECTION

The determination of the pattern of loading at the selected mounting position may be the first step but it is not perhaps the most important in the procedure for selecting vibration isolators.

The tendency amongst engineers until fairly recently was to select or specify vibration isolators on the basis of a given transmissibility or conversely the perferred percentage of isolation. However, experience has shown that there were many instances where the performance achieved in practice using this type of approach fell short of expectation. This was due in many cases to the effects of resonance conditions being set up associated with the low stiffness of the building structure.

The significance of building element stiffness is explored in greater depth later in this chapter, however, at this stage it suffices to say that specifying the performance of vibration isolators in terms of static deflection under load has become a widely accepted practice.

The use of an engineers selection chart, mentioned earlier and again shown in Figure 6.25 will normally lead to the selection of a suitable vibration isolator. The chart employs the basic parameters discussed earlier which affect the isolation efficiency, but principally these are the frequency of the disturbing vibration and the required natural frequency to achieve the preferred isolation efficiency. Also the required static deflection of the vibration isolator for a given natural frequency may be established. The chart does give some guidance in terms of the "critical state" of the installation with reference to the likely effect of inadequate vibration isolation on adjacent areas, rooms or spaces within the building.

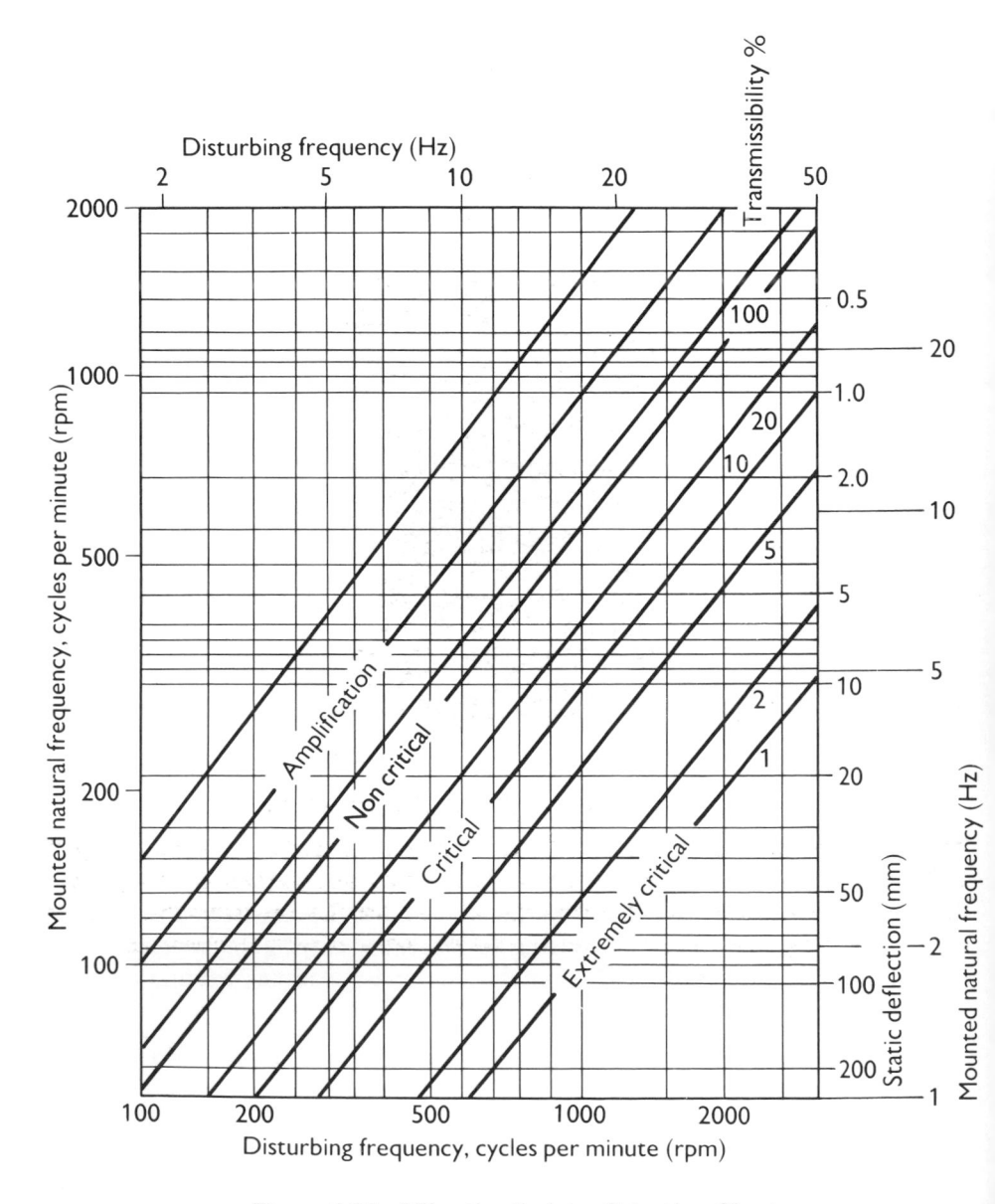

Figure 6.25 Vibration Isolator Selection Chart

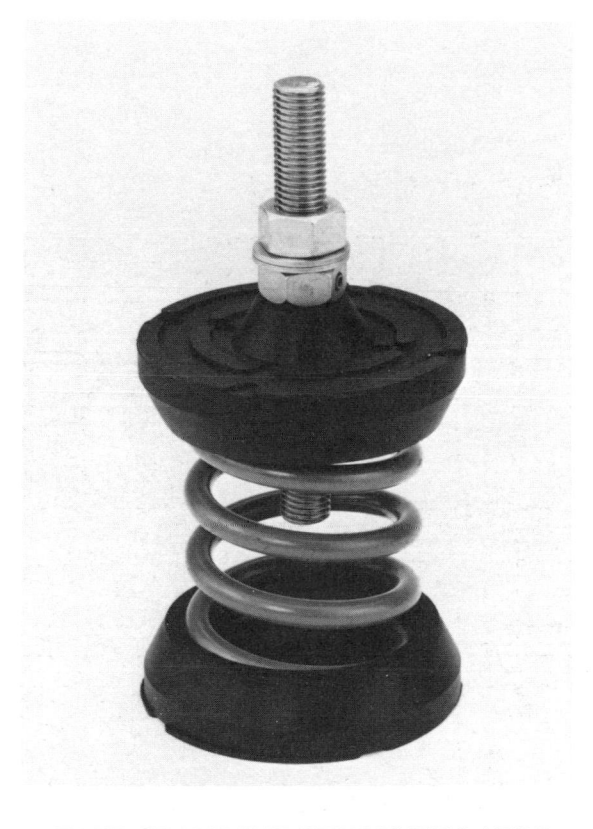

FREE-STANDING SPRING ISOLATOR

The spring coil is designed to be stable when correctly loaded and has nearly equal stiffness in all three "*x, y, z*" directions. In this sample the noise stop pads have been designed as self-captive cups incorporating the load distribution elements within the moulding. The necessary levelling bolt and lock-nut are clearly visible.
Courtesy: Sound Attenuators Ltd

CAPTIVE MOUNT

Here the free-standing coil, with its noise stop mouldings, is housed to minimise any movement by way of adjustable resilient limit stops. Such mounts are employed when the mass of an inertia base is not available to control occasional load variations such as wind loads on roof-top cooling towers, and start-up on compressors.

Courtesy: Sound Attenuators Ltd

It has become popular amongst certain acoustic engineers to cross reference the level of vibration isolation required to the preferred acoustic design criterion, on the NC or NR curves for a particular situation.

Example. "Vibration isolators shall be selected to have static deflections not less than 75mm where the mechanical equipment under consideration is to be sited on floors above/below rooms requiring an acoustic design criterion of NC30." Whilst it is generally the case that for more stringent acoustic designs there is an increasing need to ensure high levels of vibration isolation, such simply expressed cross references ignore the more important aspects relating to the flexibility of the intermediate building structure, principally the floor elements to be discussed later.

6.9 MATERIALS

The materials most widely used in vibration isolation systems for building services equipment comprise engineering glass fibre, neoprene rubber and metal coil springs. Some general comments about these materials are appropriate.

Engineering glass fibre is very different from the more familiar thermal and acoustic low density varieties. It is made by a flame attenuated drawing process such that the resultant carpet of fibres are approximately aligned, and much less random than spun products. Compression during curing produces very high density slabs — $100kg/m^3$ and more. Once in the correct state these slabs of around 20mm thickness are subjected to compressive cycling at extreme overloads, to fatigue and breakdown any seriously misaligned cross fibres. The remaining resilient slab may be considered as a series of fine fibre leaf springs bonded together to form a stable product. The final vibration isolation unit is cut as blocks from the slab and coated in a water and oil resistant flexible polymeric membrane. It possesses non-linear deflection properties which are often an applications advantage.

The rubber used in vibration isolators is normally of the neoprene or synthetic type rather than natural rubber, principally because of the greater resistance to attack from contaminants and external effects such as oils, solvents, weather, ozone and temperature. The materials are readily moulded into a wide variety of shapes and can be bonded to metal elements such as base fixing plates.

The materials employed can be compounded to have a wide range of resilience, dependent upon the "Shore hardness" factor for the synthetic rubber compound. A certain amount of damping is inherent in synthetic rubber, principally due to hysteresis effects. The damping factor may be altered by combination with other "filling" materials, but this is rarely done for reasons of economy in manufacturing costs. Again these materials are capable of supporting a wide range of applied loads, subject only to a limit of 10% for maximum permissible compressive strain. This is due to the setting up of internal stresses which, if excessive, reduce the long term stability and performance. Shear strains of 50–60% are not uncommon and for this reason rubber in shear mountings are generally capable of higher static deflections and thus higher isolation performance for a given level of disturbing vibration.

In general rubber in compression vibration isolators can support much greater loads than the rubber in shear isolators, with the limit on permissible strain largely dictating the size of the isolator.

To some extent the surface shaping of the flat surfaces of a rubber pad, by moulding in ribs or shoulders, will help to increase the static deflection capability of such an isolating material. This is due to the fact that the ribs or shoulders, are more readily compressed then the main body of the isolator and are thus capable of greater static deflection under load. However, the compressive stiffness of such a product is not linear and thus the load against static deflection peformance must be thoroughly researched by the manufacturer.

Rubber in shear vibration isolators have been widely used on building services equipment for many years with considerable success and is typically the isolator type chosen for the standard equipment format, such as fans, chillers and pumps.

The use of vertical metal coil springs is perhaps the most popular choice amongst those designers concerned with achieving high levels of vibration isolation for building services installations. Generally these products are capable of higher order static deflections up to 150mm with a wide range of load carrying capability. Increased load carrying capacity is simply achieved by clustering a number of identical springs at each of the preferred mounting points. The range of steel metals available mean that large deflections can be achieved without excessive stressing and that spring "creep" under load is kept to a minimum with high level returnability. The environmental resistance to attack can be enhanced by the use of "special" steels or surface protection coatings.

Perhaps the most significant advantage to be found by using spring vibration isolators is that their performance in practice follows very closely the predictions based upon theory, thus the design engineer can normally select springs with a good deal of confidence.

Metal spring isolators normally have very little inherent damping and whilst for many applications in building services situations this is not a problem, nevertheless damping can be added quite readily by the addition of friction dampers.

A particular problem with the use of metal spring vibration isolators is due to their ability to readily transmit higher frequency vibration. In effect acoustic frequencies simply pass down the coil without significant attenuation. The use of rubber pads on the top or bottom of springs, normally described as "noise stop pads", has been widely held as a solution to this problem. However, some research has shown that these relatively thin section rubber pads are not very efficient in reducing the transmission of acoustic frequencies. Indeed experience has shown that where the installation under consideration is adjacent to "critical areas" or areas having particularly low acoustic design criterion, that is, NC30 or below, then the use of thick section ribbed rubber pads in series with spring isolators is more likely to be successful. A further refinement of this approach is to situate the equipment on spring vibration isolators which in turn are supported on a raised plinth which is itself isolated from the floor on ribbed rubber mat or a matrix of isolator blocks such as engineering glass fibre.

When the selection of vibration isolators has required mounted resonant frequencies in the range of 2 Hz to 25 Hz, as is most often the case, then it is possible to predict the transmissibility performance at the standard first octave band frequencies, at the low end of the acoustic spectrum; in other words, at 32, 63, 125 and 250 Hz octaves. The transmissibilities illustrated in Figure 6.26 represent that percentile of the out-of-balance force at those frequencies, transferred to the base of the vibration isolator and are representative of what can be expected from either rubber or rubber and steel spring combined isolators.

Acoustic Transmissibility								
Isolation	Mounted Resonant Frequency Hz							
Frequency	2	4	6	8	10	12	14	18
32 Hz	0.5%	1.5%	4%	8%	11%	16%	25%	50%
63Hz	0.2%	0.6%	1.4%	2.6%	4.5%	7%	10%	20%
125Hz	0.05%	0.15%	0.4%	0.8%	1.1%	16%	2.5%	5%

Figure 6.26 Acoustic Transmissibility

Whilst at first reading these represent extremely small contributions, it must be borne in mind that only small vibration oscillations of large structural elements such as floors or walls can result in appreciable noise levels. There are other significant influences, not least the point(s) on the floor or wall at which these transferred effects occur.

The selection of spring isolators having large static deflection capability, typically in excess of 100mm, needs to be undertaken with some care. The engineer selecting springs for vibration isolators is always concerned to ensure that the spring element remains stable in use and to achieve this, he is careful to limit the length to diameter ratio of the spring under compressive load.

For certain manufacturers of spring vibration isolators this care in design of the compression element is qualified in their printed literature by stating a minimum limit to the ratio of horizontal to vertical stiffness, that is, "the horizontal stiffness shall not be less than 1.3 x vertical stiffness".

Where a multiplicity of springs are used in a particular spring mounting in order to increase the load carrying capacity, it can be readily seen that the horizontal stiffness may be different in the two horizontal planes. The arrangements shown in Figure 6.27 demonstrate this point for simple combination of springs, assuming that the springs are, in each case, of identical individual vertical and horizontal stiffness.

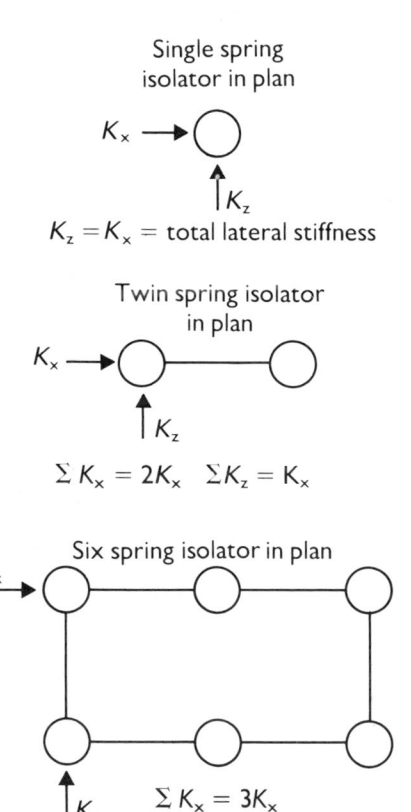

Single spring
isolator in plan

$K_x \longrightarrow$

$\uparrow K_z$

$K_z = K_x$ = total lateral stiffness

Twin spring isolator
in plan

$K_x \longrightarrow$

$\uparrow K_z$

$\Sigma K_x = 2K_x$ $\Sigma K_z = K_x$

Six spring isolator in plan

$K_x \longrightarrow$

$\uparrow K_z$ $\Sigma K_x = 3K_x$
$\Sigma K_z = 2K_x$

Figure 6.27   Total Stiffness for Multiple Spring Vibration Isolators
K_x = horizontal/lateral stiffness in the x direction
K_z = horizontal/lateral stiffness in the z direction

Thus for housed or enclosed spring mountings the risk of spring static instability is less likely than for single tall open springs. Also, by careful design of the isolator housing a controlled limit on the amount of horizontal translation of the spring elements can be incorporated. However, the use of housed spring mountings can lead to other problems such as flanking of vibration through metal parts touching and the use of such isolators should be limited to specific applications. This is particularly the case where mountings incorporating damping mechanisms or limit "stops" are to be

INERTIA BASE SKELETON

Bespoke and unitary inertia base frames are prefabricated for site location casting of the massive concrete infill. Reinforcing rods are in place and the spring location points are pocketed in and also sct to yield a 10 mm ground clearance with the selected spring mounts at their correct operating height.

Courtesy: Sound Attenuators Ltd

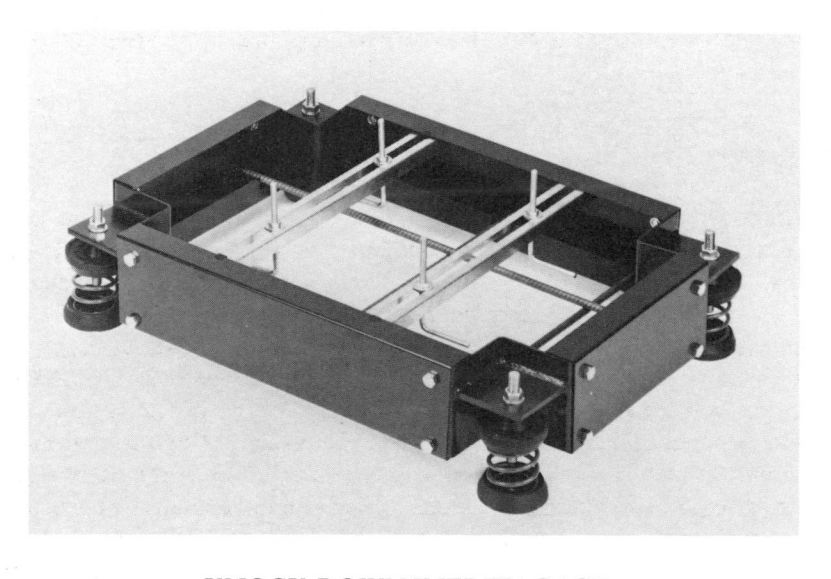

KNOCK-DOWN INERTIA BASE

Not so much a saving in portability, but one manufacturer's novel way to meet the demands of variability and dimensional flexibility in conjunction with quick delivery. Here only the corner pieces are standard and the side rails are quickly cut to length. On assembly and after hardening, the concrete becomes a necessary part of the structural integrity.

Courtesy: Sound Attenuators Ltd

used, because the equipment being isolated may otherwise undergo excessive movement resulting in metal-to-metal contact within the isolator mechanism.

Typical of such applications are reciprocating compressors and engines where dampers are incorporated in isolators to reduce machine motion at resonance during the run up and run down period. For cooling towers and chillers, the change in applied load due to varying levels of water in the pond, or draining of the chiller system, would result in reduced static deflection and hence spring height increases. This raising of the equipment and consequent strain on fitted connections such as pipes and pipe flexible connections, needs to be avoided. Hence mounts incorporating limit stops and restraints are employed.

6.10 INERTIA BASES

Our attention has already been drawn to the advantages to be gained in certain situations resulting from the use of inertia bases and foundations. Of the eight principal situations, the one which gives rise to the most interesting engineering considerations is when it is necessary to restrict the motion of the vibration isolated machine due to external or inherent out-of-balance forces. When dealing with such out of balance forces which react on the resiliently supported mass at vibration frequencies above the mounted natural frequency, the amplitude of motion of the driven system will be controlled by the mass of the machine. Stiffness of the vibration isolator will control the motion at frequencies below the mounted natural frequency. Since in order to achieve efficient vibration isolation and control we are normally concerned with the mounted natural frequency being below the forcing frequency, to increase the mass of the mounted machine is the logical step for improved motion control. This is best achieved by the introduction of an inertia base or foundation of predetermined size and weight.

Machines such as reciprocating air compressors and diesel engines widely used in modern building services systems are typical of such applications where inertia bases are used in conjunction with vibration isolators.

The calculation of the motion of a machine is both complex and time consuming, however some simplified calculation procedures have been established which are both useful and relevant. The most significant assumption made is that the mounted natural frequency does not exceed 0.33 x the frequency of the disturbing vibrationary force. This is appropriate since from earlier deductions this would result in a high degree of vibration isolation efficiency approaching 90%.

Considering firstly a vertical or horizontal force acting along an axis passing through the centre of gravity of the mounted system.

$$a_v = (28.5\, F_v)/(M f_d^2) \qquad\qquad 6.27a$$

$$a_h = (28.5\, F_h)/(M f_d^2) \qquad\qquad 6.27b$$

where a = displacement amplitude, horizontal or vertical, mm;
F = unbalanced force, horizontal or vertical, N;
M = mass of supported system, kg;
f_d = frequency of disturbing vibration, Hz.

Now considering a vertical or horizontal couple acting under the same conditions:

$$a = 341\,Cb/Mh^2f_d^2 \qquad\qquad 6.28$$

where C = unbalanced couple, Nm; b = distance between the centre of gravity and the part of the machine whose motion is to be calculated, normally an extreme point/edge, m; and h = diagonal of the equivalent rectangle enclosing most of the machine, m.

Finally considering a vertical or horizontal force acting along an axis not passing through the centre of gravity of the mounted system:

$$a = [28.5F_v(h^2 + 12xb)]/Mh^2f_d^2 \qquad\qquad 6.29$$

where x = perpendicular distance between the axis of the force and centre of gravity, m.

Where a machine is known to generate out-of-balance couples/forces at twice the rotational speed, otherwise known as the second harmonic, then the amplitude of motion can be calculated as shown above using f_d equivalent to twice the frequency of the disturbing vibratory force.

The amplitude calculated by the approach given above may then be considered in terms of acceptability by comparison to the amplitude against frequency graph given in Figure 6.28.

The amplitude of motion is effectively that which occurs at the corner of the equivalent rectangle enclosing the main bulk of the machine. In practical terms this can be regarded, in the case of a reciprocating compressor, as the motion at the cylinder head/ discharge manifold or in the case of a circulating pump at the pipe manifold.

This is an important consideration since the control of motion at such points on the machine will have a direct bearing upon the longevity and usefulness of pipework and flexible connections.

By establishing the level of vibratory motion which is acceptable from discussion with the installation designers, the required additional mass or inertia mass of the foundation base can be established. A consideration of base geometry and necessary dimensional size to accommodate the machine will normally provide sufficient options for the designer to achieve the optimum layout.

The inclusion of an inertia base will normally facilitate the selection of vibration isolators of uniform stiffness for equal loading, by virtue of reducing the eccentricity of the centre of gravity of the machine alone.

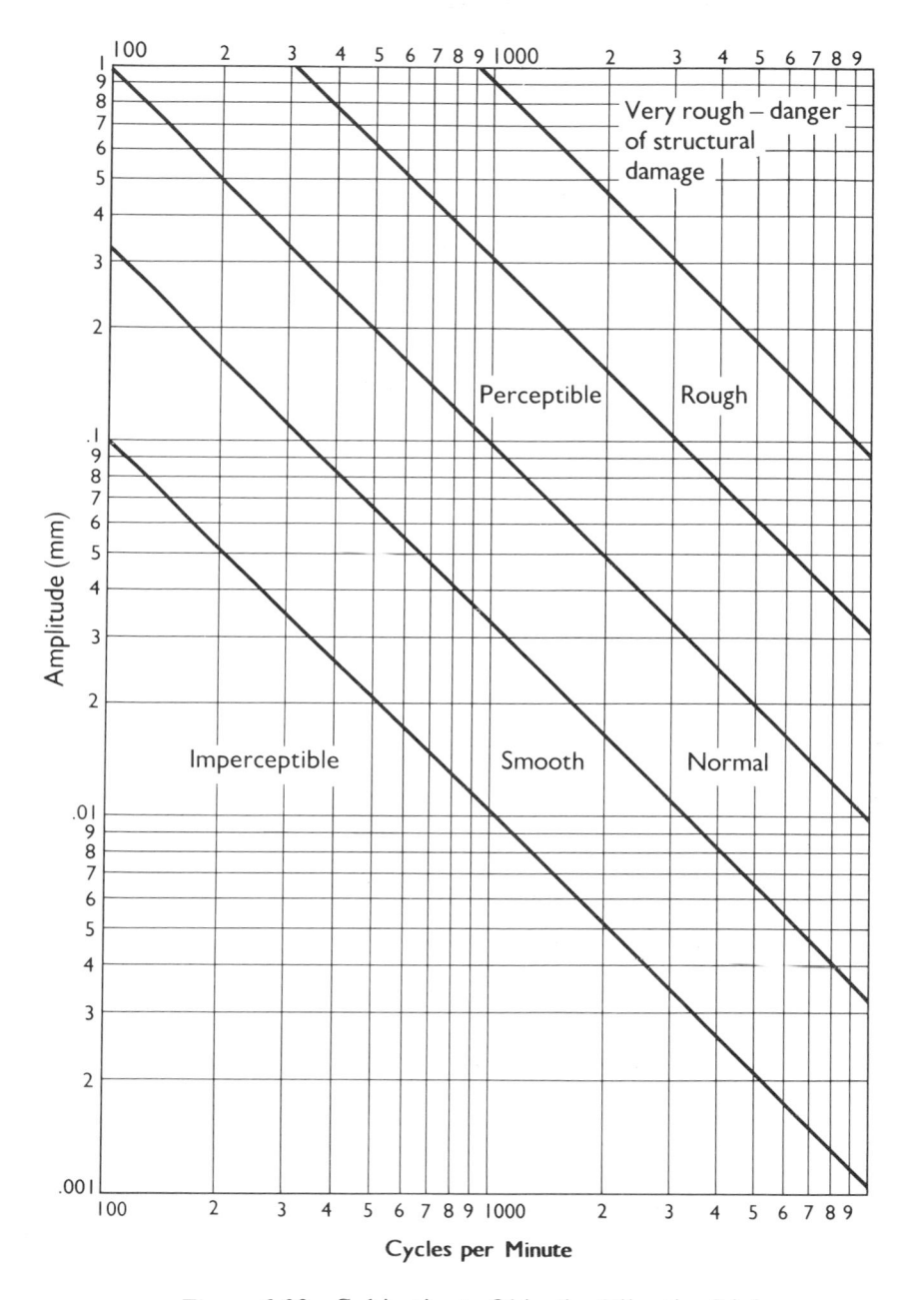

Figure 6.28 Subjective to Objective Vibration Links

The added advantage is that uniform stiffness springs will tend to provide a uniform and low level of force transfer.

A further advantage to be gained by the addition of an inertia base is the lowering of the overall centre of gravity toward the plane of support of the vibration isolators. This reduces considerably the possibility of coupling due to any horizontal vibrating forces as well as decoupling other modes of vibration, principally the rocking modes.

The use of high pressure, high velocity centrifugal fans fitted with low stiffness spring vibration isolators can lead to problems of fan tilt resulting from the effects of reaction forces due to the static and velocity pressure heads. The axis about which the reaction occurs is normally some distance above the base resulting in the fan tilting backwards on its supports.

The use of an inertia base, resulting in increased system mass and hence stiffer springs for a given mounted natural frequency, will counteract the reaction force. A consequence is that flexible connections fitted to the fan discharge will not suffer undue strain. Again the addition of an inertia base will facilitate the selection and positioning of vibration isolations of uniform stiffness.

6.11 FLANKING

The emphasis throughout the preceding considerations has been to isolate the source of vibration from the surroundings or conversely a sensitive receiver from extraneous vibrations by the introduction of vibration isolators of preferred stiffness. However, it is necessary to consider the other routes or points of contact which may exist between the source/receiver and the surroundings otherwise these may provide uncontrolled flanking paths for vibration.

The uncontrolled routes will, in all probability, occur as a consequence of a lack of care by other trades, thus adversely affecting the performance otherwise achieved by careful selection of the vibration isolators.

Typical examples include:

1. Spreading concrete across or up to bases of vibration isolated equipment.
2. The construction of walls, partitions or structural upstands against vibration isolated items, or cross bracing between the new construction and the vibration isolated equipment.
3. The lack of or improper selection of flexible connections in fluid pipelines and ducts.
4. The build up of debris against or under vibration isolated equipment, particularly where inertia bases have been included in the design.

Other flanking or bridging paths across the mounted system can occur if the selection or installation of the vibration isolator is not given careful consideration. Considering the simple spring hanger support for either ducts or pipes demonstrates most clearly the problems which can occur if this otherwise simple and low cost product is not carefully installed.

This is clearly demonstrated in Figure 6.29.

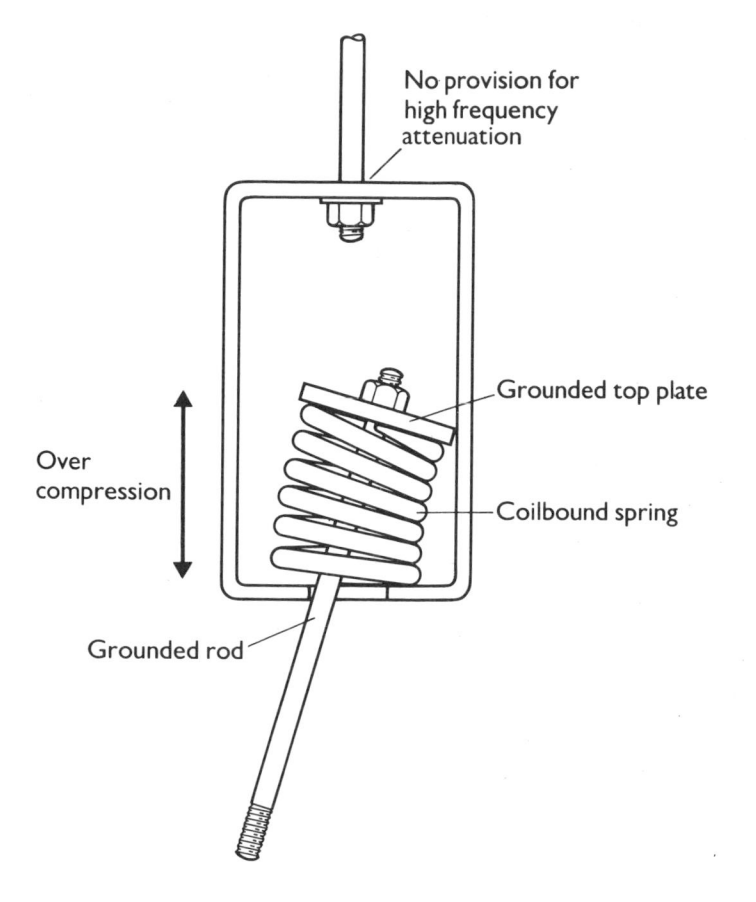

Figure 6.29   Flanking of a Spring Hanger

In addition it must be recognised that some designs of vibration isolator may well result in such problems occurring in practice if not carefully selected for each application. Fully enclosed or "captive" spring isolators may not achieve their full potential performance if installed in such a manner that the upper "isolated" metal hood comes into contact with the lower metal enclosure surrounding the springs. The addition of rubber snubbers as limit stops to control motion are more likely to "exacerbate" bridging problems than they are to restrict unacceptable levels of motion. Similarly isolators which include adjustable damping devices can result in problems of this type particularly when the dampers are so tightly adjusted that the overall isolator stiffness is dictated by that of the damping device rather than the springs.

In general, it is considered most prudent to restrict the use of either enclosed spring isolators, or isolators having damping or snubbing features to special applications identified by the specialist engineer who is, in turn, fully aware of all the aspects relating to the installation as opposed to just a working knowledge of the equipment to be isolated.

6.11.1 Ducts and Pipes

The principle services which are connected to most types of building services equipment are either in the form of duct, pipe or conduit dependant upon the nature of the service.

In ductwork, the vibration will principally be due to the inbalance of the prime mover, normally the fan impeller, or motion in the duct wall due to turbulence of the airflow. The source can be decoupled from the ductwork system by the introduction of flexible connectors fitted both to the inlet as well as outlet sides of the fan. The material used for the connectors are generally of a canvas type or possibly for highly specified acoustic considerations of mineral loaded materials. The product selected must fully satisfy the requirements of minimal air leakage, temperature and pressure stability and additionally for special applications be fire resistant, as well as corrosion and abrasion resistant.

Care must be taken to ensure that flexible connectors do not sag excessively under negative pressure and conversely that when operating under positive pressure they do not extend and deform such that the isolated equipment becomes displaced on the vibration isolators. In instances where isolators having a static deflection of 25mm or greater are to be selected then the risk of displacement due to the extension of bellows type connections must be considered as a possibility and compensated for in the design. In pipework and to a considerable extent for conduit, the main consideration is not that the walls radiate high noise levels due to vibration, but rather the "sounding board" effect which occurs if the service is rigidly connected to a large radiating surface such as a wall, floor, or ceiling slab.

There may well be two alternative methods of designing the pipe or conduit system such that vibration isolation can be included without much consequence to the dynamic performance of the pipe system.
For most situations the pipe casing is subjected to positive pressure due to the fluid and consequently any length of truly flexible piping interposed in the system will, as a consequence of its in built elasticity, try to extend or change shape.

The product must therefore be able to sustain these effects without damage or reduction in performance. However, if in changing shape/deforming, the surrounding pipework is displaced then this may result in a fracture of the system at a point of support such as a flanged connection. If the system is subjected to high operating pressure or high levels of pulsation then this adds a further complication.

The first option available to the designer is to use flexible connectors which are inherently constrained or restrained. Illustrations of two such types are given in Figure 6.30 where a unit with the bars between the flanges to control extension of the unit and also a concertina tube with spiral corrugations to control the change in

Flexible metallic hose

Figure 6.30 Restrained Flexibles

diameter under pressure are shown. In each case control of the ability to extend, twist or "balloon" tends to stiffen the connection. Thus it is normal to install two such connectors each side of an elbow so that motion in two planes is available.

The second option is to use unrestrained connectors perhaps of the bellows type, but design the pipework system such that dynamic movement is reacted against thrust plates or blocks. This approach can lead to complications and is only normally used on high pressure large pipe diameter systems.

Where flexible connectors are to be used to achieve complementary isolation of the pipework associated with the vibration isolated equipment, it will be rarely justifiable to consider the length of pipe or duct to be floated as less than 50 pipe diameters (duct minimum dimension) and rarely more than 200 pipe diameters.

In the case of more complex installations such as machinery involving reciprocating and rotating action, it must be borne in mind that the vibrational wave along the pipe involves not only transitional motion of the pipewall but also dilational motion. In addition the combination of low operating speeds and large out of balance force means that because of the large static deflections necessary, but not available from most flexible connections, then only isolation of higher acoustic frequencies is possible.

Finally, whilst considering the dynamics of piped installation, the designer must take account of the operating effect of such items as gate valves, where as a consequence of their opening and closing, large fluctuations occur which manifest themselves as pulsations. These superimposed dynamic effects have to be accounted for when considering not only the type of flexible connectors, but also the type, position and performance of flexible supports from the building structure.

6.12 STRUCTURE AND GROUNDBORNE VIBRATIONS

So far it has been assumed that the structure upon which the vibration isolated equipment is supported is in its own right rigid, or infinitely stiff.

In reality this assumption is never fully or even remotely approached and even the basement floor of a building is not rigid at vibration frequencies in the audible range. Further, building structures are a composition of elements each having its own component stiffness and natural frequency of response. Thus it is possible for parts of a building structure to be excited by vibrations emanating from machinery located in remote parts of the same building.

The situation of a resilient structure is represented in Figure 6.31 as a compound spring system, where the structure is represented as a mass on a spring.

The resonant frequency of such a system is defined as

$$f_{res} = \left[\frac{1}{2}(f_2^2 + f_1^2) \pm \frac{1}{2}\sqrt{(f_2^2 - f_1^2) + 4f_1^2 f_3^2} \right]^{1/2} \qquad 6.30$$

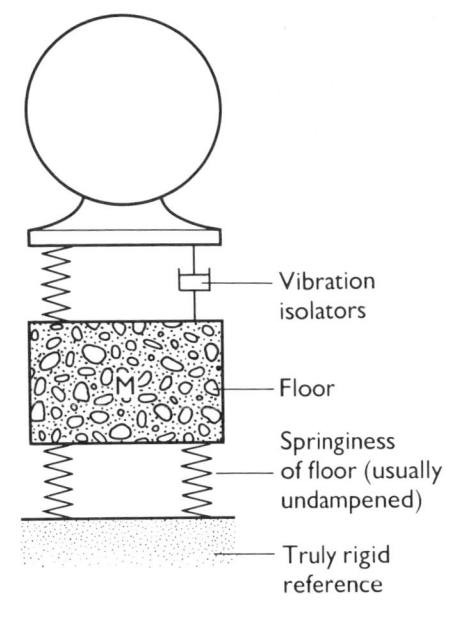

Figure 6.31 Elements of the Building Structure Modelled as a Compound Spring System

where

$$f_1 = \frac{1}{2}\pi (K_1/M_1)^{1/2} \text{ Hz} \qquad \qquad 6.31a$$

$$f_2 = \frac{1}{2}\pi [(K_1 + K_2)/M_2)]^{1/2} \text{ Hz} \qquad \qquad 6.31b$$

$$f_3 = \frac{1}{2}\pi (K_2/M_2)^{1/2} \text{ Hz} \qquad \qquad 6.31c$$

where M_1 = mass of machine, kg;

K_1 = stiffness of vibration isolator, kg force/mm;

M_2 = effective mass of supporting slab, kg;

and K_2 = effective stiffness of supporting slab, kg force/mm.

In order to establish the natural frequency of response for a structural element, it is necess..ry to consider the likely deflections due to loading that the structural engineer will permit. For the important case of floor spans, this is generally limited to a maximum deflection at the centre of the span not more than one 250th of the span dimension and allowing for a safety margin of up to 500% results in relatively small deflections. Based upon the maximum allowable deflection then correspondingly the lowest probable natural frequency of response can be established. A range of information is given in Figure 6.32 for floor spans between 3 and 15m.

	Centrally loaded span natural frequencies		
Span (m)	Allowable $\delta(1/250\text{th})$ (mm)	Maximum probable $\delta(20\%)$ (mm)	Minimum probable f_0 (Hz)
3	12	2	12
5	20	4	8
10	40	8	7
12	50	10	6
15	60	12	5

Figure 6.32 Allowable and Probable Maximum Deflections of Centrally Loaded Floor Spans

In order to avoid the problems of resonance between the isolated equipment and the loaded floor, we must ensure that the resilience of the isolator system is less than that of the floor. Thus the deflection of the vibration isolator due to the weight of the machinery must be greater than that of the floor and a multiplying factor of 5 to 6 times is usually appropriate.

To illustrate the relationships between machinery vibration and the structural vibration responses, let us consider a machine operating at 24 Hz and with a minor degree of out of balance. The machine is to be sited centre span on a floor which otherwise has an unloaded natural frequency of 70 Hz. Consequently the floor undergoes deflection due to the added machine weight. If the deflection is 0.5mm then the new natural frequency is 24 Hz, resulting in a resonance condition occurring and in turn resulting in amplification occurring at that frequency where previously none existed. If we then select a vibration isolation device to be installed between the machine and floor, the isolation offered reduces the resonance peak effect at the machine operating speed. The floor motion is consequently reduced towards that of a stiff floor. The selection of vibration isolators of even lower stiffness will result in further improvements and as previously stated ratios of 5 or 6 between isolator and floor deflection are preferable.

Figure 6.33 illustrates the effects sequence referred to above.

In general it has been found that resonances of building components fall between 5 and 200 Hz with floors in the range up to 70 Hz. With many types of building services equipment operating at speeds in the range of 3 Hz to 24 Hz there is a high degree of overlap which explains the high instance and likelihood of floor vibration problems.

As a result of these and other simple and somewhat broadbrush generalisations, the American Society of Heating, Refrigerating and Air-Conditioning Engineers (ASHRAE) laid down some guidance for the mount deflections for a wide variety of machines and floor spans, which is tabulated in Figure 6.34. The selections chosen are based on a philosophy of avoiding resonance.

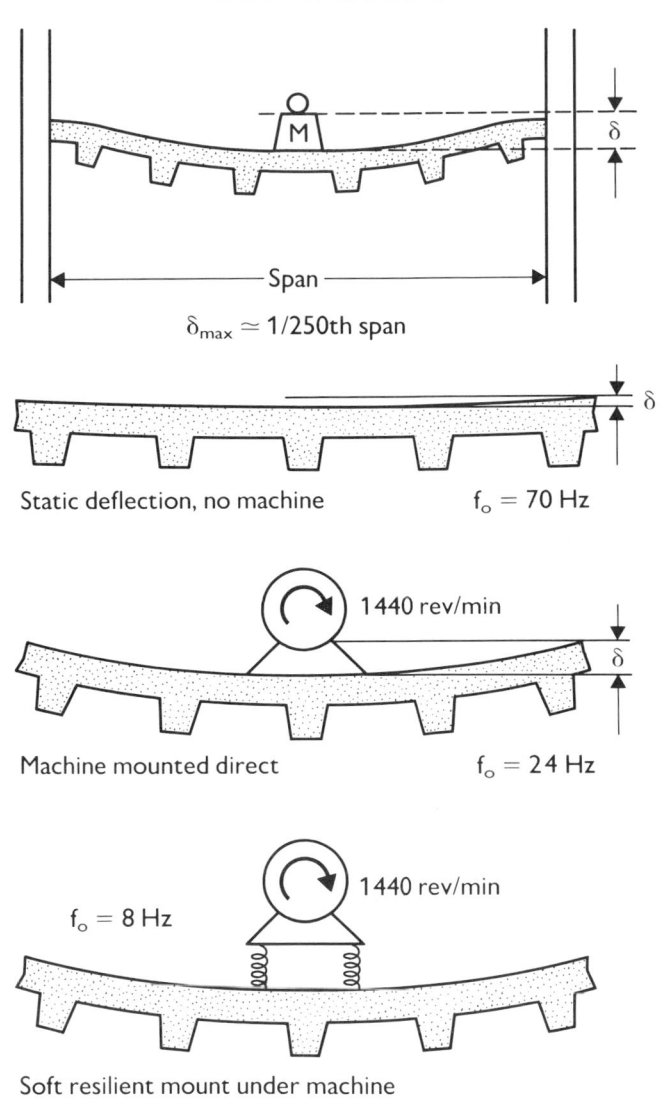

Figure 6.33 The Relationship Between Machinery Vibration and the Vibration Response of a Floor

6.12.2 Floating Floors

The increase in popularity of acoustic floating floors in the last 15 years has led to the misconception that these are included as an added vibration isolation feature. This is not so and the prime purpose of such systems is to enhance the structural sound insulation characteristic beyond that possible from considering "the mass law".

Equipment	Base-ment	Floor span			
		6m	9m	12m	15m
Refrigeration machines					
Absorption	6	12	25	50	50
Packaged hermetic	6	12	50	60	90
Open centrifugal	6	12	50	60	90
Reciprocating chillers:					
500–750 rev/min	25	50	50	60	90
751 rev/min and over	25	25	50	60	60
Reciprocating air or refrigeration compressors					
500–750 rev/min	25	40	60	70	90
751 rev/min and over	25	25	40	60	70
Boilers or steam generators	6	6	25	40	70
Pumps (water)					
Close coupled up to 5 hp	6	12	25	25	25
7½ hp and over	20	25	50	60	90
Base mounted up to 5 hp	9	12	40	50	60
7½ hp and over	25	25	50	60	90
Packaged unitary air handling units (low pressure up to 750 Pa)					
Suspended up to 5 hp	20	25	25	25	25
7½ hp and over					
up to 500 rev/min	30	40	40	50	60
above 501 rev/min	25	25	25	40	50
Floor mounted up to 5 hp	6	25	25	25	25
7½ hp and over					
up to 500 rev/min	12	40	50	50	60
above 501 rev/min	12	25	25	40	50
Axial fans (floor mounted)					
Up to 5 hp	6	25	25	25	25
6–20 hp up to 500 rev/min	12	40	50	50	60
above 501 rev/min	12	25	25	40	50
25 hp and over					
up to 500 rev/min	20	50	60	70	90
above 501 rev/min	12	25	30	40	50

Equipment	Base-ment	Floor span			
		6m	9m	12m	15m
Centrifugal fans (floor mounted)					
Low pressure (up to 750 Pa)					
Up to 5 hp	6	25	25	25	25
7½ hp and over					
up to 500 rev/min	12	40	50	50	60
above 501 rev/min	12	25	25	40	50
High pressure (above 750 Pa)					
Up to 20 hp					
175–300 rev/min	9	60	60	90	120
301–500 rev/min	12	50	50	60	90
above 501 rev/min	9	30	30	50	60
25 hp and over					
175–300 rev/min	40	60	90	120	140
301–500 rev/min	25	50	60	90	120
above 501 rev/min	12	30	50	60	90
Cooling towers					
Up to 500 rev/min	12	12	50	60	90
Above 501 rev/min	9	9	25	40	60
Internal combustion engines (standby power generation)					
Up to 25 hp	9	12	50	60	60
30–100 hp	12	50	60	90	90
Above 125 hp	25	60	90	120	120
Gas turbines (standby power generation)					
Up to 5 MW	6	6	6	9	9

The floor span refers to the largest dimension between supporting columns.
The equipment is assumed to be at mid-span.

Figure 6.34 Static Deflection: Guidance for Resilience to Provide Vibration Isolation and Avoid Resonances

However, because the floating floor is generally supported on isolators having a natural frequency of 15 Hz, it is important to avoid the problems of the compound spring installation by ensuring that the equipment sited on this floor is supported by vibration isolators having static deflections corresponding to natural frequencies of 5 Hz. Thus spring isolators with static deflection capability generally 25mm or greater are preferred in such instances.

6.13 NOISE RADIATION

The instance of major failure of buildings or building components due to vibration effects from building services equipment is unlikely in the extreme and our prime concern is normally with human comfort. New standards are available to the vibration isolation designer which give guidance to an acceptable limit for different types of environment and these are discussed elsewhere.

A much more relevant consideration, particularly where acoustic design criterion is stringent, is the structurally transmitted and re-radiated noise which occurs as a consequence of machinery vibration. Research undertaken to develop a method of determining the resultant noise level due to vibration in the surfaces enclosing a building space has resulted in a mathematical formula which allows the acoustician to establish the consequences to room noise levels and this is detailed below.

$$L_p = V + 10\log_{10}S + 10\log_{10}r - 10\log_{10}(A/4) \qquad 6.32$$

where L_p = resultant sound pressure level;
V = velocity level of the enclosing surfaces (dB re 5 x 10^{-8} m/s);
S = area of enclosing surfaces, m^2;
A = acoustic absorption of the space, m^2;
and r = radiation factor, typically 0.1 at frequencies below the critical frequency, that is, for concrete/brickwork below 70 Hz. Above the critical frequency then r can be taken as 1.

Thus the inter relationship between the careful analysis of vibration problems and specification of suitable vibration isolating systems as part of the overall acoustic design is well established on a scale which extends from individual items of plant equipment through to complete buildings. This again justifies the approach previously proposed which suggested that the responsibility for specifying the performance of vibration isolation systems should rest with the design engineer who is concerned with the installation as a whole, rather than with suppliers of any individual items of equipment.

6.14 SPECIFICATIONS AND ENGINEERED SOLUTIONS

The concept of specifying the natural frequency preferred for the vibration isolation of a given installation is not readily grasped by many engineers. Additionally and perhaps of much more concern to the engineer is that the measurement of frequency is not always directly possible and for that matter the specification of transmissibility or isolation efficiency is far from helpful in true quantitative terms. Thus the specification of preferred static deflection has been found to be much more relevant and of course may more readily be measured on site in a post mortem situation. As has also been highlighted, the structural engineer will more readily identify with specifications which, when addressed to him directly, point towards types of structure and/or typical loading and load distribution considerations. The clauses included below form the part of the overall acoustic specification favoured by many acoustic consultants. They are compatible both with the vibration and acoustic considerations and specifications attributed to both professional organisations such as CIBSE in the U.K. and ASHRAE in America. See also figure 6.34.

6.14.1 Vibration Isolators

Clause 1
Where indicated on the appropriate schedule, vibration isolators shall be provided for
the specified equipment.

Clause 2
If necessary, the isolators shall be provided with hold-down, or restraining features to
prevent changes to the equipment disposition by virtue of weight changes. They shall
also include damping features where the reduction of excessive movement is required,
even allowing for the provision of inertia bases.

Clause 3
All isolators shall provide the required static deflection indicated in the relevant
schedule, under the equipment weight. The selection of specific isolators should
account for any eccentric load distribution, dynamic reactions, so that the specified
deflections are achieved by all isolators under the equipment.

Clause 4
It is the supplier's responsibility to ensure that all isolators offered to the contract
meet the specifications and are suitable for the loads, operating and environmental
conditions which prevail.

Clause 5
Each isolator of the open spring type shall be identifiable by a colour code mark. The
spring element shall be located by a top and bottom plate, where the latter includes a
bonded ribbed neoprene "noise stop" pad to the underside of at least 6mm thickness
and the provision of hold-down fixings if so required. The spring mounting should
include a built in levelling facility with final lock nuts.

The spring elements shall be of the helical type, at least 50mm diameter and having a
horizontal stiffness not less than 130% of the vertical stiffness.

To ensure stability the outside spring diameter shall be a minimum of 0.8 times the
rated vertical operating height.

The spring elements should be rated such that a 50% overload capacity is available
before the spring becomes coil bound.

Clause 6
Resilient hangers must include noise isolation blocks of neoprene rubber or glass
fibre, but may also use helical springs as part of a sub-element design.

The elements shall be mounted within an open steel bracket, with predrilled and some
pretapped holes to permit fixing and the location of hanger rods.

The springs shall conform to the requirements of clauses 3, 4, and 5 and additionally
the spring location should facilitate at least 15 degree vertical misalignment.

HEIGHT ADJUSTMENT

Loaded spring deflections are usually between 25 and 100 mm and, hence, small
weight variances lead to comparatively large deflections of more than 2 or 3 mm.
These need to be levelled out and, hence, the inclusion of substantial levelling screws
and locking nuts in all specification coil spring mounts.
Courtesy: Sound Attenuators Ltd

Clause 7

Enclosed or restrained spring isolators shall be included for specific applications listed in the appropriate schedule. The isolators may include one or more helical springs located in a base plate. The mounting furniture shall include a telescopic top and bottom housing, incorporating built in levelling adjustments and resilient guides to the two housings.

The isolators may include adjustable dampers and shall have a ribbed rubber neoprene sound pad. The specifications for the spring elements and sound pad shall be as previously stated in Clause 5.

Clause 8

The use of neoprene and other synthetic rubber and glass fibre vibration isolators is acceptable where specified in the relevant schedule. The materials and the isolator design should render them impervious to contamination from oils and attacking chemicals and be rot and vermin proof.

6.14.2 Plant Bases

Clause 1

Where appropriate, steel rails may be applied to the equipment to provide the facility of attaching suitable vibration isolators and/or improving equipment stability. The rails should include height saving brackets, matched to the spring selection and deflection, and should be sufficiently strong to maintain the equipment stiffness.

Clause 2

Steel bases may be applied where appropriate and shall be adequate in their design and stiffness to provide adequate support of the equipment.

Clause 3

Concrete inertia bases may be used where appropriate and shall be designed with sufficient strength and rigidity to support the equipment and compensate for dynamic reactions due to the operation of the equipment.

The inertia base size should be sufficient to provide support for all integral parts of the equipment and also any overhanging components such as inlet and discharge manifolds and elbows.

A minimum clearance of 18mm shall be provided between the underside of the inertia block and the structural floor beneath the block, with the installed mountings deflected to their design value and under full plant rating.

In order to specify fully the best type of vibration isolator for a given application, a checklist of information must normally be established by the design engineer. Whilst it is possible to specify vibration isolators, ignorant of the majority of facts, nevertheless the advice of the specialist should be sought whenever possible. Listed below are both the essential and desirable facts relating to the specification of vibration isolators.

Essential Facts
1. Type of equipment to be mounted
2. Equipment weight
3. Estimation of equipment centre of gravity position
4. Number and position of mounting points
5. Operating speed and nature of the operating mechanism
6. The resulting forcing frequency of vibration

Desirable Facts
1. Fixing details for the isolator
2. Operational environmental conditions
3. Type and description of surface upon which the equipment is to be mounted
4. The required isolation efficiency
5. The permitted transmissibility at resonance
6. Cost limitations

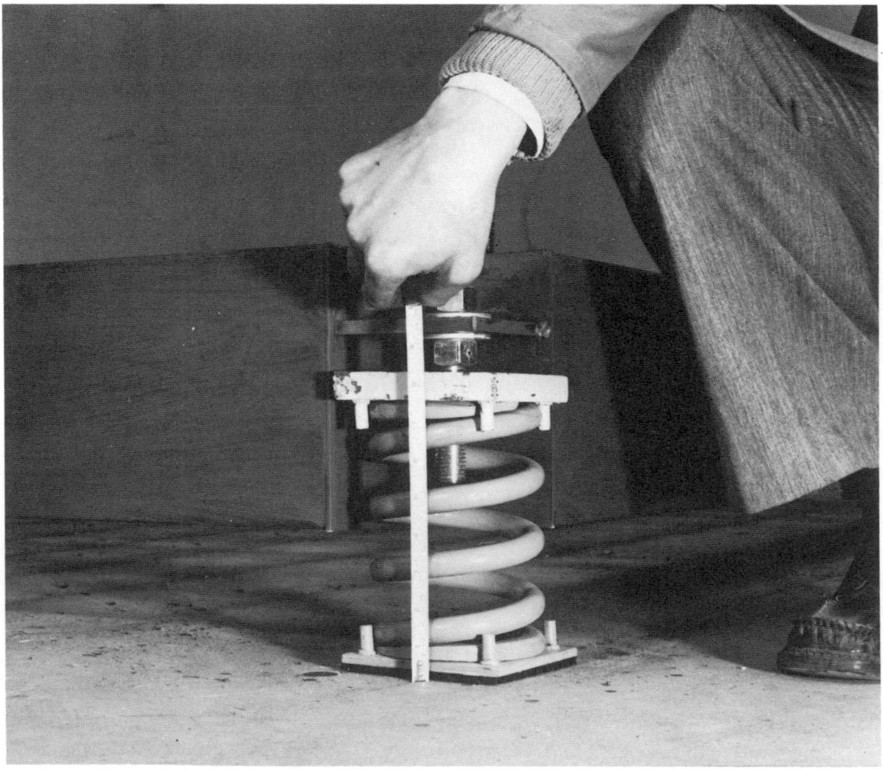

Figure 6.35   Checking the Static Deflection

Moving on towards the concluding phases of the building services contract, it is important to realise that checking and commissioning of vibration isolation systems constitutes an important part of the overall acoustic design programme. This may well be undertaken in the simplest case by actually measuring the static deflection of the spring type isolator as illustrated in Figure 6.35.

Equally, it may be necessary to adjust the levelling of certain vibration isolators beneath the equipment in order to redistribute the weight to achieve a more uniform deflection of the resilient supports. Alignment of the drop rods attached to spring isolators supporting ducts and pipes is equally important as has been demonstrated earlier. In many cases the cost of remedial work is many times that of the vibration isolator product in question thus leading to high remedial charges where, with the proper thought and attention to detail at the design and installation stage, these might be avoided.

CHAPTER 7

DUCTBORNE NOISE — TRANSMISSION

7.1 INTRODUCTION

One of the main sources of noise within a ventilation system is the fan. Hence a primary objective of the system's designer is to establish how much of the sound generated by the fan is transmitted through the ductwork system. This may be to the conditioned room via "jobside" ductwork, or to atmosphere via "atmosphere side" ductwork. Invariably, criteria are set for both areas. If the system's natural attenuation is inadequate to achieve the desired criteria, then the designer will need to consider the best means of achieving additional attenuation.

To accomplish the above objective requires a knowledge of the following parameters:

(a) Fan noise. This data should be available from the manufacturer as octave band sound power levels in the desired operating arrangement. However predictive techniques may need to be employed in many situations.
(b) System attenuation. The natural attenuation within the system provided by duct runs, bends, branches and terminations.
(c) Room or atmosphere corrections. The resultant sound power emerging from the ductwork system modified by corrections such as directivity, distance, room volume and room reverberation time.
(d) Sound reducing techniques. If required, the means of reducing the resultant sound level to the design criteria by the use of unitary attenuators, duct linings, plenum chambers or modification to the system design.

Each one of these four categories is discussed in turn.

7.2 FAN NOISE

Before looking at the prediction of fan noise it is worthwhile to discuss the design and operation of the different types of fans currently in use.

By definition, a ventilation fan is "an air moving device which continuously propels air by the aerodynamic action of a rotating impeller". The mechanical energy from the driving motor is transmitted via a shaft to the impeller and emerges as energy in the form of air velocity and pressure — the fan duty.

Different system designs and applications require different types of fan. Not only the operating characteristics, but also the noise characteristics will be dictated by the type of fan and in particular the design of impeller.

Fans may propel air centrifugally, axially, or a mixture of the two. Taking each category in turn:

7.2.1 Centrifugal Fans

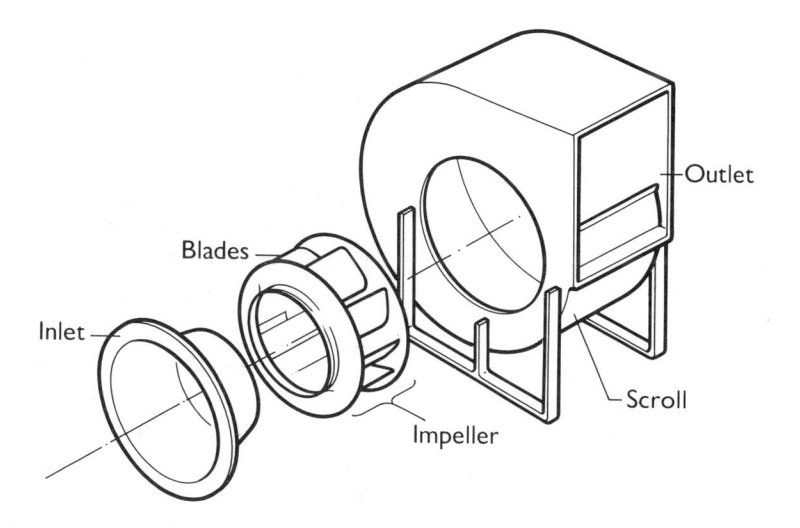

Figure 7.1 Centrifugal Fan, Exploded View

Air enters the fan via a circular aperture which is concentric with the axis upon which the impeller rotates (Figure 7.1). The rotating impeller causes the air to move radially to a point at which it passes through the impeller blades. At this point, the air is constrained by the scroll to move tangentially to the impeller until discharged centrifugally at the outlet.

There are three principal types of centrifugal fans in present use which may be characterised by the shape of the impeller blade used and are illustrated in Figure 7.2 as:

Forward curved
Backward curved
Radial or paddle blade

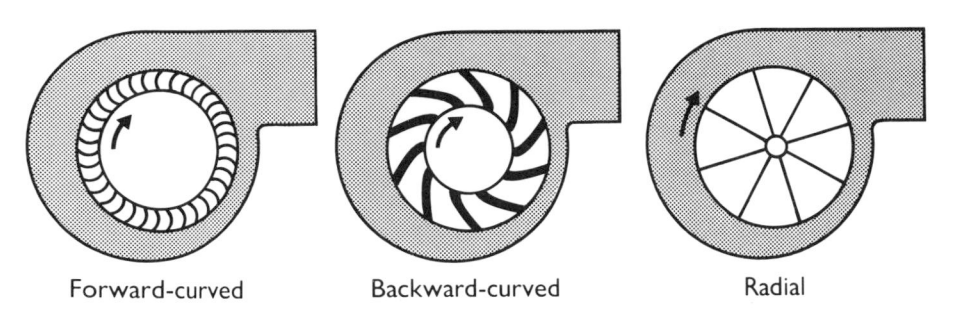

Forward-curved   Backward-curved   Radial

Figure 7.2 Types of Centrifugal Fan

Fans with forward curved blade impellers are more compact and for a given volume and pressure operate at a lower speed than other centrifugal fans. This is because the air is released into the scroll at a velocity greater than the blade tip speed. These high air velocities, generated in the impeller blade/scroll area, produce turbulent airflow. Therefore the resultant sound power level generated by these fans tends to be higher than other centrifugal fan types. Since the rotational speed is lower, the sound power level will tend to peak at a lower frequency. An advantage of the lower running speeds required for forward curved impellers is low mechanical and bearing noise.

Fans with backward curved blade impellers discharge air into the scroll at a velocity which is lower than the blade tip speed. The result is that for a given duty, backward curved impellers will generate lower sound power levels. However, to achieve the same capacity and pressure, a backward curved impeller will have to rotate at speeds nearly twice those of an equivalent forward curved impeller. This results in higher maintenance and increased mechanical noise. Backward curved impellers are preferable to forward curved impellers for systems with varying pressure and volume requirements.

Fans with radial or paddle blade impellers are seldom used in ventilation applications. Their main use is for industrial applications where high pressures are required. Radial fans may also be employed for the transportation of light materials such as: coal dust, wood/plastic chippings, etc. For this reason, their blades are often heavily constructed to reduce wear. Such fans often produce high sound levels at discrete frequencies corresponding in particular to the blade passing frequency, its harmonics and, at times, also the rotational frequency.

7.2.2 Axial Flow Fans

With axial flow fans (Figure 7.3) air enters the fan along the axis of the impeller. The impeller, contained within a cylindrical housing, is coupled to a motor either directly located in the air stream, or indirectly via a belt drive to a motor located externally on the housing. The motor imparts energy to the air via an impeller in a similar manner to that of an aeroplane propeller. In the normal course of events, this energy will be

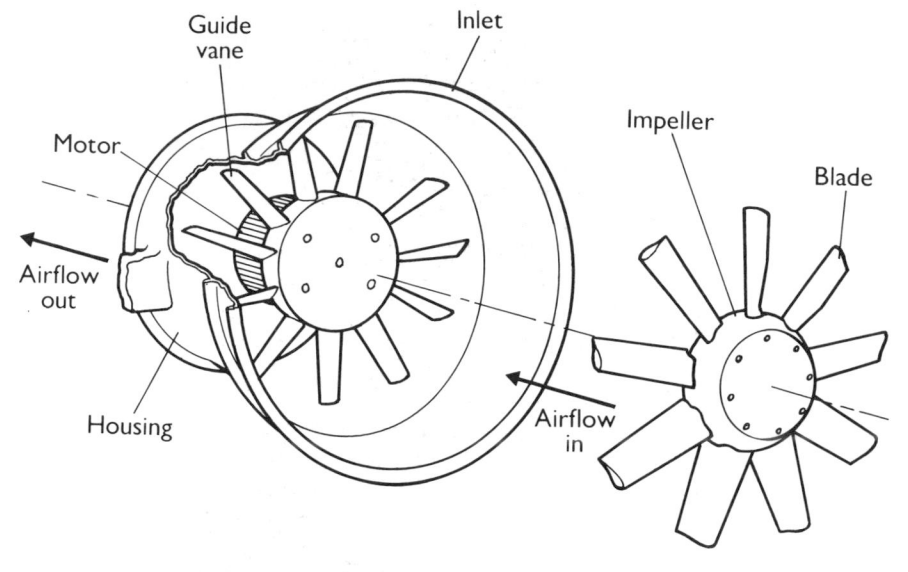

Figure 7.3 Axial Fan Exploded View

transmitted in a spiral motion along the ductwork. This simple axial-flow fan is termed "tubeaxial". Since this flow is inefficient, guide vanes are often located downstream of the impeller to straighten out the flow- "vaneaxial" fan.

In general, the design of axial flow fans is such that higher volume flows but lower delivered pressures are obtainable, when compared to an equivalent duty centrifugal fan.

Axial impellers are operated with higher tip speeds than comparable centrifugal fans. However, the overall sound power level generated is similar to that of centrifugal fans of a similar duty. Axial flow fans generate a higher proportion of high frequency noise but a lower proportion of low frequency noise than centrifugal fans of similar duty.

Axial fan blades may be either constant thickness or alternatively, aerofoil shaped. The aerofoil blade, whilst more expensive, is more efficient and quieter.

A low pressure derivation of the axial fan is the propeller fan. This fan is not capable of operating against the necessary pressures within a ductwork system. Its use is therefore confined to simple roof extract fans, or for circulation of air through heat exchangers/cooling towers.

7.2.3 Mixed Flow Fans

This type of fan has been developed with a view to obtaining the best aspects of both centrifugal and axial flow fans. To achieve this the impeller of an axial flow fan has

Figure 7.4   Mixed Flow Fan (Courtesy Woods of Colchester)

been modified to capture a characteristic of the centrifugal blade — that is a radial component of motion is added to the existing spiral motion. The result is a fan which is capable of delivering both high pressure and high volume, Figure 7.4.

The effect on overall sound power levels is a reduction over axial fans of equivalent duty. The typical sound power frequency spectrum of this type of fan lies between that of the axial flow and centrifugal fans.

All of the types of fans described generate noise as a by-product of their normal operation. As mentioned previously, one component of this noise is termed mechanical noise. This component, generated by bearings, motors etc., is of only secondary importance for airborne noise but remains significant for structure-borne noise, dealt with in Chapter 6.

The main source of airborne noise is aerodynamic noise of which there are two types:

(a) Rotational noise — arising from the interaction between the impeller blades and the surrounding air. The periodic pressure impulses from a blade produce a series of tones. The fundamental frequency of this tone will be at the blade passage frequency, b.p.f.

$$\text{b.p.f.} = \frac{n}{60} \times N \qquad\qquad 7.1$$

where

b.p.f. = blade passage frequency (Hz);

$\quad n$ = fan speed in revolutions per minute (rpm); and

$\quad N$ = number of blades in the impeller.

Lower amplitude harmonics and sub-harmonics may also be evident at frequencies which are multiples of the blade passage frequency.

(b) Vortex noise — produced by the turbulent flow across the impeller blades. Vortex noise is broad band in nature. This component of aerodynamic noise is particularly prominent with all types of centrifugal fans.

Having selected the type of fan, the operating speed of the fan and the fan duty, the system designer will require detailed information on the octave band sound power levels of the fan. Manufacturers are usually able to give this data having performed laboratory tests to relevant British and International standards.

In the event that this data is not available, and as a means of checking the performance data supplied, the approximate fan sound power level in each octave band can be estimated by using the following charts, Figures 7.5 and 7.6.

These charts were derived from empirical formulae developed by Beranek and Allen. It should be noted that the sound power level figures shown relate to the sound power at either the inlet or discharge of a fan.

Sound power levels estimated using this method will give answers to an accuracy of

63	125	250	500	1K	2K	4K	Hz
±10	±7	±5	±5	±5	±5	±7	dB

Both manufacturers sound power data and sound power figures calculated from the charts, will give sound levels which will be generated by fans operated at their representative optimum efficiency and with "ideal" intake and discharge conditions.

For example, the British Standard 848 test duct specifies a minimum free matching intake length of 4 fan diameters and matching discharge of 5.5 fan diameters. The majority of site layouts will preclude the achievement of these conditions and therefore allowance should be made for the increased sound power generated by the fan in these site situations.

Whilst many possible axial flow fan inlet and discharge configurations exist, the following Figures 7.7 and 7.8 cover some of the more common configurations and are included as guidance.

CENTRIFUGAL FAN

Courtesy: Engart Fans

AXIAL FLOW FAN

Courtesy: Engart Fans

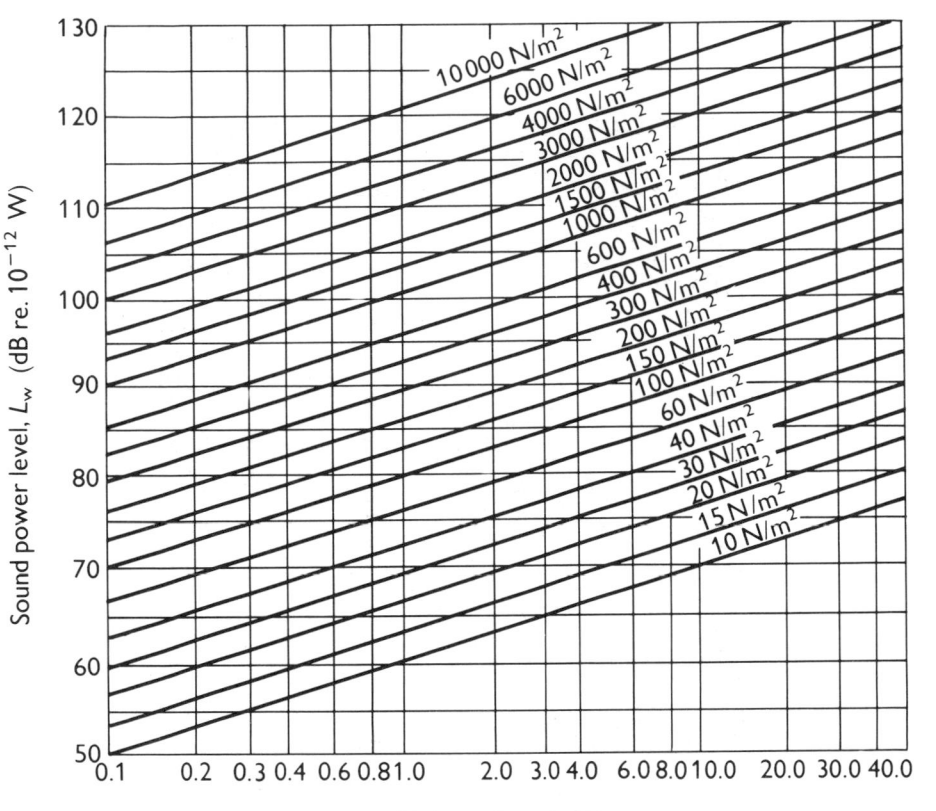

$$L_w = 10 \log Q + 20 \log P + 40 \quad \text{dB re } 10^{-12} \text{ watts}$$

where Q is flow rate in cubic metres per second
P is static pressure in N/m^2

Figure 7.5 Chart for Calculation of Sound Power

	Spectrum Corrections							
Fan Type	63	125	250	500	1 k	2 k	4 k	Hz
Forward Curved Centrifugal	−2	−7	−12	−17	−22	−27	−32	dB
Backward Curved Centrifugal	−7	−8	−7	−12	−17	−22	−27	dB
Axial Flow	−5	−5	−6	−7	−8	−10	−13	dB
Mixed Flow	−12	−11	−10	−10	−13	−17	−22	dB

Figure 7.6 Fan Sound Power, Spectrum Corrections

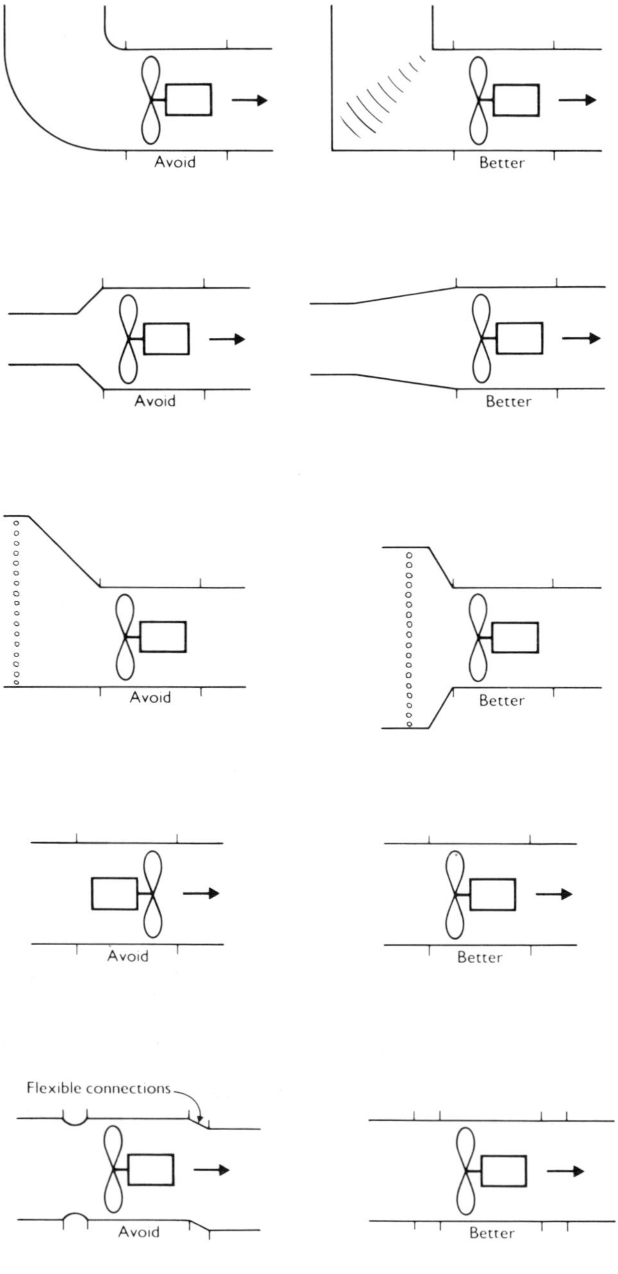

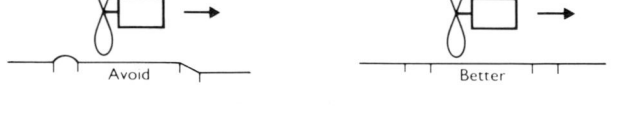

Figure 7.7   Recommended Intake and Discharge Conditions

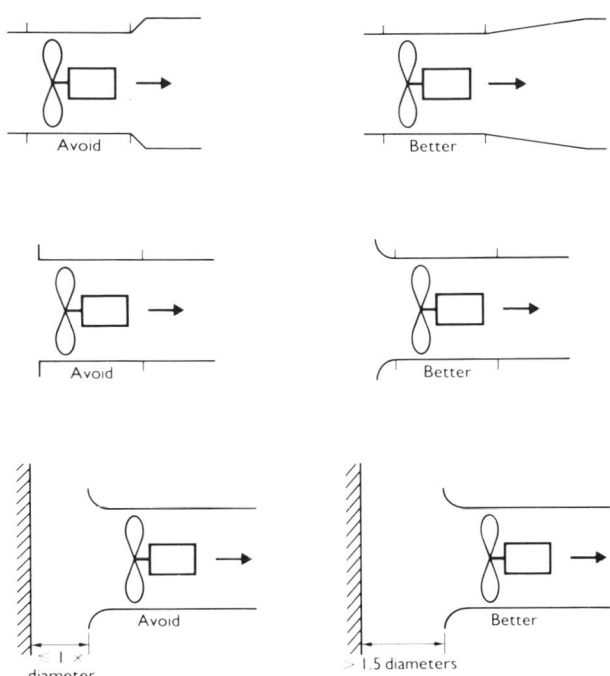

Figure 7.7 (cont.)

a) Abrupt Entry

Frequency	63	125	250	500	1k	2k	4k	Hz
Correction	+2	+5	+7	+5	+5	+5	+5	dB

b) Upstream interference for example trimming vane, idling impeller, radiused bend, acute transformations, expanders.

Frequency	63	125	250	500	1k	2k	4k	Hz
Correction	+6	+6	+6	+6	+6	+6	+6	dB

c) Flexible Connectors (misaligned or concave)

Frequency	63	125	250	500	2k	2k	4k	Hz
Correction	+6	+6	–	–	–	–	–	dB

d) Form of running (motor upstream of impeller)

Frequency	63	125	250	500	1k	2k	4k	Hz
Correction	+6	+7	+3	+8	–	–	+2	dB

Figure 7.8   Intake and Discharge Condition Corrections

It will be seen from these illustrations that when dealing with fans, and in particular axial fans, it is extremely important to ensure that for minimum noise generation air is delivered to and from the fan as "a straight line flow". Spin or assymetrical velocity profiles affect the pressure developed by the blades creating significant increases in both the broadband vortex and pure tone blade passage frequency related noises.

Having now established the fan sound power generated within the octave bands, it is now possible to determine how much of that power will be transmitted to the conditioned area through the ductwork distribution system.

7.3 SYSTEM ATTENUATION

7.3.1 Attenuation in Straight Duct Runs

Sound generated by the fan will travel along the ductwork both upstream and downstream of the fan. The velocity of sound is so great compared to the velocity of air in the duct, that for mechanical services applications, the sound can be said to travel equally well upstream and downstream.

The attenuation of sound due to molecular absorption of air is extremely small for the distances encountered even in the longest of ductwork systems. The attenuation, in sheet metal ductwork, is therefore purely due to the acoustic pressure within the duct in exciting the duct wall into vibration. A component of this vibration will be radiated and emerge as duct breakout noise. This component is dealt with in Chapter 9. The ever-reducing remainder will be retained within the duct and be transmitted along the duct run.

If the ductwork through which the sound travels is circular, or masonry builders work duct, then the walls will tend to be relatively rigid. In this case, more of the sound energy will pass along the duct efficiently. High frequency sound will be slightly attenuated in contrast to the minimal attenuation of low frequency sound.

On the other hand, if sheet metal ductwork is rectangular or oval in section, the walls will be less rigid and have less damping. This is particularly true with large unstiffened areas of ductwork. This greater degree of flexibility will result in more loss of sound along the ductwork run.

The sound passing through and out of the duct, "noise breakout", may however cause problems of another kind.

Whatever the cross sectional shape of ductwork, attenuation will be involved. That is, the original sound power generated by the fan will reduce linearly along the length of the duct. It is therefore convenient to express this reduction of sound power as "attenuation per metre run".

This data is given in Figure 7.9.

Attenuation in dB/metre run

Straight Duct Circular/oval or Rigid Walled (unlined)	x (mm)	Octave Band Centre Frequency (Hz)						
		63	125	250	500	1k	2k	4k
	75–200	0.07	0.10	0.10	0.16	0.33	0.33	0.33
	200–400	0.07	0.10	0.10	0.16	0.23	0.23	0.23
	400–800	0.07	0.07	0.07	0.10	0.16	0.16	0.16
	800–1500	0.03	0.03	0.03	0.07	0.07	0.07	0.07
Straight Duct rectangular (unlined)	x (mm) side dimension	Octave Band Centre Frequency (Hz)						
		63	125	250	500	1k	2k	4k
	75–200	0.16	0.33	0.49	0.33	0.33	0.33	0.33
	200–400	0.49	0.66	0.49	0.33	0.23	0.23	0.23
	400–800	0.82	0.66	0.33	0.16	0.16	0.16	0.16
	800–1500	0.66	0.33	0.16	0.10	0.07	0.07	0.07

Figure 7.9   In Duct Attenuation in Unlined Straight Ducts

Note:
When calculating attenuation in rectangular ductwork, it is necessary to take both duct dimensions into account. The attenuation for each dimension should be calculated separately and then added.

Limit attenuation to a maximum of:

40 dB in the 63 Hz octave band
30 dB in the 125 Hz octave band
35 dB in the 250 Hz octave band
40 dB in the 500 Hz octave band
45 dB in the 1000 Hz and above octave bands

The external lagging of ductwork, although reducing duct breakout, also increases low/medium frequency attenuation within the duct. Rather than totally retaining the sound within the ductwork, external lagging still allows the excitation of the ductwork, albeit it now damped, and primary radiation of noise remains. However this radiation of noise is retained and absorbed between the ductwall and external barrier sheeting.

The attenuation figures shown for unlined ducts (Figure 7.9) may be doubled for lagged ductwork in the 63, 125, 250 and 500 Hz octave bands as indicated in Figure 7.10.

Attenuation in dB/metre run

Circular/oval	x (mm)	Octave Band Centre Frequency (Hz)						
		63	125	250	500	1k	2k	4k
	75–200	0.14	0.20	0.20	0.32	0.33	0.33	0.33
	200 – 400	0.14	0.20	0.20	0.32	0.23	0.23	0.23
	400–800	0.14	0.14	0.14	0.20	0.16	0.16	0.16
	800–1500	0.06	0.06	0.06	0.14	0.07	0.07	0.07
Rectangular	x (mm) side dimension	Octave Band Centre Frequency (Hz)						
		63	125	250	500	1k	2k	4k
	75–200	0.33	0.66	1.00	0.66	0.33	0.33	0.33
	200–400	1.00	1.32	1.00	0.66	0.23	0.23	0.23
	400–800	1.64	1.32	0.66	0.32	0.16	0.16	0.16
	800–1500	1.32	0.66	0.32	0.20	0.07	0.07	0.07

Figure 7.10   In Duct Attenuation within Externally Lagged Straight Ducts

The attenuation of ductwork lined internally with acoustic absorptive material will depend on the following factors:

the thickness of the lining
the sound absorbing characteristics of the lining
the dimensions of the duct passage
the frequency or wavelength of sound.

In general, the smaller the duct passageway, the greater the attenuation per metre run. At the same time, the greatest effect will be experienced with the high frequencies. Figure 7.11 gives an approximated attenuation per metre run for various shapes and sizes of internally lined ductwork.

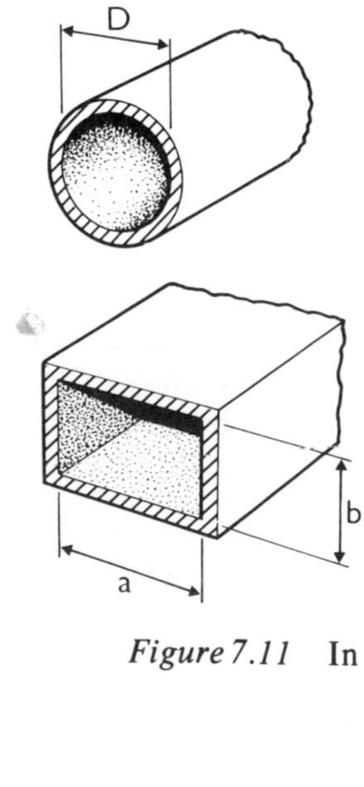

$$\text{Attenuation} = 1.07 \left(\frac{P}{S} \right) \alpha^{1.4} \text{ dB/metre} \quad \text{...............} 7.2$$

where P = perimeter (inside) of lining , m
S = cross sectional area of duct, m^2
α = absorption coefficient

Constraints:

accuracy $\pm$ 10%
frequency range 250 to 2000 Hz
$\alpha \leqslant 0.8$
for circular duct $D > 0.15$ m
for rectangular ducts a or b $\leqslant$ 900 mm
and $0.5 < \dfrac{a}{b} < 2$

Figure 7.11   In Duct Attenuation within Internally Lined Ducts

7.3.2 Attenuation Due to Bends and Duct Fittings

Where airflow changes direction or is expanded/restricted, attenuation is obtained by the sound being partially "reflected back towards the source", by the corresponding geometry changes.

Useful high and medium frequency attenuation may thus be obtained from mitre bends of a sufficient cross-sectional area. Attenuation achieved with such bend fittings will be substantially reduced if large chord turning vanes are employed. Turning vanes when used should be small in width — less than 1/8 of the wavelength of sound in the octave where full potential attenuation is required. That is, for 1000 Hz and below, turning vanes should be no greater than 50 mm in width.

Radiused bends provide very little attenuation and are similar in performance to mitre bends with long chord turning vanes.

Figure 7.12 shows attenuation performances for bends with the following geometry:

(a) mitre or mitre with small chord turning vanes
(b) mitre with long chord turning vanes or radiused bends.

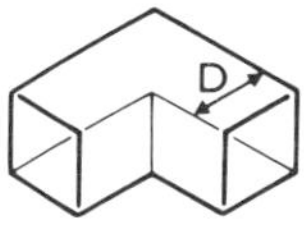

Duct Width	Octave Band Centre Frequency							
D mm	63	125	250	500	1k	2k	4k	Hz
75–100	–	–	–	–	1	7	7	dB
100–150	–	–	--	–	5	8	4	
150–200	–	–	–	1	7	7	4	
200–250	–	–	–	5	8	4	3	
250–300	–	–	1	7	7	4	3	
300–400	–	–	2	8	5	3	3	
400–500	–	–	5	8	4	3	3	
500–600	–	–	6	8	4	3	3	
600–700	–	1	7	7	4	3	3	
700–800	–	2	8	5	3	3	3	
800–900	–	3	8	5	3	3	3	
900–1000	–	5	8	4	3	3	3	
1000–1100	1	6	8	4	3	3	3	
1100–1200	1	7	7	4	3	3	3	
1200–1300	1	7	7	4	3	3	3	
1300–1400	2	8	7	3	3	3	3	
1400 1500	2	8	6	3	3	3	3	
1500–1600	3	8	5	3	3	3	3	
1600–1800	5	8	4	3	3	3	3	
1800–2000	6	8	4	3	3	3	3	

(a)

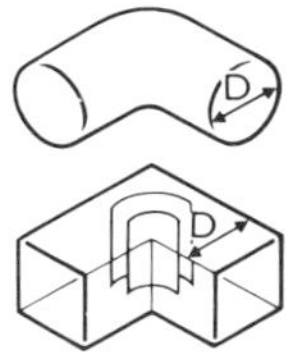

Duct Width/Diameter	Octave Band Centre Frequency							
D mm	63	125	250	500	1k	2k	4k	Hz
150–250	–	–	–	–	1	2	3	dB
250–500	–	–	–	1	2	3	3	
500–1000	–	–	1	2	3	3	3	
1000–2000	--	1	2	3	3	3	3	

(b)

Figure 7.12 Attenuation of Unlined Bends

Very little test data is available for bends of other than 90 degrees in angle. One may, by way of guidance, approximate the attenuation of a bend in ratio to its subtended angle, that is for a 45 degree bend the attenuation will be approximately half of that for a 90 degree bend. For bends greater than 90 degree only the attenuation figure for a 90 degree bend should be employed.

In the event that two bends are located close together, then it has been found that to take both bend losses gives an overestimation of the attenuation attained. This results from sound reflected back at the second bend being reflected forward down the system again from the first bend and thus becoming partially retained within the system. With two bends close together it is recommended that only one bend loss be taken into account.

The attenuation of an acoustically lined mitre bend is shown in Figure 7.13.

Duct Width D mm	Octave Band Centre Frequency							
	63	125	250	500	1k	2k	4k	Hz
75–100	–	–	–	–	2	13	18	dB
100–150	–	–	–	1	7	16	18	
150–200	–	–	–	4	13	18	18	
200–250	–	–	1	7	16	18	16	
250–300	–	–	2	11	18	18	17	
300–400	–	–	4	14	18	18	17	
400–500	–	1	5	16	18	16	17	
500–600	–	1	8	17	18	16	17	
600–700	–	2	13	18	18	17	18	
700–800	–	3	14	18	17	16	18	
800–900	–	4	15	18	18	17	18	
900–1000	–	5	16	18	17	17	18	
1000–1100	1	7	17	18	16	17	18	
1100–1200	1	8	17	18	16	17	18	
1200–1300	1	10	17	18	16	18	18	
1300–1400	2	11	18	18	16	18	18	
1400–1500	2	12	18	18	16	18	18	
1500–1600	3	14	18	18	17	18	18	
1600–1800	4	15	18	18	17	18	18	
1800–2000	5	16	18	17	17	18	18	

Lining thickness = $\dfrac{D}{10}$

Lining to extend distance 2D or greater

Figure 7.13 Attenuation of Lined Bends

7.3.3 Attenuation at Ductwork Contractions and Expansions

As mentioned earlier in this chapter, sound attenuation is also obtained at points of duct-work contractions or expansions. The mechanism involved is similar to that of a bend in that medium and high frequency sound is reflected back along the duct. The amount of attenuation achieved will be a function of the steepness of the area change — the more acute, the higher the attenuation — and the ratio of the cross sectional areas of ductwork.

LINED BEND

The narrow chord tuning vanes have little effect on the acoustic attenuation, but may well prove a source of flow-generated noise.

Figure 7.14 gives attenuation figures for typical examples.

Ratio of areas A1/A2	5	4	3	2.5	2	1	0.5	0.4	0.33	0.25	0.20
Attenuation dB	2.6	1.9	1.3	0.9	0.5	0	0.5	0.9	1.3	1.9	2.6

Figure 7.14   Sound Power Loss at Abrupt Changes of Duct Cross Sectional Areas

7.3.4 Attenuation Due to Branches

On reaching a branch, the total sound energy will divide between the two duct runs. This division will be in direct proportion to the individual branch duct areas. Figure 7.15 shows the attenuation in each duct as a percentage of the total areas of the branches.

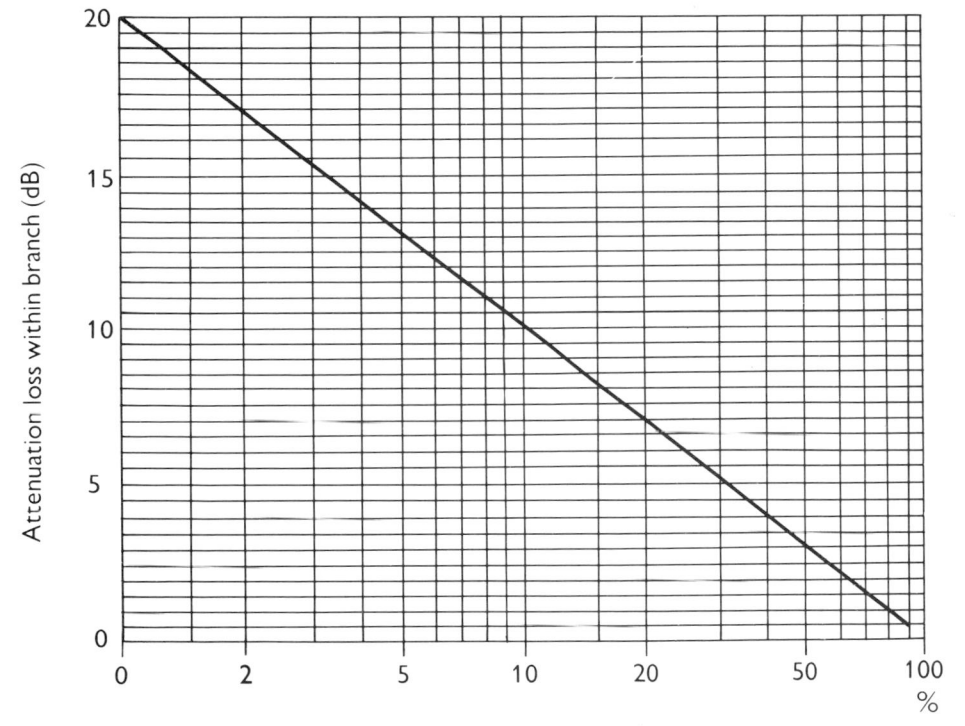

Figure 7.15   Attenuation at Duct Branches

It should be noted that when a branch divides, the cross sectional area of the branches is generally slightly higher than that of the original duct. Calculations should therefore be based on the ratio of the new branches and not the ratio of a branch with that of the primary duct.

When the volumes divide and the velocities in the branches remain approximately equal, then the air volume ratios may be used instead of the duct area ratios. This is usually the case for low velocity systems. In this case, the calculation is simplified significantly, since it is then only necessary to know the ratios of air volume reaching the particular room or grille as a percentage of the total fan volume. The fan sound power may then be divided in the same ratio. Figure 7.15 may be employed similarly when calculating the attenuation at branches based on this volume flow rate percentage.

When analysing the sound emerging from a side branch into a much larger main duct, for example when looking at cross-talk between rooms, there will be a similar division of sound power as previously described. However, in this instance, there will also be a large loss due to the abrupt change in cross sectional area. The attenuation may therefore be calculated by adding the "attenuation due to an abrupt change of area" (Figure 7.14) to the "loss at a branch" (Figure 7.15). An example of a cross talk calculation is shown at the end of this chapter.

7.3.5 Attenuation Due to Duct Terminations

When the ductborne sound energy reaches the end of the duct, the duct termination is usually a diffuser (supply side), or grille (extract side). At the point where the sound emerges into the room, the extreme increase in volume creates similar losses as those previously described when expansions occur within the duct system itself. This attenuation at a duct termination is called "end reflection loss". It will be seen when looking at Figure 7.16 that end reflection losses are particularly prominent at low frequency and for smaller areas of termination.

In an air-conditioning system, this end reflection effect is very useful in reducing the amount of low frequency noise which enters the conditioned areas and which, in the absence of the end reflection, would be very difficult to control. In some cases therefore it may be possible to use this effect to advantage. A larger number of small outlets, spaced well apart, will transmit less low frequency noise into a conditioned area than a single large one.

The illustration shown in Figure 7.17 demonstrates this effect.

Extract or supply stub duct terminations are often located within a ceiling void or below a raised floor system. The void is either pressurised or evacuated which in turn forces air through light fittings, grilles and diffusers distributed over the floor or ceiling area. When such a system is employed, then care should be exercised at the point of duct termination. In many cases the cross-sectional area of such a

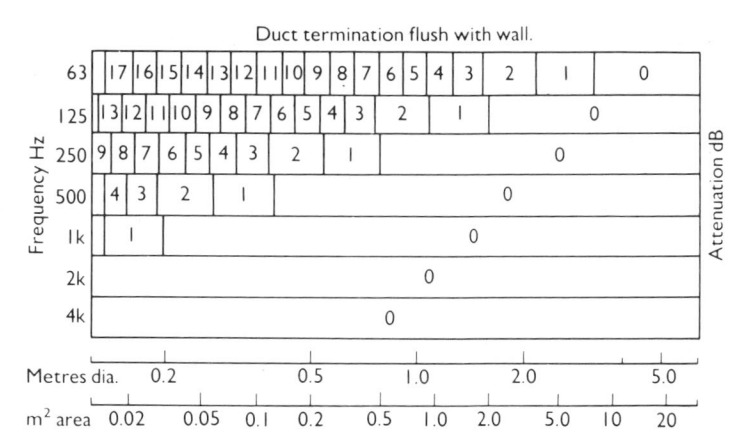

Figure 7.16 Attenuation at Duct Termination Due to End Reflection Loss. Dimensions or areas shown, are "gross, overall" figures and ignore the presence of grilles, louvres etc., which at low frequencies do not significantly effect end reflection losses.

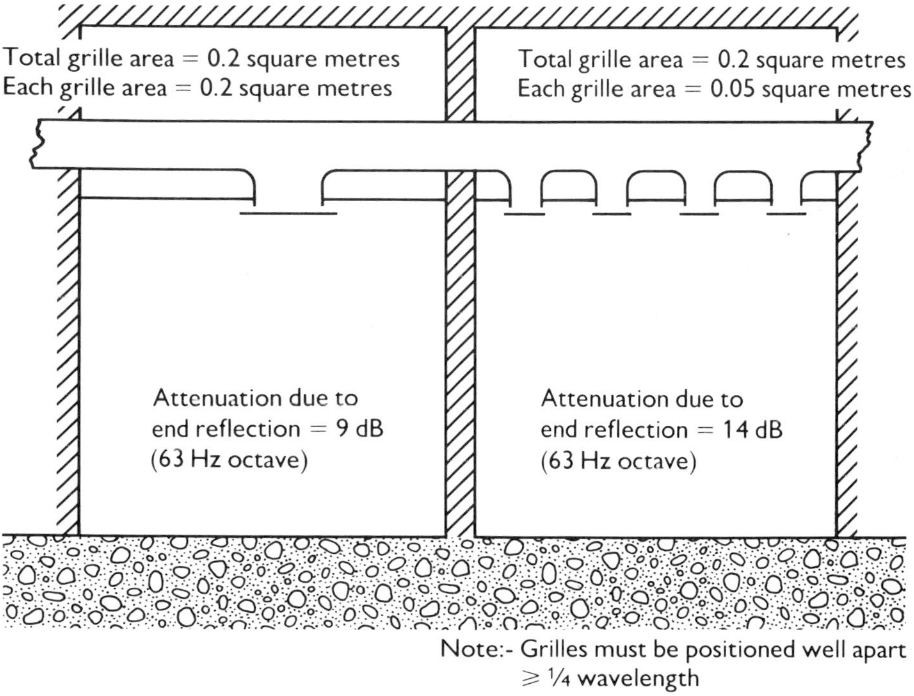

Total grille area = 0.2 square metres
Each grille area = 0.2 square metres

Total grille area = 0.2 square metres
Each grille area = 0.05 square metres

Attenuation due to
end reflection = 9 dB
(63 Hz octave)

Attenuation due to
end reflection = 14 dB
(63 Hz octave)

Note:- Grilles must be positioned well apart
≥ ¼ wavelength

Figure 7.17   End Reflection Losses from Large and Small Grilles

termination is much greater than would be the case if a fully ducted floor or ceiling diffuser distribution system had been employed. The resulting lack of end reflection loss may create localised areas of low frequency noise which will not be attenuated by a lightweight ceiling/floor system.

To overcome potential problems from supply or extract stub terminations, both good in-duct attenuation and a heavier ceiling/floor construction should be provided.

7.3.6 Overview

This discussion on "system attenuation" has so far concentrated on the ductwork system itself. It has been shown that ductwork acts as quite an efficient carrier of sound. Nevertheless it can also be seen that in any typical run of ductwork there are many mechanisms for the reduction of in-duct noise, whether this noise is from the fan or from other sound sources.

Other items of equipment located within the ductwork (for example: terminal units, filters, coils etc.) will also provide some attenuation. In this case however, the insertion loss of the unit is normally known from test data and so it is simply a matter of adding this attenuation to the natural system attenuation. If this data is not available, then the item of equipment should be tested in an acoustic test laboratory. Typical values of attenuation for such items are shown in Figure 7.18.

Item	Octave band frequency							
	63	125	250	500	1k	2k	4k	Hz
Bag or Roll Filter	0	2	2	2	4	7	12	dB
Absolute Filter	3	5	5	5	7	8	12	dB
Humidifier	3	3	3	4	4	4	5	dB
Heating/Cooling Coils								
1–2 rows	1	1	1	1	1	2	2	dB
> 2 rows	2	2	2	2	2	4	5	dB
Spray washer	2	3	2	3	3	4	4	dB
Thermal wheel	–	–	–	2	2	3	4	dB

It should be appreciated that many of these items of equipment will generate noise from air flow.

Figure 7.18   Attenuation of Miscellaneous Items of Equipment

7.3.7 Summary

Commencing with the octave band sound power levels generated by the fan, and factoring for efficiency and inlet/discharge conditions, it is now possible to

NATURAL DUCT LOSSES

A complex duct arrangement gives rise to an accumulation of acoustic and aerodynamic losses at bends, take-offs and cross-section changes, all in a small region of ceiling void.

Courtesy: Lindab

systematically calculate the transmission of noise through the ductwork system, deducting the energy lost by natural attenuation from:

(a) Straight duct runs — rectangular/circular/oval (Figure 7.9)
(b) Lagged duct runs — rectangular/circular/oval (Figure 7.10)
(c) Lined duct runs — rectangular/circular/oval (Figure 7.11)
(d) Mitre bends — (Figure 7.12a)
(e) Mitre bends with short chord turning vanes — (Figure 7.12a)
(f) Radius bends/mitre bends with long chord turning vanes — (Figure 7.12b)
(g) Lined mitre bends — (Figure 7.13)
(h) Changes in cross sectional area — (Figure 7.14)
(i) Branch losses — (Figure 7.15)
(j) Cross talk attenuation due to changes in cross sectional area (Figure 7.14) and branch losses (Figure 7.15)
(k) Duct terminations due to end reflection loss — (Figure 7.16)
(l) Miscellaneous items — (Figure 7.18)

It should be noted that noise generated at fittings by the flow of air in the system — "flow generated noise", should be added at the same time. This subject is discussed in the next Chapter—Chapter 8.

The sound emerging from the duct termination will be expressed as sound power level over a range of octave bands. In order to compare with design criteria it is necessary to convert this to sound pressure level. To do this requires the application of certain "corrections".

7.4 ROOM AND ATMOSPHERE CORRECTIONS

7.4.1 Room Corrections

As previously described in Chapter 3 the sound pressure level at any point in a room emerging as sound power, from a grille or grilles, may be considered to made up of two components:

(a) The direct sound — reaching a position in a direct line from a single source
(b) The reverberant sound — reaching a position after successive reflections and re-reflections from the internal surfaces of the room. This is also referred to as the reflected sound.

The direct sound pressure level component will be influenced by:

the sound power emerging from a single grille
the distance between the source (grille) and receiver (listener)
the directivity factors — surface and source

The reverberant sound pressure level component will be dependent on:

the room volume
the total absorption available from the rooms surface/contents

or

the reverberation time of the room

The reverberation time method is simpler to use and is accurate for most applications. It is therefore used in this instance. For a more detailed explanation of direct and reverberant or reflected fields, see Chapter 3 — "Sound In Rooms".

The two components of direct sound pressure level and reverberant sound pressure level are logarithmically combined to give the total sound pressure level at a particular position — hence without due regard to direction of arrival:

$$L_p \text{(total)} = L_p \text{dir} + L_p \text{rev} \qquad\qquad 7.3$$

$$L_p \text{(dir)} = L_w \text{(per grille)} + 10 \log_{10} Q - 10 \log_{10} 4\pi r^2 \qquad\qquad 7.4$$

$$L_p \text{(rev)} = L_w \text{(to room)} + 10 \log_{10} T - 10 \log_{10} V + 14 \qquad\qquad 7.5$$

where L_p (total) = total sound pressure level, dB;
　　　　L_p (dir) = direct sound pressure level, dB;
　　　　L_p (rev) = reverberant sound pressure level, dB;
　L_w (per grille) = sound power level from each grille, dB;
　L_w (to room) = sound power level entering room, dB;
　　　　　　Q = directivity factors;
　　　　　　r = distance from source to receiver, m;
　　　　　　T = reverberation time, s; and
　　　　　　V = volume, m³.

To obviate the need for a lengthy calculation, a set of tables (Figures 7.19 and 7.20) have been formulated for each of these terms ($10 \log_{10} Q / 10 \log_{10} V$ etc.) enabling the easy determination of both reverberant and direct room corrections.

Figure 7.19 shows the corrections necessary to convert from sound power level of the source to *reverberant* sound pressure level.

$$L_p \text{(rev)} = L_w \text{(to room)} + A + B \qquad\qquad 7.6$$

where $A = 10 \log_{10} T$ and
　　　$B = -10 \log_{10} V + 14$

Figure 7.20 shows the corrections necessary to convert from sound power level to *direct* sound pressure level:

$$L_p \text{(dir)} = L_w \text{(per grille)} + A + B \qquad\qquad 7.7$$

where $A = 10 \log_{10} Q$ and
　　　$B = 10 \log_{10} 4\pi r^2$

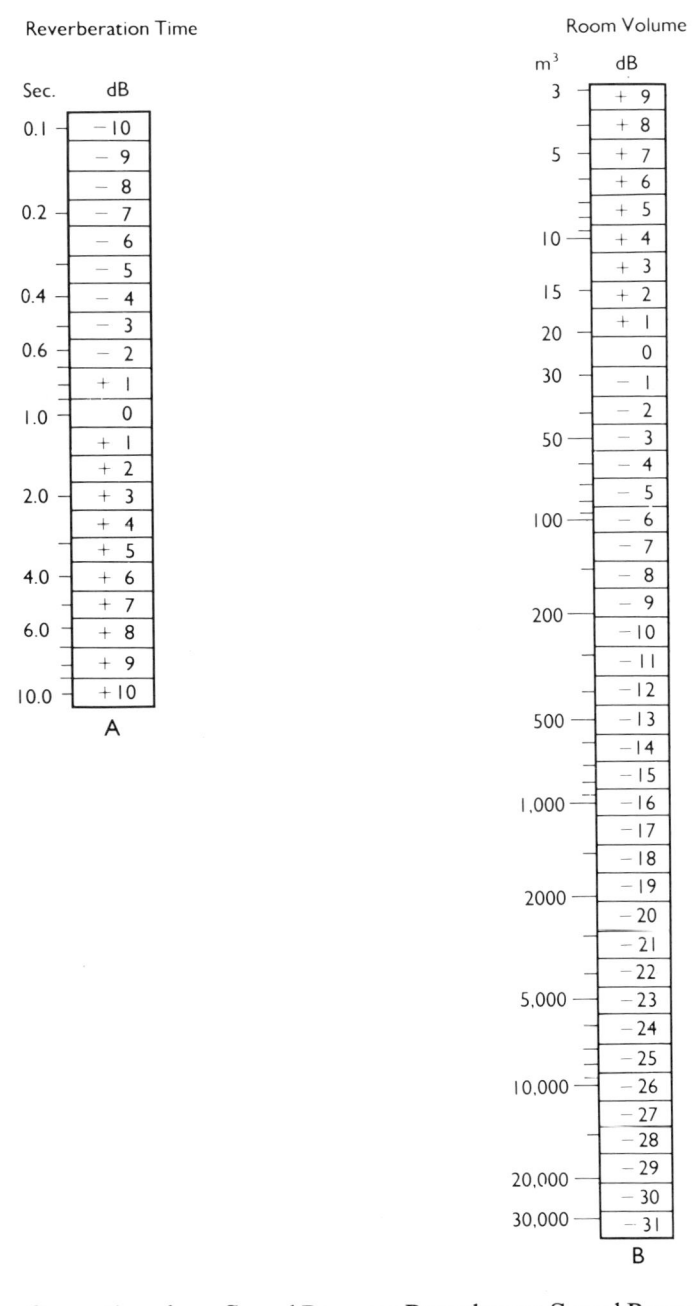

Figure 7.19 Corrections from Sound Power to Reverberant Sound Pressure Level

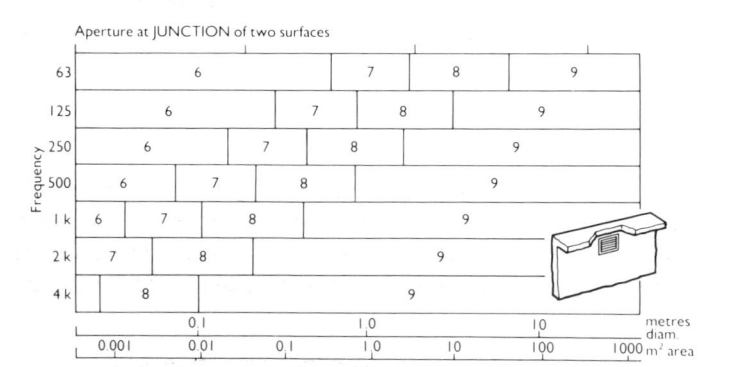

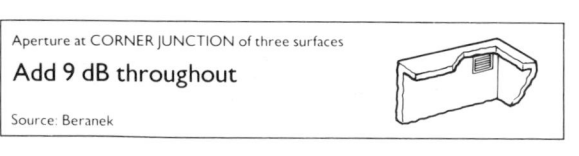

A

Figure 7.20   Corrections from Sound Power to Direct Sound Pressure Level

The direct sound pressure level and reverberant sound pressure level should then be added logarithmically to determine the total sound pressure level by using the chart shown in the chapter on Physics of Sound.

7.4.2 Noise To Atmosphere

If concerned with noise to atmosphere, a similar process is followed. In this case, there will not be a reverberant component and only the direct sound pressure level should be calculated from directivity and distance corrections. However, the large areas of louvres often employed on the atmosphere side require special consideration. This subject is covered more fully in the chapter "Noise to the Exterior" - Chapter 11.

The use of calculation sheets is illustrated at the end of this chapter.

The total sound pressure level predicted for the conditioned room or to atmosphere may then be compared directly with the design criteria. In the event that the targeted design criteria is exceeded, natural attenuation within the system must be increased or supplemented in some way.

7.5 SOUND REDUCTION TECHNIQUES

In practice, it is unlikely that the design of the system can be changed to a degree that would allow "natural attenuation" to have any fundamental effect on noise levels. The usual engineering approach when designing for fan noise is therefore to add sufficient attenuation into the system to meet the noise criterion.

Apart from unitary attenuators, this may be achieved through the internal lining of ductwork walls with an absorbent material (Figure 7.11). Other valuable attenuation may be achieved by the lining of mitre bends (Figure 7.13). An alternative may be the use of plenum chambers.

7.5.1 Plenum Chambers

If a plenum chamber is already part of the ventilation systems aerodynamic considerations, high attenuation may be gained by lining the internal walls. An internally lined plenum chamber is a good attenuator of the difficult low frequencies and so therefore there may be justification for its insertion into the system on purely noise attenuation grounds.

Several formulae have been derived for calculating plenum attenuation including one developed by Wells.

$$\text{Attenuation dB} = -10 \log S_{\text{out}} \left[\frac{\cos\theta}{2\pi d^2} + \frac{1}{R} \right] \qquad 7.8$$

where S_{out} = area of outlet, m²;

θ = angle between centre of inlet and centre of outlet (Figure 7.21);

d = distance between centre of inlet and centre of outlet, m;
$R = S\overline{\alpha}/(1-\overline{\alpha})$ — the room constant;
S = total internal area of plenum, including inlet and outlet areas, m²;
$\overline{\alpha}$ = average absorption coefficient of the plenum lining; assume inlet and
     outlet = 1.

and
Note:

$$\overline{\alpha} = \frac{(S_1\alpha_1 + S_2\alpha_2 + S_3\alpha_3 + \dots S_n\alpha_n)}{S}$$

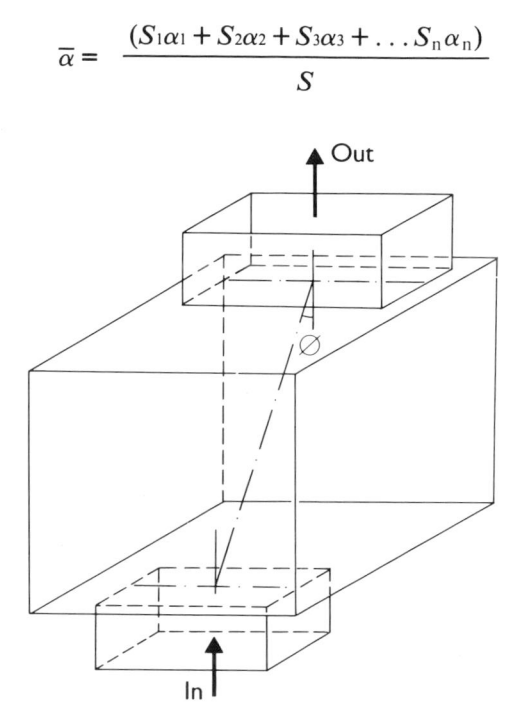

Figure 7.21   Attenuation from Plenum Chambers

For frequencies sufficiently high that the wavelength of the sound is less than the plenum dimension the equation predicts accurately to within 2 dB. At lower frequencies, where the wavelength is longer than the dimension of the chamber, calculated values may be an underestimation of performance of 5 to 10 dB, since no account is taken of the end reflection loss of the inlet and discharge apertures.

7.5.2 Purpose Made Unitary Attenuators

To achieve high levels of attenuation, both lined ductwork and plenum chambers require large areas of absorption lining. A primary advantage of purpose made attenuators therefore is high levels of attenuation in a small length. Another advantage is that the performance of the unit may be accurately predicted by the manufacturer through laboratory testing. Figure 7.22 illustrates a range of performances for typical cylindrical and rectangular attenuators under conditions of no flow-static insertion loss.

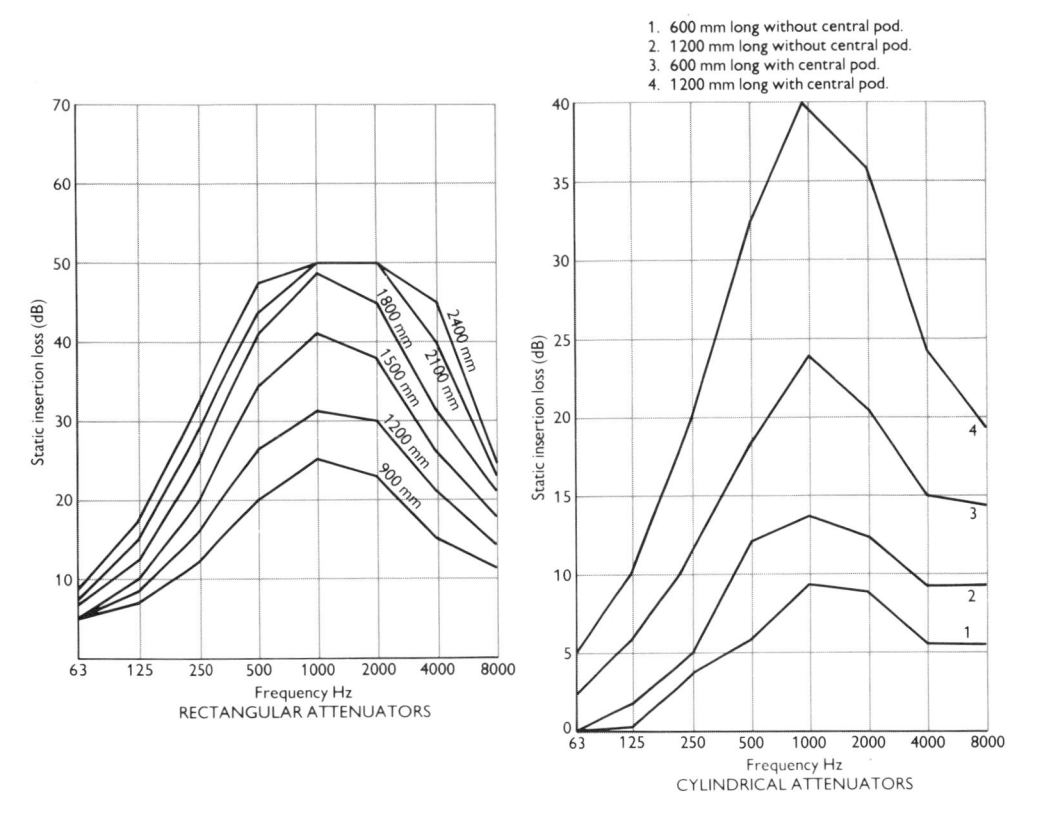

Figure 7.22 Static Insertion Losses of Rectangular and Cylindrical Attenuators

For many applications the cylindrical attenuator will be bolted directly to an axial flow fan, and the dynamic insertion loss will depend on pitch angle. The insertion loss will be reduced as indicated in Figure 7.23.

For centrifugal fan systems, or axial fan systems where a high degree of attenuation is necessary, it is usual to consider a purpose-made rectangular attenuator in conjunction with an appropriate transition. This type of attenuator consists of a sheet metal duct housing several sound absorbent splitters.

If required, the noise reduction available from a device of this type can be estimated for design purposes, using the Sabine formula (equation 7.2 on Fig. 7.11). It should be noted, however, that in computing the attenuation of the individual airway, it is

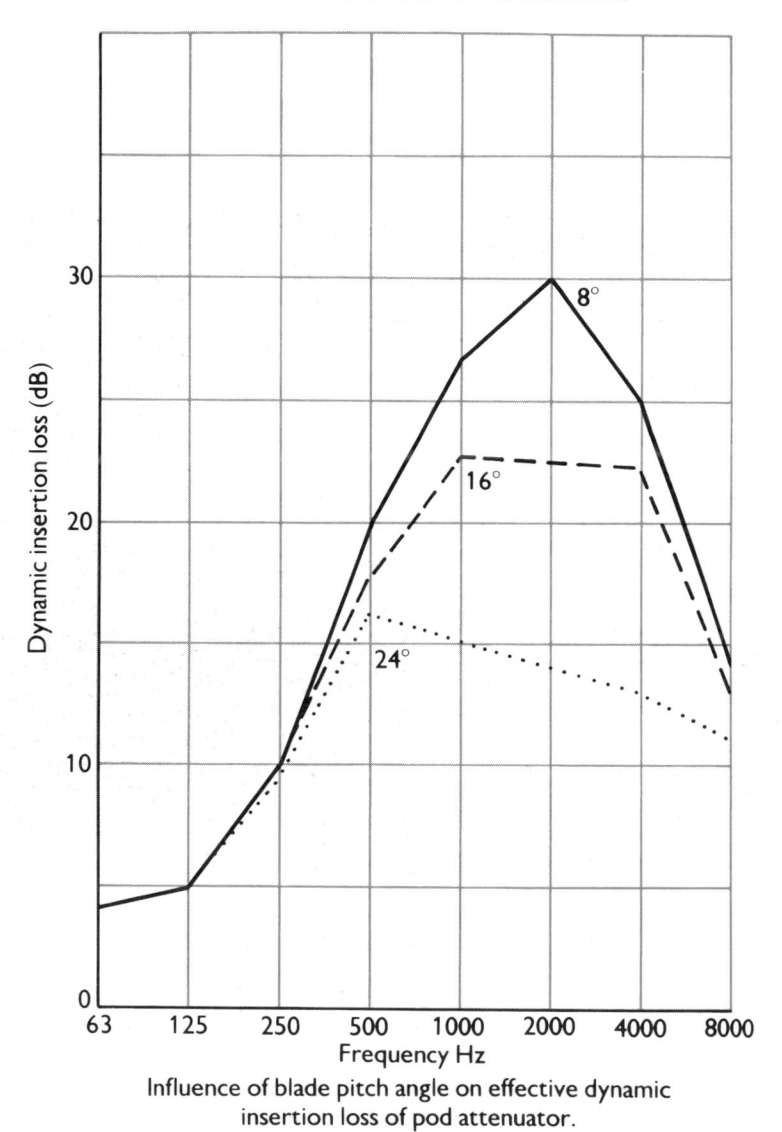

Influence of blade pitch angle on effective dynamic
insertion loss of pod attenuator.

Figure 7.23   Dynamic Insertion Losses of Cylindrical Attenuator under Different
Fan Pitch Angles

necessary to use the absorption coefficient for a thickness of material equal to one half
of the thickness of the common splitters between the airways. In order to be equally
effective in absorbing sound, the splitters should therefore consist of a material twice
as thick as that used for lining the duct wall. This consideration is particularly

RECTANGULAR ATTENUATOR

Splitters or baffles of glass fibre faced Fibreglass — Eurolon — being slid into a multisection ductwork casing manufactured to HVCA specifications.
Courtesy: Sound Attenuators Ltd

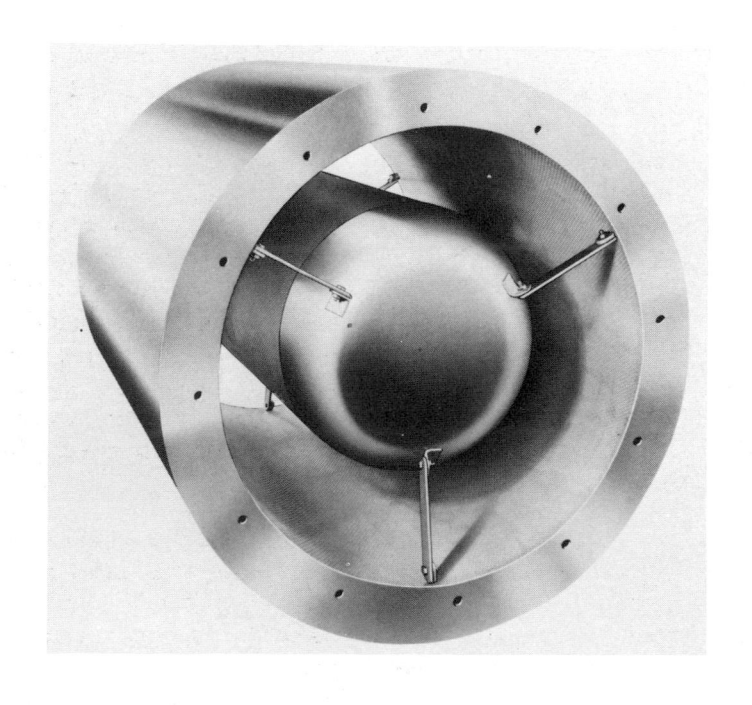

CYLINDRICAL ATTENUATOR

The central pod gives a substantial lift to the insertion loss over that of simply lined
ductwork. Its location is well suited to bolt-on axial fan applications, being adjacent
to the motor and blank fan hub.

Courtesy: Sound Attenuators Ltd

important at the lower frequencies where the absorption depends on the thickness of the acoustic material. Where the number of splitters is large and the open area decreases to a fraction of its original size, an additional attenuation occurs due to the end reflection at the end of the airways. The end reflection loss is calculated from the ratio of the open area of the splitter to the area of the ducts connected to the splitter.

It is usual for the splitter attenuator to have a greater cross-sectional area than the duct in which it is located. This is to limit the velocity through the airways, and hence keep the pressure drop through the unit to a reasonable limit. In a restricted space, however, it may be that the airway velocities are rather high.

In most applications the requirement to keep the pressure drop across the attenuator to a reasonable level automatically ensures that the flow noise generated within the attenuator is insignificant compared with the permissible sound power which emerges. If however, extremely low levels have to be obtained, or if the sound power from the fan is relatively low, the flow noise generated by the attenuator can be significant and can reduce its effective insertion loss. It is for this reason that when an acoustic consultant specifies the attenuator performance he will normally specify the dynamic insertion loss which is required, together with sound power level entering the attenuator. This then enables the attenuator manufacturer to select a unit of such a size that the flow noise generated within it will not reduce the effective insertion loss below the required level.

7.5.2.1 Location of Attenuators Having decided the magnitude of the attenuation required in a particular system, and the method of providing it, a further point should be considered. If either duct lining or a purpose-made unitary attenuator is to be installed, it is important that its position should be correct (Figure 7.24). Ideally, primary attenuation, that which is necessary to prevent fan noise, should be located at the point at which the distribution ductwork leaves the plantroom. In this position noise generated within the plantroom will be prevented from entering the "quiet" side of the attenuator. If the attenuator is positioned remotely from the plantroom, noise radiating from the "noisy" side of the attenuator will be prevented from creating problems in areas adjacent to the plant room. Obviously, in a complex plant room, it is not always possible to position all attenuators in this ideal position, either due to the numbers of attenuators involved, or if zoning is required, when it is preferred to provide a single attenuator adjacent to the fan, prior to splitting the system into various zones.

Invariably, therefore, the need to prevent noise entering the "quiet" side of the attenuator must be considered. External insulation for thermal effect is an advantage in preventing this, and in plantrooms it is more than likely that the thermal insulation will have an external covering which provides a good additional acoustic shield.

In short, primary attenuation should be located in the plantroom as near as possible to the point at which the ducts penetrate its walls. If this is not possible, suitable acoustic insulation must be considered.

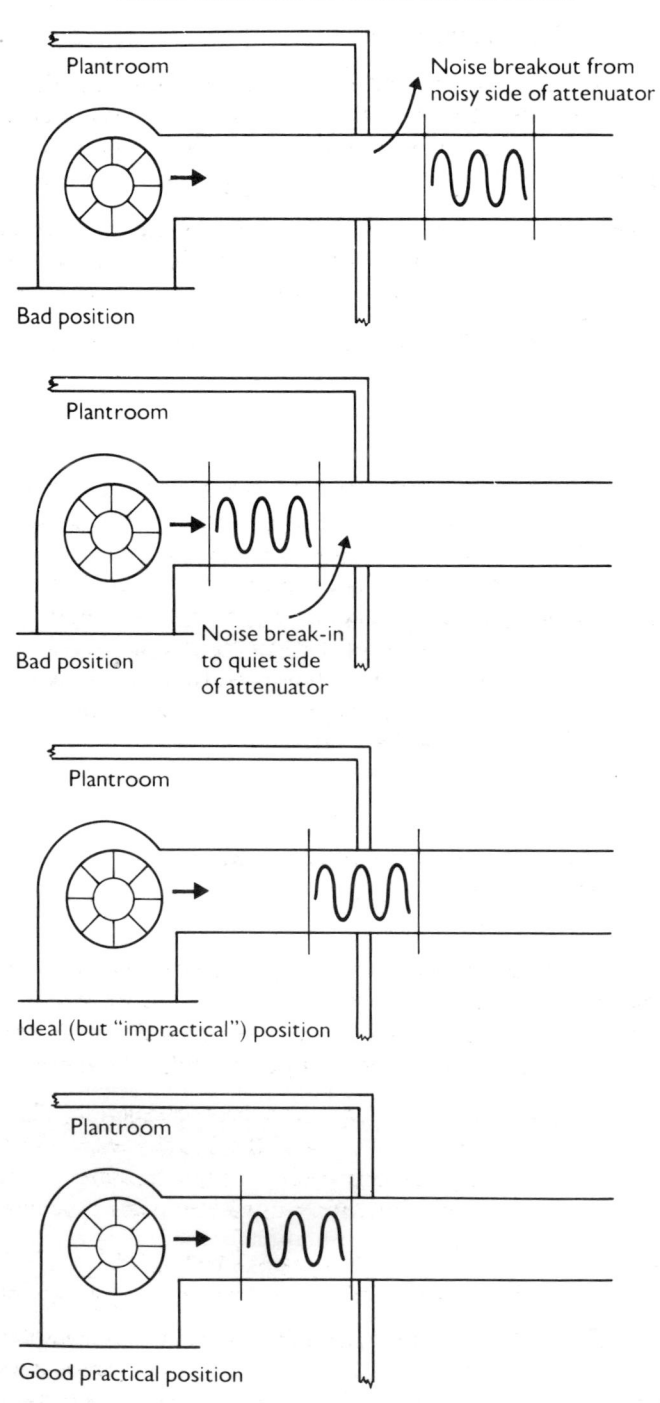

Figure 7.24   Positioning the Attenuator in System

7.5.3 Control of Sound Within the Room

In certain applications it may be possible to control both the direct field and reverberant field within the room itself. To reduce the direct field it is necessary to re-locate the grilles further from the occupants. Obviously this is only practical if the ventilation requirements will allow it, and only if it is physically possible to do so.

It may be possible to reduce the reverberant field by reducing the reverberation time of the room. This can be done by acoustic ceiling tiles, carpets, curtains, wall panelling and the like. However, this again may be impractical and expensive.

It can be seen from Figure 7.19a that a halving of reverberation time will result in a decrease of reverberant sound pressure level of 3 dB.

7.5.4 "Crosstalk"

The fan is not the only source of airborne sound power which can be radiated along the ductwork. Just as sound power can come out of the duct openings in rooms, it can also enter. If two rooms are served by branches on a common duct system the possibility arises of sound from one room passing into the ductwork and out into the adjoining room. This sound source can take many forms, for example, a raised voice in one room may be transmitted via the ducting to the adjacent room, or machinery noise in that room could be similarly transferred. A suitable method of providing attenuation within the duct between the adjacent rooms is therefore necessary, and once again this attenuation can take the form of:

lined ductwork
increased ductwork length
increased end reflection — by making the grilles smaller
fitting a purpose-made attenuator
modification to the rooms themselves

Figure 7.25 indicates in sketch form a typical system where crosstalk transmission has been avoided by a revised ductwork configuration, effectively increasing the natural ductwork attenuation as a result of length increase. This would be especially powerful if internally lined ductwork were being employed, as in studio applications.

To calculate the amount of attenuation required to prevent crosstalk it is necessary to know the sound power level of the source, and the background noise level in the receiving room. The lower the background level in a room, the more chance there is of hearing speech or other noise from adjacent rooms via connecting ductwork.

Figure 7.26 indicates in graphical form the sound power level of various sources including different intensities of speech and the usual office type of machinery.

If speech privacy is to be considered for example, the sound power level indicated in Figure 7.26 by the line "D" can be substituted into "Sound power level leaving system" in the calculation sheet (Figure 7.29 line No 23). Room corrections may then

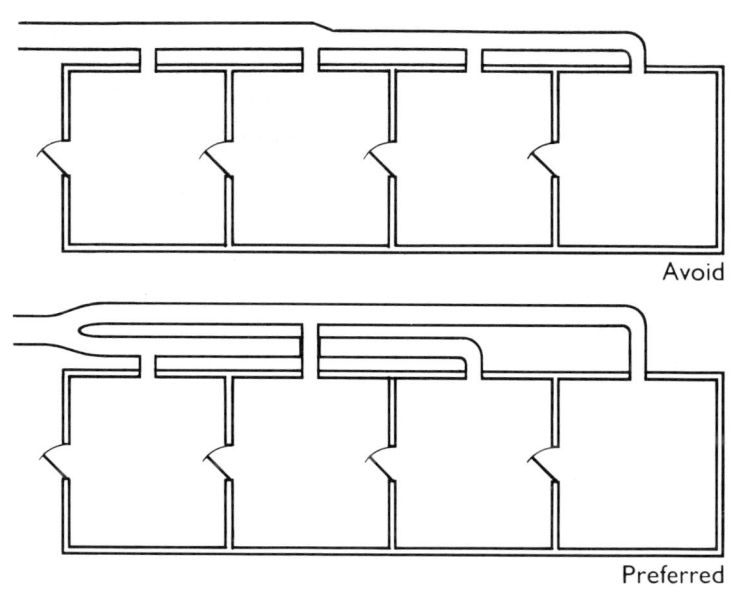

Avoid

Preferred

Figure 7.25   Ductwork Layouts to Reduce Crosstalk

A. 2 tabulators or 4 teletype machines   C. Loud as possible without strain
B. 4 typewriters   D. Raised
                                           E. Normal   } Male voice

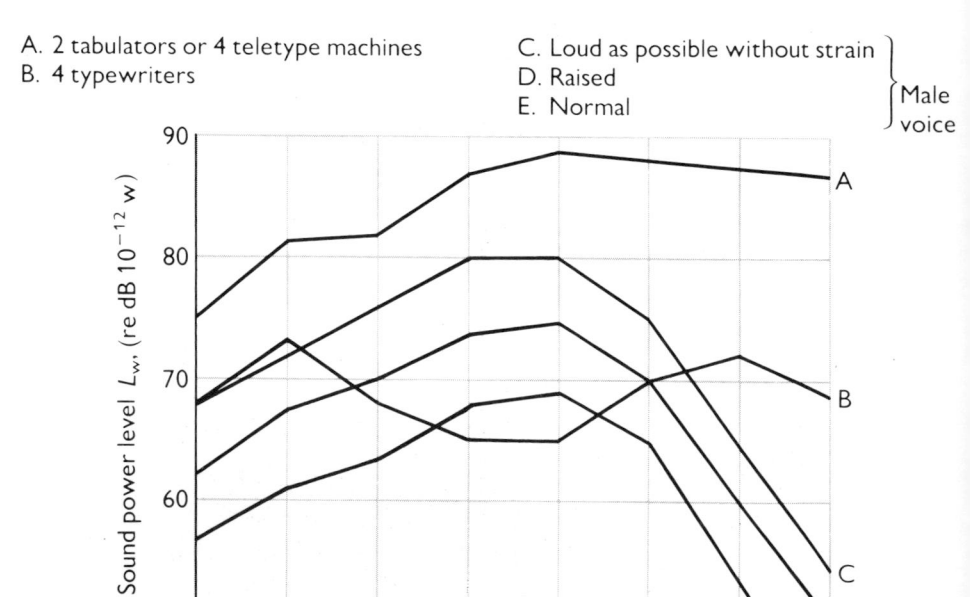

Figure 7.26   Sound Power Levels of Speech and Office Equipment

be applied to obtain the resultant sound pressure level in the source room at the grille. This, then, is the sound pressure level endeavouring to enter the grille in the source room. For further calculations it is necessary that this be converted to sound power level. The sound power level entering the grille is given by:

$$L_w(\text{total}) = L_w(\text{reverberant}) + L_w(\text{direct}) \text{ dB} \qquad 7.9$$

$$\text{where } L_w(\text{reverberant}) = L_p(\text{reverberant}) + 10 \log A - 6 \text{ dB} \qquad 7.10$$

$$\text{and } L_w(\text{direct}) = L_p(\text{direct}) + 10 \log A \text{ dB} \qquad 7.11$$

where A = grille area, m².

This total sound power level may now be inserted directly into a calculation sheet under "unit L_w" (Figure 7.29 line No 4) and the attenuation requirement can be calculated in the normal way for a duct system.

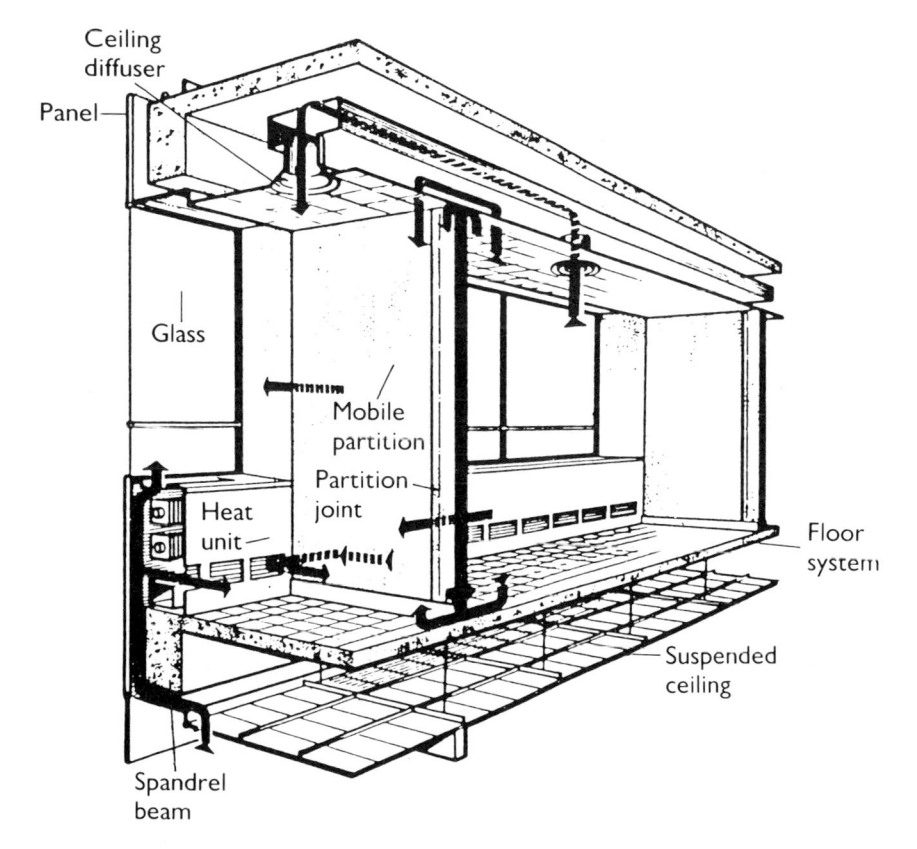

Figure 7.27   Flanking Paths Between Rooms

7.5.5 Flanking

It is of course important that the transmission loss of the partition dividing the two rooms is taken into consideration.

There is little point in providing adequate crosstalk attenuation in the ventilation system if the partition is of a lower acoustic insulation. Even if the partition is adequate in itself, it is important to check that there is not an easier flanking path for the sound to travel between one room and the next. This can occur for instance if the partition is only taken up to the false ceiling line, and lightweight false ceilings are provided. Similarly, if there is perimeter heating fitted in the form of a continuous sheet metal casing, the sound is more likely to travel around the partition than through the ductwork system. These possible flanking paths are illustrated in Figure 7.27.

7.6 CONCLUSIONS

In the course of this chapter the various types of fans used have been illustrated along with the octave band sound power level generated by each. The inlet/discharge conditions and operating efficiency affecting fan sound power levels have been detailed.

It has been shown how the sound power of the fan, as with other sound sources, will be attenuated by the duct work system.

On reaching a conditioned area, the effect of room corrections has been discussed thus allowing the computation of sound pressure level from the sound power level emerging from the system.

Lastly, the available sound reduction techniques have been detailed.

It will be seen that although the acoustic design of a ventilation system involves a detailed understanding of all aspects of the system, it also involves a considerable amount of simple addition and subtraction of a number of source/attenuation elements. This is particularly pertinent when considering sources of flow generated noise or when "shaping" the spectrum to a required noise criteria curve. It is for this reason that a number of computer programmes have been designed.

In the absence of computer programmes, it is of assistance to the system designer to use one of a number of standard calculation sheets (Figure 7.29). These sheets may be structured to work either from the desired criterion back through the system to the fan, or vice versa from the fan to criterion within the conditioned room.

An example of fan noise transmission to a conditioned room and an example of a cross-talk calculation are illustrated.

7.7 EXAMPLES

7.7.1 Example 1: Supply System

The supply system shown in Figure 7.28 has been designed to the following specification:

Axial fan flow, duty — volume 2.5 m³/s, pressure 40 N/m²

Ductwork primary 900 x 600 mm, secondary 600 x 300 mm and 600 x 600 mm

Bend 90° mitre

Diffusers flush mounted 0.04 m² gross outlet area each. Handling 0.25 m³/s each. Diffusers mounted in ceiling away from influence of walls

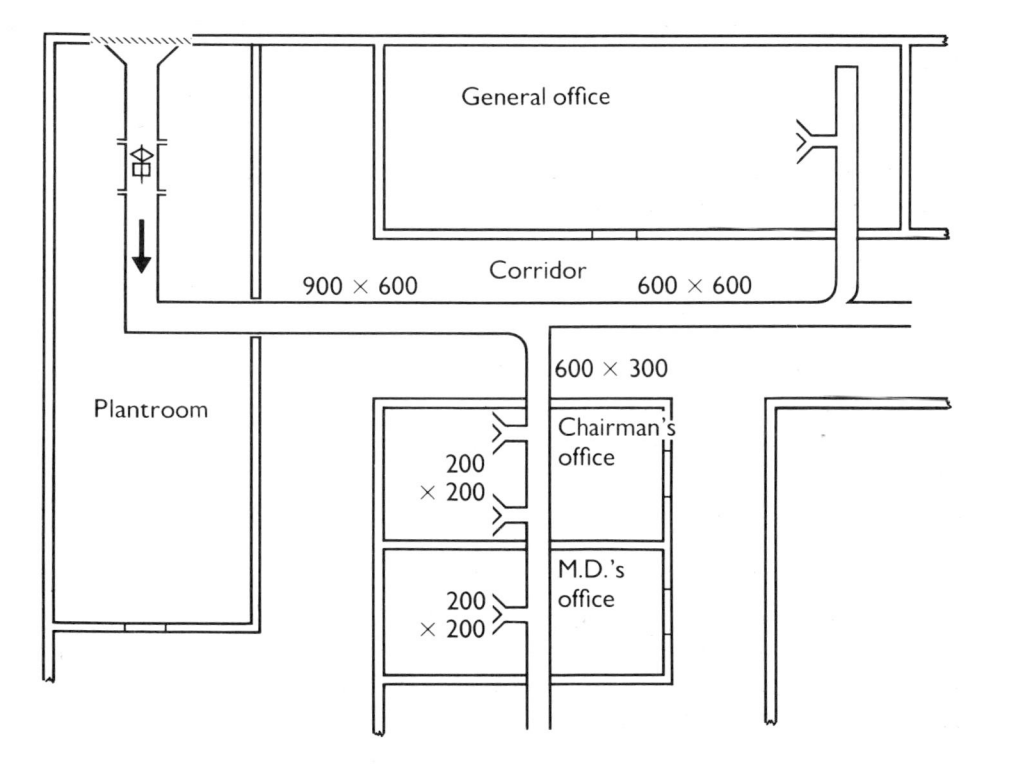

Figure 7.28 Layout of Office Supply System

Architectural data the chairman's office must have a background sound level due to the ventilation system of NC30. Partitions and ceiling are to a very high sound insulation performance. No component of duct breakout or noise flanking through the structure need therefore be consider.

Calculation Sheet

The calculation sheet (Figure 7.30) shows the sound pressure level that will be generated in the chairman's office by the fan and the corresponding additional attenuation required to meet NC30.

7.7.2 Example 2: "Crosstalk""

The Chairman is a quiet, mild man who enjoys relaxing in his office with a low background sound level (NC30). On the other hand the Company Managing Director is prone to talking in a loud voice to his staff. The sound power level of the M.D.'s voice is shown in Figure 7.26, Curve C.

The calculation sheets (Figures 7.31 and 7.32) shows the calculation process undertaken to determine the sound pressure level generated in the Chairman's office by the M.D.'s voice. Since the partitioning and ceiling is of a high acoustic insulation, only transmission via the ventilation system need be considered.

	CONTRACT TITLE			C/C		SHEET NO		
SRL	SYSTEM DESIGNATION			REF. NO.			HIGH/LOW VEL	
	SOUND SOURCE			UNIT REF NO		PATH TO		
ENG	MAKE & TYPE		DUTY		running at			rpm
				m^3/s	N/m^2			

DATE		OCTAVE BAND CENTRE FREQUENCY Hz						
		63	125	250	500	1k	2k	4k
OVERALL Lw ①	SPECTRUM CORRECTION							
② CORRECTIONS TO Lw DATA								
③								
④ STRAIGHT DUCT	UNIT Lw							
⑤ LENGTH m	SIZE mm	TREATMENT						
⑥								
⑦								
⑧								
⑨								
⑩ BENDS & TAKE OFFS								
⑪ NO.	TYPE	SIZE mm	TREATMENT					
⑫								
⑬								
⑭								
⑮								
⑯								
⑰								
⑱								
⑲								
⑳ OTHER ATTENUATION								
㉑								
㉒ END REFLECTION	REF	x = m^2						
㉓ Lw LEAVING SYSTEM								
㉔ SOUND POWER TO ROOM	m^3/s %							
㉕ ROOM SIZE	x x = m^3							
㉖ MID FREQUENCY REVERBERATION TIME	(secs)							
㉗ REVERBERANT Lp								
㉘ SOUND POWER TO OUTLET	m^3/s %							
㉙ DISTANCE TO LISTENER	mm							
㉚ DIRECTIVITY (FLUSH/CORNER/)	m^2							
㉛ DIRECT Lp								
㉜ RESULTANT Lp								
㉝ cf DESIGN CRITERION								
㉞ ADDITIONAL ATTENUATION REQUIRED								
㉟ Lw ENTERING ATTENUATOR								

© Sound Research Laboratories Limited

Calculation sheet No1 Ductborne noise

Figure 7.29 Blank Calculation Sheet

CONTRACT TITLE XYZ CORP PLC
SYSTEM DESIGNATION OFFICE SUPPLY   **REF NO** CC 1234   **SHEET NO** 1/1
SOUND SOURCE FAN   **PATH TO** CHAIRMAN'S OFFICE   **HIGH LOW LEVEL**
MAKE & TYPE AXIAL FLOW   **UNIT REF NO**   **DUTY** 2.5 m³/s   40 N m²   turning at — rpm
DATE 6.2..86   **ENG** DFS

Calculation sheet No1 Ductborne noise   © Sound Research Laboratories Limited

					OCTAVE BAND CENTRE FREQUENCY Hz						
					63	125	250	500	1k	2k	4k
(1) OVERALL Lw		77dB									
(2) CORRECTIONS TO Lw DATA				SPECTRUM CORRECTION	-5	-5	-6	-7	-8	-10	-13
(4)				UNIT Lw	72	72	71	70	69	67	64
(5) STRAIGHT DUCT	LENGTH m	SIZE mm	TREATMENT								
(6)	10	900 × 600	None		-15	-10	-5	-3	-2	-2	-2
(7)	5	600 × 300	None		-7	-7	-4	-2	-2	-2	-2
(9) BENDS & TAKE OFFS	NO	TYPE	SIZE mm	TREATMENT							
(10)	1	M.B.	900 × 600	No T/vanes	0	-5	-8	-4	-3	-3	-3
(21) OTHER ATTENUATION											
(22) END REFLECTION	REF	—	200 × 200 = .04 m²		-14	-10	-5	-2	0	0	0
(23) Lw LEAVING SYSTEM					36	40	49	59	62	60	57
(24) SOUND POWER TO ROOM 0.5 m³	20%	30 m³			-7	-7	-7	-7	-7	-7	-7
(25) ROOM SIZE 4 × 3 × 2.5 = 30 m³											
(26) MID FREQUENCY REVERBERATION TIME 0.5 (secs)					-3	-3	-3	-3	-3	-3	-3
(27) REVERBERANT Lp					25	29	38	48	51	49	46
(28) SOUND POWER TO OUTLET 0.25 m³	10%				-10	-10	-10	-10	-10	-10	-10
(29) DISTANCE TO LISTENER 0.9 m				DIRECT Lp	19	24	34	45	49	48	46
(30) DIRECTIVITY FLUSH		-.04 m²			+3	+4	+5	+6	+7	+8	+9
(31) RESULTANT Lp					26	30	39	50	53	52	49
(33) DESIGN CRITERION NC30					57	48	41	35	31	29	28
(34) ADDITIONAL ATTENUATION REQUIRED					0	0	0	0	0	0	0
(35) Lw ENTERING ATTENUATOR Fan-unit Lw					72	72	71	70	69	67	64

COMMENTS:-

(1) see Fig. 7.5 for Fan Lw
(2) See Fig. 7.6 for Spectrum Corrections
(3) Note ideal entry discharge conditions. No correction required
(4) See Fig. 7.9 for attenuation in rectangular straight duct.
(5) Mitre bend (MB) with no treatment or turning vanes. See Fig. 7.12a
Note: Don't take a bend loss at the branch
(6) See Fig. 7.16 for end reflection loss figures
(7) Since the system is low velocity. Rather than take area losses at each branch, the air volume reaching the room or outlet has been taken as a percentage of total — see Fig. 7.15
(8) see Fig. 7.19
(9) See Fig. 7.19
(10) See comment (7) above
(11) See Fig. 7.20
(12) See Fig. 7.20 for directivity loss (flush, away from walls)
(13) Add Rev + Dir Lp using chart in Physics of Sound Chapter. Note: Sound dominated by Rev Lp.
(14) see NC levels in Chapter on Design Targets
(15) NC30 – resultant Lp
(16) Required for calculation of attenuator dynamic insertion loss.

Figure 7.30   Calculation Sheet Example 1 — Supply System

Calculation sheet No1 Ductborne noise © Sound Research Laboratories Limited

CONTRACT TITLE XYZ CORP PLC	REF NO C.C 1234	SHEET NO 1/2
SYSTEM DESIGNATION CROSSTALK M.D. TO CHAIRMAN		
SOUND SOURCE M.D. VOICE	PATH TO CHAIRMAN'S OFFICE	HIGH LEVEL / LOW LEVEL
MAKE & TYPE NOT APPLIC.		
UNIT REF NO	DUTY	
ENG / DFS		
DATE 6.2.86		
OVERALL SWL	Lp	

NO						SPECTRUM CORRECTION / UNIT Lw / TREATMENT	63	175	250	500	1k	2k	4k
(2)	CORRECTIONS TO Lw DATA												
(3)	STRAIGHT DUCT												
(4)	LENGTH m												
(5)													
(6)													
(7)		SIZE mm		TREATMENT									
(8)													
(9)													
(10)	BENDS & TAKE OFFS	NO	TYPE	SIZE mm	TREATMENT								
(11)													
(12)													
(13)													
(14)													
(15)													
(16)													
(17)													
(18)													
(19)													
(20)	OTHER ATTENUATION												
(21)													
(22)	END REFLECTION	REF ×											
(23)	Lw LEAVING SYSTEM						69	72	77	80	80	75	66
(24)	SOUND POWER TO ROOM	100 %					0	0	0	0	0	0	0
(25)	ROOM SIZE 4 × 3 × 2.5 = 30 m³						-1	-3	-1	-1	-1	-1	0
(26)	MID FREQUENCY REVERBERATION TIME 0.5 secs						-3	-3	-3	-3	-3	-3	-3
(27)	REVERBERANT Lp						65	68	73	76	76	71	62
(28)	SOUND POWER TO OUTLET	100 %					0	0	0	0	0	0	0
(29)	DISTANCE TO LISTENER 0.9 m						-10	-10	-10	-10	-10	-10	-10
(30)	DIRECTIVITY (FLUSH CORNER) 1						+3	+3	+3	+3	+3	+3	+3
(31)	DIRECT Lp						62	65	70	73	74	70	63
(32)	RESULTANT Lp												
(33)	Lw (rev) = Lp (rev) + log A - 6 =						45	48	53	56	56	51	42
(34)	Lw Dir = Lp (dir) + 10 log A =						48	51	56	59	60	56	49
(35)	Lw (Tot) = Lw (rev) + Lw (dir) =						50	53	58	61	61	57	50

OCTAVE BAND CENTRE FREQUENCY Hz

COMMENTS –

(1) M.D.'s voice Lw see Fig. 7.26 Curve C
(2) are taking 'losses' at individual branch therefore ignore
(3) volume of source room see Fig. 7.19
(4) rev time of source room see fig. 7.19
(5) see comment (3)
(6) distance attenuation see fig. 7.20
(7) may be taken as a directivity factor constant of the human voice
(8) do not need total Lp
(9) to find Lw into grille add 10 log A – 6 to Lp (rev)
(10) to find Lw into grille add 10 log A to Lp (dir)
(11) add Lw (rev) and Lw (dir) using Fig. 1 in Physics of Sound chapter
  – substitute t is line into unit Lw on sheet 2

Figure 7.31 Calculation Sheet Example 2—Crosstalk

SRL

CONTRACT TITLE	XYZ CORP PLC	REF NO	C.C. 1234	SHEET NO 2/2
SYSTEM DESIGNATION	CROSSTALK M.D. TO CHAIRMAN			
SOUND SOURCE	M.D. VOICE	UNIT REF NO		HIGH LOW VEL
MAKE & TYPE	NOT APPLIC.	DUTY	PATH TO	
ENG	DFS			
DATE	6.2.86			

Calculation sheet No1 Ductborne noise   © Sound Research Laboratories Limited

No.	Item		63	125	250	500	1k	2k	4k
(1)	OVERALL SWL	Lp							
(2)	CORRECTIONS TO Lw DATA	SPECTRUM CORRECTION							
(4)	STRAIGHT DUCT — UNIT Lw		50	53	58	61	61	57	50
	LENGTH m / SIZE mm / TREATMENT								
(10)	BENDS & TAKE OFFS — NO / TYPE / SIZE mm / TREATMENT								
(13)	1st brnch 600x300/600x300 (50%)		-3	-3	-3	-3	-3	-3	-3
(14)	2nd brnch 600x300/200x200 (18%)		-7	-7	-7	-7	-7	-7	-7
(15)	1st branch 200x200/600x300 (1:4)		-2	-2	-2	-2	-2	-2	-2
(16)	2nd branch 600x300/200x200 (4:1)		-2	-2	-2	-2	-2	-2	-2
	END REFLECTION REF								
	OTHER ATTENUATION								
(23)	Lw LEAVING SYSTEM		36	39	44	47	47	43	36
(24)	SOUND POWER TO ROOM 100%		0	0	0	0	0	0	0
(25)	ROOM SIZE 4 x 3 x 2.5 = 30 m³								
(26)	MID FREQUENCY REVERBERATION TIME 0.5 secs								
(27)	REVERBERANT SPL		32	35	40	43	43	39	32
(28)	SOUND POWER TO OUTLET 100%		0	0	0	0	0	0	0
(29)	DISTANCE TO LISTENER 0.9 m		-10	-10	-10	-10	-10	-10	-10
(30)	DIRECTIVITY (FLUSH CORNER)		+3	+4	+5	+6	+7	+8	+9
(31)	DIRECT Lp .04 m²		0	0	0	0	0	0	0
	DIRECT Lp		29	33	39	43	44	41	35
(32)	RESULTANT Lp		34	37	43	46	47	43	37
(33)	DESIGN CRITERION NC30		57	48	41	35	31	29	28
(34)	ADDITIONAL ATTENUATION REQUIRED		—	—	2	11	16	14	9
(35)	SWL ENTERING ATTENUATOR								

OCTAVE BAND CENTRE FREQUENCY Hz

Notes:

(12) brought forward from sheet 1 Figure 7.31

(13) attenuation due to loss at a branch. In this case due to split of areas in system see Fig. 7.15 (note 2nd branch taken as nearest to source)

(14)

(15) attenuation due to abrupt change of cross section see Fig. 7.14

(16) see Fig. 7.14

(17) calculated by taking losses from unit Lw

(18) Receiver room volume see Fig. 7.19

(19) Receiver room rev. time see Fig. 7.19

(20) Receiver room distance attenuation see Fig. 7.20

(21) Grille flush in ceiling away from walls — see Fig. 7.20

(22) Add Lp (rev) to Lp (Dir) using Fig. 1 in Physics of Sound chapter

(23) See NC levels in chapter 4 on design targets

(24) will be cross talk problem

Figure 7.32   Calculation Sheet Example 2 — Crosstalk

CHAPTER 8

DUCTBORNE NOISE — FLOW GENERATED

8.1 INTRODUCTION

Having studied, and possibly attenuated, the sound power levels travelling down the ductwork as acoustic waves which started, for instance, at the fan, the engineer must turn his attention to noise resulting from the flow of air through the duct work system.

The accurate prediction and control of flow-generated noise of duct fittings is important because at long distances from noise source, flow noise from duct fittings can be the primary noise source. If the fittings are located near terminal devices, there may not be enough duct length available between the duct fitting and terminal device to provide sufficient attenuation of the flow generated noise, as is usually the case for the fan noise.

Flow generated noise is the result of air flow turbulence which has usually been concentrated into a discrete region. It must be remembered that all air flow in heating and ventilating ducts is turbulent air flow and not the smooth streamlined flow which can easily be illustrated with liquid in small glass tubes. Smooth flow of the air is not possible because it becomes unstable and instead of sliding layer over layer, the layers, in fact, roll on circular "eddies" rather like ball bearings. Additional eddies are shed from any duct irregularities as illustrated in Figure 8.1 to form what is called a "Karman vortex street". However, such whirlpools can also be created by any discontinuities or even the rough sides of the airways. Below are listed the most likely sources of this turbulence.

1. The wake behind a moving object, for example, rotating fan blades.
2. Flow over objects in duct, for example, tie bars, dampers, grilles.
3. Constriction noise, for example, orifice plates, changes of section, attenuator splitters.
4. Jet noise as the air emerges finally from a duct, for example, induction unit noise, duct outlets.
5. Boundary layer turbulence, as the air passes over apparently smooth surfaces, for example, duct walls.
6. Flow around corners, for example, bends and take-offs.

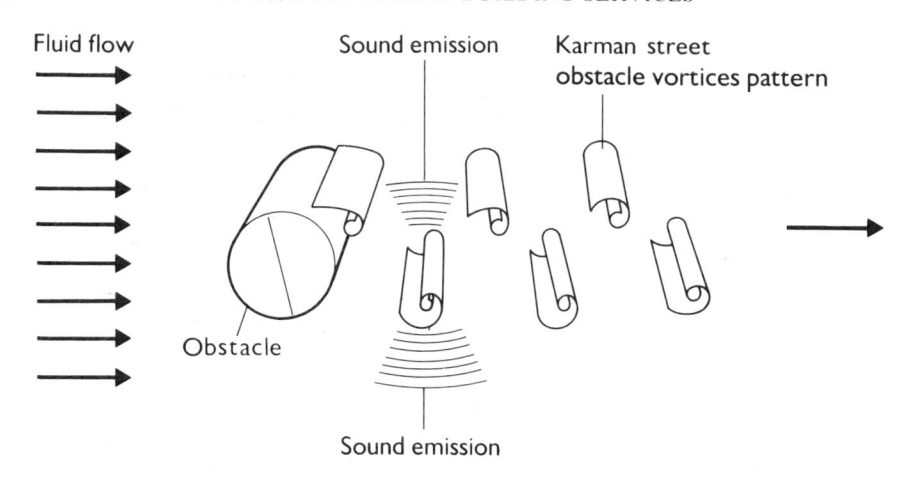

Figure 8.1   Formation of Turbulence

The intensity of the noise generated from the air flow will be determined primarily by the magnitude of the velocity increase created by the obstruction etc.

The spectrum of the noise produced will depend mainly on the size of the obstacle or discontinuity but also on the maximum velocity. Figure 8.2 illustrates the turbulence round a large single damper blade and a small tie rod. Large objects producing large turbulence will be associated with a spectrum peak at low frequencies, whilst small objects will be correspondingly biased towards higher frequencies. In general, the peak in the spectrum is determined by a simple equation involving a representative dimension of the obstruction or discontinuity d, and the maximum flow rate involved round the obstacle or through the discontinuity, "V". This formula is:

(peak frequency x representative dimension)/maximum flow velocity = 0.2

$$\frac{fd}{u} = 0.2$$

or

$$\text{Peak Spectral Frequency} = \frac{0.2u}{d} \qquad 8.1$$

The turbulent eddies themselves are relatively harmless noise sources. They simply consist of little whirlpools of air and in free space they are generally very inefficient generators of noise at the velocities familiar in building services systems. A possible exception to this generalisation is the jet outlet of induction units. Efficient generation of noise occurs mainly when the turbulence is shed from the object or discontinuity or when it impinges subsequently on an obstruction. Then the eddies and corresponding velocity variations are converted into pressure fluctuations. The interaction or back reaction on the bluff object, for example, fan blades, leads to a

Butterfly damper

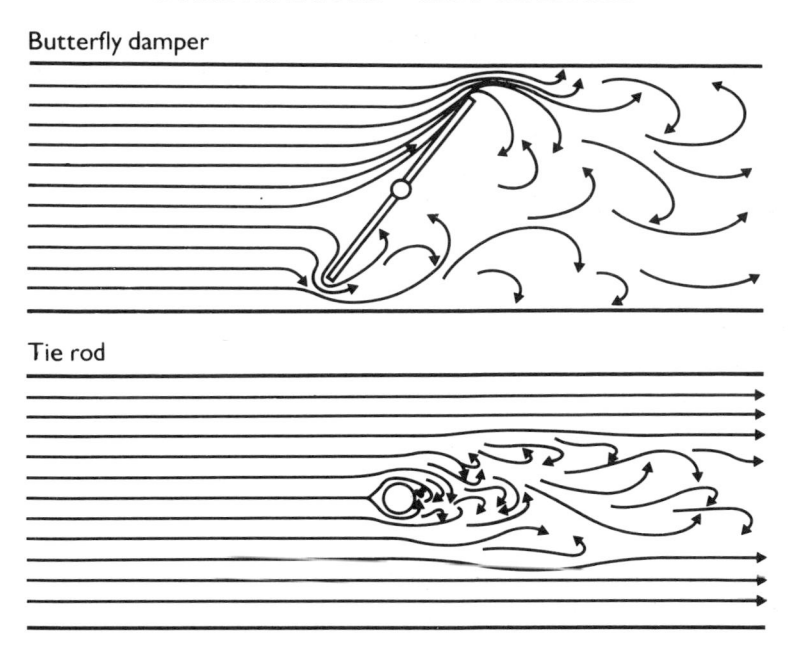

Tie rod

Figure 8.2   Turbulent Flow Behind Obstacles

further link between the spectrum distribution and the size of the object. Large objects become progressively more efficient at radiating low frequencies. Hence the reaction radiated sound from large objects will be enhanced at low frequencies compared to that from small objects (Figure 8.2.)

This turbulent reaction noise on solid objects may be easily demonstrated. If the necessary turbulent air discharges straight from a circular duct, it will usually do so quietly, even though the jet velocity may be fairly high — 20 m/s. If, however, there are obstructions, such as a grille or sheet of metal, placed in the air stream, there is an increase in the radiated noise level. This is due to interaction between the turbulence and the solid objects.

Although extensive research on flow generated noise has been carried out over the past fifteen years, this has unfortunately had a strong emphasis on jet engine noise. The air flow speeds involved in this situation are generally near to the speed of sound and not mainly applicable to the situation of the building services engineer. In general, comparatively little is known about the situation of noise produced by turbulent flow in conventional ducted systems. The situation becomes complicated by the presence of duct resonances and flow instabilities. Some generalised results will be presented as a result of existing systematic studies on flow generated noise, but, of necessity, they lack the precision of measurements on specified duct units. The major problem is the spectral distribution of the turbulent generated noises. Hence, much of the data presented here is the result of studies on specific types of duct elements coupled with reliable extracts from manufacturers' tests. Further to this, these isolated features and fitting studies completely avoid the interaction of turbulence on subsequent downstream obstructions and hence may be considered as minimum or certain flow generated noise levels.

The following flow generated noise sources will be considered:

1. Plain straight ducts
2. Bends
3. Take-offs
4. Abrupt and gentle transformation pieces
5. Tie rods
6. Butterfly dampers
7. Opposed blade dampers
8. Damper-damper interaction
9. Attenuators
10. Terminal units
11. Diffusers and associated dampers
12. Grilles
13. High speed jet outlets and nozzles.

8.2 PREDICTION OF FLOW GENERATED NOISE

In most cases the experimental data base, on which the prediction of flow generated noise is based is very limited and there is usually not enough variation of the key geometric parameters. Consequently the prediction methods presented in this chapter are subject to the shortcomings of the underlying data base. Even if the ± 4 dB accuracy of the predictions is less than satisfactory they have been included where they represent an improvement over previously used methods.

The calculation of the Strouhal Number is, in general, the starting point for predicting flow generated noise.

$$\text{Strouhal Number (St)} = \frac{fd}{u} \qquad 8.2$$

as touched upon previously re Peak Spectral Frequency,

where d is a characteristic dimension,
   u is the flow velocity, and
   f is the octave band centre frequency.

These quantities must be in compatible units, for example m, m/s and Hertz but not mm, m/s and Hertz.

The nomograph in Figure 8.3 may be used as a convenient way of calculating the Strouhal Number. Place a straight edge joining the scales of the selected dimension and air velocity and mark off a point on the reference line. From this reference point, place the straight edge across to any chosen octave band centre frequency to read off the corresponding Strouhal Number.

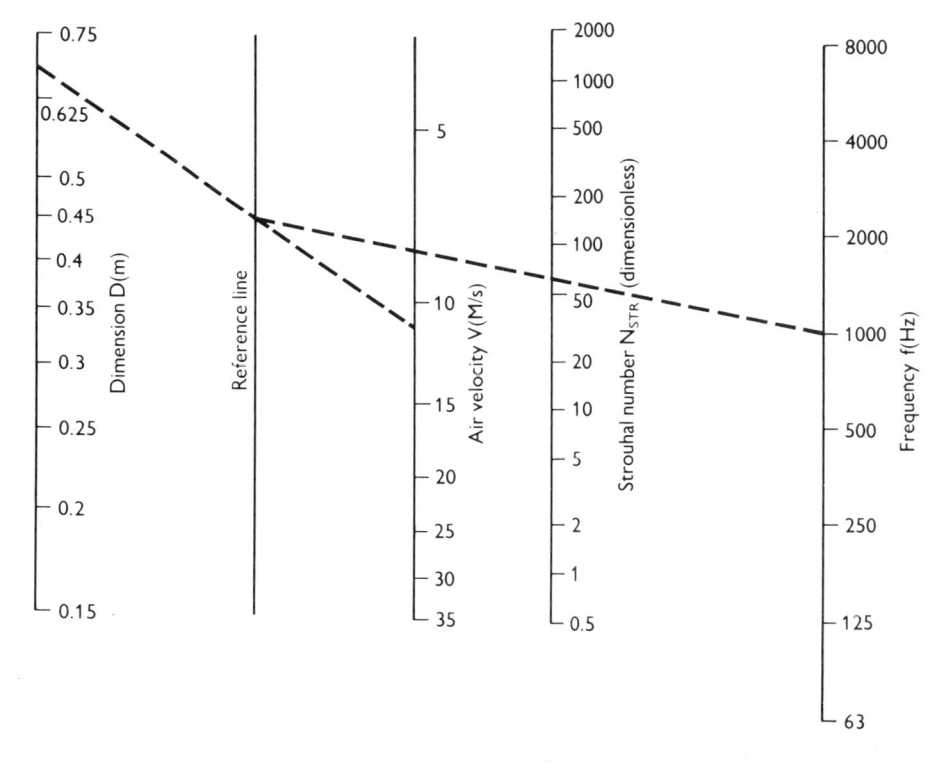

Figure 8.3   Nomograph for Calculating Strouhal Number

8.2.1 Flow Generated Noise in Straight Empty Ducts

Measurements have shown that contributions from this flow, when correctly "flow staightened" are very small compared with any other minor misalignment factors in a duct run. Air entry conditions are likely to give more important noise power contribution.

An indication of the sound power levels to be expected is shown in Figure 8.4 for 600 mm square ductwork. This figure is consistent with an increased power level of 18 dB per doubling of speed and also shows an increased (16 dB) output at a frequency (500 Hz) corresponding to one wavelength fitting across the duct's width (500 Hz is approximately equivalent to a wavelength of 600 mm).

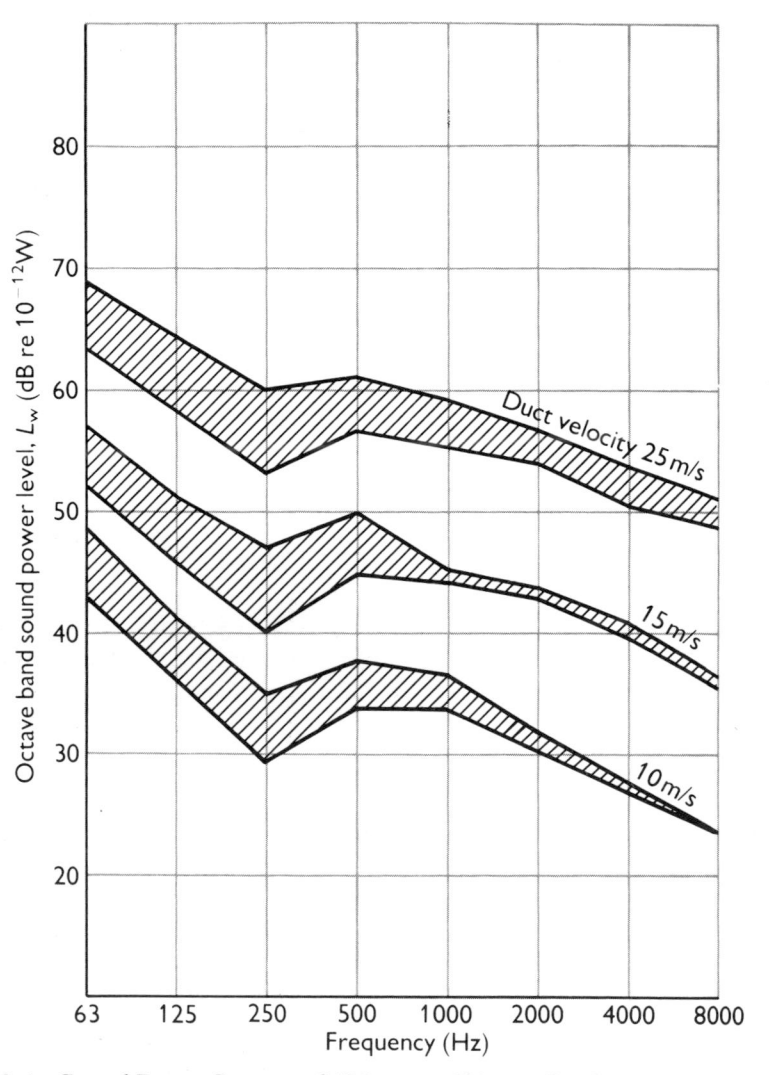

Figure 8.4 Sound Power Spectra of 600 mm x 600 mm Straight Empty Steel Duct for Various Air Flow Velocities

As a general design guide the following maximum duct velocities for conventional low velocity systems should not be exceeded.

NC design level	Main ducts	Branch ducts	Final run outs
20	4.5m/s	3.5m/s	2.0m/s
25	5.0m/s	4.5m/s	2.5m/s
30	6.5m/s	5.5m/s	3.25m/s
35	7.5m/s	6.0m/s	4.0m/s
40	9.0m/s	7.0m/s	5.0m/s

These figures are only intended as a guide. The position of the duct relative to the conditioned space, and the geometry of fittings etc will determine whether the velocity can be increase or reduced.

8.2.2 Flow Generated Noise in Bends

The noise regeneration of the following four 90° bends are presented in conjunction with their aerodynamic "k" factors.

(Pressure drop = k factor x velocity head):

(i) Simple radiused bend
(ii) Plain 90° bend
(iii) 90° bend with single long central splitter turning vane
(iv) 90° bend with short-chord multiple turning vanes

Simple radiused bends of aspect ratios up to 3:1 make no more noise than corresponding straight ductwork.

Figure 8.5 compares the flow generated noise in a rectangular duct 600 mm x 200 mm.

Here a single long chord central splitter turning vane has been introduced, without an appreciable change occurring in the regenerated noise levels despite a 50 per cent pressure drop reduction from a k factor of 1.2 to 0.65. The regenerated sound power levels increase by about 20 dB for each doubling of flow rate for this duct size and configuration.

Figure 8.6 compares the regeneration in a 200 mm square bend with and without multiple short chord turning vanes. A reduction of some 10 dB occurs at 125 Hz from the introduction of the turning vanes, probably resulting from their mild action as flow straighteners. However, over the middle octave bands an increased noise level is observed, even though the pressure loss decreases from a k factor of 1.2 to 0.22. The generated sound power levels increased by about 15 dB for each doubling of flow rate for this particular duct size and configuration.

The octave band sound power level of the noise generated by bends fitted with curved-blade-type turning vanes can be predicted provided the total pressure drop across the blades is known or can be estimated. This method is applicable to any bend angle between 60° and 120°. The octave band sound power levels generated by the bend with turning vanes is predicted by:

$$L_{\text{w oct}} (f_0) = K_T - 18 + 10\log(f_0) + 50\log(U_c) + 10\log(S) + 10\log(CD) + 10\log n \quad 8.3$$

The symbols are defined as follows:

f_0 = central frequency of the octave band, Hz
U_c = flow velocity in the constricted part of the field between the blades determined from

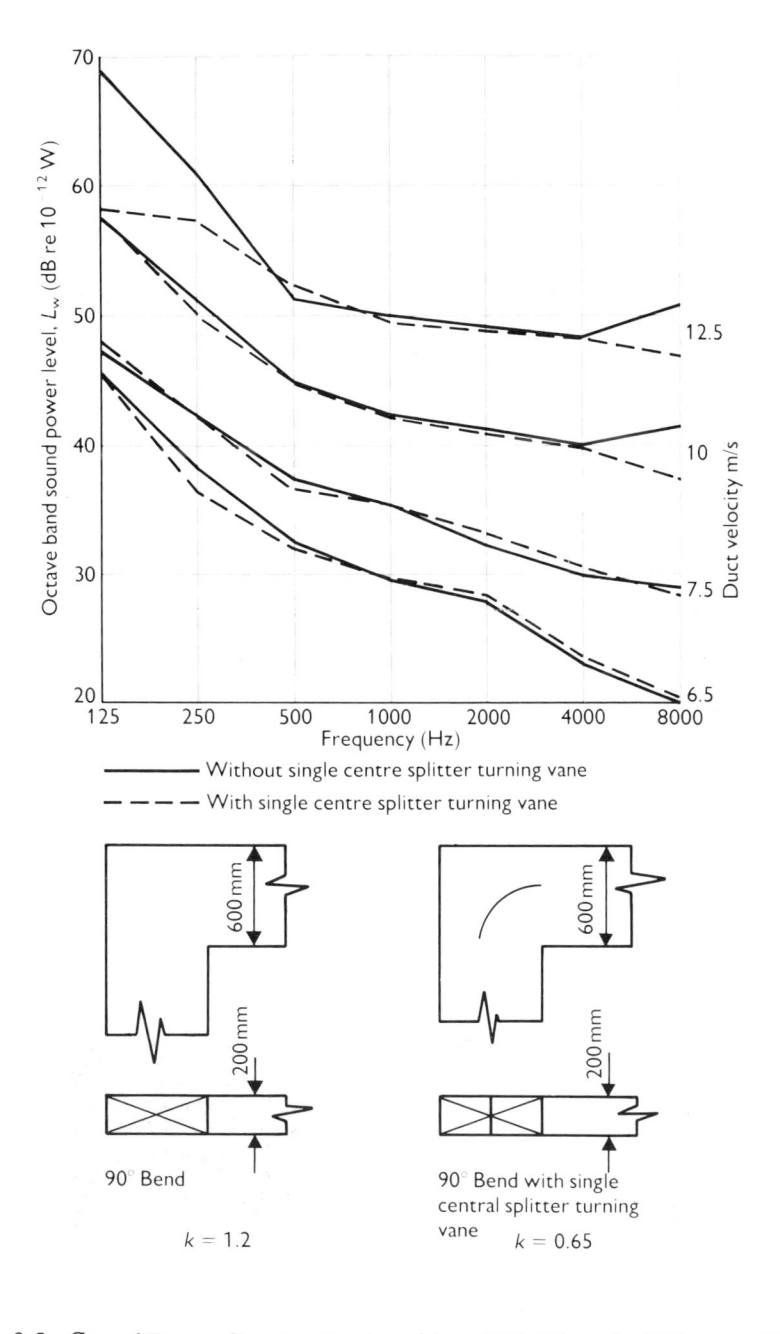

Figure 8.5 Sound Power Spectra Produced by a 90° Elbow in 600 mm x 200 mm Duct with and without a Single Splitter Turning Vane.

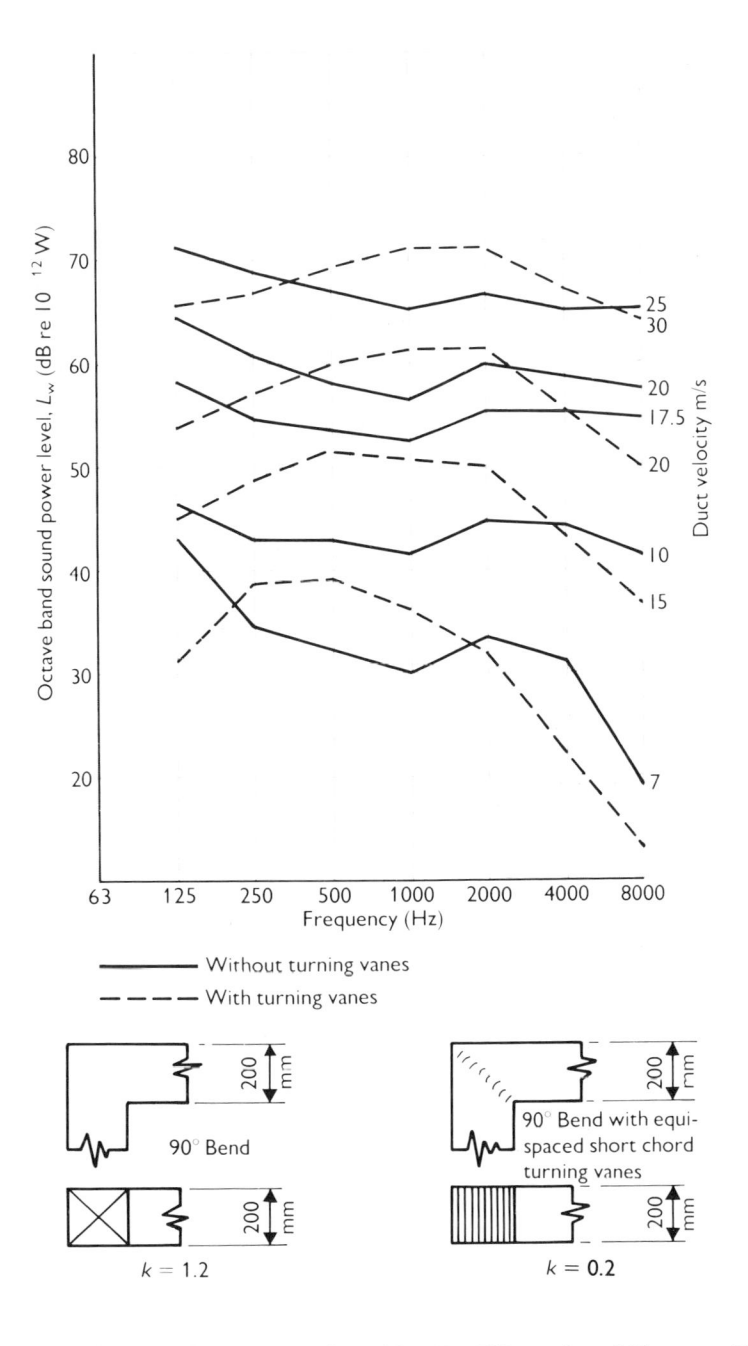

Figure 8.6 Sound Power Spectra Produced by 90° Elbows in a 200 mm x 200 mm Duct System with and without Short Chord Turning Vanes.

$$U_c = \frac{Q}{S(\text{BF})} \text{ m/s} \qquad\qquad 8.4$$

where BF is the Blockage Factor and this is given by

$$\text{BF} = \frac{(C^{1/2} - 1) \text{ if } C \ne 1}{(C - 1)} \qquad\qquad 8.5$$

or

$$\text{BF} = 0.5 \text{ if } C = 1 \qquad\qquad 8.6$$

where C is the total pressure loss coefficient which is given by

$$C = \frac{1.67 \Delta p}{[(Q/S)^2]} \qquad\qquad 8.7$$

Q = volume flow rate in m^3/s
Δp = total pressure loss in Pa
S = cross sectional area of the duct in m^2
CD = chord length of the typical vane in m
n = number of turning vanes

K_T = characteristic spectrum which is given by

$$K_T = 4.5 - 7.69 [\log_{10}(\text{St})]^{2.5} \qquad\qquad 8.8$$

St = Strouhal Number which is given by

$$\text{St} = \frac{f_0 D}{U_c} \text{ where } f_0 \text{ is the octave band centre frequency and } D \text{ is the duct dimension perpendicular to the vane length.} \qquad 8.9$$

The prediction proceeds according to the following steps:

1. The following information must be acquired

Information	Symbol	Units
Volume flow rate	Q	m^3/s
Total pressure loss	Δp	Pa
Duct cross sectional area	S	m^2
Chord length of vane	CD	m
Duct height, normal to vane length	D	m
Number of vanes		
	n	—

2. Determine the total pressure loss coefficient, C
3. Determine the blockage factor, BF

4. Determine the flow velocity in the constriction, U_c
5. Determine the Strouhal Number, St
6. Determine the characteristic spectrum K_T
7. Calculate the octave band sound power levels of the flow generated noise.

8.2.3 Flow Generated Noise in Take-offs and Junctions

Figure 8.7 shows the results of an experiment in which the sound power leaving the end of a 100 mm branch duct for two main flow duct velocities 9 m/s and 18 m/s and two junction designs, sharp and coned. These results include "end reflection" and hence indicate a substantial fall off below 250 Hz. Doubling the main duct flow velocity gives approximately a 10 dB increase in the radiated sound levels.

The advantageous result of employing a more streamlined coned junction is clearly illustrated, indicating a reduction of about 5 dB at the higher flow velocity, the advantage decreasing as the main flow duct velocity is reduced.

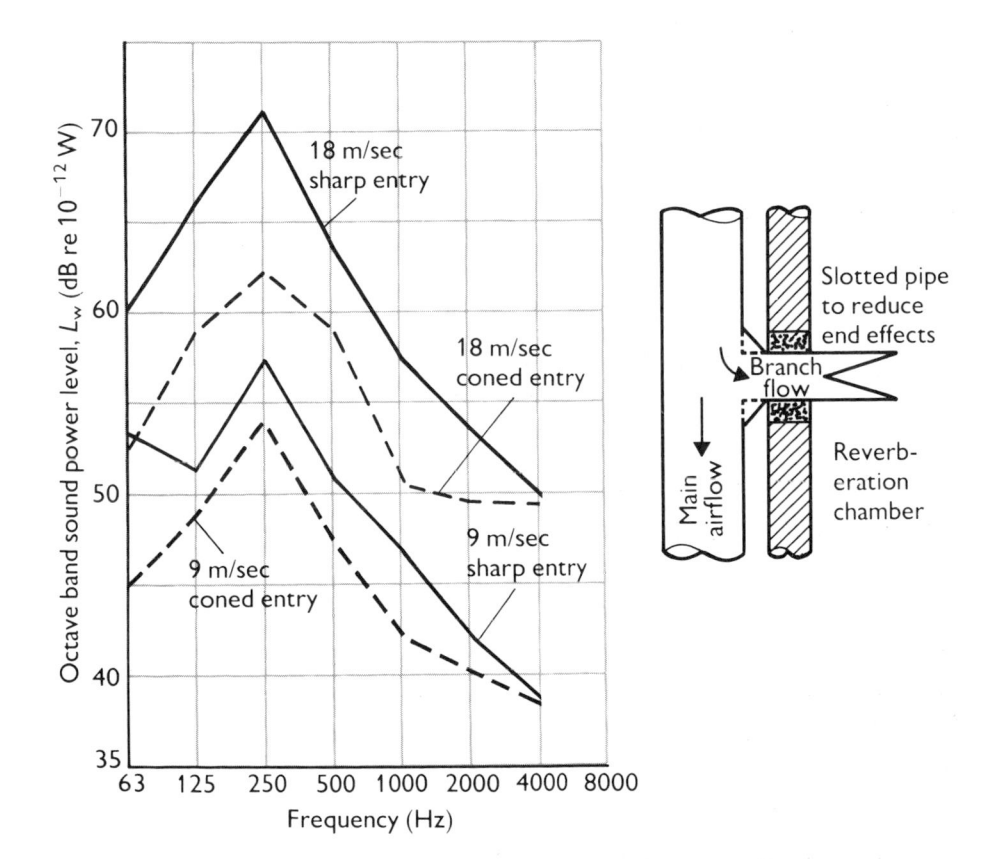

Figure 8.7   Flow Generated Noise at 90° Take-off — abrupt and coned

The general increase in flow generated noise with increased branch flow is shown. Typical power distribution spectra are shown in Figure 8.8.

The effect of flow distribution between the main and branch ducts is shown in Figure 8.9 for the 63, 125 and 250 Hz octave bands, both for the main duct noise and the branch duct noise.

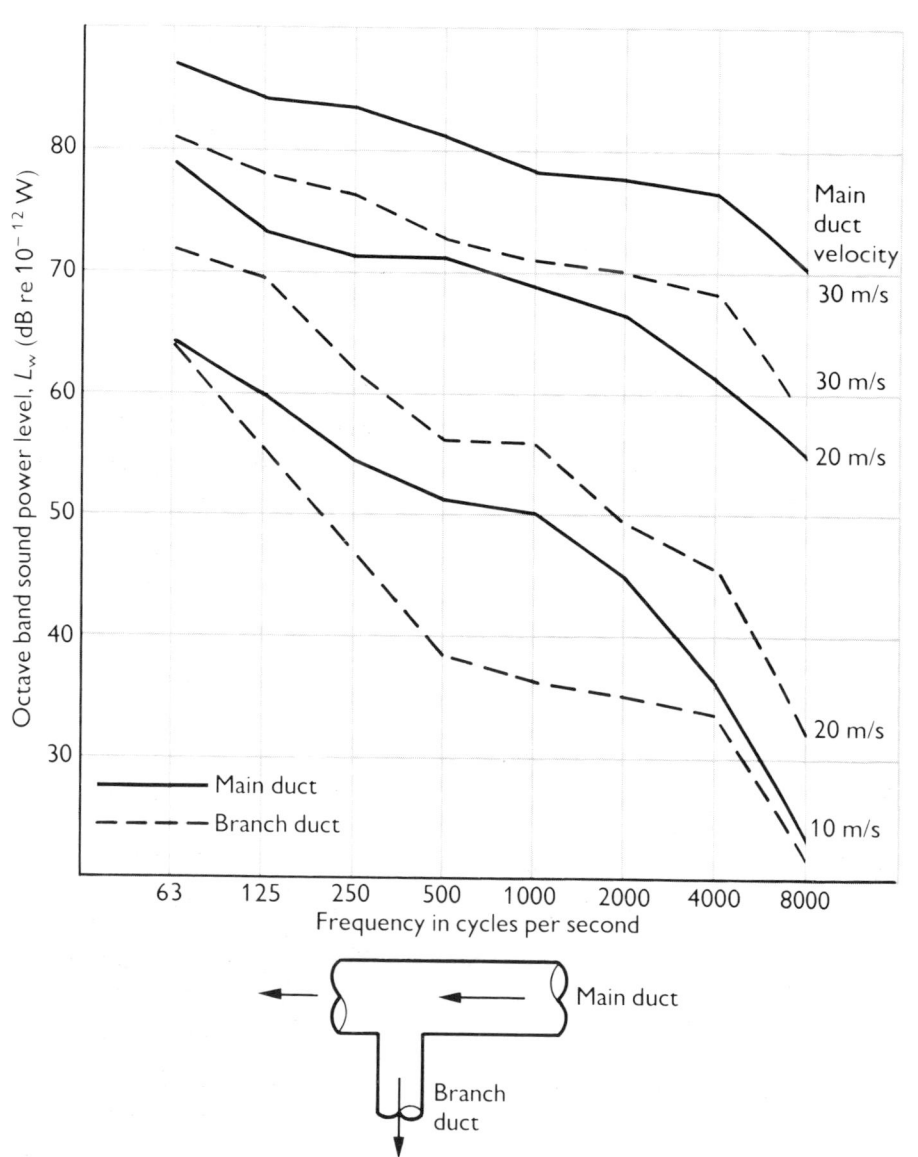

Figure 8.8   Sound Power Spectra from a Branch Take-off System with a 90° Tee Comparing the Branch and Main Duct Sound Power Levels.

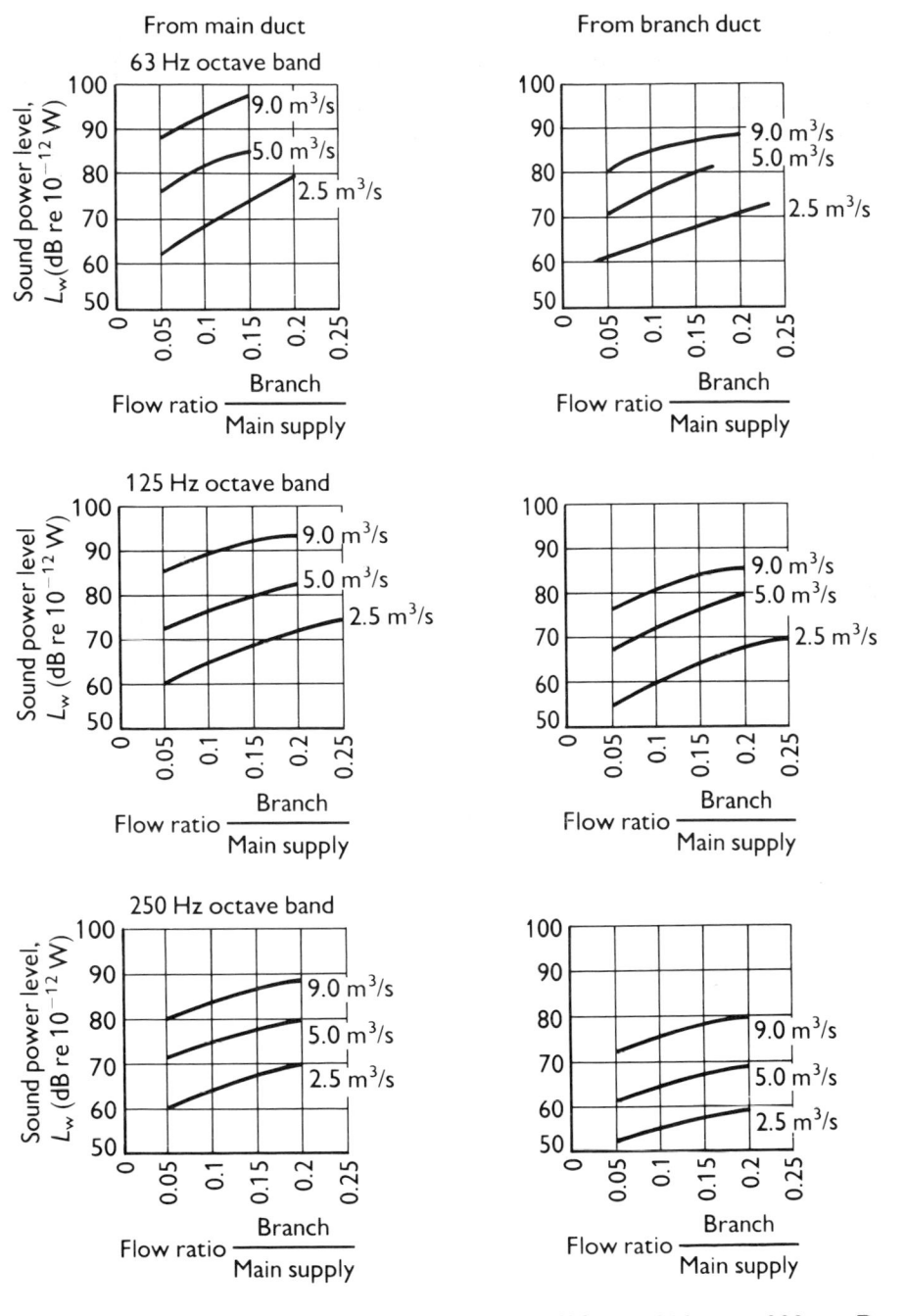

Figure 8.9   Sound Power Levels of Branch Take-off from a 900 mm x 300 mm Duct System Plain Tee. Volume Flow Rates are for Main Duct.

8.3 PREDICTIVE TECHNIQUES

8.3.1 Junctions and Turns

The flow generated noise of turns and junctions can be predicted for diverging flow conditions. Converging flow generates considerably less noise than diverging flow and consequently it is seldom of concern.

Figure 8.10 shows the various junctions and turns and gives information on how the sound power level is obtained by modifying the equivalent spectrum, L_{we} which is

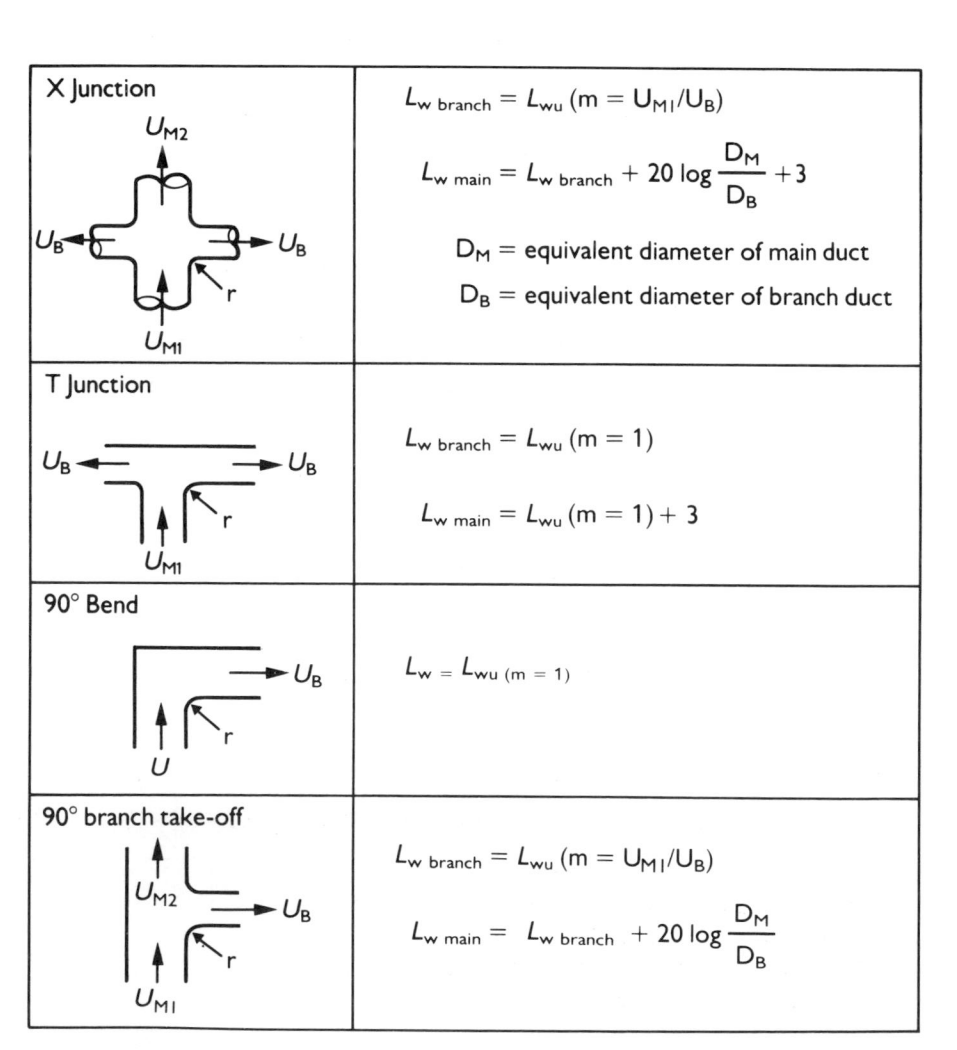

Figure 8.10   Prediction of Flow Generated Noise for Junctions and Turns

defined as:

$$L_{we} = L_{wu} + \Delta r + \Delta T \qquad 8.10$$

where Δr is a correction term for the extent of rounding the corner of the turn or junction

and ΔT is a correction factor for upstream turbulence given by Figure 8.11.

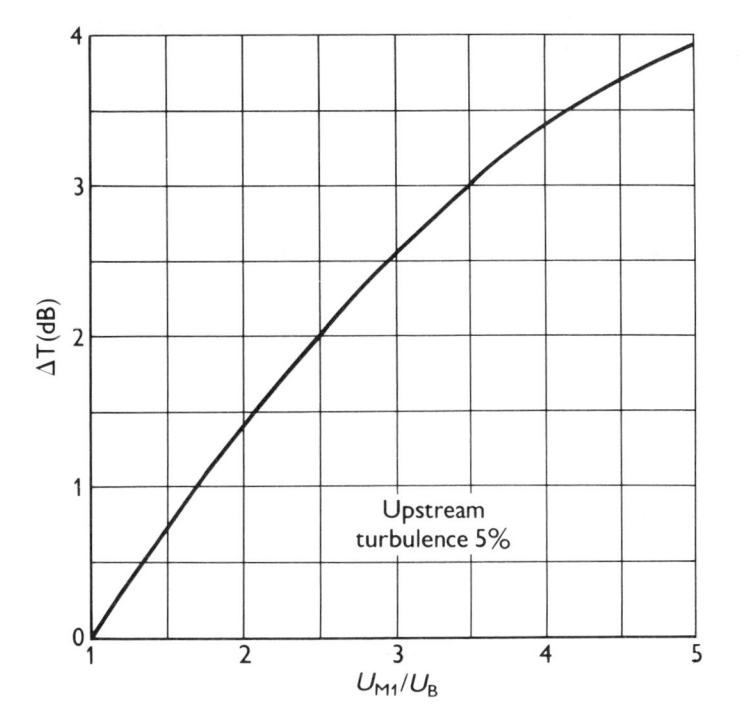

Figure 8.11   Correction for Upstream Turbulence

The spectrum L_{wu} may be obtained from:

$$L_{wu}(f_0) = K_J + 10\log_{10}(f_0) + 50\log_{10}(U_B) + 10\log_{10}(S_B D_B) - 18 \qquad 8.11$$

where f_0 = octave band centre frequency, Hz;
  D_B = equivalent diameter of the branch duct, m;
  U_B = flow velocity in the branch duct, m/s;
  S_B = cross sectional area of the branch duct, m²;
 and K_J = characteristic spectrum.

The characteristic spectra K_J are plotted in Figure 8.12 as a function of the Strouhal number with the velocity ratio U_{MI}/U_B as parameter.

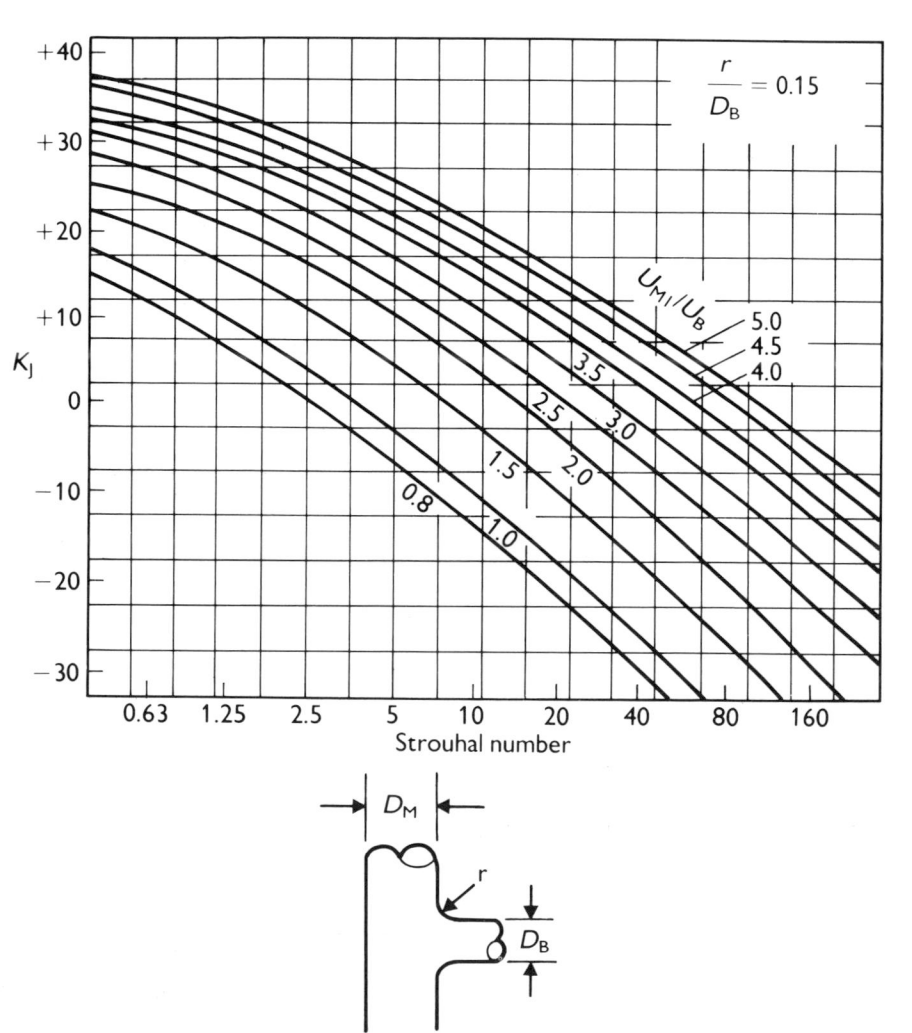

Figure 8.12 Formalized Spectra for Flow-Generated Noise of Take-Offs and Junctions.

$$\text{Rounding of corner } \frac{r}{D_B} = 0.15 \text{ from Figure 8.13.}$$

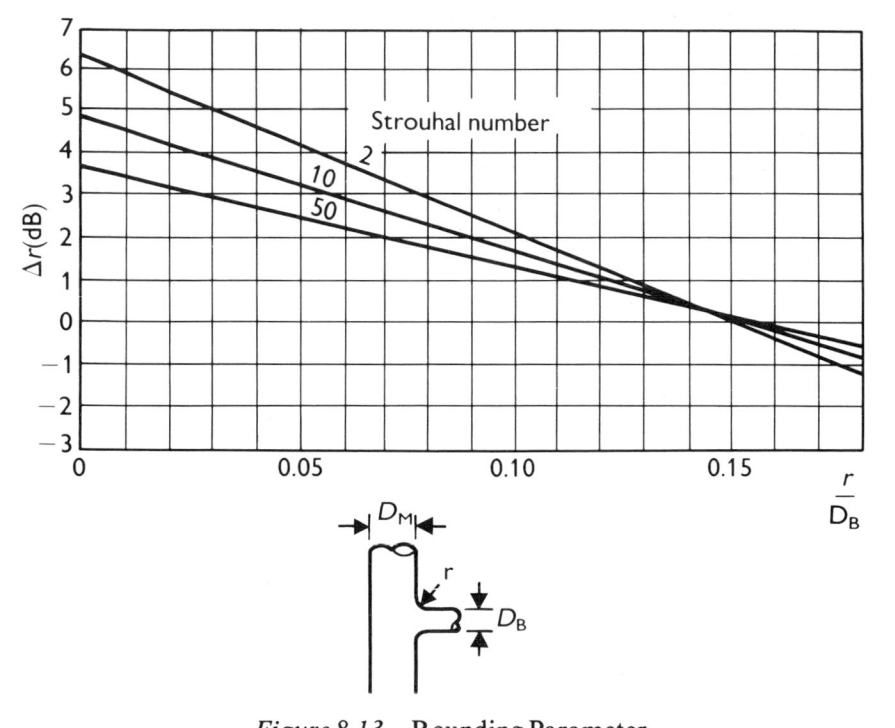

Figure 8.13   Rounding Parameter

The prediction requires the following information:

Information	Symbol	Units
Flow velocity in branch	U_B	m/s
Flow velocity in main duct upstream of junction	U_{MI}	m/s
Cross sectional area of branch duct	S_B	m²
Cross section area of main duct	S_M	m²
Rounding parameter	$\dfrac{r}{D_B}$	—
Turbulent inflow conditions	No*	Yes**

* In the case of long straight duct with no dampers or branches 5 main duct diameters upstream, no inflow turbulence correction is required

**If the duct upstream has dampers, turns, take-offs within length of 5 main duct diameters, apply turbulent inflow correction.

The following additional parameters need to be calculated:

$$\text{Velocity ratio } m = U_{MI}/U_B \qquad\qquad 8.12$$

$$\text{Area ratio } \frac{S_M}{S_B} \qquad\qquad 8.13$$

$$\text{Strouhal number St} = \frac{f_0 D_B}{U_c} \qquad\qquad 8.14$$

where U_c is the corresponding value for U_B or U_{MI}

8.3.1.1 Worked Example Predict the octave band sound power level in the branch duct, L_{WB} and the octave band level in the main duct, L_{WM} for a branch take off geometry sketched in the calculation sheet (Figure 8.14). The rectangular main duct is 0.305 m x 0.915 m and the volume flow in the main duct, Q_{MI} is 5.7 m³/s. The cross dimensions of the rectangular branch duct are 0.254 m x 0.254 m and the volume flow rate in the branch duct. $Q_B = 0.1 Q_{MI}$. The 90° turn from main duct to branch has no rounding.

Step 1 Determination of additional parameters

Area ratio: $S_M/S_B = (0.305 \text{ x } 0.915)/(0.254 \text{ x } 0.254) = 4.32$
Velocity ratio: $U_{MI}/U_B = (Q_{MI}/S_M)/(0.1 Q_{MI}/S_B) = 10 S_B/S_M = 2.3$
Velocity in branch: $U_B = Q_B/S_B = (0.1 \text{ x } 5.7)/(0.254 \text{ x } 0.254) = 8.835 \text{ m/s}$
Equivalent diameter: $D_B = (4 S_B/\pi)^{1/2} = 0.286 \text{ m}$
Strouhal Number: $\text{St} = D_B f_0/U_B = (0.286/8.835) f_0 = 0.0324 f_0$

Step 2 Determining branch L_{WB}

Enter the St, Strouhal Number, values corresponding to each octave band centre frequency into the calculation sheet in Figure 8.14. Complete all the rows, using the figures indicated in the calculation sheet. The last row yields the octave band sound power level of the flow generated noise in the *branch duct*.

Step 3 Determining Main Duct L_{WM}

According to Figure 8.10, the octave band sound power level of the flow generated noise in the main duct is given by:

$$L_{wM} = L_{wB} + 20\log_{10} \frac{D_M}{D_B} = L_{wB} + 10\log_{10} \frac{S_M}{S_B} \qquad\qquad 8.15$$

therefore

$$L_{wM} = L_{wB} + 10\log_{10} 4.32 = L_{wB} + 6.3 \text{ dB} \qquad\qquad 8.16$$

Consequently, the octave band sound power level in the main duct in both directions, upstream and downstream of the junction, is obtained by adding 6.3 dB to the octave band sound power level of the side branch listed in the last row of the calculation sheet, Figure 8.14.

Junction Type (Sketch)

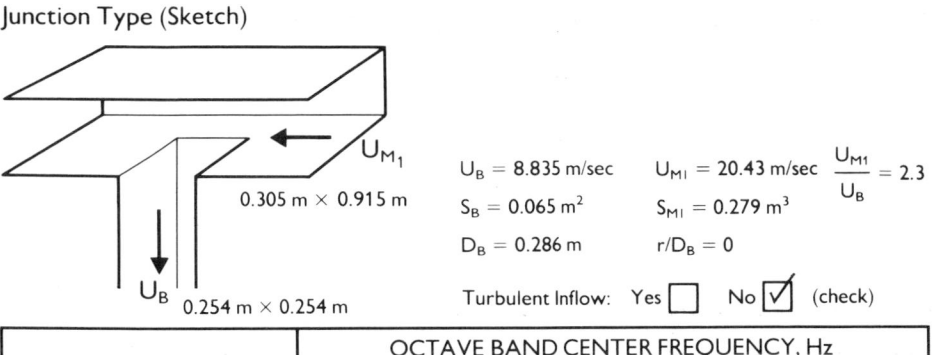

$U_B = 8.835$ m/sec $U_{MI} = 20.43$ m/sec $\dfrac{U_{MI}}{U_B} = 2.3$

0.305 m × 0.915 m $S_B = 0.065$ m^2 $S_{MI} = 0.279$ m^3

$D_B = 0.286$ m $r/D_B = 0$

U_B 0.254 m × 0.254 m   Turbulent Inflow: Yes ☐ No ☑ (check)

	OCTAVE BAND CENTER FREQUENCY, Hz								
	31	63	125	250	500	1K	2K	4K	8K
St	1	2	4	8	16	32	64	128	256
10 log (f_0)	15	18	21	24	27	30	33	36	39
K_J (St) From Fig. 8.12	27	23	18	11.5	5	−2	−10.5	−18	−27.5
50 logU_B − 18	29	29	29	29	29	29	29	29	29
15 log$S_B D_B$	−1.8	−1.8	−1.8	−1.8	−1.8	−1.8	−1.8	−1.8	−1.8
Δr(St) From Fig. 8.13	7	6.5	6	5	4.5	4	3.5	0	0
ΔT From Fig. 8.11	−	−	−	−	−	−	−	−	−
L_{WU} (f_0)*	76.2	74.7	72.2	67.7	63.7	59.2	53.2	45.2	38.7
Corr. from	−	−	−	−	−	−	−	−	−
$L_{W\,OCT}$ (f_0) dB re 10^{-12} Watt	76.2	74.7	72.2	67.7	63.7	59.2	53.2	45.2	38.7
Rounded L_W	76	75	72	68	64	59	53	45	39

* Sum of all rows below the second emboldened line.

Figure 8.14 Calculation Sheet for Predicting Flow Generated Noise of Duct Junctions and Turns without Turning Vane.

8.3.2 Transformation Pieces

Figure 8.15a, b, c and d show the sound power levels generated for four transition combinations using abrupt and gradual transformations. The most striking point is the general failure to indicate a trend. More air flow does produce more noise in each case, but the law of change is very different throughout the spectrum and from case to case. This is most probably the result of preferential acoustic coupling between the turbulence and the duct's acoustic properties and indicates a very complex aerodynamic/acoustic situation for analysis. Also the gradual transformation achieves various degrees of success. In case b, with an area transition of 3:1, an improvement of only a few dB is indicated from the use of a gradual transition, while for a smaller area ratio change of only 1.5:1 in case d, about 20 dB improvement is gained. Case a, with a large change of 9:1 indicates improvements approaching 30 dB from the use of the gradual transition.

8.3.3 Rods and Tubes

Figure 8.16 shows the effect of placing 4 off 13mm diameter rods in a duct flow of 25 m/s. The increased noise from regeneration has here been preferentially amplified by coupling to a resonant mode of the 600 mm square duct in the 500 Hz octave band (500 Hz is approximately equivalent to a wave length of 600 mm).

8.3.4 Butterfly Dampers

For guidance, Figure 8.17a, b and c show the sound power levels generated by a single blade damper plate in a 600 mm square duct for various flow rates from 4 m/s to 30 m/s and three blade angles, $0°$, $15°$ and $45°$. The corresponding aerodynamic k factors are included. Being a variable geometry device, it is extremely difficult to devise a general purpose formula to predict total sound power levels and spectra. Figure 8.17d shows an example of a butterfly damper.

8.3.5 Predictive Technique for Dampers

The octave band sound power levels of the flow noise generated by single or multi-blade dampers of both parallel and opposed blade configurations can be predicted by:

$$L_{w_{OCT}}(f_0) = K_D - 18 + 10\log_{10}(f_0) + 50\log_{10}(U_c) + 10\log_{10}(S) + 10\log_{10}(D) \quad 8.17$$

where f_0 = centre frequency of the octave band, Hz;
and U_c = flow velocity in the most constricted part of the flow field determined according to:

$$U_c = Q/S(BF) \text{ (m/s)}$$

where BF is the blockage factor given by:

(1) For multi-blade dampers

$$BF = 0.5 \text{ if } C = 1 \quad\quad\quad 8.18$$

or

$$BF = (C^{1/2} - 1)/(C - 1) \text{ for } C \neq 1 \quad\quad\quad 8.19$$

(2) For single blade dampers

$$BF = (C^{1/2} - 1)/(C - 1) \text{ for } C < 4 \quad\quad\quad 8.20$$

or

$$BF = 0.68C^{-0.15} - 0.22, \text{ for } C > 4 \quad\quad\quad 8.21$$

where

$$C = \text{total pressure loss coefficient} = \frac{1.67\Delta P}{(Q/S)^2} \quad\quad\quad 8.22$$

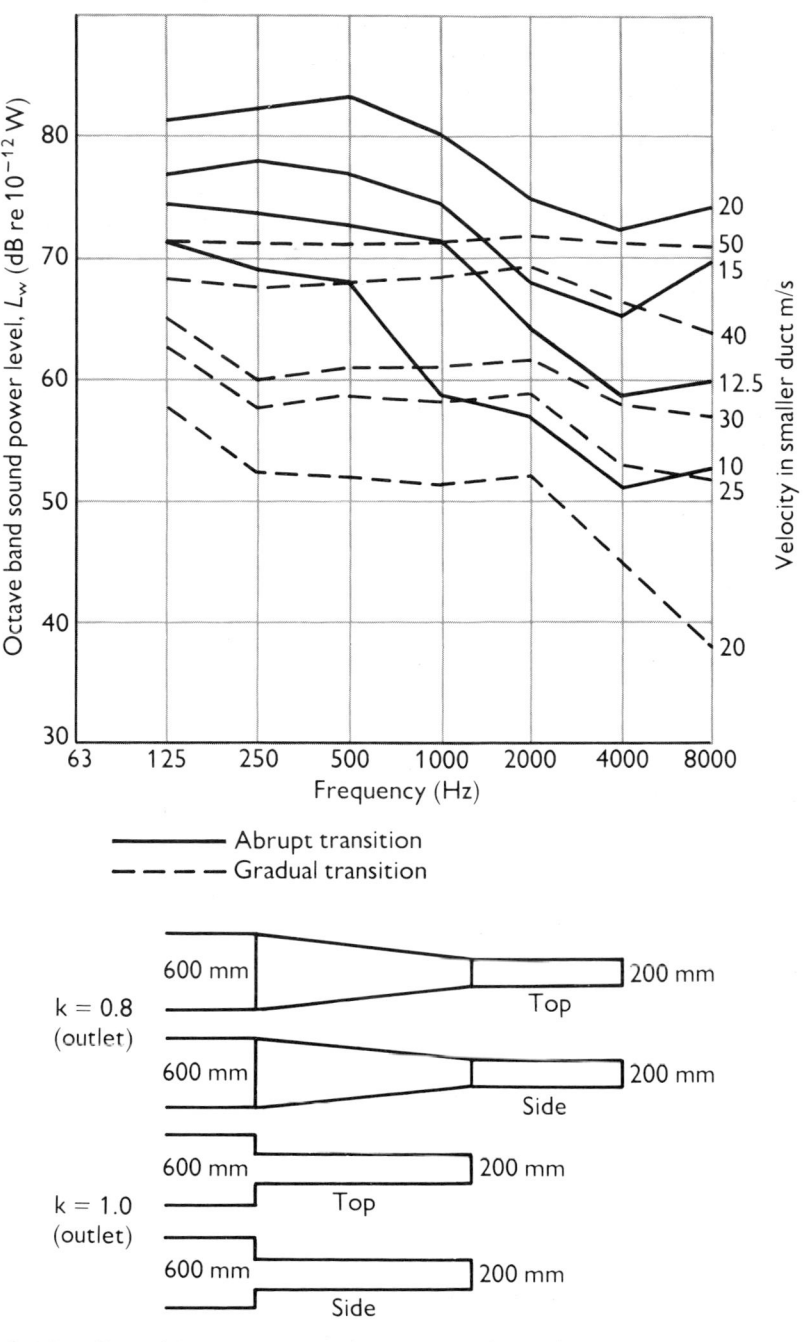

Figure 8.15a   Sound Power Levels of Abrupt and Gradual Area Transitions from
600 mm x 600 mm to 200 mm x 200 mm Duct Cross Sections (9:1)

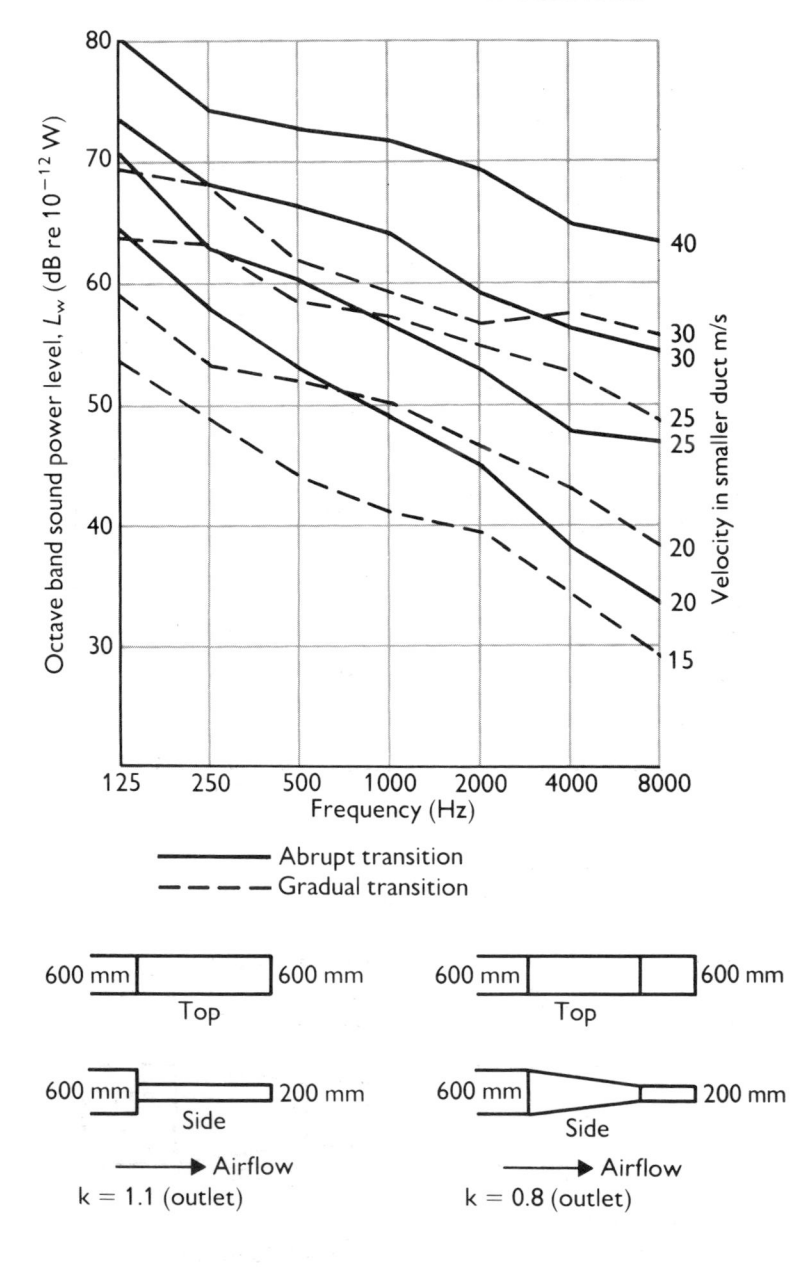

Figure 8.15b   Sound Power Levels of Abrupt and Gradual Area Transitions from 600 mm x 600 mm to 600 mm x 200 mm Duct Cross Section (3:1).

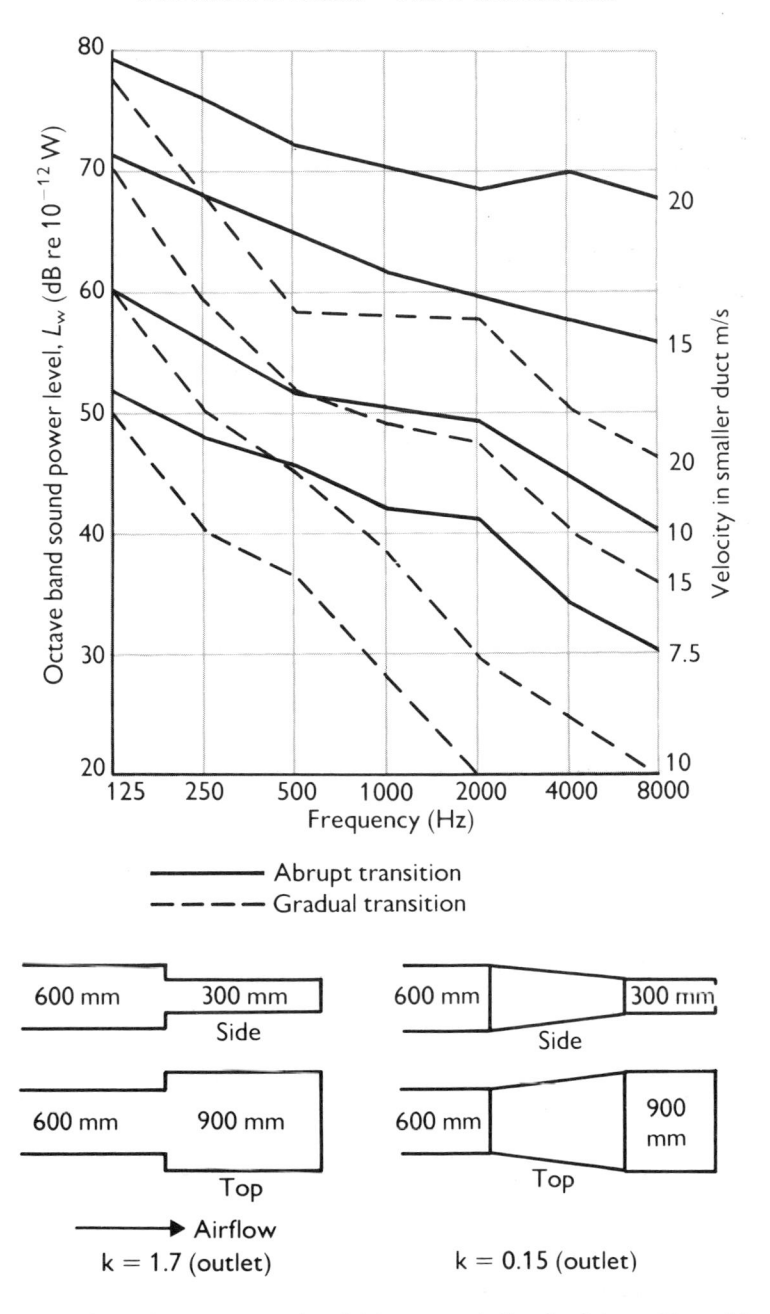

Figure 8.15c   Sound Power Levels of Abrupt and Gradual Area Transitions from
600 mm x 600 mm to 900 mm x 300 mm Duct Cross Section (4:3).

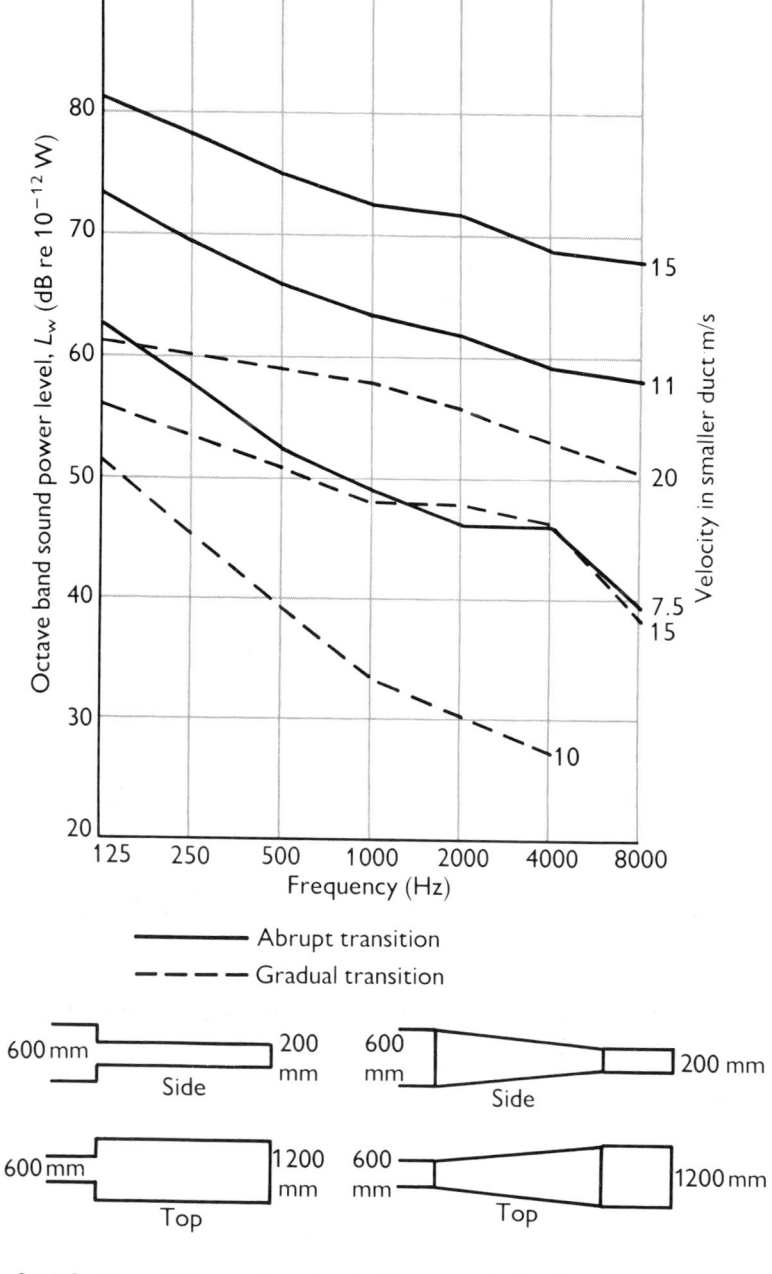

Figure 8.15d Sound Power Levels of Abrupt and Gradual Area Transitions from 600 mm x 600 mm to 1200 mm x 200 mm Duct Cross Section (1.5:1).

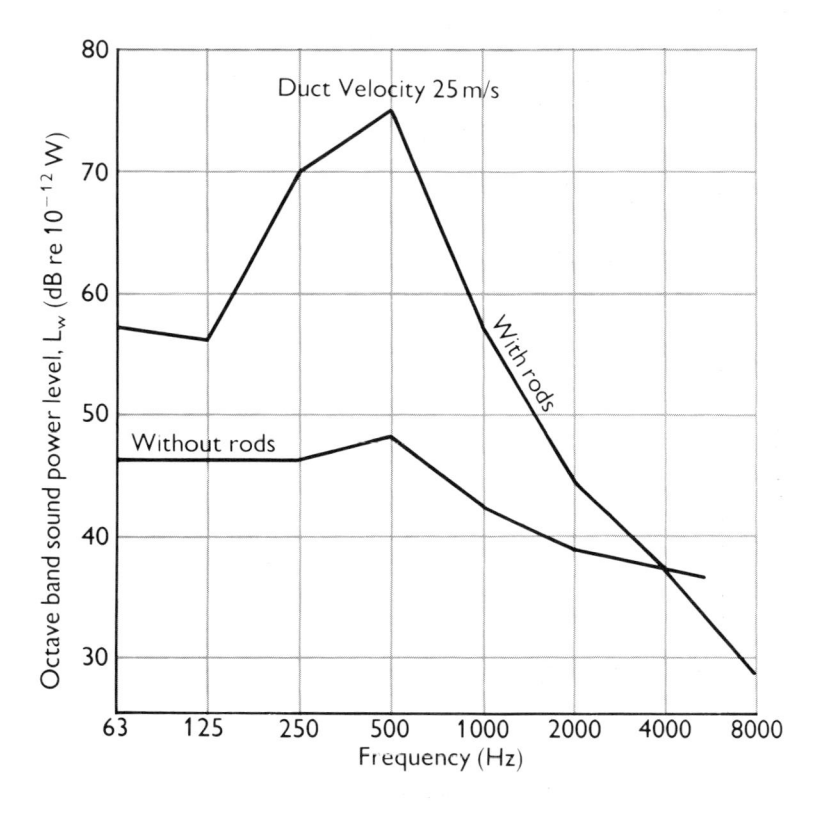

Figure 8.16   Sound Power Spectra of Noise from a 600 mm x 600 mm Duct with and without 4 off 13 mm Dia. Rods

Q = volume flow rate, m^3/s S = cross sectional area of the duct, m^2
ΔP = total pressure loss, Pa D = duct height normal to the damper axis, m
K_D = the characteristic spectrum which is given by:

$$K_D = 5.0 - 10.7\log_{10}(\text{St}) \text{ for St} < 25 \qquad\qquad 8.23$$

$$K_D = 40.2 - 35.9\log_{10}(\text{St}) \text{ for St} > 25 \qquad\qquad 8.24$$

(N.B. a function of band centre frequency f_0)

where St is the Strouhal Number given by St = $f_0 D/U_c$

8.3.5.1 Example To illustrate the use of the prediction scheme to determine the sound power level of a multi-blade damper positioned in a 0.305 m x 0.305 m duct to drop the pressure by 125 Pa at a volume flow rate of 1.9 m^3/s.

Input data	Symbol	Values
Volume flow rate	Q	1.9m^3/s
Total pressure loss	ΔP	125Pa
Duct cross section	S	0.093m^2
Duct height	D	0.305m

Step 1 Total Pressure Loss Coefficient, C
$C = (1.67 \times 125)/[(1.9/0.093)^2] = 0.479 \, C \neq 1$

Step 2 Blockage Factor, BF
$\text{BF} = (0.479^{1/2} - 1)/(0.479 - 1) = 0.615$

Step 3 Constricted Velocity, U_c
$U_c = 1.9/(0.093 \times 0.615) = 33.2 \, \text{m/s}$

Step 4 Strouhal Number
St = $f_0 \times 0.305/33.2 = 0.0092f_0$

Step 5 Characteristic Spectrum

Calculate the characteristic spectrum and enter it into the calculation sheet as shown below.

f_0, Hz	31	63	125	259	500	1k	2k	4k	8k
St	0.28	0.58	1.15	2.3	4.6	9.2	18.4	36.7	73.5
K_D	10.8	7.5	4.3	1.1	−2.1	−5.3	−8.5	−16.0	−26.8
10 log (f_0)	15	18	21	24	27	30	33	36	39
50logU$_c$	76	76	76	76	76	76	76	76	76
10log(S) + 10logD	−15.5	−15.5	−15.5	−15.5	−15.5	−15.5	−15.5	−15.5	−15.5
Constant	−18.	−18.	−18.	−18.	−18.	−18.	−18.	−18.	−18.
$L_{w_{oct}}$ (f_0), dB*	68.3	68.0	67.8	67.6	67.4	67.2	67.0	62.5	54.7
Rounded $L_{w_{oct}}$	68	68	68	68	67	67	67	63	55

* Add all the rows from below the double line.
i.e., a fairly flat spectrum of pink noise as per the grille alone in Figure 8.26.

OPPOSED BLADE DAMPER

Another potential source of flow generated noise which is very dependent on the flow duty required. High frequency whistles can be produced near to the fully shut-off setting.

Courtesy: Trox

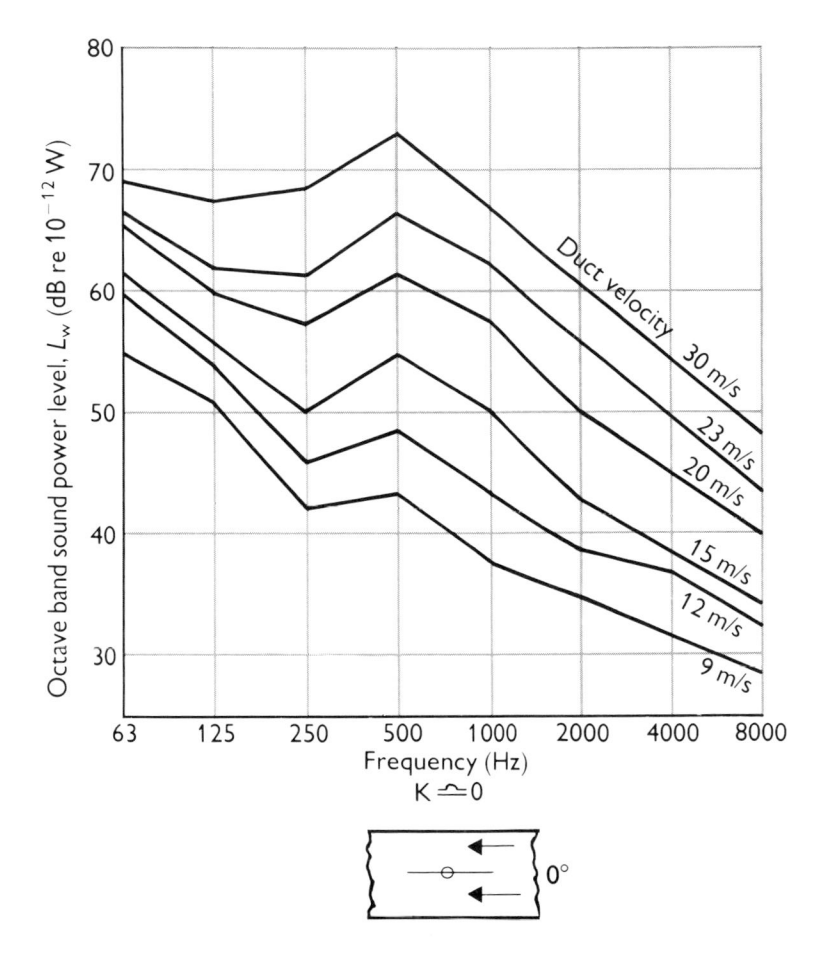

Figure 8.17a   Sound Power Spectra Produced by a Single Blade Butterfly Damper in a 600 mm x 600 mm Duct at Three Angles, *k* is the Aerodynamic Loss Factor

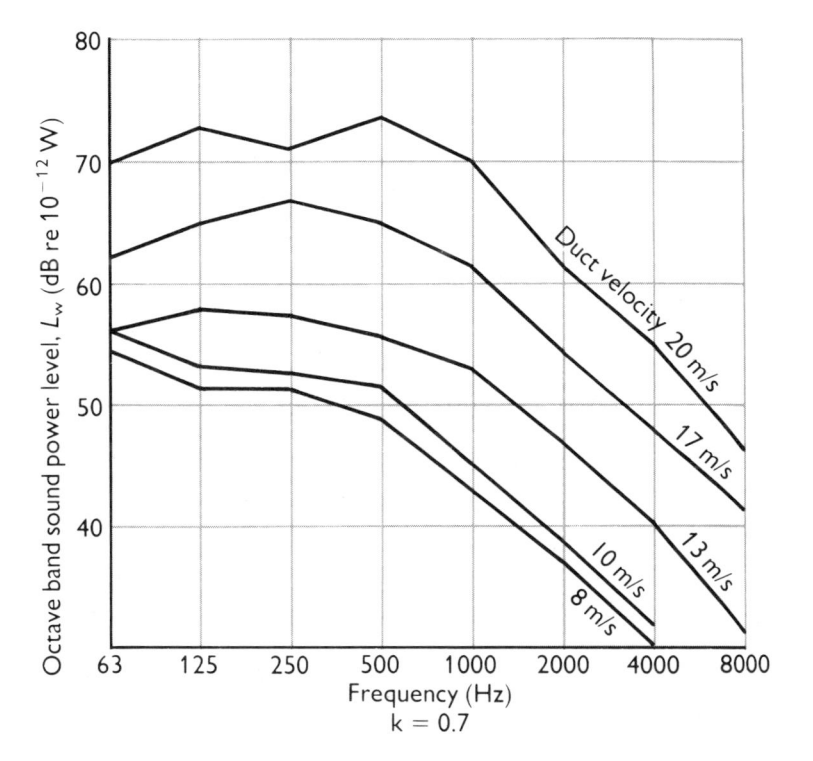

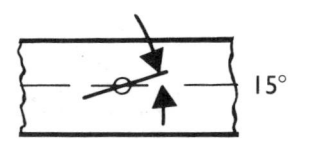

Figure 8.17b Sound Power Spectra Produced by a Single Blade Damper Butterfly
Damper in a 600 mm x 600 mm Duct at Three Angles, *k* is the Aerodynamic Loss
Factor

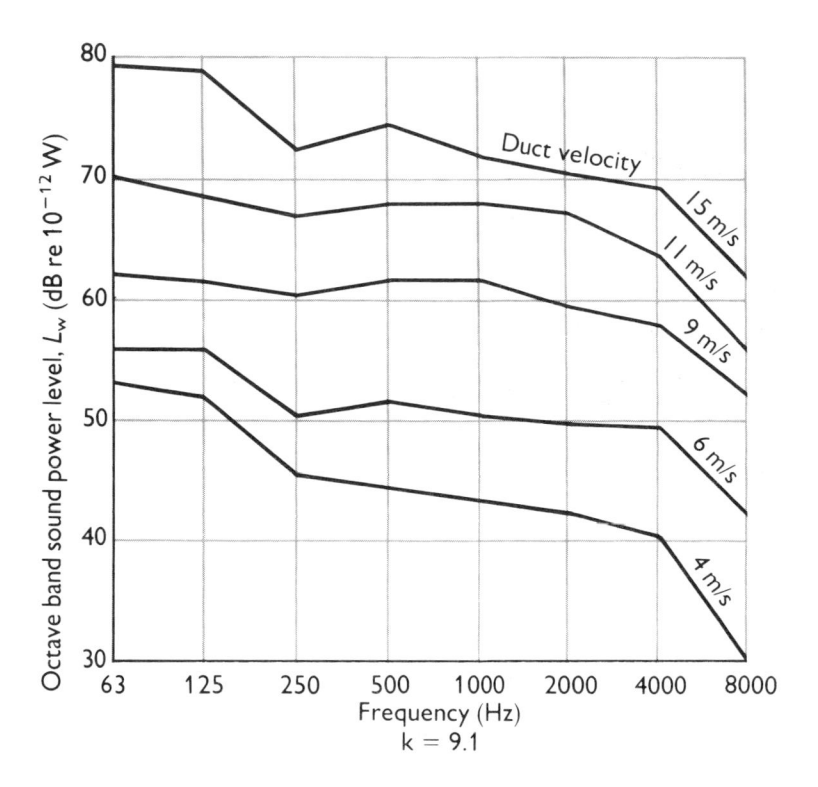

Figure 8.17c Sound Power Spectra Produced by a Single Blade Damper Butterfly Damper in a 600 mm x 600 mm Duct at Three Angles, *k* is the Aerodynamic Loss Factor

Figure 8.17d   A Single Blade Butterfly Damper

8.3.5.2 Damper/Damper Interaction Aithough only limited data is available, the following observations may be taken as being generally true:

1. The level of flow-generated noise increases when two duct elements are closely spaced rather than being in relative isolation.
2. The increase in flow generated sound power level is frequency dependent. The effect is enhanced at frequencies for which the acoustic wave length is similar to the element separation distance. The effect is suppressed for frequencies for which the acoustic wave length is equal to or twice the duct diameter.
3. The magnitude of the increase in the flow generated sound power level is a function of the pressure loss coefficient of the upstream element. This means that the increase is almost independent of the speed of airflow.
4. The increase in flow generated sound power level is a function of both the upstream element pressure loss coefficient and the separation distance. For element separation of more than five duct diameters only the most severe upstream obstructions will have any effect.

8.3.6 Attenuators

Figure 8.18a indicates that a duct mounted splitter attenuator is a series of transitions which may be abrupt and a source of flow generated noise. Figure 8.18b shows an example of an attenuator. Manufacturers of these unit attenuators supply catalogued information in terms of the face velocity (Figure 8.19) or in terms of the air way velocity together with a spectrum correction (Figure 8.20).

Individual manufacturers are the best suppliers of figures for flow generated noise of their own products. For the preliminary design calculation, however, a good approximation can be calculated from the following equation for the overall or total sound power level.

$$L_w \text{ (total)} = 55\log_{10}V + 10\log_{10}N + 10\log_{10}H - 45 \text{ dB} \qquad 8.25$$

where V is the local velocity in the splitter airway, m/s;
　　　N is the number of airways; and
　　　H is the airway height (rectangular) or circumference (cylindrical), mm.

Distribution of the energy over the frequency spectrum can be obtained by subtracting the following from the overall sound power level.

Octave band centre (Hz)	63	125	250	500	1k	2k	4k	8k
Frequency correction for octave band sound power level, dB	−4	−4	−6	−8	−13	−18	−23	−28

8.3.7 Grilles, Diffusers and Associated Dampers

Flow noise generation at terminal units, grilles, diffusers, induction units and the like, is of critical importance. Unlike the noise of in-duct elements, there is no opportunity to attenuate energy produced by the terminal unit before it reaches the occupant of the same room.

Two main types of diffuser exist:

1. Circular ceiling diffusers supplied through circular ductwork with a neck area of A_N with or without plenum boxes.
2. Linear and rectangular grilles, registers, or slots fed from plenum boxes and having a free discharge area A_F.

For both designs the radiated sound power level is given by the general equation:

$$L_w = \text{(Constant)}_1 + 57\log_{10}Q - 47\log_{10}A \text{ dB} \qquad 8.26$$

where Q = volume flow rate, m^3/s
and A = area, A_N or A_F, m^2.

This will apply for the overall total radiated sound power level or to the corresponding octave bands. The "Constant" term will vary from design to design but for each "family" of grilles may be determined by experiment (N.B. different values for each octave band).

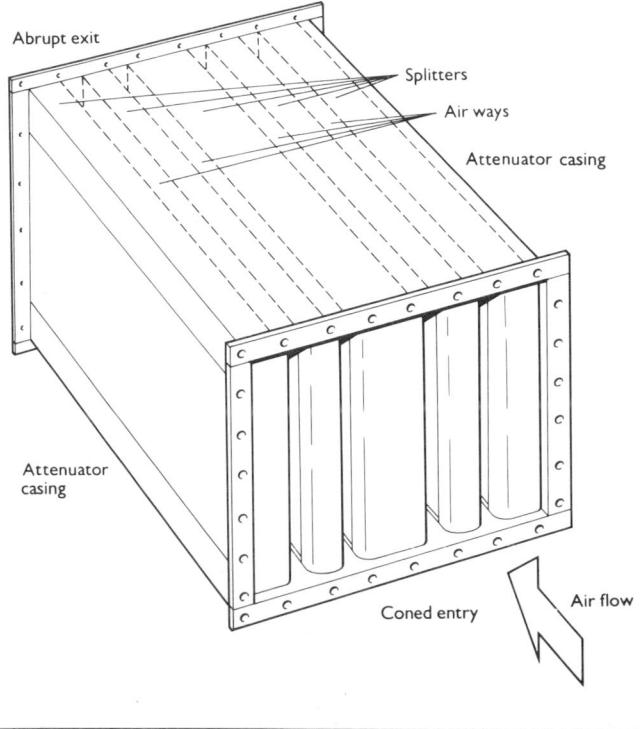

Abrupt exit

Splitters

Air ways

Attenuator casing

Attenuator casing

Coned entry

Air flow

Figure 8.18 Rectangular Attenuator

A typical value for the total radiated sound power level is given by:

$$L_w \text{ (total)} = 32 + 57\log_{10}Q - 47\log_{10}A \text{ dB} \qquad 8.27$$

where Q = volume flow rate, m^3/s
and A = area, A_N or A_F, m^2.

The formula indicates that the radiated sound power level is very sensitive to changes of flow rate or grille area and approximately proportional to:

(i) (Flow rate)^6, that is 18 dB per doubling of flow rate
(ii) $1/(area)^5$ for a fixed volume flow rate, that is 15 dB reduction in L_w from a grille of doubled area.

A similar expression relates the total radiated sound power level to neck flow rate V.

$$L_w \text{ (total)} = 32 + 60\log_{10}V + 13\log_{10}A \text{ dB} \qquad 8.28$$

where V is the neck (or maximum) flow velocity in m/s.

A general spectral distribution is indicated in Figure 8.21 for circular diffusers, which may be applied directly to this over all sound power level.

When the duty of a diffuser is specified by the volume flow rate Q and the pressure drop Δp the general equation becomes

$$L \text{ (total)} = (\text{Constant})_2 + 10\log_{10}Q + 26\log_{10}\Delta p \text{ dB} \qquad 8.29$$

Octave band Hz	125	250	500	1000	2000	4000	8000
Attenuator face velocity m/s	Sound power levels in decibels						
5	32	36	34	31	32	29	21
7.5	47	45	43	40	42	40	34
10	54	52	50	47	48	47	44
15	64	60	58	56	58	59	57
20	72	67	65	63	64	67	67

Figure 8.19 Attenuator Flow-generated Noise

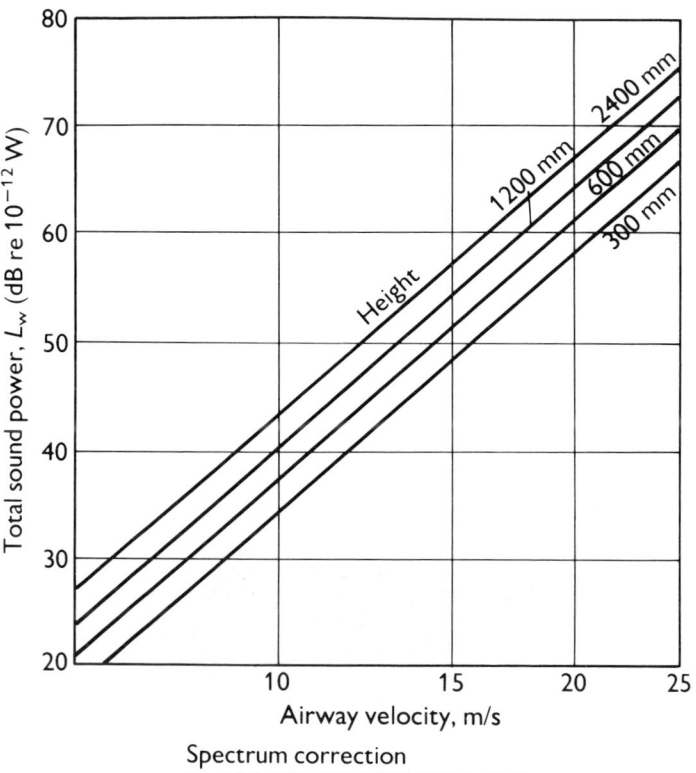

Figure 8.20   Attenuator Flow-generated Noise

When allowance has been made for the spectral shape of a typical rectangular or linear grille the equation may be expressed in terms of the "NC sound power level" (NC_{LW}).

$$NC_{LW} = 26\log_{10}\Delta P + 10\log_{10}Q \qquad\qquad 8.30$$

where ΔP = grille pressure drop in Pa
and Q = volume flow rate in m³/s.

JET NOZZLE

Long throws can be achieved quietly with correctly selected diffusers of this type.
Courtesy: Trox

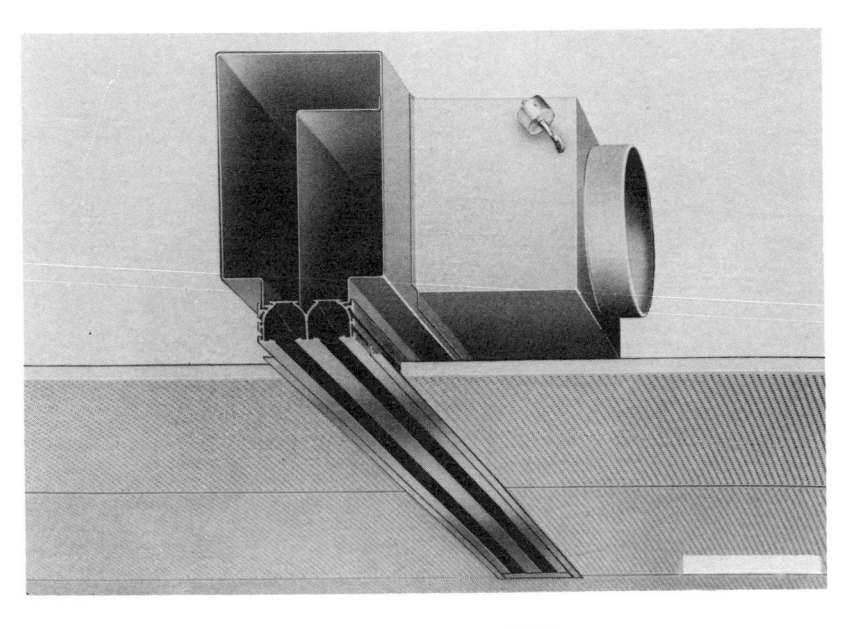

LINEAR DIFFUSER

Catalogue selection from reliable data will ensure that flow noise is not a problem with linear diffusers. However, good header box design is essential to ensure that these data are achieved in practice. This design incorporates a novel biased flap valve arrangement, which will demand access to catalogued performance data.

Courtesy: Trox

FLOOR DIFFUSER

Clever designs of floor grilles send air spinning up into the room with enhanced secondary air entrainment and with potentially low noise levels.

Courtesy: Trox

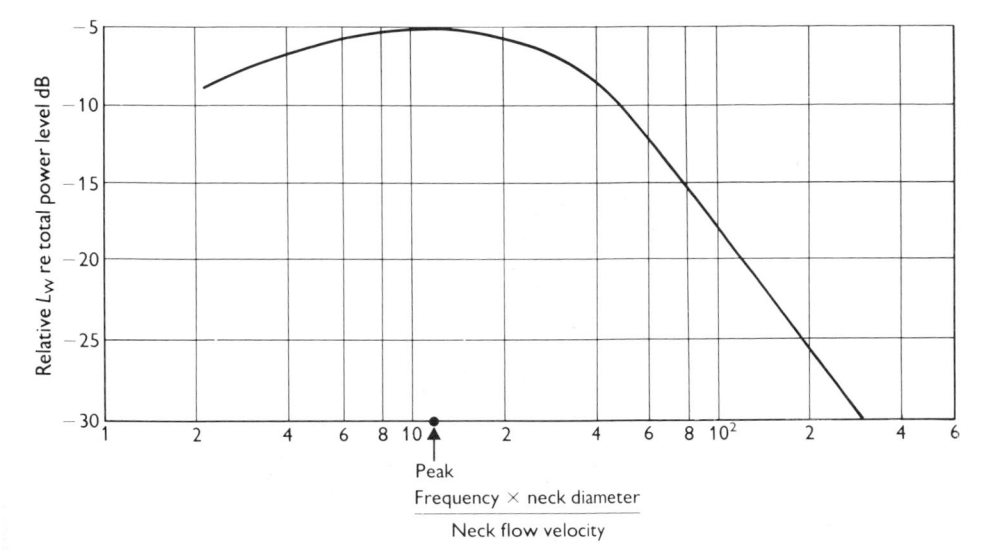

Figure 8.21   Spectral Distribution for Circular Diffuser

For site applications, the room correction must be deducted from this NC_{LW} to obtain a representative NC sound pressure level.

A measured representative spectrum shape is shown in Figure 8.22 indicating the comparatively stable shape at various flow rates, the noise generation here being mainly from the diffuser grid, bars or blades.

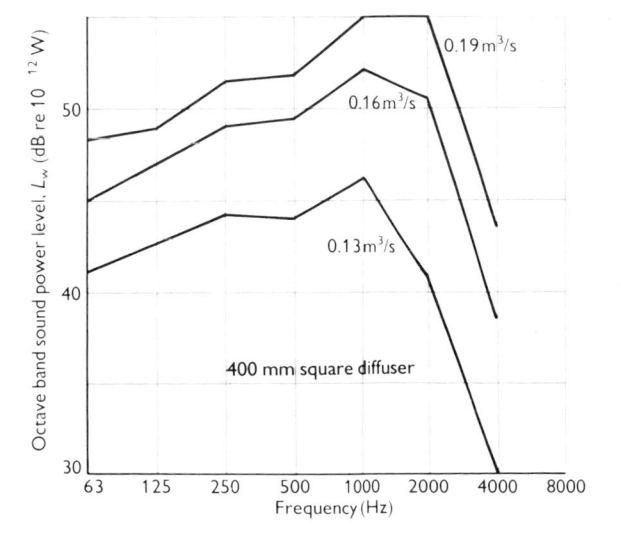

Figure 8.22   Diffuser Sound Power Level

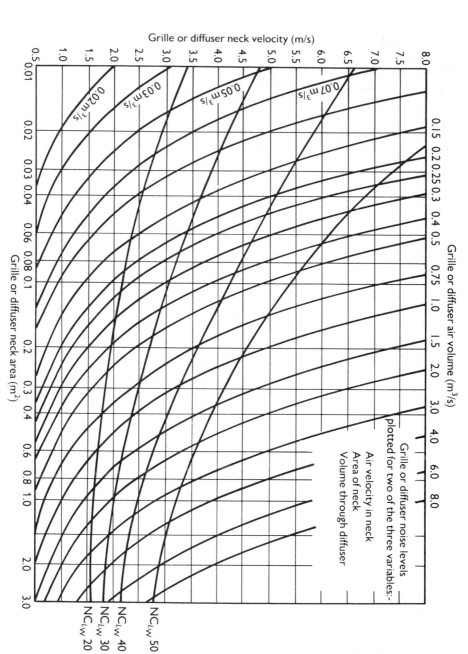

Figure 8.23 Diffuser/grille — Flow-generated Noise Guidance

Figure 8.23 indicates the radiated sound power levels with respect to the NC curves (NC_{LW}) for representative circular ceiling diffuser units. These figures require deduction of the room correction to obtain representative NC sound pressure levels.

8.3.7.1 Effect of Throw. Figure 8.24 indicates the change in radiated sound level as diffusers are designed to deflect the air to produce a wider distribution. It indicates the increased level from wide throw grilles, an increase in the region of 20 dBA resulting from a demand for wide distribution, albeit with increased pressure loss.

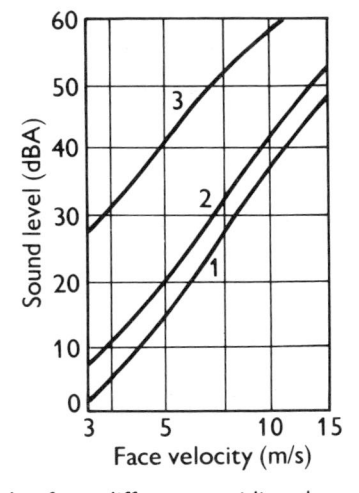

Curve 1 is for a diffuser providing almost no spread
Curve 2 is for a type giving a small spread
Curve 3 is for a wide throw diffuser

Figure 8.24   Effect of Throw on Sound Level.

8.3.7.2 Effect of Dampers. Fitting any sort of control damper to a grille or diffuser will add two more noise sources. First there will be the vortex shedding noise component from the damper blades themselves.

Secondly, the impact of the damper blade turbulence wake, see Figure 8.1, upon the grille or diffuser vanes, will generate noise as if the outlet grille is the prime source, as illustrated in Figure 8.25. The second is by far the most important and whilst removal of the grille in such situations may reduce the noise level, this grille should not always be identified as the prime "culprit".

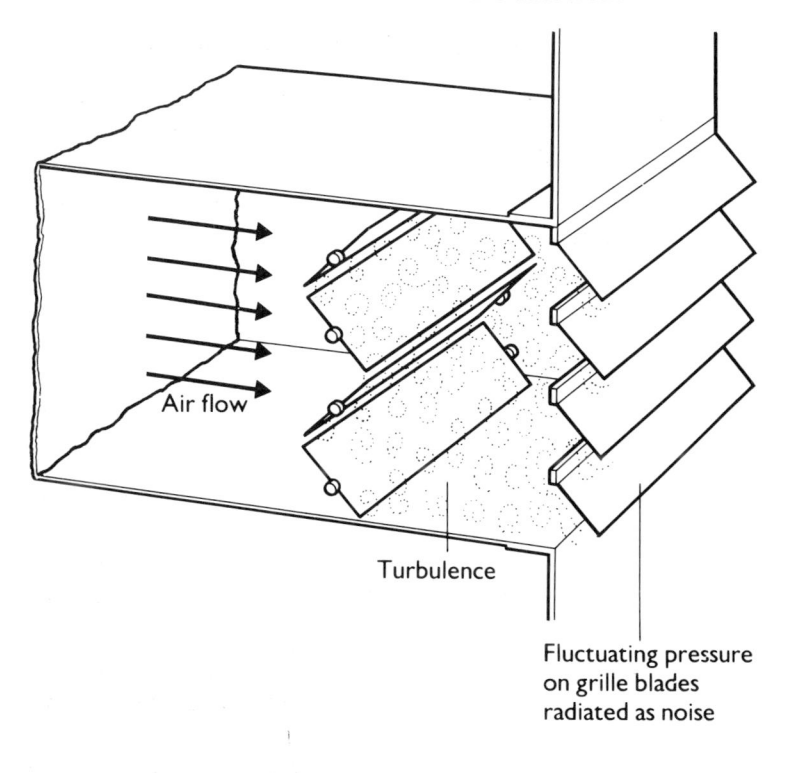

Figure 8.25　Noise Generated by Damper/grille Combination

In current grille design, it is the practice to manufacture the damper as a multi-leaf opposed-blade pack which clips on to the rear of the grille. The system has a number of advantages, of course, including the ability to balance flow through a number of grilles easily, adjusting the damper by turning a screw reached through the front face of the grille. Unfortunately, the system also pays a penalty, sometimes a heavy one, in additional noise generation.

Nearly all manufacturers of grilles and diffusers can supply quite detailed figures for the sound power level produced for their grille/damper assemblies, as a function of either face velocity or volume flow rate, or more usually, as a function of pressure drop across the unit.

In the absence of such information at the design stage, however, one can assume an increase in the above estimated sound power level, of 5 dB in each octave band for the addition of an integral damper vane set fully open.

For subsequent flow adjustment, that is an increase in pressure drop across the damper, it can be assumed that the overall sound power level will increase approximately according to the following equation.

$$L_w = 33\log_{10}(\Delta p / \Delta p_0) \text{ dB (correction term)} \qquad 8.31$$

where Δp is the new static pressure drop across the unit and
Δp_0 is the initial pressure drop with damper pack added but vanes fully open, with the same volume flow rate.

The overall and octave band radiated sound power levels of an isolated damper approximately obey the following equation:

$$L_w \text{ (total)} = \text{(constant)} + 10\log_{10}Q + 26\log_{10}\Delta p \qquad 8.32$$

where Q = volume flow rate, m³/s; and
Δp = working throttled pressure drop, Pa.

However, as it is a variable geometry device, its radiated sound spectrum distribution is not constant. See previous section (8.3.5).

The frequency of maximum sound intensity will rise as the damper is throttled, the increased velocity between the partly closed vanes controlling the shift to a greater extent than the larger "bluff" area presented to the air flow by the blades. This is shown in Figure 8.26 for a fixed supply rate as the damper is closed against increased duct pressure.

Figure 8.26 also indicates the interaction which occurs between the fully open damper and grille due, in this case, to the damper blades being in the wake of the turbulence from the grille vanes. The shift in the spectral maximum towards higher frequencies modifies the effect on an NC rating system. The NC level becomes more sensitive to pressure drop and the correction term must be applied to the above equation, 8.32.

For a variable damper a representative NC equation becomes:

$$NC_{L_w} = 31 + 33\log_{10}(\Delta P / \Delta P_0) + 26\log_{10}\Delta P + 10\log_{10}Q \text{ dB} \qquad 8.33$$

where Q = volume flow rate, m³/s;
ΔP_0 = minimum open damper pressure drop, Pa; and
ΔP = working throttled pressure drop, Pa.

Manipulating this equation it becomes:

$$NC_{LW} = 31 - 33\log_{10}\Delta P_0 + 59\log_{10}\Delta P + 10\log_{10}Q \qquad 8.34$$

For any given design, the term $-33\log\Delta P_0$ is a constant and the equation reduces to

$$NC_{LW} = \text{constant} + 59\log_{10}\Delta P + 10\log_{10}Q \qquad 8.35$$

This shows the relative importance of pressure drop and the volume flow rate and demonstrates that the flow generated noise is very sensitive to changes in pressure drop.

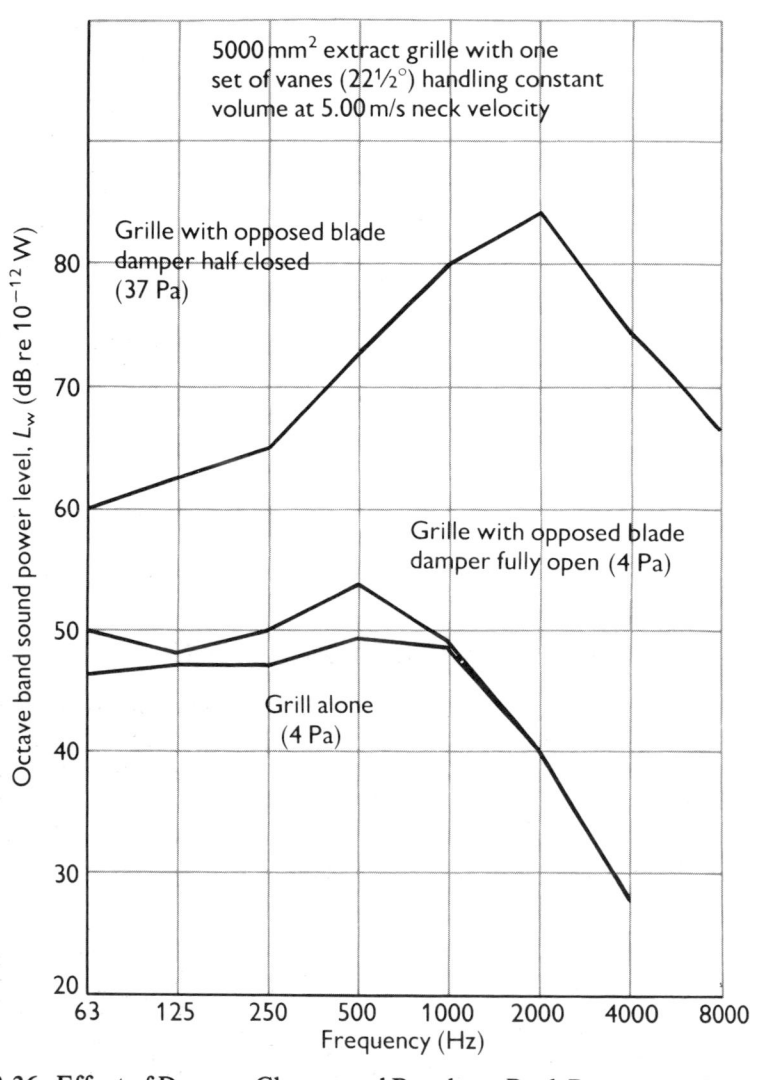

Figure 8.26 Effect of Damper Closure and Resultant Back Pressure on Noise Level

This equation indicates that a damper, suitably designed with a large minimum pressure loss, may generate a lower NC_{LW} for a given air flow duty. This basic principle can be carried across to single plate dampers, constant flow rate controllers and variable flow rate controllers.

Figure 8.27 shows a set of typical data produced by a manufacturer for a ceiling diffuser with various volume flow rates and percentage open areas. Many manufacturers present this type of data in the form of nomograms and these vary widely from manufacturer to manufacturer.

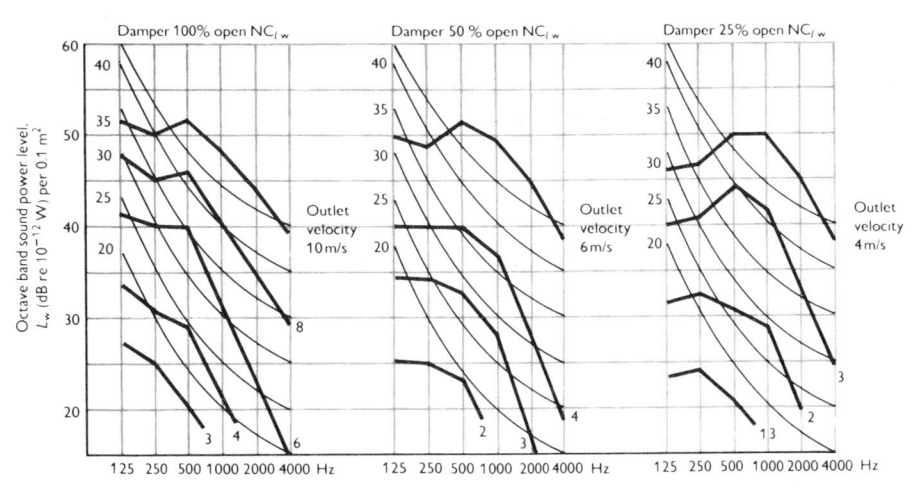

Figure 8.27   Ceiling Diffuser Damper — Sound Power Spectra

8.4 MULTIPLE AIR DISTRIBUTION DEVICES

When more than one air distribution device serves a room, the sound power levels generated by multiple sources and transmitted through multiple outlets must be accounted for in determining the room sound pressure level. The total sound power level for multiple air distribution devices in small rooms can be determined by adding the factor $10\log_n$, as shown in Figure 8.28, where n is the number of devices with equal sound power levels, to the sound power level for one device.

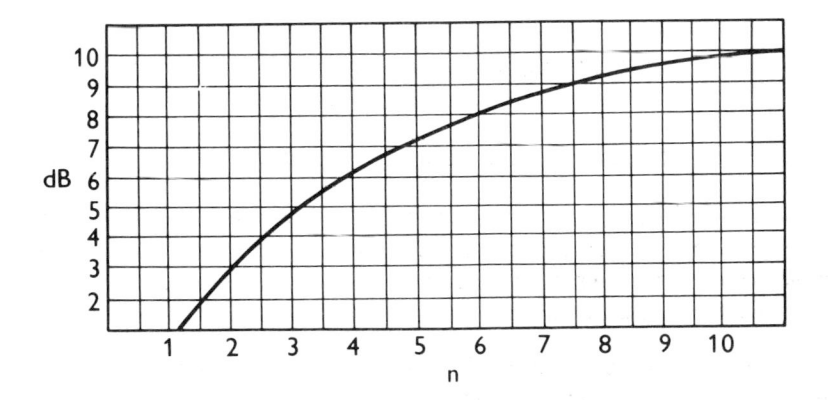

Figure 8.28   Graph of $10\log n$

This factor is valid for the reverberant field in small rooms only, not for large low ceiling rooms where the receiver may not be affected by all the devices in the room.

In multiple outlet air distribution systems, great care has to be taken to ensure that flow generated noise from one outlet does not travel along the duct to a more noise sensitive area. Figure 8.29 shows a situation where a store room which requires little

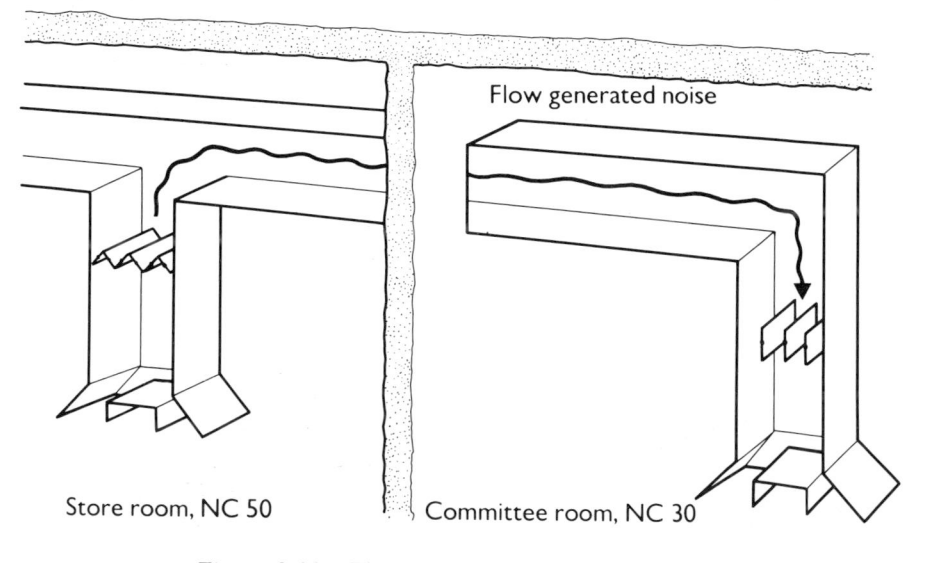

Figure 8.29   Flow-generated Cross-talk noise

air, and has its dampers virtually closed, generates a high level of flow generated noise that is not a problem with regard to NC50. However, this noise can travel down the duct to the committee room, which has open dampers and becomes a problem in the light of the NC30 criteria.

8.5 ROOM TERMINAL INSTALLATION FACTORS

The sound level output of an air diffuser or grille depends not only on the air quantity and the size of the outlet, but also on the air approach configuration as has already been illustrated and analysed for close damper to grille configuration. Manufacturers' ratings apply only to outlets installed as recommended, that is with a uniform air velocity distribution throughout the neck of the unit. Poor approach conditions can easily increase sound levels by 10 to 20 dB above the manufacturers' ratings, see Figures 8.30 and 8.31. The consequences of poor approach conditions can be mitigated by the use of properly adjusted accessories such as turning vanes or equalizing grids. However, using these in critical, low noise level projects should be avoided.

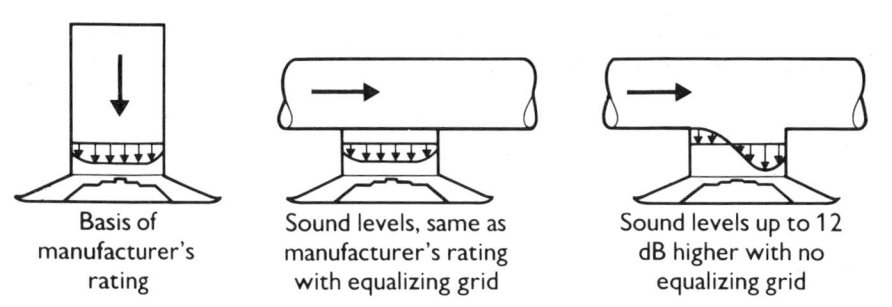

| Basis of manufacturer's rating | Sound levels, same as manufacturer's rating with equalizing grid | Sound levels up to 12 dB higher with no equalizing grid |

Figure 8.30   Correct and Incorrect Air Flow Conditions to an Outlet

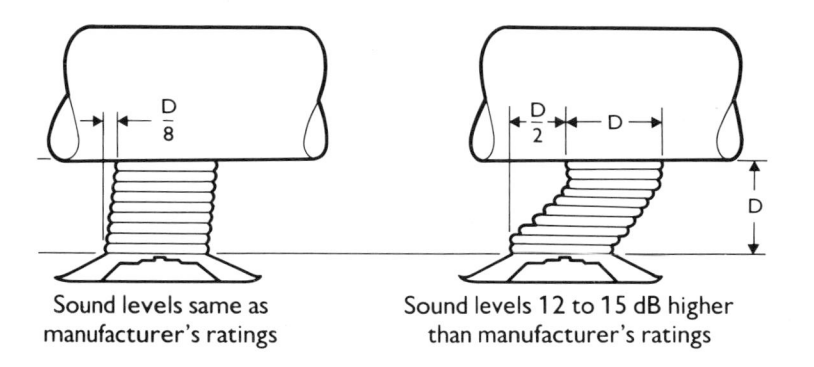

| Sound levels same as manufacturer's ratings | Sound levels 12 to 15 dB higher than manufacturer's ratings |

Figure 8.31   Effect of Correct and Incorrect Alignment of Flexible Duct Connector

Flexible duct is often used to correct misalignment between the supply duct and the diffuser ceiling location. A misalignment or offset that exceeds one-fourth of a diffuser diameter in a diffuser collar length of two diameters significantly increases the diffuser sound level, see Figure 8.31. There is no appreciable change in diffuser performance with an offset less than one-eighth the length of the collar.

In acoustically critical spaces, such as concert halls, it is necessary to avoid using balancing dampers, equalizers and so forth directly behind terminal devices or open end ducts. They should be located five to ten duct diameters from the opening, followed by lined duct to the terminal or open duct end. Linear diffusers are often installed in distribution plenums that permit installing the damper at the plenum entrance. The further a damper is installed before the outlet, the lower the resultant interaction sound level will be.

8.6 INDUCTION UNITS

Induction units are rather less amenable to a general estimate of the noise they produce. The generating mechanism itself is basically one of free jet noise from the nozzles, but with the possibility of additional output from turbulent flow impingement on solid surfaces. If free jet noise is the only source, its sound power outlet may be estimated from a design chart such as Figure 8.32. Note here that the

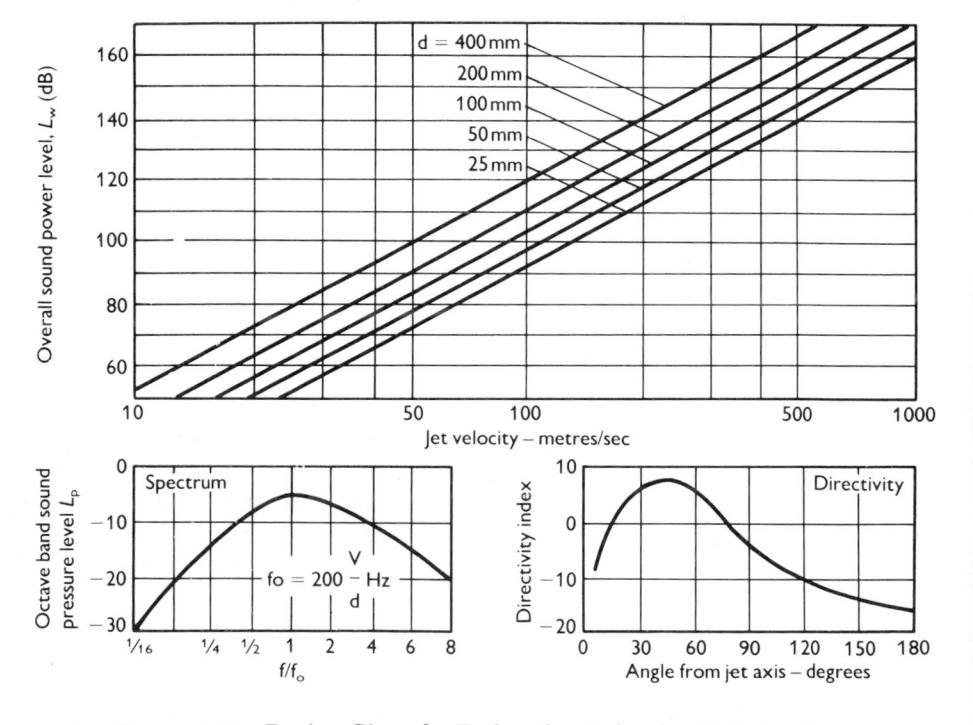

Figure 8.32   Design Chart for Estimating Induction Unit Jet Noise

power level estimated applies to one jet or nozzle. For the total jet noise from assembly within a unit, an addition of $10\log_{10}N$ dB should be made, where N is the number of nozzles. Note also that the directivity of the free jet will not apply in general, being affected by the casing of the unit to a very marked degree.

In addition to the noise of the free jets, additional sources will be present in most induction units. Any damper into the unit plenum will produce noise as discussed earlier in this chapter, and secondary generation at the air entry is to be expected. Sound power from both of these sources, however, may be expected to be reduced by some 15 dB across the spectrum before emerging from the unit to the room as a result of nozzle attenuation.

Additional noise of the turbulence interaction type may be expected if the primary nozzle jets are allowed to impinge upon any part of the casing of the unit, or if there is a protective grille placed directly in line. Under these conditions about 5 dB should be added across the spectrum.

It should always be the aim to obtain, from the manufacturer, induction unit noise levels which appertain to the particular design to be used, and which are based upon test figures for that design.

8.7 GENERAL GUIDELINES FOR SELECTING DUCT FITTINGS FOR LOW FLOW-GENERATED NOISE

Flow-generated noise in duct fittings can be reduced substantially by choosing geometries which minimise the generation of vortices and the interaction of the vortices with sharp edges. The information provided below is aimed to guide the duct designer in choosing the less noisy alternative.

8.7.1 Cross Sectional Area Changes

As depicted in Figure 8.33 one should, whenever possible, avoid cross sectional area changes where flow separation occurs. The transition angle should not be larger than 10° to minimise the risk of flow detachment.

If the available transition length L_{max}, is insufficient to obtain less than a 10° transition angle in critical situations, it is advantageous to apply a perforated plate which provides enough resistance to diffuse the flow evenly into the larger duct. This is usually accomplished if the pressure drop across the perforated plate is in the range of one quarter to one half of the local velocity head. The high frequency noise generated by the perforated plate is easily attenuated by duct lining downstream of the area change.

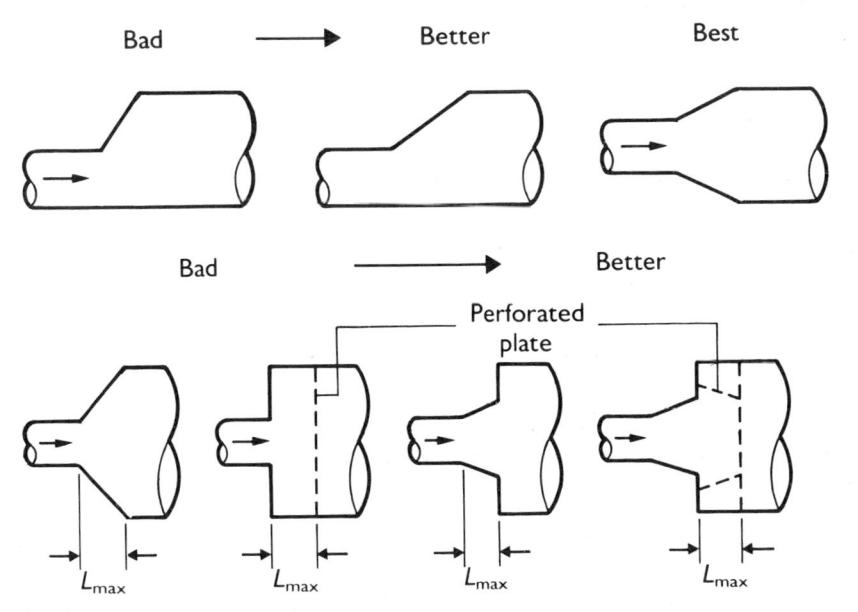

Figure 8.33 Guidelines for Minimizing Flow-generated Noise in Cross Section Changes

8.7.2 Bends

As shown in Figure 8.34, mitred 90° bends are noisier than radiused bends without turning vanes. Accordingly, in critical situations, narrow chord and large radius bends are preferable (see pages 232 and 233).

8.7.3 Branchings and Take-offs

As shown in Figure 8.35, flow generated noise at branchings and take-offs can be reduced by using duct fittings which provide a smooth flow and by avoiding a change of velocity pressure, especially a reduction of the velocity pressure which is usually accompanied by flow separation. Consequently, branches of a Tee should have the same or less cross sectional areas as the main duct, that is, $d_1 \geqslant 1.4d_2$.

Avoid Cross and Tee junctions where the volume flow in one branch, V_3 is substantially smaller than the volume flow in the other branches V_2. Instead, use the construction depicted in Figure 8.35d. The branch take-off should be more than five duct diameters downstream of the main junction so that the turbulence generated in the main junction decays sufficiently until it reaches the downstream branching with the small volume flow.

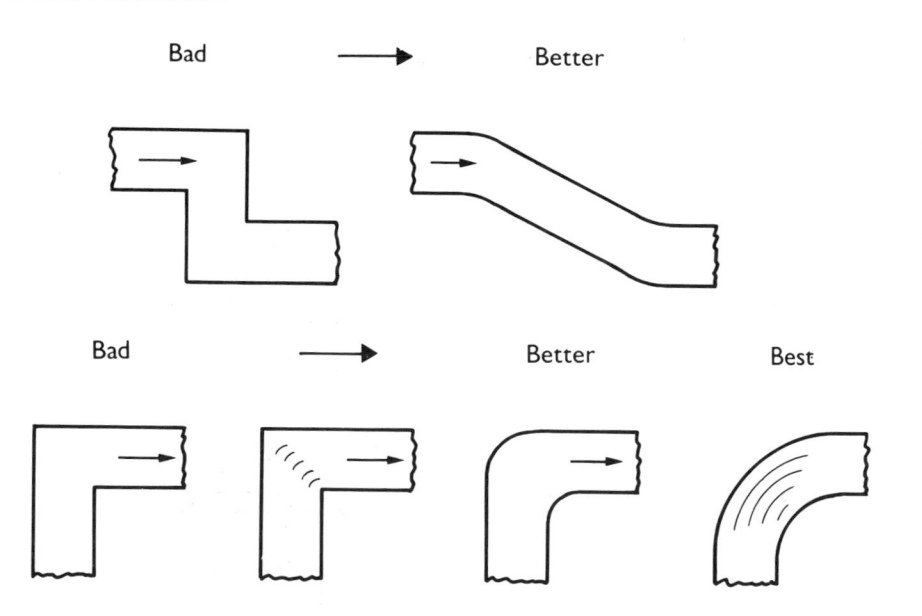

Note however, that the mitred bend without turning vanes yields the highest reflection loss for the incident sound.

Figure 8.34   Guidelines for Minimizing Flow-generated Noise in Duct Bends.

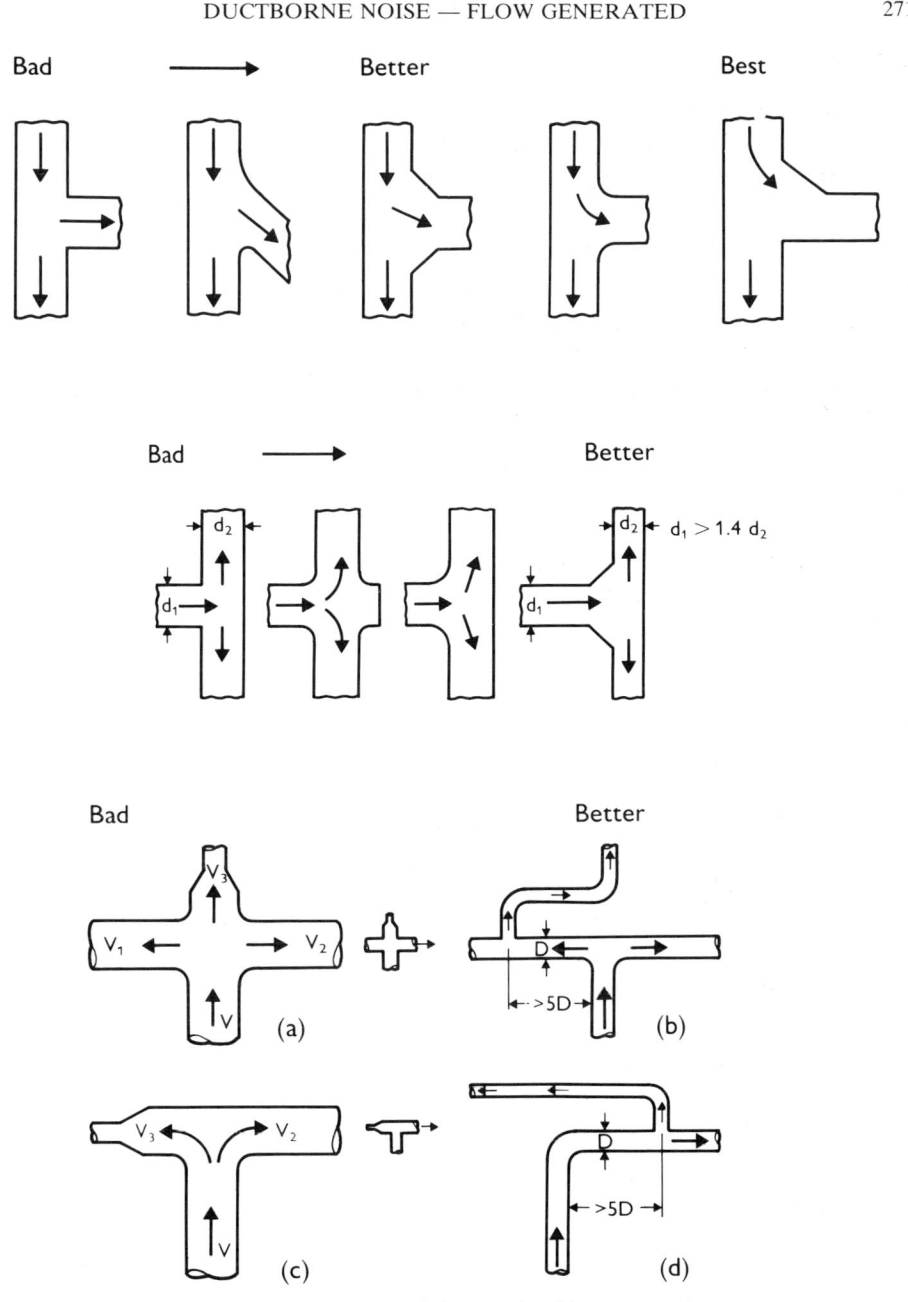

If branch volume flow V_1 or V_2 is much larger than V_3
then use a T-piece as illustrated above.

Figure 8.35 Guidelines for Minimizing Flow-generated Noise at Branchings and
Take-offs

8.7.4 Duct Buffeting

As illustrated in Figure 8.1 the whirlpools or vortices produced by objects or duct discontinuities, are alternately in opposite directions. If these are generated, or develop, to completely fill the duct dimensions a situation illustrated in Figure 8.36 will exist.

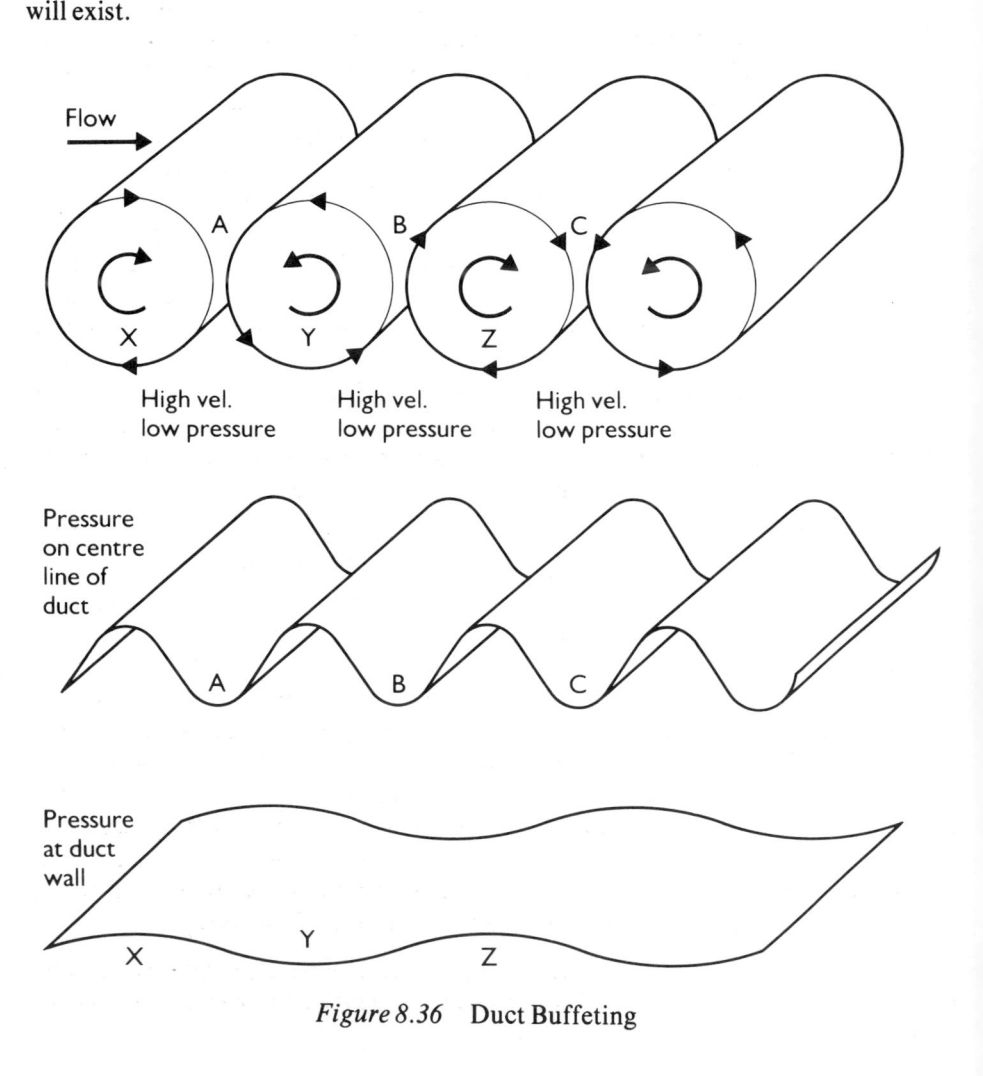

Figure 8.36   Duct Buffeting

At points A, B, C etc., there will be local maxima in the velocities which will be associated with pressure minima as shown for the pressure on the centre line. At points X, Y, Z, etc., the flow will be alternately speeded up and slowed down, producing pressure peaks at X and Z and minima at Y, as shown for the wall pressure. Both these pressure patterns will move along at the speed of flow and interact with the duct walls to produce buffeting motions of the ducts and low frequency sounds with harmonics. When changes of direction, corners or take offs are encountered, even more intense interactions are generated as the tangential whirlpool velocities attempt to impinge at right angles on the duct wall.

For a duct 500 mm (20 in) square, the distance between the pressure dips A, and B, will correspond to one fully developed whirlpool and hence be 500 mm or 0.5 m. If the flow speed is 20 m/s (4000 ft/min) these pulses will pass at a rate of 40 per second, i.e., 40 Hz. XY will correspond to 20 Hz and in both cases, there will be higher frequency harmonics, also from smaller whirlpools. Hence, the generation of intense low frequency "break out" from direct excitation of the duct walls by turbulence. The frequency will depend on the flow velocity and low velocities will carry less intense whirlpools, which will react much below the audible frequency range, e.g., for 2 m/s (400 ft/min) the frequency becomes 4 Hz and also carries less kinetic energy. This 4 Hz can be detected as vibration by the hand. Audible buffeting break out can be avoided by the use of low flow velocities carrying less turbulent energy and exciting low frequencies. For higher velocities, the adoption of adequate flow straighteners can minimise the build up of excessive turbulence. Multiple long chord turning vanes as splitters can act usefully in the prevention of turbulent excitation at bends. Take-offs, bends and transformations, may also excite turbulence when high flow rates (10 m/s) occur and smooth air flow paths will minimise excitation.

Duct fittings in high aspect ratio rectangular ducts often result in drumming if the velocity of the flow is high enough so that the localised turbulence generated by the duct fitting vibrates the duct wall with an amplitude which exceeds the bulging of the duct walls. Round ducts, which have a much higher form stiffness, are less subject to drumming. Drumming in high aspect ratio rectangular and oval ducts can be eliminated by partitioning the duct downstream of the fitting with plates orientated parallel to the smaller side wall, that is making many parallel small aspect ratio passages. The length of the partitioning should be five times the smaller side of the duct or longer, as shown, schematically in the top sketch of Figure 8.37. This may be an appropriate retrofit measure in cases where additional pressure drop cannot be tolerated. If there is sufficient reserve capacity in the system to overcome additional pressure drop, a perforated plate, as shown in the lower sketch in Figure 8.37, which breaks down the large size vortices, would be a preferable alternative.

The use of tie rods in high velocity situations, more appropriately in high aspect ratio oval ductwork, runs the risk of exciting resonant tones in the 200 to 600 Hz frequency range as illustrated earlier in Figure 8.16.

8.7.5 Duct Lining Downstream of Duct Fittings

Noise generated by duct fittings in both the main and branch ducts can be reduced effectively if the ducts connected to the duct fitting are lined as in Figure 8.38.

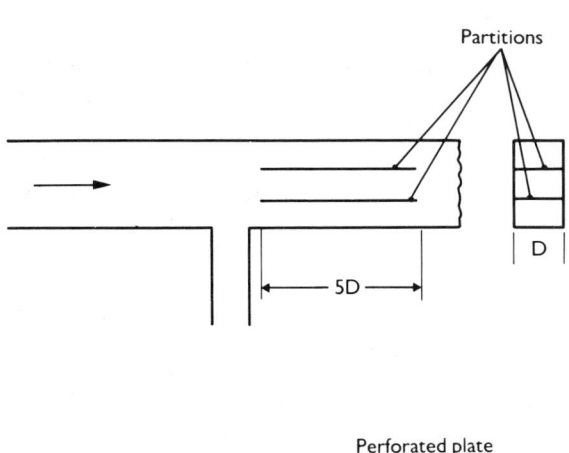

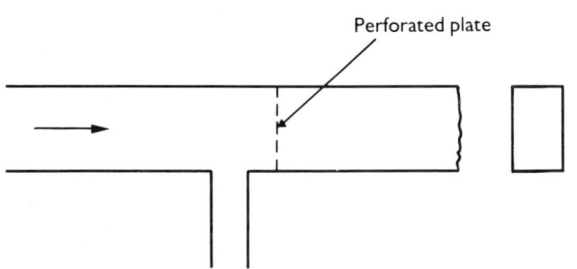

Figure 8.37   Measures to Reduce the Probability of Drumming in Large Aspect Ratio Rectangular Ducts

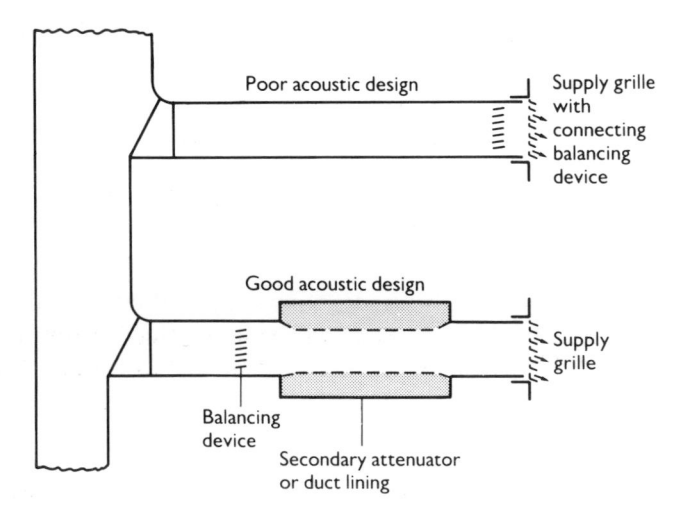

Figure 8.38   Secondary Attenuators

8.7.6 Organ Pipe-Type Resonances

Ducts that are terminated with a high sound reflection coefficient at both ends, such as branch ducts of small cross sectional area that are fed from a large main duct and tied to terminal devices, represent a highly resonant acoustic system. Such branch ducts have resonance frequencies where their length equals one half of the wavelength or its multiple. These longitudinal acoustic resonances can be excited by the turbulent flow at their junction with the main duct. This is a situation similar to the excitation of an organ pipe or flute.

To avoid the occurrence of organ pipe-type resonances, branch ducts of small cross sectional areas that originate from a large main duct and lead to terminal devices should be lined. The acoustic lining provides sufficient losses to prevent resonant amplification. The danger of resonant excitation can also be reduced by reducing the turbulent excitation of the junction by choosing the branch duct size so that the velocity in the branch is the same or greater than that in the feeder duct as shown schematically in Figure 8.39.

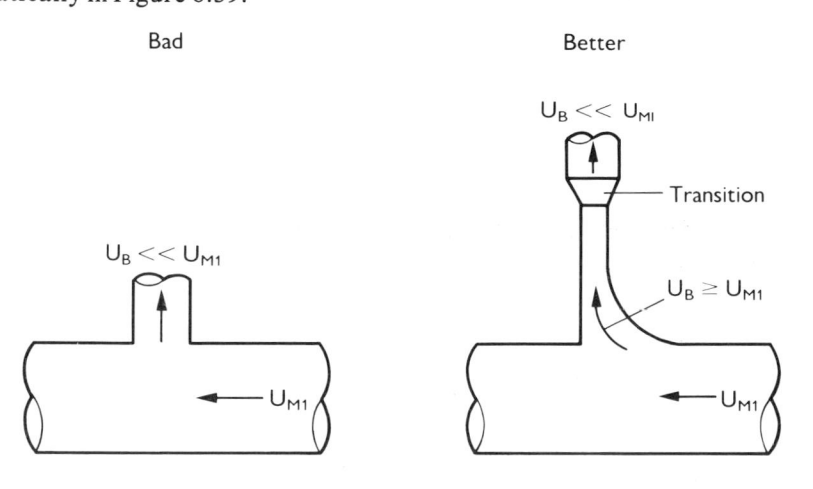

Figure 8.39 Guidelines for Reducing the Danger of Resonant "Organ Pipe" Excitation of Small Side Branch Ducts.

8.7.7 Flow Rates

The flow rates through grilles and diffusers must be kept sufficiently low. Accurate values for the radiated sound power levels should be obtained from specific manufacturers' data sheets.

Most important is to ensure that the approach conditions to the grille or diffuser supply a uniform flow distribution across the face of the unit. Local high flow spots will generate a disproportionate quantity of noise. If plenum boxes are employed, corresponding noise data must be employed for the entry conditions envisaged.

CHAPTER 9

DUCTBORNE NOISE — BREAKOUT

9.1 INTRODUCTION

Methods of selecting attenuators to control fan noise transfer through ducted systems have become fairly well established over the years and so problems from this source are comparatively rare. Rather more common, and unfortunately more difficult to solve, are problems arising from the "breakout" of noise through the walls of the ductwork along which it is passing.

Noise breakout may be due to three causes:

1. High levels of ductborne noise, most commonly fan noise.
2. Aerodynamic noise from the interaction of airflow on duct fittings.
3. Turbulent airflow interacting with duct walls and causing the duct walls to vibrate, and to radiate airborne noise.

9.1.1 Fan Noise Breakout

Conventional calculations (Chapter 7) to assess the level of primary attenuation to reduce fan noise to an acceptable level at the nearest conditioned area mainly assume that the duct system itself is capable of containing the noise on its path to the room.

Duct walls, like any acoustic barrier, will provide a level of sound insulation which is primarily dictated by the superficial mass of the wall material and, at low frequency, by the flexural stiffness of the walls themselves. Since these two factors are small in the case of rectangular sheet metal ductwork, it is frequently found that the total level of noise breaking out of duct systems into areas through which the system passes is higher than that transferred via the duct system itself to its final room outlet.

When selecting primary attenuation, therefore, it is essential that any possible noise breakout from duct systems should be considered so that noise from this source may be reduced to acceptable levels.

The problem generally occurs when a ductwork system serves areas remote from the plant room and passes over lightweight false ceilings in intervening rooms. If the duct run is considerable, and hence its natural attenuation is high, it may not be necessary to provide any primary or secondary attenuation for the rooms served by the system in order to meet design requirements. If this requirement is taken as the basis of the primary silencing need, the sound power levels in the first few feet of connecting ductwork may be sufficiently high to cause noise problems in the rooms below. In other words, the critical path for attenuator selection is not the path along the ductwork to the conditioned area but through the duct walls to the intervening rooms (Figure 9.1). Even if a primary attenuator is provided the breakout can still occur if the duct runs to critical areas are long.

Figure 9.1   Breakout where Little or No Primary Attenuation is Required

Breakout from ductwork can also occur if secondary attenuation at the outlet of the system, usually provided to control noise from balancing devices, is also relied on to control noise from fans and so reduce the primary attenuation requirement. This can lead to high in-duct sound power levels and consequent breakout problems (Figure 9.2).

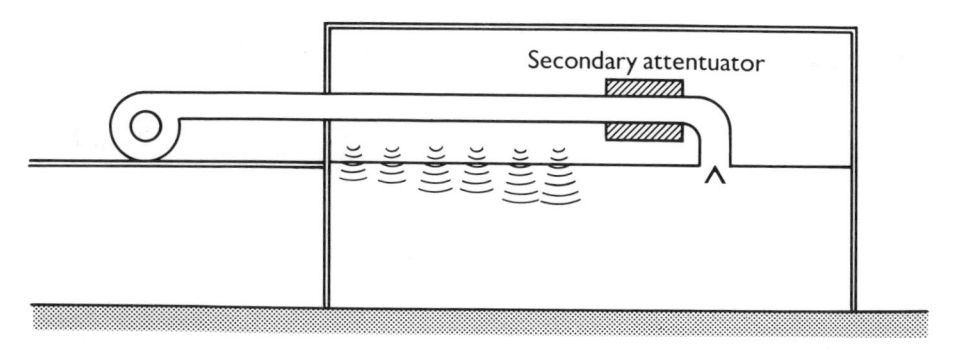

Figure 9.2   Breakout where Secondary Attenuators are Used to Control Fan Noise

TERMINAL UNITS

Great strides have been made in reducing the casing radiation from flow control terminal units, but they must be evaluated in the total breakout noise calculation. This is much assisted by the manufacturers' data, which nowadays includes casing-radiated sound power levels. Substantial false ceilings, corridor locations and special "low breakout boxes" all help guard against problems from such necessary control units.
Courtesy: Sound Attenuators Ltd

Hence, when defining attenuation requirements, the weakest acoustic path to the conditioned areas must be assessed and this may well be via breakout from the ductwork.

Location of the primary attenuator is also a factor to be taken into account: a poor location can result in break-in to the duct system and breakout further along the ducting (Figures 9.3 and 7.23).

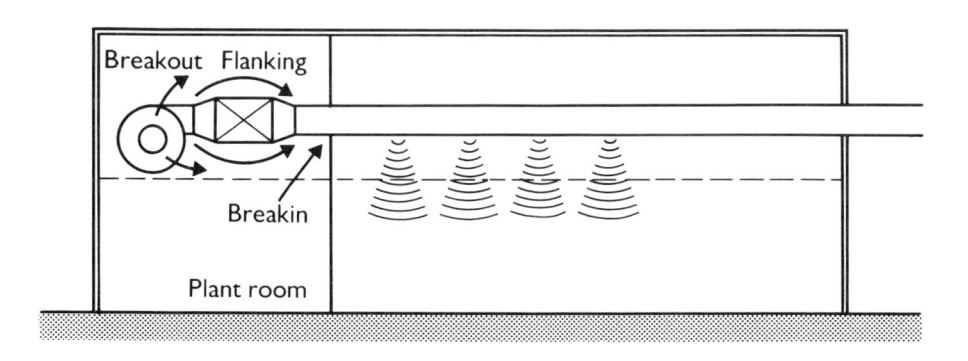

Figure 9.3   Breakout where Attenuator is Incorrectly Located

In some cases, it may be found that space restrictions do not permit the incorporation of sufficient primary attenuation to reduce the level of noise breakout to an acceptable level. Under these circumstances, the use of external lagging for the ductwork to improve sound insulation is recommended.

9.1.2 Flow-generated Noise Breakout

In-duct noise may also be due to the interaction of high velocity airflows on duct fittings and dampers and, in high velocity systems, by flow rate regulation devices such as terminal units. This aerodynamically generated noise will propagate in all directions along duct systems and may break out from the ductwork at an acoustically weak duct section some distance from its source.

It is necessary, therefore, to calculate the in-duct noise level due to generated noise for each section of the duct system and to compute the level of noise breakout and additional insulation required prior to selection of primary attenuation. For terminal units the upstream or inlet sound power levels from the flow control device should be available from the manufacturer.

9.1.3 Turbulence-induced Breakout

Finally, noise may be radiated from ductwork due to turbulent airflow directly interacting with the walls of the ductwork themselves. Although this is not noise breakout as generally recognised, this phenomenon has similar characteristics and treatments to the previous types of breakout.

Turbulence induced breakout is almost impossible to predict quantatively, but can be avoided by good ductwork design and layout (see Chapter 7).

9.2 CALCULATION OF NOISE BREAKOUT

The level of noise breaking out of a given section of ductwork depends on the following factors:

1. The sound intensity within the duct
2. The area of ductwork exposed within the conditioned area
3. The construction and shape of the duct — rectangular, oval, round.

The calculation method can be developed from first principles taking these factors into account.

Sound power transfer tends to be by plane wave propagation for duct systems which have cross sectional dimensions less than the acoustic wavelength considered. At 63 Hz, the wavelength of a sound wave is over five metres and a wavelength as small as one metre does not occur until a frequency of about 330 Hz is reached.

Hence, plane wave propagation occurs for most lower frequency sound within commonly used duct sizes.

The sound power level within the ducting system is available for fan or regenerated noise calculations and this can be related to the sound intensity (I) incident on the duct walls as follows:

$$I \text{ (in duct)} = \frac{W \text{ (in-duct)}}{A} \qquad 9.1$$

where A is the duct cross sectional area in m^2
and W (in-duct) is the in-duct sound power.

The total sound power incident within the duct system of the conditioned area is equal to this intensity multiplied by the surface area of the duct system, that is:

perimeter $\times$ length $= S \, m^2$

$$W \text{(incident on duct wall)} = \frac{W \text{(in-duct)} \times S}{A} \qquad 9.2$$

and since the duct wall will attenuate sound depending on its Sound Reduction Index (SRI) the total sound power emerging from the duct system may be found by taking logarithms and subtracting the Sound Reduction Index.

$$L_w \text{ (radiated)} = L_w \text{ (in-duct)} - \text{SRI} + 10\log_{10} \frac{S}{A} \qquad 9.3$$

where L_w (radiated) = total sound power level radiated as "breakout"
and L_w (in-duct) = sound power level within the duct.

This is known as the ALLEN formula and has been used for many years. It is reasonably accurate for short lengths of ductwork and medium/high frequencies but is inaccurate for long lengths of ductwork and low frequencies for the following reasons:

1. As sound power passes along the duct system it is dissipated through the duct walls and is not uniform from one end of the duct system to the other. This leads to over-prediction for long duct lengths.
2. The Sound Reduction Index (SRI) of materials is measured under reverberant test conditions and therefore will give a different sound insulation under the totally different incidence conditions encountered within ducts.
3. For large duct areas, the calculated transmitted sound power level can exceed the in-duct sound power level. This is clearly impossible and it is often assumed that the worst conditions will be that half the in-duct sound power level will be re-radiated, that is 3 dB below the in-duct level.

To compensate for the over-prediction of noise breakout due to long duct lengths, a formula can be derived which takes into consideration the decay of sound along a finite duct length, L.

This is as follows:

$$L_{wB} = L_{wD} - 10\log_{10} \frac{1}{T} + 10\log_{10}L + 10\log_{10} \frac{[1 - \exp(-TL)]}{TL} \qquad 9.4$$

where L_{wB} = sound power breakout from the duct;

L_{wD} = sound power entering the duct;

T = duct transmission coefficient (not SRI); and

L = duct length.

To use this formula it is necessary to measure values of T for the appropriate duct, design, gauge and size.

By way of illustration, some values of duct transmission loss are given in Figure 9.4 and the correction due to duct length can be estimated from Figure 9.5.

Having established the total sound power breaking out of a duct length it is a fairly simple matter to predict the reverberant sound pressure level in a given room by entering this power level into the conventional room calculation (Chapter 7, Figure 7.28).

Duct Size (mm)	Octave Band Hz							
	63	125	250	500	1 k	2 k	4 k	Hz
150 × 150	16	16	12	18	21	24	28	dB
150 × 300	12	17	13	19	22	25	29	dB
150 × 600	18	18	14	20	23	26	30	dB
300 × 300	14	19	20	21	24	27	31	dB
300 × 300	15	20	21	22	25	28	32	dB
300 × 1200	16	21	21	23	26	29	33	dB
600 × 600	22	22	23	24	27	30	34	dB
600 × 1200	23	23	24	25	28	31	35	dB
600 × 1800	24	24	25	26	29	32	36	dB
1200 × 1200	25	25	26	27	30	33	37	dB
1200 × 1800	26	26	27	28	31	34	38	dB

Figure 9.4   Duct Tranmission Loss per Metre

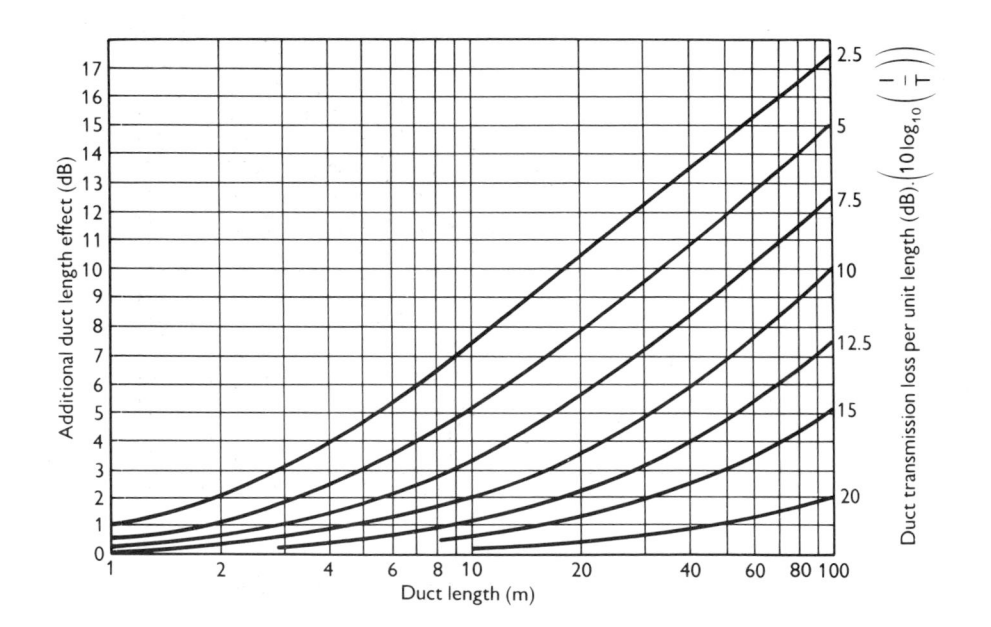

Figure 9.5   Breakout Duct Length Correction

However calculation of the direct sound field emitted from a duct system must realise that the duct system is a line source and that the listener receives contributions from each point on the duct system that are related to the distance to the listener.

The direct free field sound pressure level, L_p (direct), at a distance, r, from a duct of length, L, can be calculated from the formula:

$$L_p \text{ (direct)} = L_{wB} + 10\log_{10} \left(\frac{I}{4\pi r^2} \right) + 10\log_{10} \left[\left(\frac{2r}{L} \right) \tan^{-1} \frac{L}{2r} \right] \qquad 9.5$$

The last correction factor can be conveniently expressed graphically as Figure 9.6.

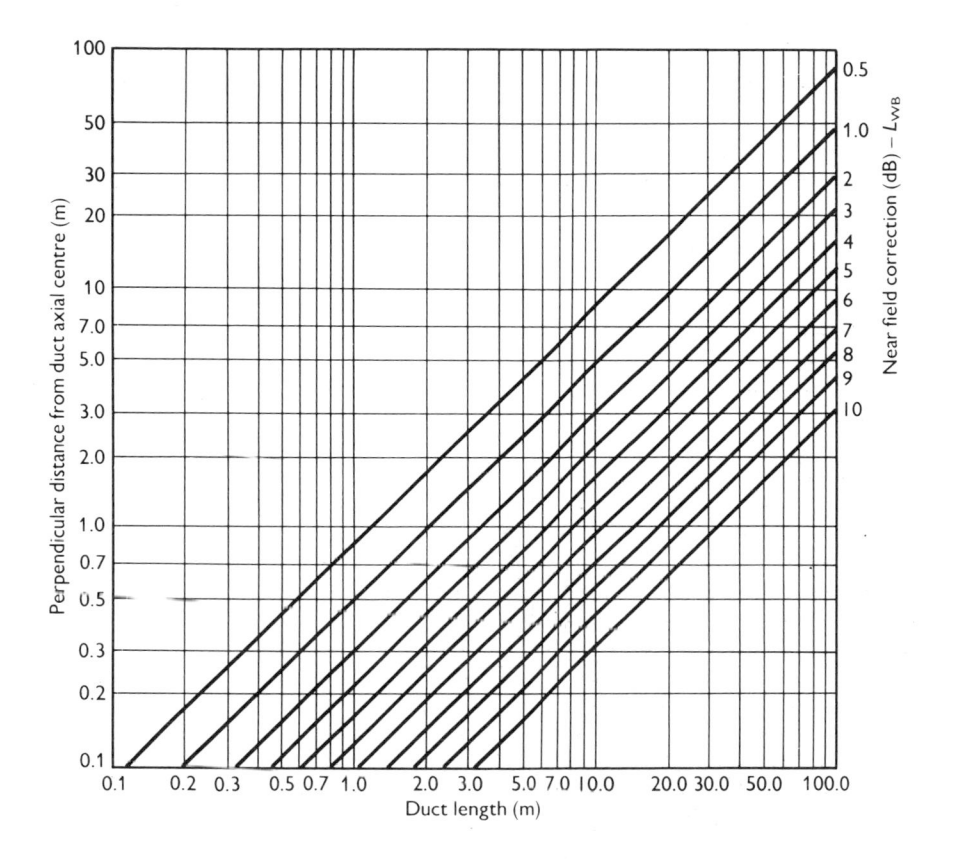

Figure 9.6   Near Field Correction for a Line Source

Where a duct passes above a suspended ceiling, the sound insulation of the ceiling may be added to the total attenuation of breakout noise for mid and high frequencies if the ceiling void is of normal size and if the ducting does not pass closely above it. Typical sound insulation values for ceilings are given in Chapter 5.

The calculation assumes that the duct is in free space, and when the duct is located close to a reflecting surface the conventional directivity corrections shown in Figure 9.7 apply.

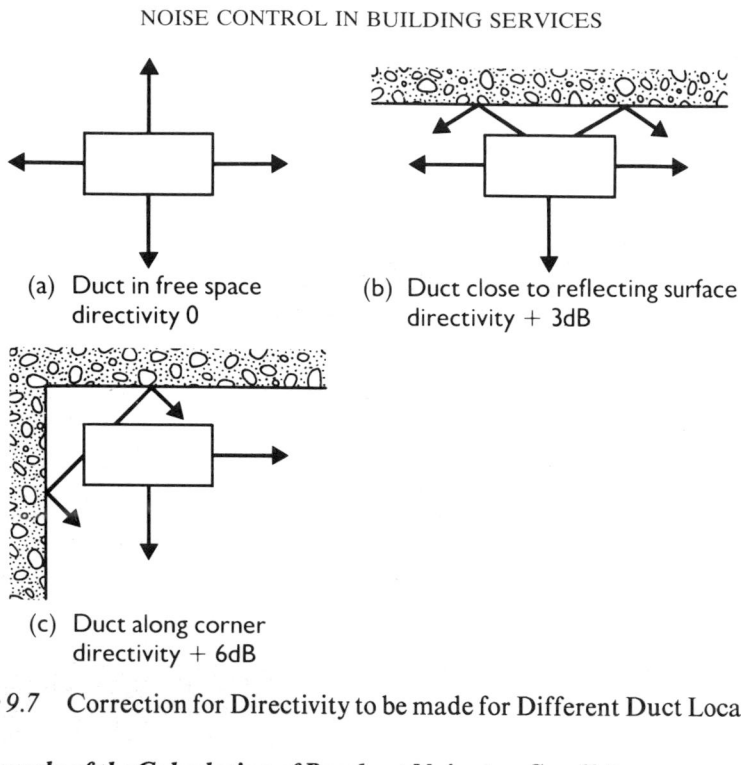

(a) Duct in free space
directivity 0

(b) Duct close to reflecting surface
directivity + 3dB

(c) Duct along corner
directivity + 6dB

Figure 9.7   Correction for Directivity to be made for Different Duct Locations

9.2.1 Example of the Calculation of Breakout Noise to a Small Room

A room measuring 5 m × 5 m × 2.5 m contains a duct having dimensions 5 m long, 300 mm high and 300 mm wide. The duct passes through the room but does not serve the room. The room has a reverberation time of 1.0 s at all frequencies.

	Octave Band (Hz)						
	63	125	250	500	1k	2k	4k
Duct sound power level, L_{wD}	84	84	83	79	76	71	71
$10\log_{10} \dfrac{l}{T}$	14	19	20	21	24	27	31
− Length effect (Figure 9.5)	0	0	0	0	0	0	0
+ $10\log_{10}L$	+7	+7	+7	+7	+7	+7	+7
Breakout sound power level L_{wB}	77	72	70	65	59	51	47
Reverberant Field							
Room volume 62.5m³	−4	−4	−4	−4	−4	−4	−4
Reverberation Time, 1s	0	0	0	0	0	0	0
Reverberant L_p	73	68	66	61	55	47	43
Direct Field							
Directivity (Figure 9.7)	+3	+3	+3	+3	+3	+3	+3
Near field correction (Figure 9.6)	−3	−3	−3	−3	−3	−3	−3
Direct L_p	77	72	70	65	59	51	47
Resultant L_p	78	73	71	66	60	52	48

MASSIVE DUCTWORK

Whilst in this photograph outlet grilles are included, it serves to warn us that large
lengths of large cross-section ductwork may progressively and cumulatively allow all
the inlet sound power level out into the area below. Only upstream attenuation can
sensibly combat this situation.

RECTANGULAR DUCTWORK

Large areas of rectangular ductwork are a potential source of low frequency noise breakout, be it duct noise or flow noise in origin. Here, in a studio complex, secondary attenuators are incorporated just to reduce in-duct noise levels in ductwork passing above critical areas.

In the previous example and Figure 9.4, the duct transmission loss figures were extrapolated from laboratory test data. Where this data is not available, the architectural random incidence Sound Reduction Index can be applied with less certainty for rectangular ducts.

Different conditions apply with circular ducting. Because of its shape, an ideal circular duct is virtually infinitely stiff at low frequencies and thus has a very high resistance to low frequency breakout. This is because circular ductwork, being completely symmetrical, cannot change its cross-sectional area and remain symmetrical unless the metal itself actually stretches.

Low frequencies, which are the main area of concern, have relatively long wavelengths. As noted earlier, at 63 Hz the wavelength of sound in air is approximately 5 m while at 250 Hz it is still over 1.2 m. Clearly wavelengths of this magnitude will only fit into ductwork of for example 600 mm diameter if aligned along the axis of the duct (Figure 9.8). At any given instant, therefore, pressure at any point along the duct is either uniformly high or uniformly low across the ducts cross-section. As a result, the air is either trying to push the duct wall out equally in all directions or, alternatively, to contract in all directions, as also shown in Figure 9.8 —

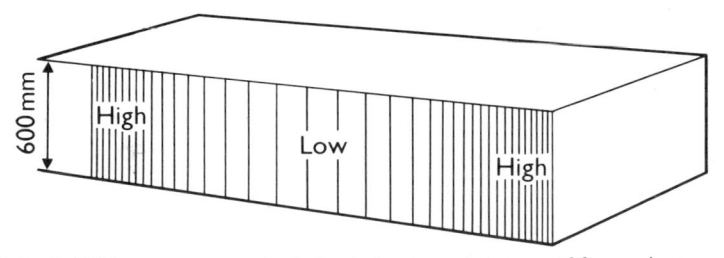

(a) A 63 Hz wave can only fit its 5.3 m length into a 600 mm duct along the duct axis. As a result at any section the duct must expand or contract like a balloon for breakout to occur.

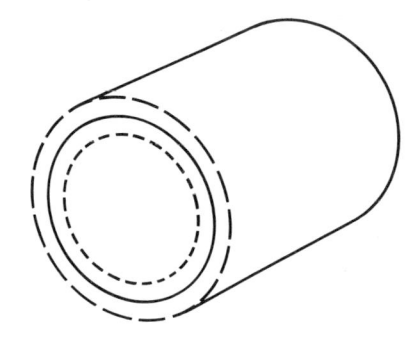

(b) Virtually impossible in practical circular ducting

Figure 9.8   Low Frequency Noise in Ducts

a phenomenon which can only occur if the ductwork can stretch like a balloon. Because circular ducts are unable to do this, low frequency breakout from such ductwork is extremely rare.

However, in practical ducting systems, small deviations from a true circular shape will degrade a duct's sound insulation performance. Simple prediction methods are not accurate enough to take these subtle but important differences into account and only indicative results are given in this chapter (Figure 9.9).

As the frequency of the sound in the duct increases, and correspondingly the wavelength decreases, cross modes are formed where changes of area can occur simply by the bending of the duct walls. This effect becomes significant when the wavelength of the sound approaches the diameter of the ductwork as illustrated in Figure 9.10. This would be above 500 Hz for a 600 mm diameter duct.

Even at these frequencies, the spiral seams in rolled circular ductwork enable it to remain stiffness controlled up to quite high frequencies. At high frequencies, circular ductwork is at least as resistant to breakout as rectangular systems.

Spiral wound oval ducting falls somewhere between the two extremes of rectangular and circular systems. It has greater stiffness than rectangular ductwork because it incorporates rolled seams and, in larger sizes, stiffening rods. However, it still does not possess the unique radial stiffness at low frequencies of circular ductwork.

9.3 BREAKOUT DUE TO TURBULENCE DOWNSTREAM OF FITTINGS

Complex problems arise when investigating noise breakout downstream of fittings, a phenomenon which is mainly associated with rectangular ducting. In general, the problem is likely to arise with rectangular ducts as soon as average air speeds reach about 4.5 m/s.

Although this is a relatively low speed, systems often have inherent design constraints which mean that a low average speed may conceal areas where the maximum velocity is much higher than the average.

There are cases where velocities close to 10 m/s have been measured in straight sections of ductwork where the nominal average velocity was under 5 m/s. The fact that the velocity is within limit in over half the duct cross-section does not help to ease turbulence-induced breakout problems in the remainder.

Similarly, commissioning may reveal that it is not possible to balance the system as designed and air quantities may need to be increased.

A design which follows the guidelines for airflow generated noise given in Chapter 8 would not cause problems until higher average speeds are reached.

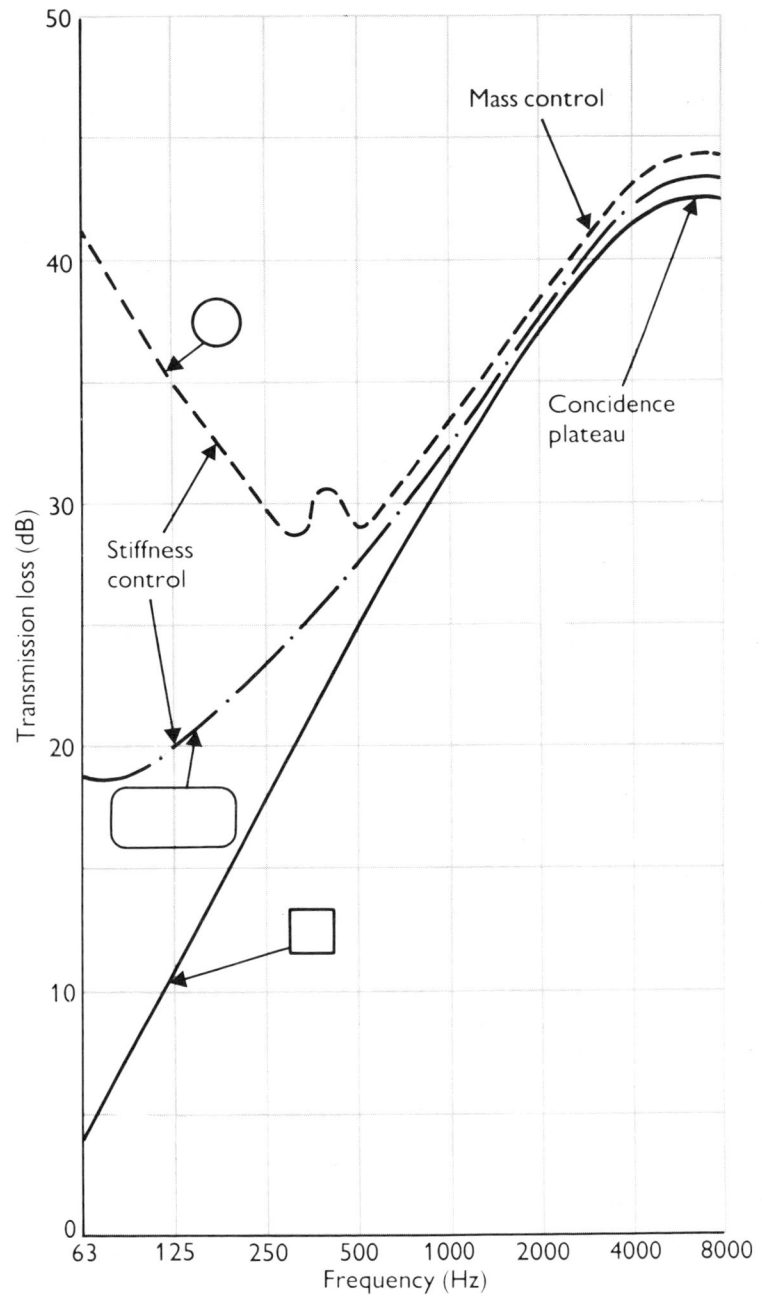

Figure 9.9 Typical Transmission Loss of Rectangular, Oval and Circular Steel Ducts

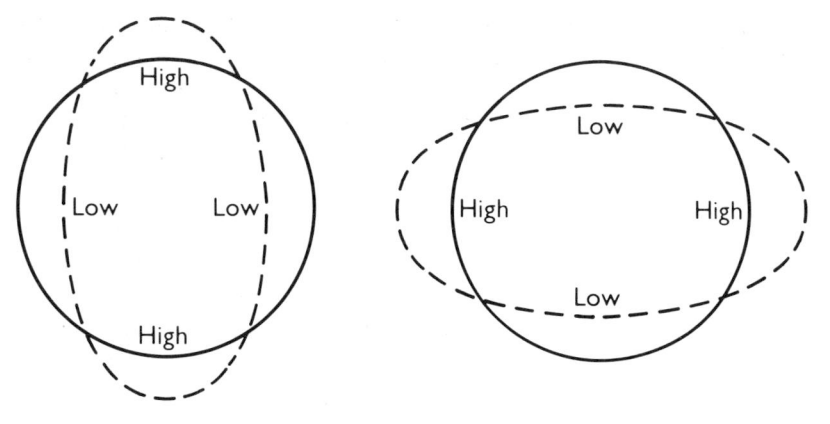

At high frequencies, when the wavelength approaches the duct diameter the second cross mode occurs and the duct can respond by becoming elliptical. The duct walls follow the mass law for sound insulation.

Figure 9.10   Sound Insulation of Circular Duct Walls at High Frequencies

Breakout problems are likely to occur with large rectangular metal ductwork panels as soon as velocities reach 4.5 m/s for normal conditioned areas where, say, NC35 is required. If the system is very "clean" it may be possible to reach much higher velocities without trouble. Velocities approaching 20 m/s have been reported with rectangular ductwork without problems but such cases are rare. This indicates that the fast airflow has been well "guided" uniformly through the system, and has not encountered flow restrictive fittings. Velocities of this order can be accepted without breakout problems with builders work ducts.

No reliable figures exist to enable breakout due to turbulence downstream of all fittings to be estimated, owing to the vast amount of work that would be necessary in view of the large variety of possible configurations. Information is available, however, about a number of fittings in isolation which regularly lead to trouble and some of these are touched on in Chapter 8.

9.3.1 Internal Flanges

These fittings not only reduce the effective cross section of the duct but are also efficient generators of vortices, which then vibrate the duct walls.

9.3.2 Mitred Bends

Air travels in straight lines, continuing its motion at constant speed and in a constant direction unless acted on by an outside force. As a result, when air reaches a radiused bend, it tends to carry straight on until sufficient quantity has been moved over to the outer wall to generate the necessary pressure gradient to force the remainder round the corner. This leads to very high velocities on the outer surface of the bend and accounts for the aerodynamic loss coefficient encountered at such a bend. Any take-offs or other fittings on the outside of a bend or immediately downstream of the bend will therefore be operating in a very much faster air stream than expected, with the usual breakout implications (Figure 9.11).

Plain mitred bends produce a similar but more severe problem. However, if narrow chord turning vanes are fitted, the air can be guided and turned more evenly and the distribution pattern preserved. The resulting distribution is better than in a radiused bend. The energy loss takes the form of smaller eddies shed from the individual blades, rather than large eddies from the inside of the bend. This has the effect of converting the problem from one of low frequencies to one of higher frequencies,

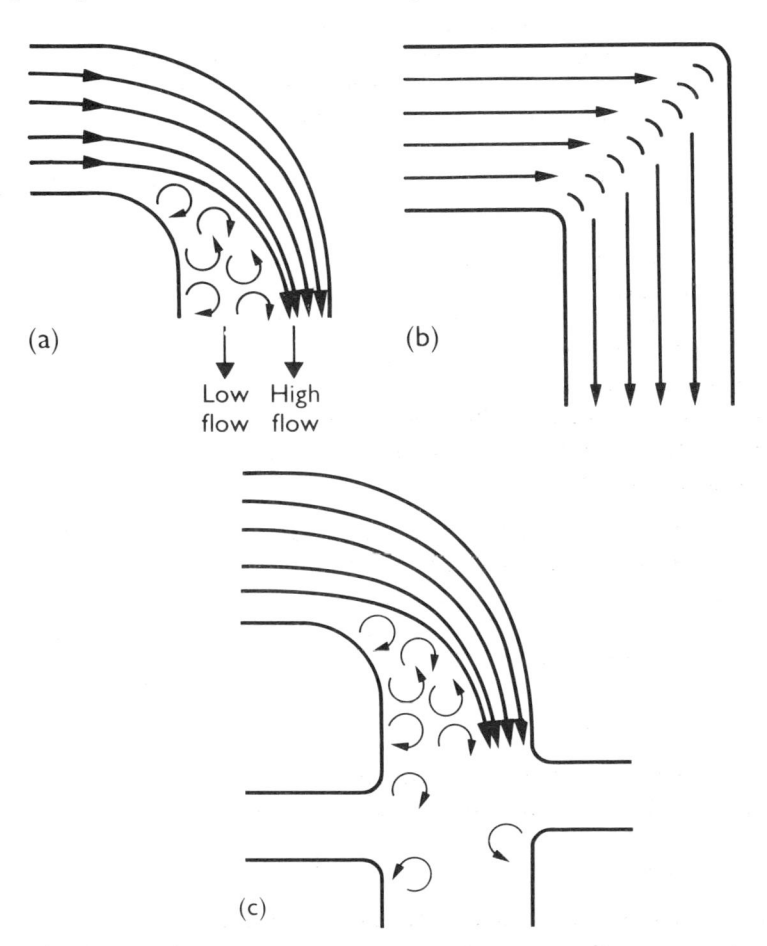

Figure 9.11   Airflow Patterns in Radiused and Mitred Bends

which is less difficult to solve. A mitred bend has the additional advantages of occupying less space than a radiused bend and providing more fan noise attenuation.

9.3.3 Square Branch Connections

The inability of air to pass round sharp corners without assistance also leads to severe ductwork vibration and consequent noise breakout at square branch connections. Coned, shoed or angled take-offs are therefore preferable.

9.3.4 Other Fittings

In general, any fitting which constitutes a pressure loss device or is a source on in-duct flow generated noise can cause breakout problems. A fire damper is a common example.

9.3.5 Breakout due to Interaction between the Airflow and the Duct Walls

In theory, the natural frequencies of vibration of the duct wall can interact with the natural instabilities of the air stream so that eddies are produced and the walls "drum", even if the system itself is aerodynamically "clean". In practice, however, the problems caused by this source are at a very low level and need not generally be dealt with. They become important in the design of very low noise level systems such as those for broadcasting studios and audiometry rooms.

9.4 CONTROLLING BREAKOUT FROM DUCTWORK

It is usually easier to prevent breakout by using aerodynamically clean fittings and circular ducting than to attempt a cure by employing insulation techniques. However, if it is necessary to establish a cure, there are four possible ways to increase the resistance of ductwork to noise breakout, namely:

– stiffening the ductwork, 9.4.1
– increasing the mass of the duct walls, 9.4.2
– increasing the damping of the duct walls, 9.4.3
– providing acoustic lagging, 9.4.4

Not all of these methods are effective in practice and in any case they are difficult and can be expensive to apply.

9.4.1 Stiffening the Ductwork

An obvious technique used quite extensively in the past is to fit external bracing to the ductwork. Unfortunately, this also improves the radiation efficiency of the ductwork at low frequencies and partly cancels out any benefits derived from increased stiffness.

TAKE-OFFS

Low frequency turbulence and in-duct flow noise is inevitable at take-offs, but here the total use of cylindrical ductwork will contain this noise and prevent it creating a breakout noise problem.

Courtesy: Lindab

ACOUSTIC LAGGING

Proprietary user-friendly lagging treatments now offer a degree of secondary break-out noise control when problems are predicted or found to occur. Here a limp intermediate layer of lead is resiliently supported on a glass fibre layer. The outer foil cladding not only acts as a containment and fire barrier, but more especially gives an integrity and surface finish which makes the concept into a finished unitary product. Courtesy: Salex Acoustic Materials Ltd

9.4.2 Increasing the Mass of the Duct Walls

By using a thicker gauge material or applying a proprietary mass barrier compound to the duct walls, their mass can be increased. Either method may be effective though, as any increase in mass that can be achieved is relatively small, any increase in insulation obtained is correspondingly modest.

9.4.3 Increasing the Damping of the Duct Walls

Damping compounds applied for their sound absorbing properties as well as to increase mass can reduce breakout at low frequencies by up to 10 dB as the combined result of increasing the mass and the damping.

9.4.4 Acoustic Lagging

By far the most effective method of reducing breakout is acoustic lagging but it can prove both difficult to apply and expensive. It is important to distinguish between acoustic and thermal lagging. For example, though both expanded polystyrene and uncovered glass fibre mat are effective thermal insulators, they have little or no acoustic lagging properties.

An effective acoustic lagging requires an impervious, heavy and preferably limp outer covering which completely encloses the duct, but is mechanically separated from it. The intervening space should contain a sound absorbing material such as mineral wool to damp out noise build-up within the space.

One of several methods of achieving these requirements is to erect a studwork enclosure round the duct and to cover this with plasterboard while filling the intervening space with loose glass fibre or rockwool. The joints between the duct and the enclosure are best sealed with a soft rubber seal or similar material. This method is relatively simple to make and fairly inexpensive, provided that space is available (Figure 9.12).

As an alternative to this, the duct can be wrapped with 75 mm of soft heavy density glass fibre or rockwool mat so that the flanges, if any, are covered and then encase the whole in a 12.5 mm thick layer of Keene's Cement, 1mm thick lead foil or one of the proprietary heavy loaded barrier mats. These treatments are effective but can be very difficult to apply in practice, when it is necessary to negotiate duct hangers and access panels, etc.

It is possible to suggest an approximate idea of the effectiveness of these lagging techniques. For example, if a studwork partition can be spaced 150 mm or more from the ductwork, it is possible to achieve an additional reduction nearly comparable with the expected Sound Reduction Index of the partition material itself.

In the case of the plaster or lead on mineral wool treatments, the following table indicates the likely improvements in Sound Reduction Index which can be anticipated:

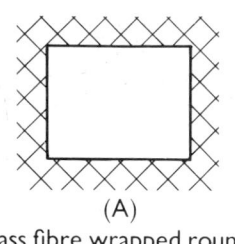

(A)
Glass fibre wrapped round a
duct is NOT a very effective
acoustic lagging

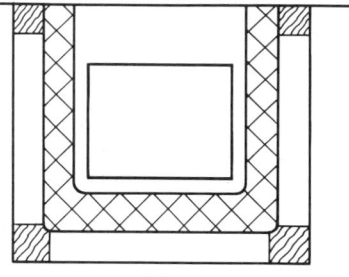

(B)
Studwork enclosure covered
with plasterboard and loosely
filled with glass fibre or
rockwool, makes a good enclosure.
Seal joints on to ductwork
with soft rubber.

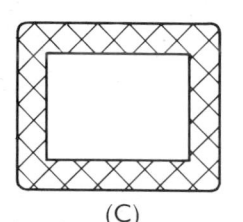

(C)
Heavy cladding (lead or
plaster) over soft glass fibre

Figure 9.12 Methods of Cladding Ducts to Reduce Breakout Noise

				Octave Band (Hz)					
Frequency	63	125	250	500	1k	2k	4k	8k	Hz
Plaster (30kg/m²) on mineral wool	9	11	13	12	12	12	12	21	dB
Lead (12kg/m²) on mineral wool	6	7	8	7	7	7	7	7	dB

Though circumstances may dictate the need to carry out remedial measures, it must always be remembered that it is almost certainly cheaper to avoid generating noise breakout problems than to try to contain them.

FALSE AND SUSPENDED CEILINGS

Such ceilings can offer a degree of extra noise control against breakout noise, but the magnitude will depend primarily on the mass involved and the acoustic integrity of the assembled construction. Generally they are light and have various penetrations, so caution and vigilance are called for. Also the overall performance will depend on the depth of the void and the nearness of the ductwork to the ceiling's rear surface, so generalised calculation techniques of sufficient sophistication have not been developed.

CHAPTER 10

NOISE FROM ROOM UNITS

10.1 ROOM UNITS FOR VENTILATION AND AIR-CONDITIONING

While centralised mechanical ventilation and air-conditioning systems will provide the basic requirement for a building, the ever-increasing demand for more sophisticated control in each individual room calls for room units and terminal units which will process incoming air to specific requirements. Also, in certain cases improvement of existing rooms may only be achievable by fitting air conditioners in each room. In all these cases the equipment is a potential noise source which requires careful consideration, selection and perhaps treatment if it is not to be intrusive.

10.2 MAIN TYPES OF ROOM UNITS

10.2.1 Air-conditioning Units

Packaged air-conditioning room units do not require any supplies, other than electricity, from a central source, the plant being installed in the room to be conditioned. There are two types commonly available: the through-the-wall or window type (Figure 10.1); and the split unit, which consists essentially of a fan coil unit mounted inside the room with its coil connected by two small pipes to the condenser outside (Figure 10.2). The condenser is normally located externally and no more than a 15 m pipe run from the fan coil unit itself. These types of packaged unit are also available with heat pump facilities to extract additional heat from the external air.

10.2.2 Air-handling Room Units

In order to air-condition larger room volumes such as computer rooms, packaged air handling units are often used. These feed air to the room via under-floor and ceiling void plenums, the units themselves being built into the walls of the room or into immediately adjacent plantrooms and supplied with hot and chilled water from remote sources. Make-up fresh air is provided from an independent fan system.

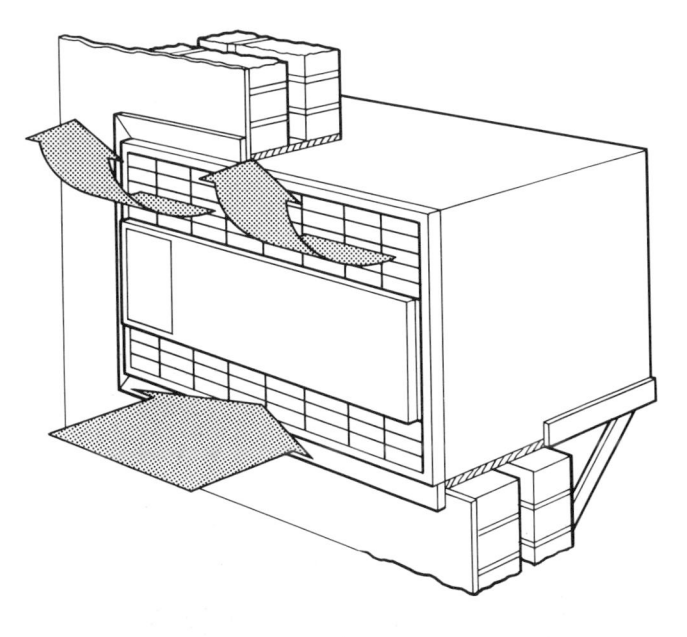

Figure 10.1 A Through the Wall Type of Packaged Air-conditioning Room Unit (courtesy Andrews Industrial Equipment Ltd)

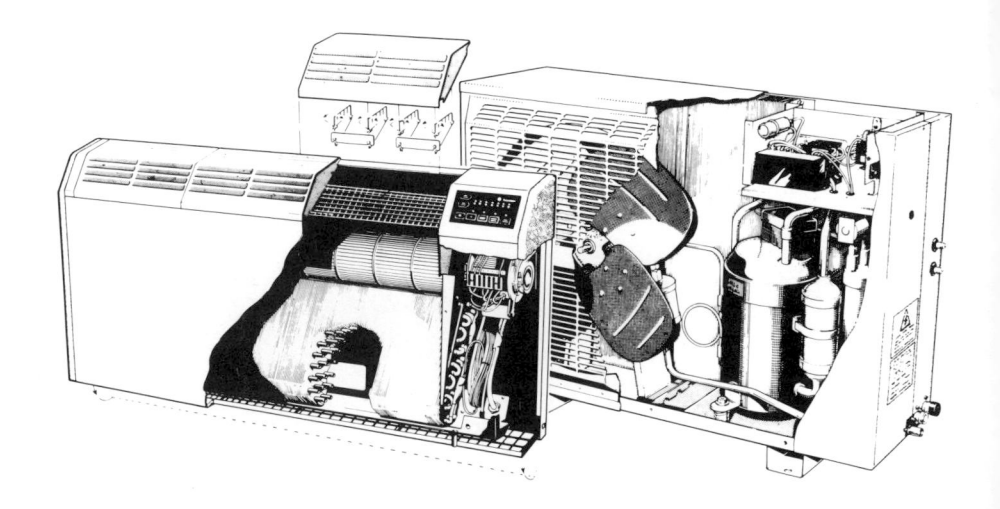

Figure 10.2 Split System Room Air Conditioner/Heat Pump (courtesy Temperature Ltd)

INDUCTION UNIT

An under-window unit.

Courtesy: Carrier Distribution Ltd

TERMINAL UNIT

A new concept is illustrated in that each function of the compound terminal unit is broken down and manufactured separately. This obviously yields greater freedom, and probably the most useful acoustic feature is the ability to place the sometimes "hissy" water coil before the secondary attenuator, or even between two short secondary attenuators.

Courtesy: Sound Attenuators Ltd

10.2.3 Fan Coil Units

The fan coil unit is a simple system usually involving heating or cooling of only recirculated air but with a degree of control over the volume flow of air output usually achieved by varying the speed of the fan. Fan coil units may be wall-mounted at low or high level, below a ceiling or in the ceiling void itself, although in the latter case adequate provision for airflow through the ceiling must be provided.

10.2.4 Induction Units

The principle on which induction units work is to be fed with primary air from a remote source, this air being directed through an arrangement of nozzles or jets. The jet action causes room air to be entrained in the stream (Figure 10.3), thus creating recirculation of room air up to as much as 10 times the primary air supply rate. Heat exchanger coils are incorporated over which the recirculated air passes. Both heating and cooling coils may be used or, if a single coil only is fitted this may have a seasonal change-over between the heating and cooling functions.

Induction units are usually installed under windows but ceiling void installations are also available. Where multiple units are used in series they are often fed from a series plenum system which accommodates the primary air, nozzle dampers ensuring that all induction units are appropriately balanced.

10.2.5 High and Low-Velocity Ducted Air-Systems

High-velocity and low-velocity ventilation systems both have their share of noise problems and there is no clear advantage in using one system over the other in terms of quietness. In practice it is increasingly the case that precise control of criteria from room to room will dictate the choice of a high-velocity system which can incorporate heating/cooling and controllable supply rate at each of its terminal units.

Another very important factor which may decide the selection of a high-velocity system is the economic use of building space which can be obtained. Figure 10.4 shows how an extra floor can be obtained in a given building height by selecting a high-velocity system. The building shown in Figure 10.4a requires a large ceiling void to accommodate large section low-velocity ductwork. On the other hand, as can be seen in Figure 10.4b, a high-velocity system uses much smaller section ducts and pipework and hence the space saving in ceiling voids can be used to add to "people space".

10.2.6 Terminal Units

Whereas the basic control of air in high-velocity systems may be carried out at the plantroom, individual room requirements dictate that fine tuning of air-conditioning systems has to be carried out at each room as required. To do this the end of the ducted system is provided with a "terminal unit" which is capable of meeting varying

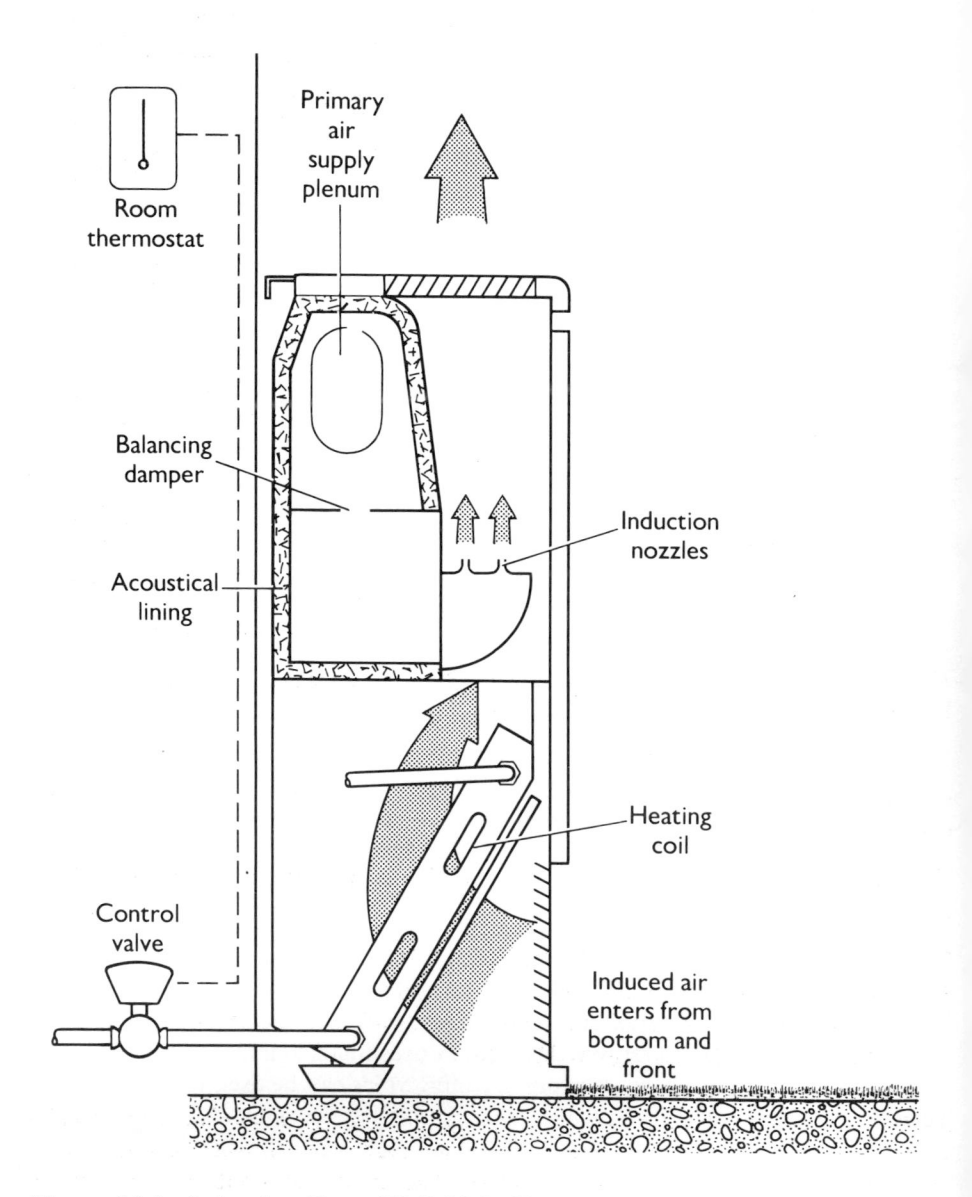

Figure 10.3 Induction Type Wall Unit Showing Method of Air Entrainment (courtesy Weathermaker Equipment Ltd)

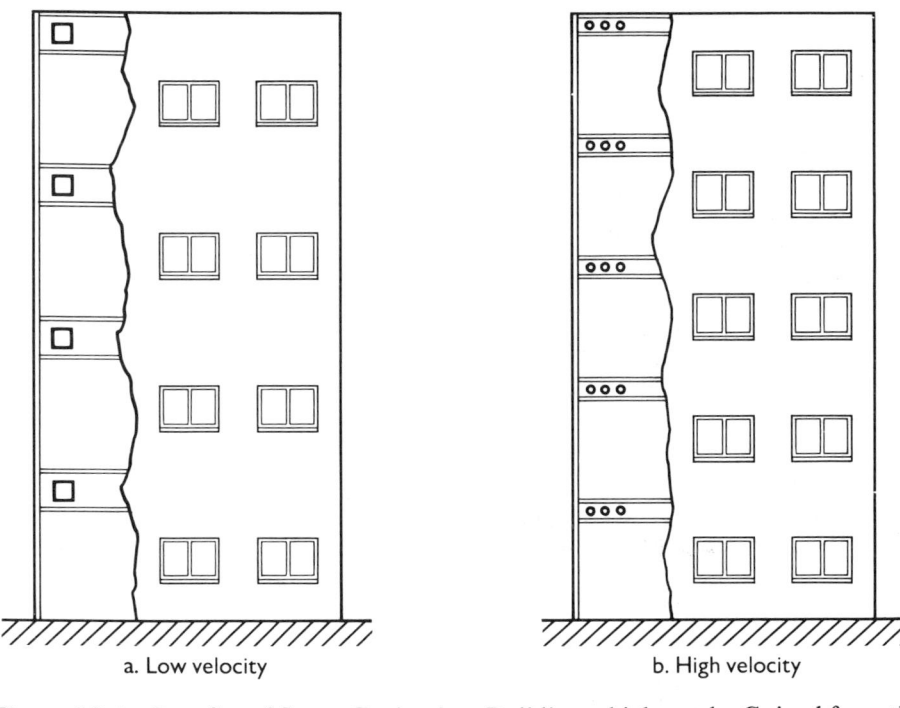

a. Low velocity   b. High velocity

Figure 10.4   Benefits of Space Saving in a Building which can be Gained from the Use of High-velocity Air-conditioning Systems

demands by filtering, controlling output air-flow, air temperature and draught-free distribution.

There are five basic categories of terminal unit in common usage. These are:

Induction terminal units (Figure 10.3)
Constant flow rate control units, CFRC (Figure 10.5)
Variable flow rate control units, VFRC (Figure 10.6)
Slot-type variable flow rate diffuser units (Figure 10.7)
Fan assisted terminal units (Figure 10.11)

Terminal units are either of single-duct or dual-duct type. A single-duct unit typically consists of a box with one high velocity air-inlet, a flow-rate controller feeding into an air distribution arrangement to discharge air at acceptable draught-free velocity into the room (Figure 10.5). A re-heat facility, either by hot water coils or electric element may be provided. Volume flow-rate may be controlled electrically, pneumatically or self-powered by means of the system pressure itself. Dual-duct terminal boxes work on generally similar principles to single-duct units but they have two independent inlet ducts, one carrying high pressure chilled air and the other carrying high pressure

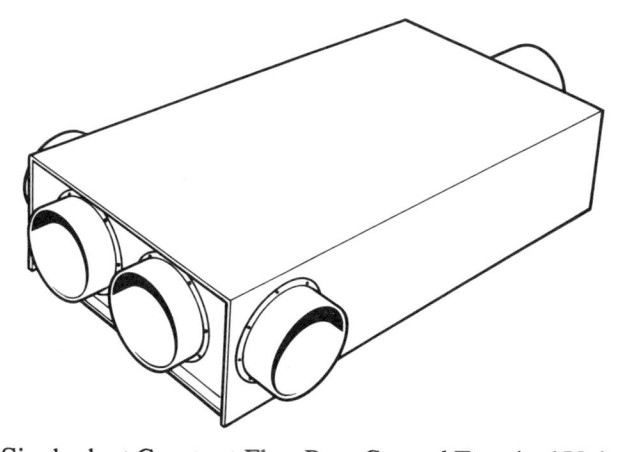

Figure 10.5   Single-duct Constant Flow Rate Control Terminal Unit with Multiple Outlets

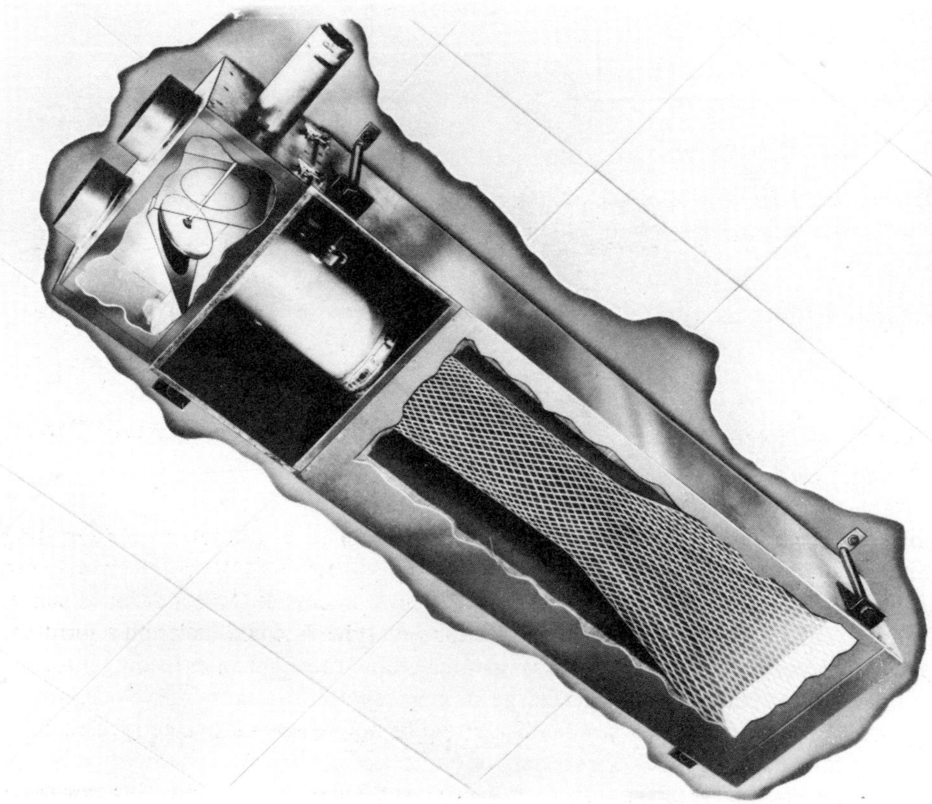

Figure 10.6   Dual-duct Variable Flow Rate Control Unit with Air Blending System (courtesy Sound Attenuators Ltd.)

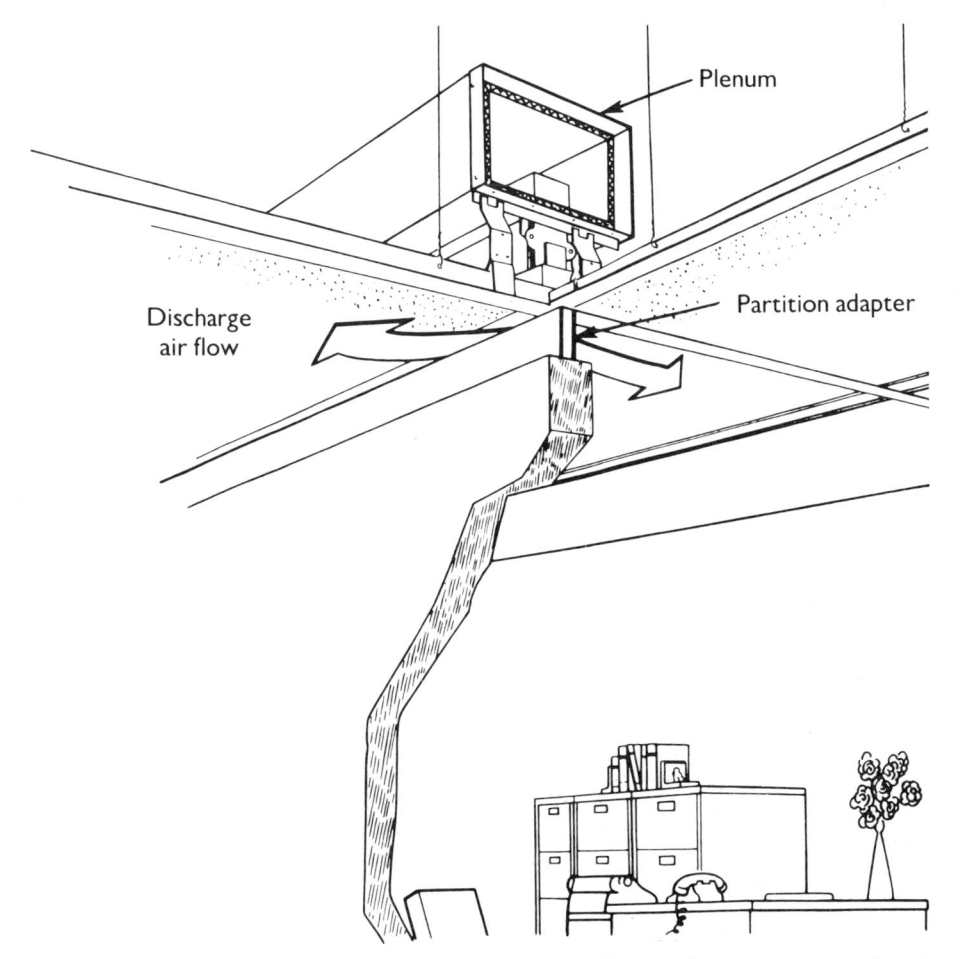

Figure *10.7* Slot Type Variable Flow Rate Diffuser Units (courtesy Carrier Moduline)

warmed air, both from a remote source. The two inputs enter an air-blending chamber which permits mixing of the tempered air in proportions which are regulated, dependent upon the temperature requirements of the conditioned space. Figure 10.6 shows a typical blending section and flow-rate controller on a terminal unit for mounting in a ceiling void.

10.3 ACOUSTIC CHARACTERISTICS OF ROOM UNITS

The generation of noise from the various types of room unit is a major factor to be considered when making a choice of system. For general guidance the acoustic characteristics of several units are shown in Figures 10.8, 10.9 and 10.10, although

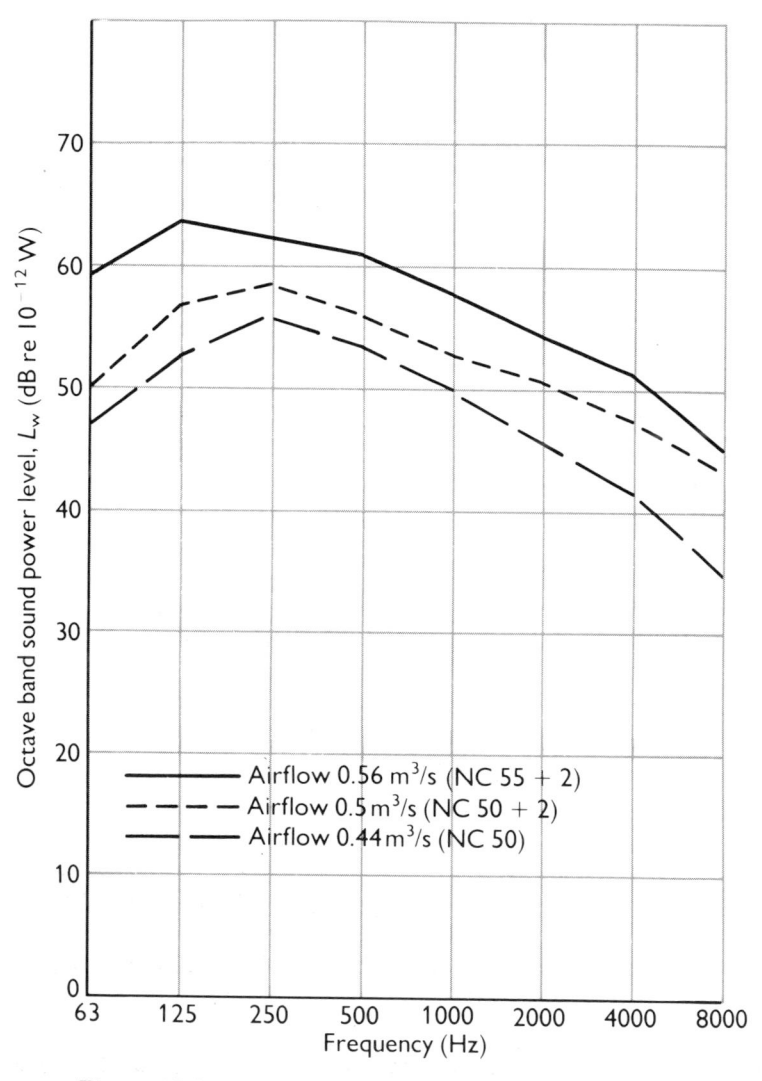

Figure 10.8   Acoustic Characteristics of Fan Coil Units

variability has been found between units of the same type but of different makes. For design purposes it is imperative to determine sound pressure or power levels of the actual type of unit to be used and, indeed, the unit selection may well be based upon the specific acoustic performance required.

10.4 SOURCES OF NOISE

There are several distinct sources of noise from room units, each of which needs to be identified and, if excessive, treated in the manner most appropriate to that source —

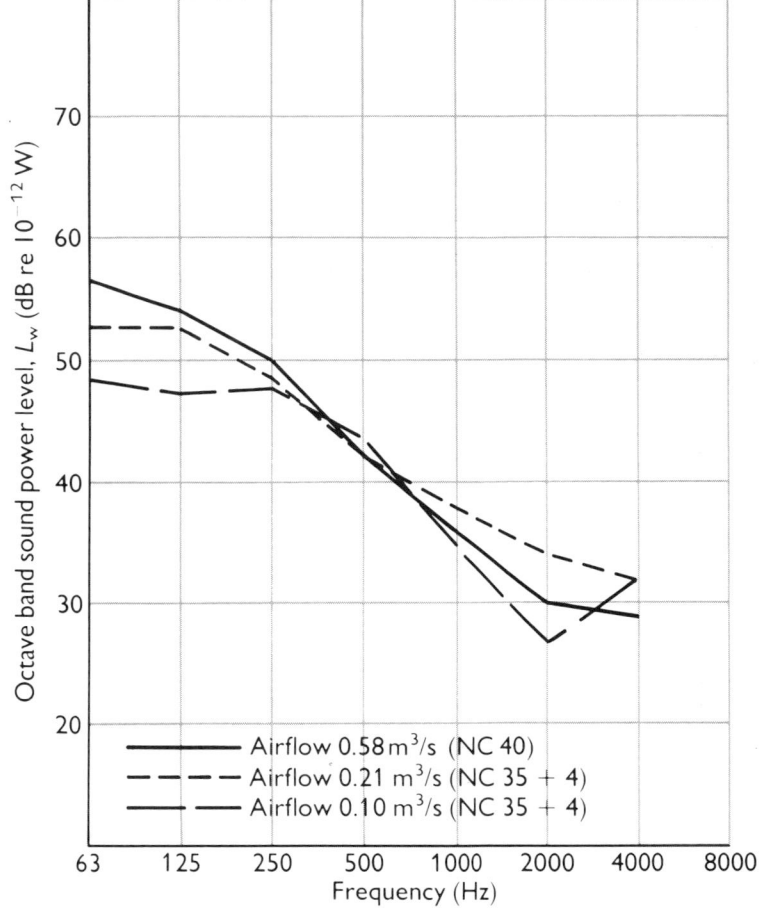

Figure 10.9 Acoustic Characteristics of Variable Flow Rate Terminal Units — Discharge Sound Power Level. Differential Pressure 250Pa

where possible. Figures 10.11 and 10.12 show the most common causes of noise from room units.

It should be mentioned in passing that some equipment manufacturers claim that their units, far from being sources of noise, do actually provide a certain degree of natural attenuation — while sometimes true this claim needs to be examined with some care, particularly if there is no evidence of acoustic treatment having been applied.

10.4.1 Duct-borne Noise

Duct-borne noise should not be a problem in a room unit since attenuation at remote sources of primary air should be designed to overcome the problems of plantroom fan

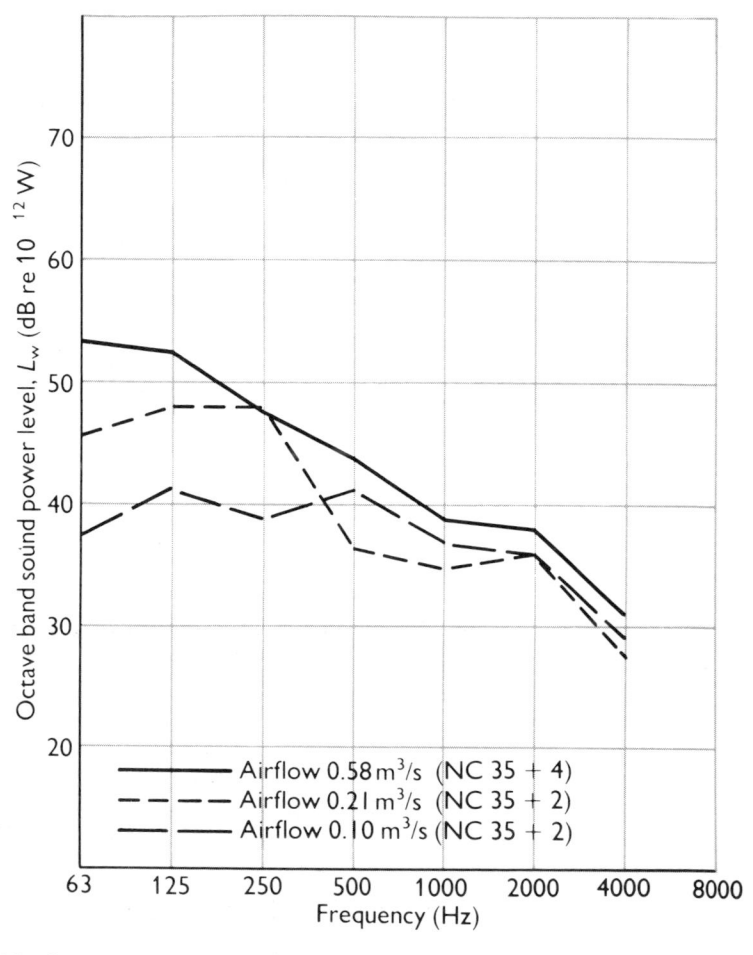

Figure 10.10   Acoustic Characteristics of Variable Flow Rate Terminal Units —
Radiated Sound Power Level. Differential Pressure 250Pa

noise (Chapter 7). However, breakout from high-velocity ducting, notably through
extended lengths of flexible connector is a common cause of high frequency noise,
while oval spiral ducting can be a weak link for low frequency noise breakout. It must
be remembered that noise generated by control valves at the terminal unit will also
travel back up the supply ductwork against the air flow.

10.4.2 Fan Noise

Any room unit which incorporates a fan — either for introducing fresh air, as in a
through-the-wall type of air-conditioning unit, for example, or for recirculation as
with a fan coil unit — is likely to require attenuation to meet the more stringent noise

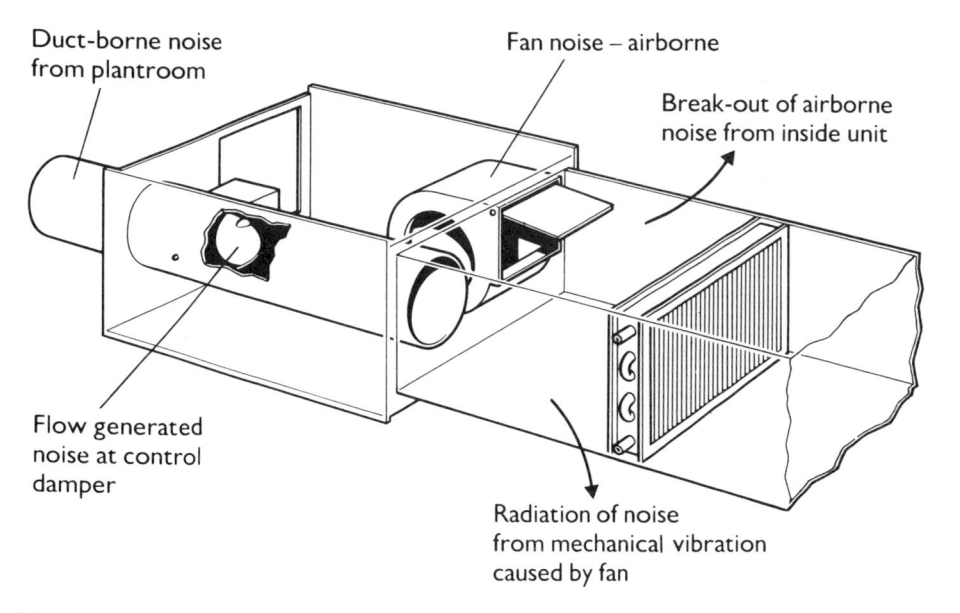

Duct-borne noise
from plantroom

Fan noise – airborne

Break-out of airborne
noise from inside unit

Flow generated
noise at control
damper

Radiation of noise
from mechanical vibration
caused by fan

Figure 10.11 Potential Sources of Noise Produced by a Fan-assisted Terminal Unit
(courtesy Senior Colman Ltd.)

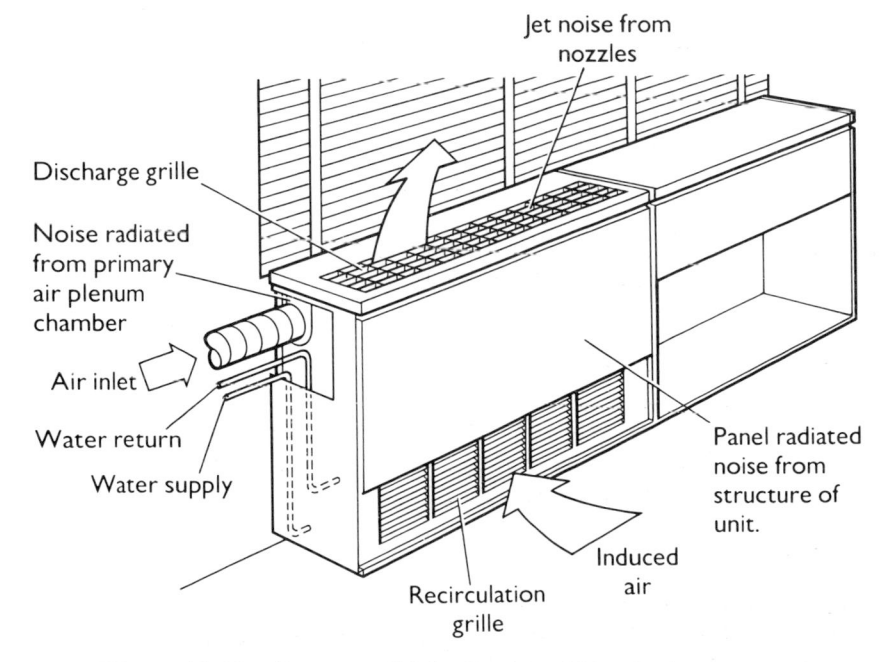

Jet noise from
nozzles

Discharge grille

Noise radiated
from primary
air plenum
chamber

Air inlet

Water return

Water supply

Panel radiated
noise from
structure of
unit.

Induced
air

Recirculation
grille

Figure 10.12 Sources of Noise Produced from Induction Unit

criteria. The first step towards this is to obtain manufacturers' data from which the best running speed is chosen to give the required ventilation performance, hopefully without exceeding the noise criterion. If the unit selected is likely to produce excessive noise it is then necessary to examine the sound power level in octave bands, both the discharge levels and case-radiated sound power levels, in order to establish how best to achieve the required noise reduction. Unfortunately, however, there is a marked lack of consistency in the form in which manufacturers present their acoustic data and while sound power levels are required, often information is limited to the room noise levels, usually expressed in NC terms and occasionally qualified by a room effect correction factor. In the case of units to be installed in a ceiling void, an acoustic insertion loss figure for the ceiling itself is sometimes included on the room rating data.

Alternatively, test data may be limited to A-weighted sound pressure levels at a prescribed distance from the unit,perhaps with reference to testing in free field conditions. While these latter NC or dBA expressions of unit output give a general indication of acoustic performance, they do not enable system design calculations to be made and consequent optimised frequency dependent noise attenuation procedures to be implemented.

Assuming that the discharge and case-radiated sound power levels can be obtained,it is possible to determine the appropriate form of acoustic treatment. Discharge noise will be reduced either by lining the unit with an acoustically absorptive material or by adding secondary attenuation at the outlet of the unit, although it should be remembered that the fan is unlikely to tolerate much additional flow resistance and so the pressure loss created by the attenuation must be kept to a minimum. Reference can be made to the fan curves to estimate the maximum additional pressure drop which can be tolerated in each case.

It should further be appreciated that useful reductions in low frequency noise levels can be obtained by using multiple small outlets since these will create greater natural attenuation due to end reflection losses, thereby minimising this noise contribution to the room, particularly at low frequencies. This comment applies equally well to the damper noise of terminal units, where the outlet noise may be divided by an octopus outlet to several small outlets.

10.4.3 Casing Sound Radiation and Break-out

Terminal and other types of room unit will radiate noise from their casings in two ways. First, the casing will have a limited sound insulation and airborne sound within the unit will break out, the amount of break-out depending upon sound power level, casing material, dimensions and thicknesses, stiffening effect of flanges or ribs, and possible case resonant behaviour. The second cause of casing-radiated noise is the transmission of vibration from a fan or, perhaps, flutter of internal components such as dampers which are excited into vibration by the airflow itself.

10.4.4 Jet Noise

The major noise source in a simple single induction unit is the "hissing noise" from the inducing nozzles themselves. Primary air travels through the nozzles at high velocity and generates mid to high-frequency noise (Figure 10.13). Induction units are the only type of room unit in which jet noise occurs but since they can be installed either under windows or in ceiling voids, the best solution in one case may not be ideal for the other situation.

Because higher frequencies respond more readily to acoustic screening it will be found that the detailed casing (often an architectural feature) around an induction unit has a large effect on the noise output and its directivity, the highest levels being measured in line with the nozzle outlets — usually vertically above.

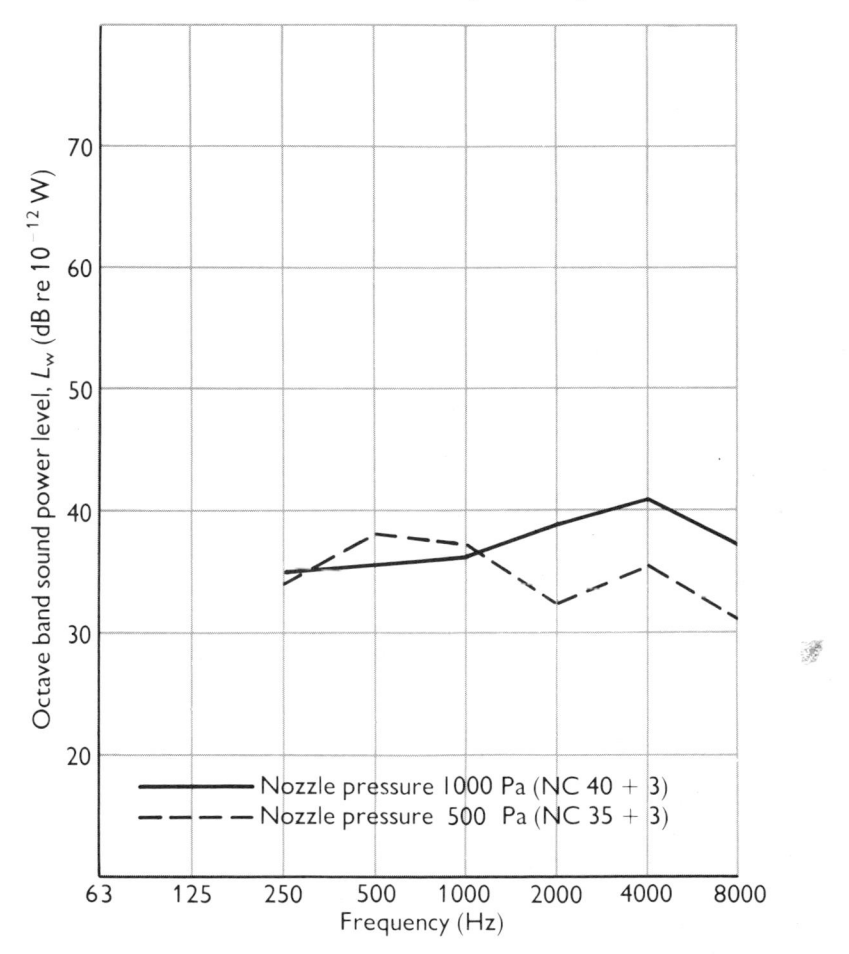

Figure 10.13 Acoustic Characteristics of Induction Units. Constant Primary Air Flow Rate 0.02 m³/s

10.4.5 Flow Noise

Flow noise is the result of local turbulence and may occur in the supply/return ducts or in a terminal or induction unit itself. In practice, many noise problems are attributable to this cause (see Chapter 8) and it should never be overlooked when designing a system which involves room units. The essential factors to be considered are:

Air velocity
Flow pattern
Fire and control damper design
Changes in cross-section of ducts or terminal boxes
Bends
Turning vanes
Header boxes
Diffuser design

10.5 HOW NOISE FROM A ROOM UNIT AFFECTS SOUND PRESSURE LEVELS IN THE ROOM

It is important to understand how noise from a room unit affects the room itself. New system designs will need to be checked to ensure that the acoustic design criterion will be met, while remedial measures may be necessary to achieve the same objective in the case of an existing room system. Figure 10.14 shows the factors which influence

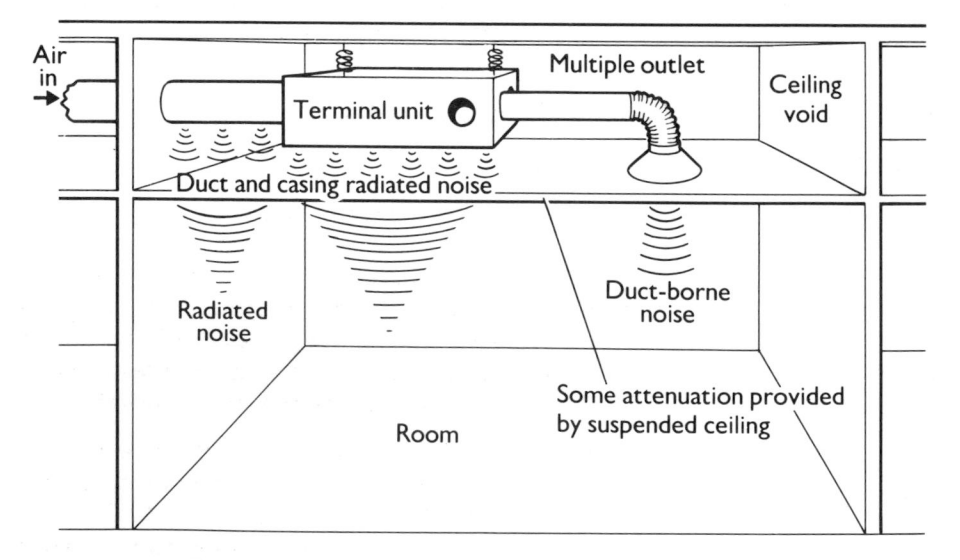

Figure 10.14   Transmission of Sound from Ducting and Terminal Unit in Ceiling Void into the Room Below

the resulting background noise in a room supplied by conditioned air from a terminal unit installed in the ceiling void. The unit has a finite sound power which is distributed via several paths into the void space and directly into the room space itself. The radiated sound pressure in the void will be attenuated as it passes through the suspended ceiling into the room, while ductborne noise feeds directly as a virtual point source at each grille position. The number of outlets has the effect of distributing the ductborne sound power as shown in Figure 10.15. Also, one is reminded that a greater number of small outlets will have less low-frequency noise output than a single large outlet because of the improved end reflection losses (Chapter 7).

Number of outlets	1	2	3	4	5	6	8	10	12
Duct outlet area, % of total	100	50	30	25	20	15	12	10	8
Sound power reduction dB per outlet	0	3	5	6	7	8	9	10	11

Figure 10.15   Outlet Sound Power Division at Multiple Outlets

The resulting noise in the room is a combination of the direct airborne sound from grilles, radiated noise from the terminal unit casing and inlet duct break-out noise.

The room itself has an absorptive effect, partly because of its volume and partly due to the acoustic absorption of wall, floor and ceiling finishes as well as room furnishings. Most technical data supplied by manufacturers of room units assumes a standard room absorption effect of either 10 dB or 8 dB. The precise value for a specific case, however, should be obtained from the correction tables in Chapter 7.

10.6 TECHNIQUES FOR NOISE CONTROL

10.6.1 Location or Positioning of Units

While room units are selected on the basis that they are required to serve local areas, and hence may be close to the occupants of those areas, every opportunity should be taken to physically locate and orientate the units away from critical areas. Even in an office no more than 5 metres square by 3 metres high, arranging for a desk and a single room unit to be as far apart as possible can save up to 8 dB in direct radiated sound level at the ear. If units can be positioned over less critical areas such as corridors or store-rooms, noise breakout from their casing may well be tolerable.

In the situation where terminal units or induction units are installed in the ceiling void, the ceiling itself can be selected to provide some useful attenuation; the Sound Reduction Index of typical ceilings being illustrated in Figure 10.16. Since the performance of the ceiling improves at higher frequencies this technique of noise control is particularly suited to induction units with their strong upper frequency output (Figure 10.13).

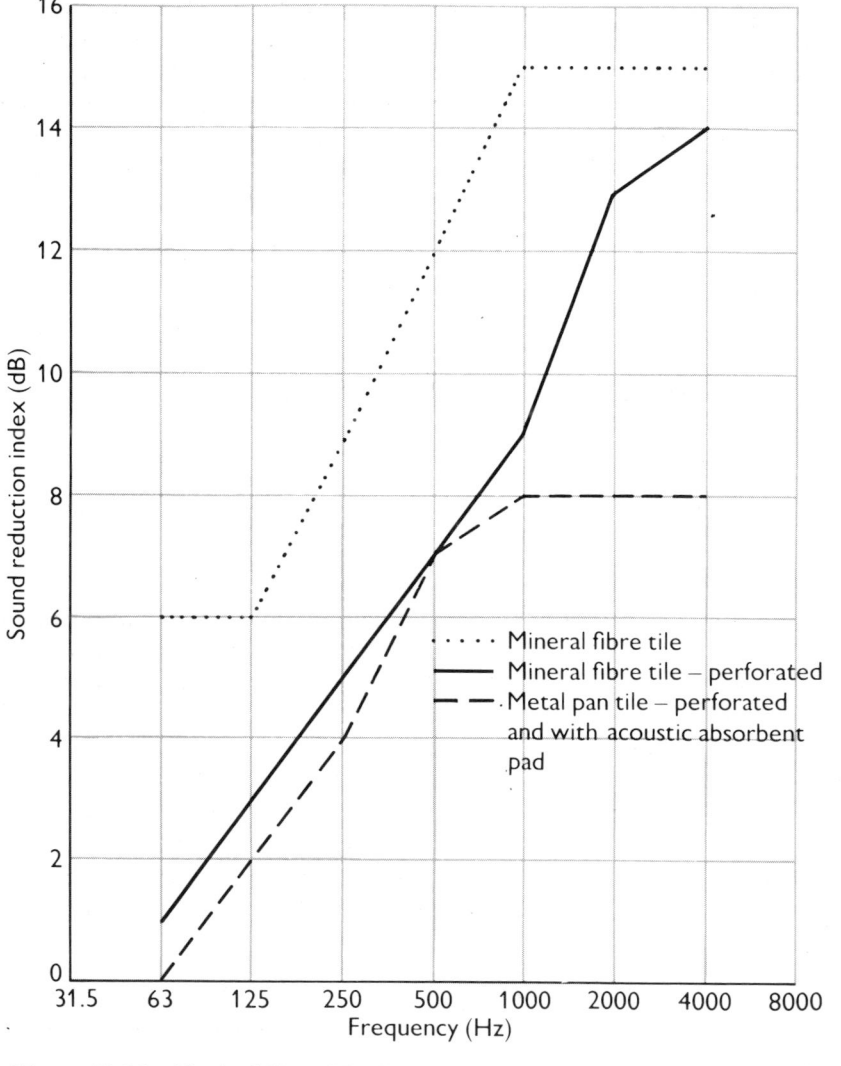

Figure 10.16   Typical Sound Reduction Index of Suspended Ceiling Tiles

It must be appreciated that in rooms which are particularly live (mid-frequency reverberation times in excess of 2s) there will be markedly less benefit from precise location of room units since the sound field is predominantly reverberant and the direct sound component is masked.

10.6.2 Selection of a Quiet Unit

Much manufacturers' literature is sadly lacking in definitive acoustic performance data and while the ventilation performance can be easily assessed, it is difficult to make comparisons of the noise produced by competitive units. Hence one is advised to choose only from units for which acoustic data is available. In order to design a system and select the room furnishings to meet a particular noise criterion sound power levels are currently the essential basis. Equipment suppliers should, when asked, provide laboratory test data from which it will be possible to select the unit most suited to the individual requirements. The minimum data required for calculation purposes is as follows:

Room-side outlet sound power levels in octave bands from 63 Hz to 4 kHz
Casing noise radiated sound power levels in octave bands from 63 Hz to 4 kHz.

Care should be taken in interpreting data which is quoted as on-site or room noise level in NC or dBA terms. The precise conditions of tests are rarely stated, may well be favourable to the unit under consideration, and inevitably have a major influence on the values produced.

The outlet sound power levels from terminal units of American origin often include "10 feet" of acoustically lined downstream ductwork in the test arrangement.

10.6.3 Ductwork Design

The fundamental requirements to design ductwork for good aerodynamic behaviour, thereby ensuring low noise levels, is every bit as important in close proximity to terminal units as it is in main and branch systems. Because the terminal unit is often of simple box construction with large flat resonant sides, it is capable of vibrating in response to low-frequency duct-borne noise and turbulence which may have been generated some distance up-stream by, for example, a poor take-off. Figure 10.17 shows the preferred form of take-off by comparison with designs which in practice often cause extra flow generated noise and poor entry conditions into the terminal unit, thus setting up the potential for further noise being generated in the terminal unit itself. An effect on terminal unit noise generation after a sharp 90° take-off is shown in Figure 10.18.

More general guidance on ductborne noise control is given in Chapters 7 and 8.

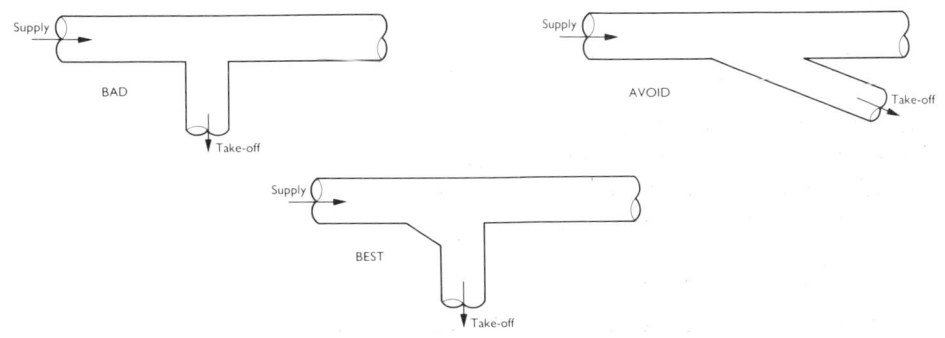

Figure 10.17 Examples of Take-off Design to Minimize Flow Generated Noise

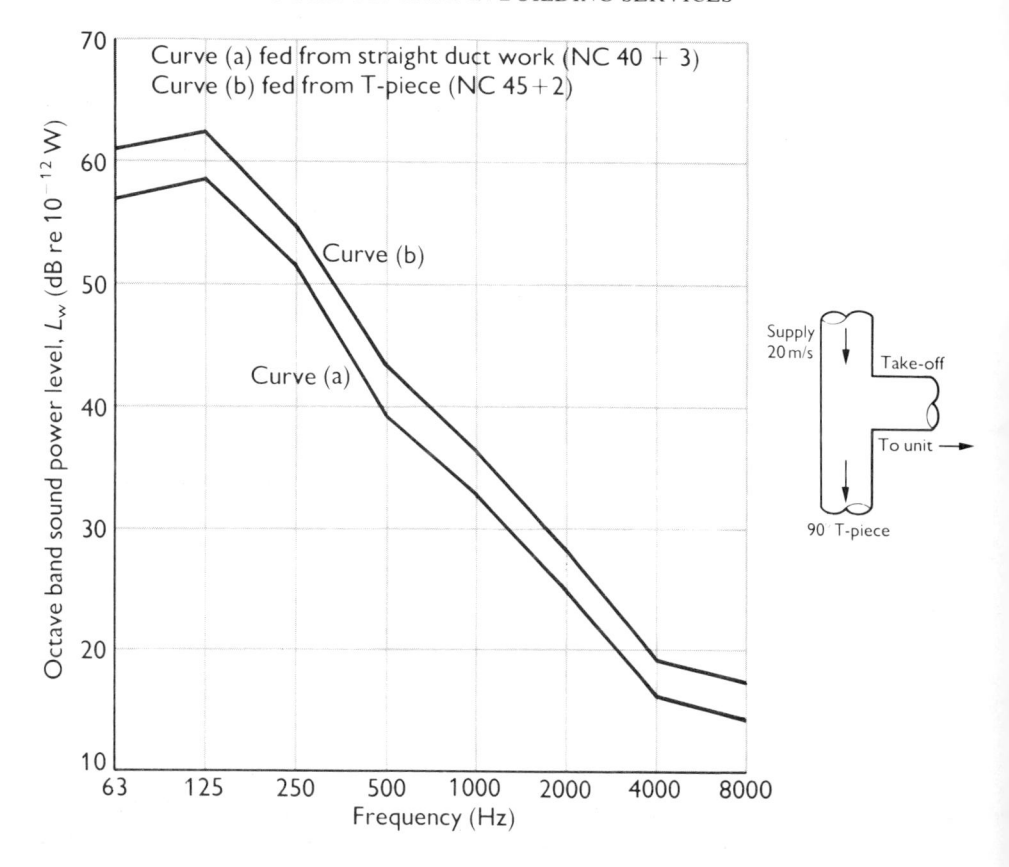

Figure 10.18   Terminal Unit Excess Outlet noise due to 90° take-off

10.6.4 Lagging of Terminal Units and Ducts

Before contemplating the control of noise breakout from ducts or terminal boxes, the availability from the manufacturer of suitably treated quiet options should be explored. The major manufacturers of terminal units now offer low noise breakout versions of their standard high velocity boxes, Figure 10.19 showing the substantial benefits in room or area NC levels that can be achieved. Typical frequency characteristics of break-out noise with different lagging or skin treatment are shown in Figure 10.20.

Where it is not possible to obtain low noise breakout room units then acoustic barrier materials should be wrapped round the casing to contain the sound. It may be necessary to use spaced barrier material where maximum sound insulation performance is needed. Similar materials are used to prevent duct breakout and are discussed in Chapter 9.

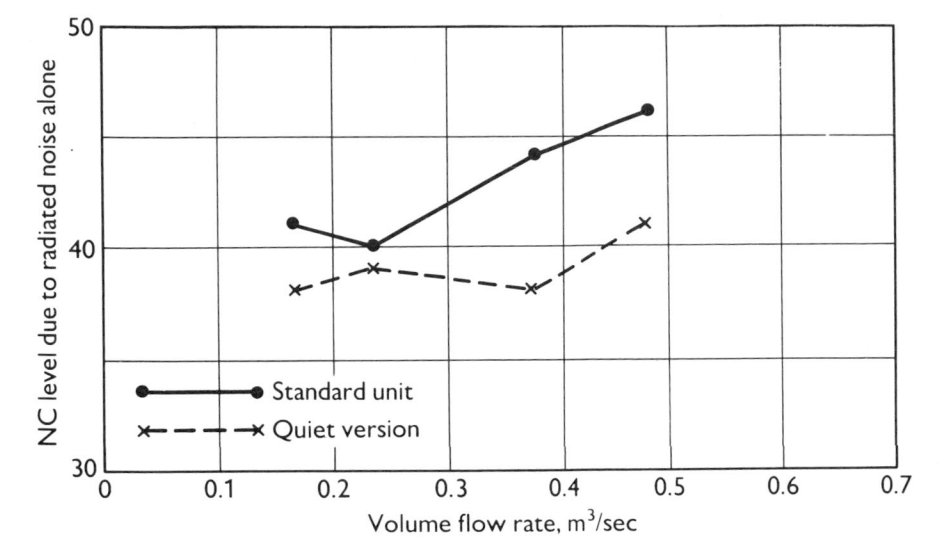

Figure 10.19 Room Sound Levels of Standard and Quiet Terminal Units as Affected by Volume Flow Rate

10.6.5 Secondary Silencing and Acoustic Lining

Discharge of outlet noise from a ducted room unit may be controlled by means of a small secondary attenuator, often supplied by the terminal unit manufacturer as a clip-on module to be fitted between the main body and the diffuser outlet.

Alternatively, a length of acoustically lined duct, preferably with the same cross-section as the discharge, can be attached to the discharge itself. Typical values of sound attenuation obtained by 1.5 m of lined duct (lining thickness 25 mm) are shown in Figure 10.21.

Attenuator section modules are usually manufactured with mineral fibre lining materials retained behind perforated metal. Lining of downstream ducting, however, relies sometimes on polyurethane foam materials with necessary fire-retardant additives.

With any secondary attenuation, it is very important to ensure that the additional flow resistance can be handled by the fan. All too often the fan is selected before the need for secondary attenuation has been identified and no allowance made for an increase in back pressure. This is typified in induction units which cannot stand any significant back pressure. Yet another aspect of "discharge" noise is the emission via the recirculation grille, where fitted. If this inlet to the unit proves to be a source of excessive noise, it will be necessary to provide supplementary attenuation as appropriate.

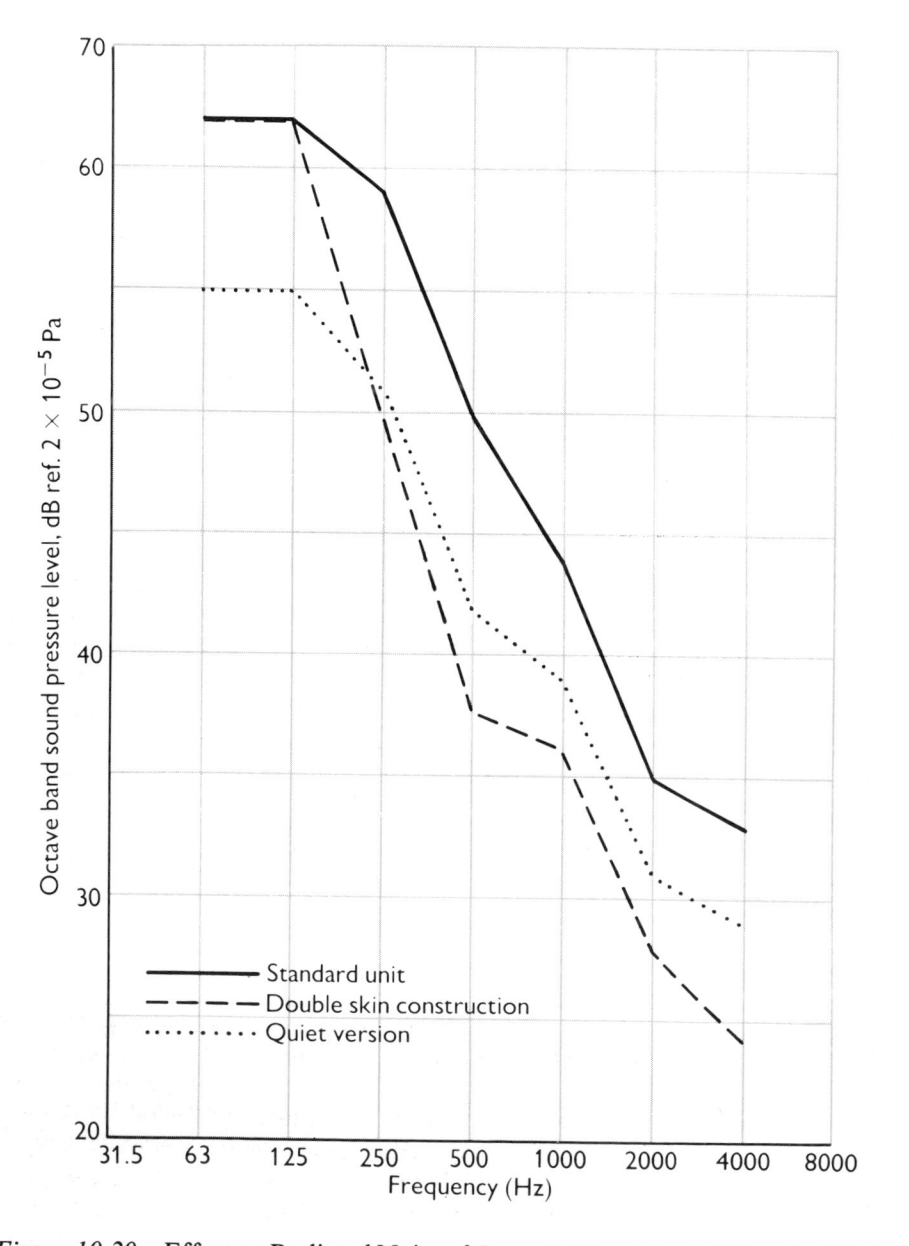

Figure 10.20   Effect on Radiated Noise of Acoustic Treatment of Terminal Units

| Internal dimensions of duct, mm | Octave band centre frequency, Hz | | | | | | |
(lining to lining)	125	250	500	1 K	2 K	4 K	8 K
250 × 200	5	10	20	38	40	18	9
400 × 250	4	7	15	30	38	15	7
600 × 250	3	7	13	28	35	15	7
950 × 250	3	6	12	23	32	12	7
950 × 300	2	5	9	18	16	8	6

Figure 10.21   Sound Attenuation of Lined Duct, 1.5 m long

Most terminal units, including single and dual-duct high velocity units and induction-type units, are lined for two reasons; to aid heat retention and to guard against a build-up of reverberant sound within the unit which would lead to excessive noise break-out.

10.7 TREATMENT OF NOISE FROM ATMOSPHERE

Fan coil units which introduce fresh air from the exterior facade of the building or which discharge to atmosphere can be a serious problem in that traffic noise can break-in to the room through the unit itself. While the coil system itself may provide a little attenuation, it is safe to assume that the unit represents a large open aperture in the wall and that it is necessary to incorporate a rectangular splitter attenuator which will effectively prevent break-in. In calculating the size of attenuator required, the site survey data should yield typical traffic noise levels at the facade and appropriate level of the building, or based on source data (Figure 10.22). The total level is then considered as the source noise in a calculation of the form used for determining break-out through a louvre (see Chapter 11).

10.8 MOUNTING OF ROOM UNITS

Ceilings and walls which are subjected to vibration excitation become potentially very effective radiators of sound. It is of paramount importance, therefore, to design room unit installations with comprehensive vibration isolation between the units themselves, ductwork and the building structure, suspended ceilings, lightweight partition walls, etc. The principles and methods by which vibration isolation is obtained can be found in Chapter 6. However, it is important to appreciate one particular detail. While flexible duct connectors (Figure 10.23) should be used extensively as vibration breaks, they should be kept to short lengths with the adjacent ducts or spigots being carefully aligned. Failure to do so will cause excessive noise generation due to air turbulence within the duct at this point, together with break-out through unnecessarily long flexible sections.

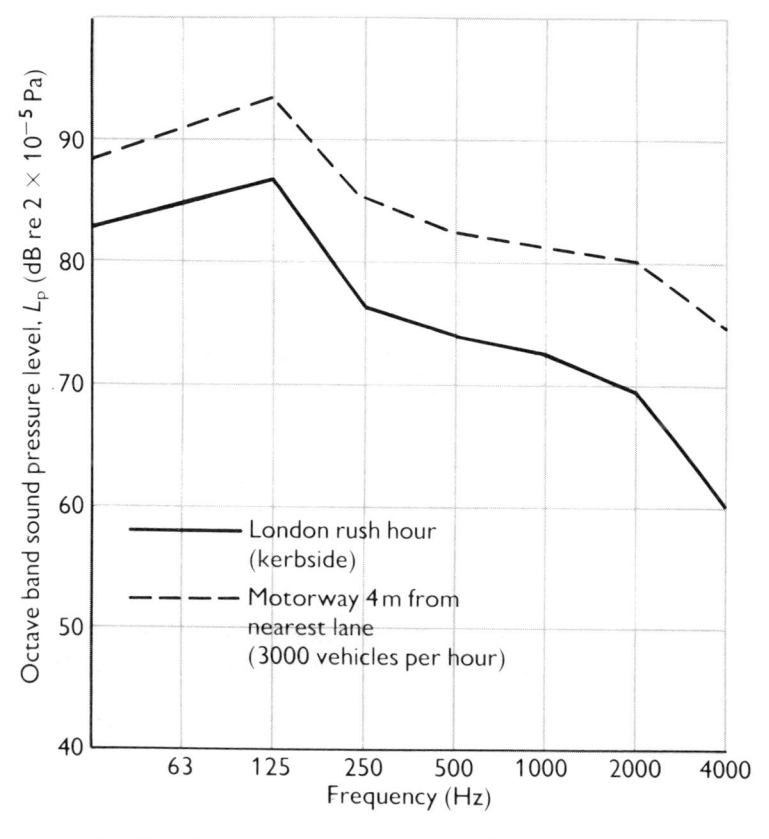

Figure 10.22 Traffic Noise Spectra (courtesy Road Research Laboratories)

It must also be remembered that "flexible" is to be considered in relation to the items being linked. Some installations which utilise "flexible" connectors onto lightweight ductwork derive no benefit from the relatively stiff isolation.

Flexible connectors are manufactured from, typically, acoustic barrier material of surface density around 5 kg per m², giving a mean Sound Reduction Index (100–3150 Hz) of about 22 dB. They have a limited working pressure of 2500 N/m² which is ample for room systems. Leakage rates through flexible connectors are lower than leakage at corresponding pressures of sheet metal ductwork, while maximum operating temperatures should usually not exceed 60°C.

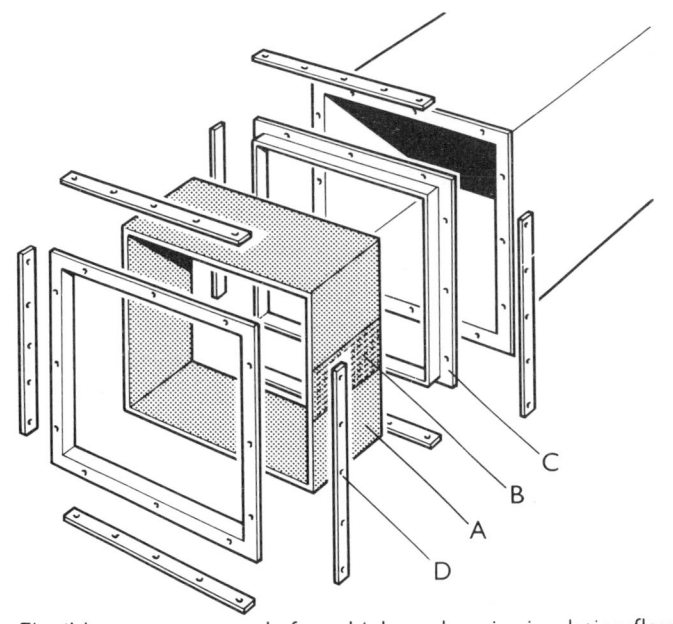

A. Flexible connector made from high grade noise insulation flexible barrier materials, jute backed and with tough anti-scuff PVC facing on outer surface.

B. Bonded ends for minimum air leakage and maximum working pressure. Suitable for H & V high velocity ductwork.

C. Pregalvanised matching end flanges, drilled as standard or to match customer requirements.

D. Steel fixing strips for rectangular connectors having pre-drilled holes and a sufficient quantity of self-tapping screws.

E. Steel banding clamps for circular connectors

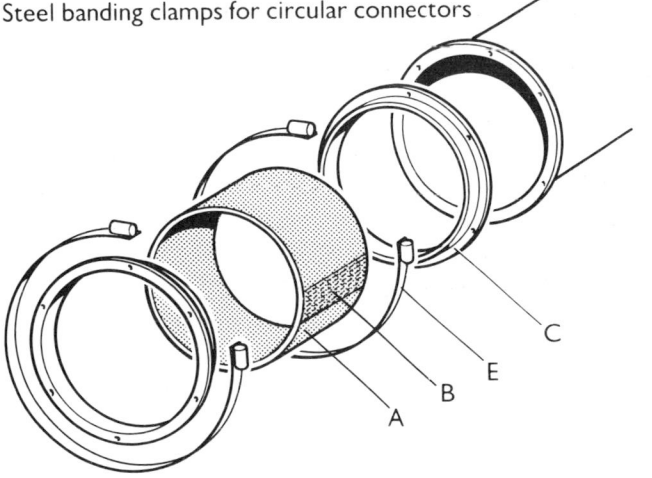

Figure 10.23　Flexible Duct Connectors

10.9 TROUBLE SHOOTING OF NOISE PROBLEMS FROM TERMINAL UNITS

Despite all these guidance comments for good noise control, site application problems will occur. These will first need to be correctly diagnosed and subsequently alleviated. To this end a flow chart is presented (Figure 10.24) to aid the systematic diagnosis of noise problems arising in high-velocity terminal units. The following supplementary notes should be read in conjunction with the flow chart.

NOTE 1 When the extract system is turned off, room doors should be opened to prevent a build-up of supply pressure. Ensure that transfer of internal noise from adjacent rooms having a higher noise criterion or external noise are not contributing to the problem.

NOTE 2 If, having eliminated all other possible noise sources, the room design criterion is met, this eliminates the air supply and control system as a source of excessive noise.

NOTE 3 This test is carried out to discover whether lower noise levels occur on floors further away from the plantroom. (It is assumed that the layout of air control boxes and their design volumes are similar on each floor).

NOTE 4 Excessive noise is most likely to be due to plantroom breakout or in-duct regeneration close to the plant, that is where volume flows are highest.

NOTE 5 This would tend to eliminate plantroom breakout and suggest that the boxes are the likely source of excessive noise. However, it could conceivably be caused by local regeneration at badly tailored branches from riser ducts to each floor.

NOTE 6 The reverberation times in critical areas should be measured at octave intervals from 63 Hz to 8 kHz and, using the correction values given in Chapter 3, true room corrections for the reverberant field can be derived. Additionally, the actual Sound Reduction Index performance of the acoustic ceiling should be measured. From this information the room correction factor from sound power level to reverberant sound pressure level can be obtained.

NOTE 7 If room correction factors are not in accordance with design, rectification will be necessary. However, all previous results should first be corrected for the difference between design and actual acoustic characteristics to check whether correction of this discrepancy would result in the design criterion being met.

NOTE 8 This test will positively remove any source of flow generated noise at points of air supply to the room. It is frequently found, particularly with large units feeding several different air outlets, that to achieve a balance between these outlets the dampers are highly constricted, thus creating excessive flow noise.

NOTE 9 If the source of the problem is flow generated noise at grilles, etc., the solution is to reduce flow velocity through the grilles.

NOTE 10 This test will eliminate airflow through the system and thus prevent localised flow generated noise. Noise from the boxes will be eliminated whereas fan noise will remain essentially the same.

NOTE 11 If noise levels measured in the room remain constant the requirement is for additional primary attenuation of the fan.

NOTE 12 If the noise levels measured in the room are reduced the problem is either localised flow generated noise or noise associated with the terminal unit itself.

NOTE 13 This test differentiates between the two possible causes identified in Note 12. By reducing the fan pressure until the controllers are just handling their design volume but at reduced total static pressure, a reduction in corresponding noise level will indicate a fault within the terminal unit.

NOTE 14 If the problem is one of local flow generation there is unlikely to be a change in noise level when the fan pressure is reduced but volume remains essentially constant.

NOTE 15 Having now found that the problem lies with the terminal unit, it remains to discover whether this is causing radiation, breakout of flow controller noise in the upstream ductwork, or in-duct noise from the discharge of the unit.

NOTE 16 A sound level meter is used to diagnose which of the three sources mentioned in Note 15 is contributing to excessive noise. Octave band analysis adjacent to each of the sources in turn will locate the cause.

NOTE 17 If highest levels are measured at the air outlet then secondary attenuation is required on the terminal unit. This assumes that there is no likely cause of downstream flow generated noise.

NOTE 18 If highest levels are measured at the casing of the unit then radiation is the problem. Check that the sound insulation of the ceiling is adequate (Ref Note 6) before lagging the casing.

NOTE 19 Excessive noise breakout from flexible supply ducting is not uncommon. If this is the case, then either lag the supply duct/s or up-rate the sound insulation given by the ceiling or introduce secondary attenuation after the terminal unit.

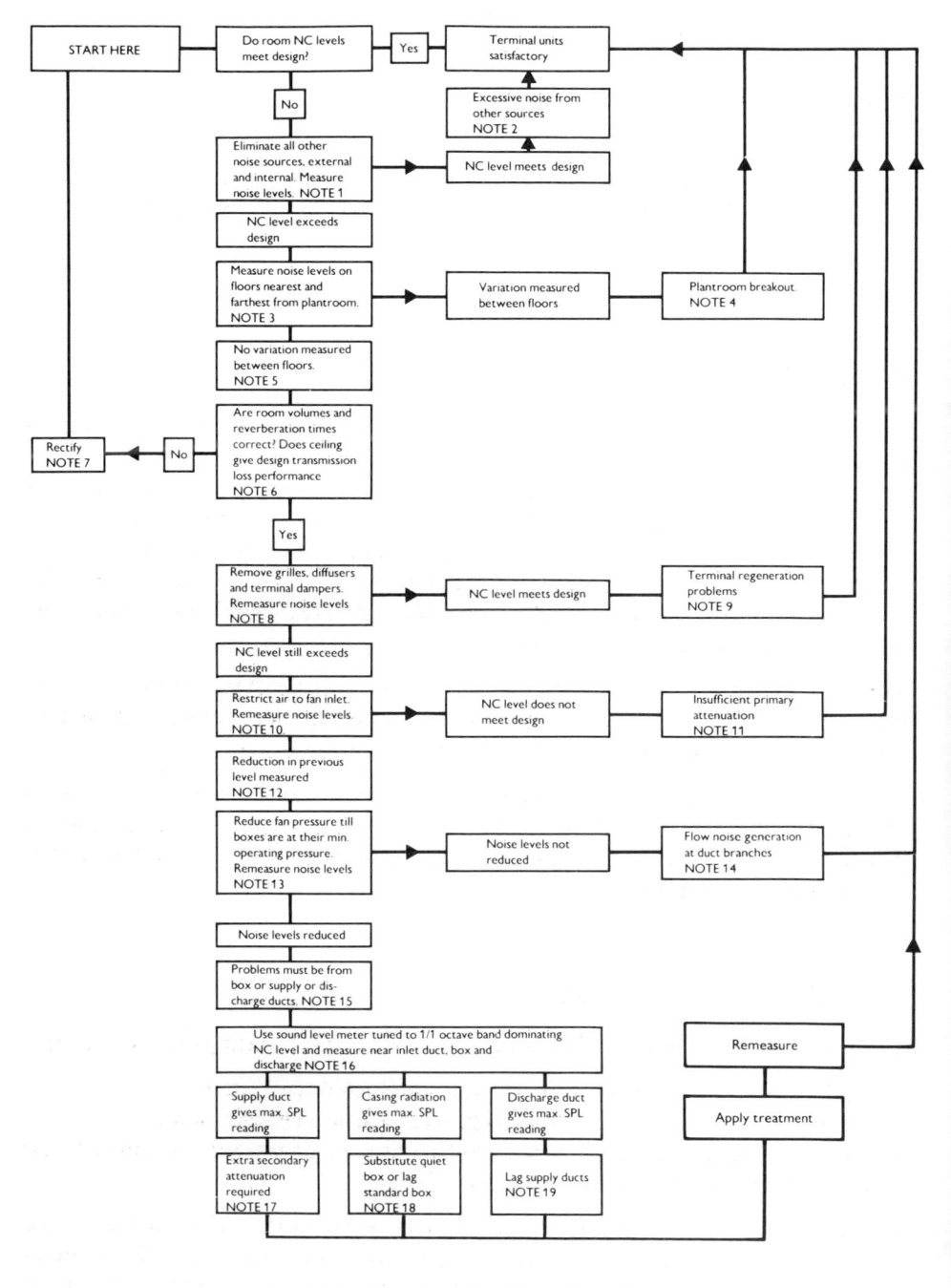

Figure 10.24   Trouble Shooting Chart

CHAPTER 11

NOISE TO THE EXTERIOR

11.1 INTRODUCTION

Although building services are usually provided for the benefit of the internal environment much of the noisy equipment is also linked to outside. Ventilation equipment requires air intake and exhaust whilst many items of equipment require their own ventilation to avoid overheating. Furthermore some items of equipment are actually located externally. It is therefore necessary to consider whether the noise transmitted to atmosphere will affect the nearby community. It is also possible for noise transmitted to atmosphere to break back into the building and cause the internal design criteria to be exceeded.

To assess whether either of these two cases could occur, noise to the exterior has to be evaluated and compared to either existing external background sound levels and/or the noise criteria selected for the internal environment of the building.

This chapter discusses typical sources of noise, how the noise is transmitted to atmosphere and how this can affect the nearby community and the building itself. The method of calculating noise to the exterior is dealt with and typical methods of noise control are discussed.

11.2 SOURCES OF NOISE

There are three basic sources of building services noise propagating to the exterior:

Plant situated externally e.g., cooling towers, fan cooled condensers.
External plantrooms e.g., housing diesel generator sets or chilling plant.
Ventilation openings serving internal plantrooms e.g., air intakes and exhausts, air handling plant or louvres for plantroom ventilation.

Cooling towers are normally mounted on the roof of a building and are therefore in a good position to radiate noise to the exterior. Fan cooled condensers are often situated on the roof but are, also, frequently located in lightwells or similar parts of buildings which results in a build up of noise in a reverberant area.

Other plant which is often installed externally includes compressors and packaged chiller units. Some of these items of plant, especially cooling towers or condensers, have to operate at night which becomes more of a nuisance to the neighbourhood because the night time background noise levels are, generally, significantly lower than the daytime levels.

External plantrooms housing such plant as diesel generators or chilling plant are normally located either at roof level or at ground level. Noise in these situations can emanate from ventilation openings serving the plantroom, via the plantroom structure itself — particularly from lightweight walls, doors or roofs — and also from engine exhaust pipes.

Figure 11.1 shows how noise sources can be located at various positions around a building.

Figure 11.1   Location of Noise Sources to the Exterior

STANDBY GENERATOR SET

Here the power unit is housed and serviced by a complete acoustic package, where-upon predicting the surrounding noise levels obeys the conventional rules starting from given sound power levels.

11.2.1 Frequency Content

A breakdown of the sound level should be known in order to establish the relative amounts of energy throughout the frequency range. Invariably this breakdown is given in terms of the sound levels in each octave frequency band between 63 Hz and 8 kHz. The actual units supplied can vary. In many cases sound power levels will be supplied. Sometimes, however, sound pressure levels will be stated. When sound pressure levels are supplied it is necessary to know at what distance the measurements have been made and, also, in what conditions — reverberant room, free field, etc.

11.2.2 Source Directivity

The directivity of a source is dependent on the dimensions of the source, the components of the equipment and the particular frequency under investigation. Sources of a constant size become more directive as the frequency increases, alternatively if the frequency remains constant then the directivity increases as the source size increases.

Some cooling tower and chiller manufacturers can supply sound data at various locations around the equipment and at various distances from the equipment. As sound levels in some directions can be 5–10 dB greater than in other directions this type of information is extremely useful when the equipment is to be located outside. Although in most situations it is most convenient to be supplied with sound power data of equipment, in the case of externally located equipment it is best to have sound pressure levels around the equipment. This is because overall sound power levels give no indication of directivity of the equipment whereas sound pressure levels at various locations can give useful information as to which way round to position the equipment and, as such, can save on noise control measures.

In the case of openings such as a louvre and open window–plane sources — the directivity of these apertures can be obtained by reference to Figure 11.2. Normally since these apertures are rectangular then the directivity in the horizontal direction will be different from that in the vertical direction because of the effective dimensions of the source width and height.

11.2.3 Surface Directivity

This is totally dependent on the location of the noise source and for omni directional sound sources has been defined earlier in Chapter 3 as the factor Q in the direct field sound pressure level equation

$$L_p = L_w + 10\log_{10}\left(\frac{Q}{4\pi r^2}\right) \qquad 11.1$$

for the direct sound calculation.

Frequency Hz	Width or Height of Louvre														
(m)	0.5	1	1.5	2	2.5	3.5	4.5	5.5	6	7.5	9	10.5	12	15 m
(ft)	1.5	3	5	7	9	12	15	18	20	25	30	35	40	50 ft
63	2	2.5	3	3	3.5	3.5	4	4	4	4	4.5	4.5	4.5	4.5
125	2.5	3	3.5	3.5	4	4	4	4.5	4.5	4.5	4.5	4.5	4.5	4.5
250	3	3.5	4	4	4.5	4.5	4.5	4.5	4.5	4.5	4.5	4.5	4.5	4.5
500	3.5	4	4.5	4.5	4.5	4.5	4.5	4.5	4.5	4.5	4.5	4.5	4.5	4.5
1 K	4	4.5	4.5	4.5	4.5	4.5	4.5	4.5	4.5	4.5	4.5	4.5	4.5	4.5
2 K	4.5	4.5	4.5	4.5	4.5	4.5	4.5	4.5	4.5	4.5	4.5	4.5	4.5	4.5
4 K	4.5	4.5	4.5	4.5	4.5	4.5	4.5	4.5	4.5	4.5	4.5	4.5	4.5	4.5
8 K	4.5	4.5	4.5	4.5	4.5	4.5	4.5	4.5	4.5	4.5	4.5	4.5	4.5	4.5

Table gives directivity at $\theta = 0$ for either horizontal bearing or vertical bearing of observation point from centre of louvre

Directivity correction for the normal axis of atmospheric louvres and grilles

D.I. KEY $\theta°$							
0	20	40	60	80	100	120	140
2	2	1.5	1.5	1	1	0.5	0
2.5	2.5	2	1.5	1	0.5	0	−1
3	3	2	1.5	0.5	−0.5	−1.5	−3
3.5	3	2.5	1	−0.5	−2	−4.5	−7.5
4	3.5	2.5	1	−2	−6.5	−12	−15
4.5	4	3	0	−15	−20	−20	−20

Directivity correction for a bearing angle away from the normal axis of atmospheric louvres and grilles

Figure 11.2   Plane Source Directivity Corrections

If the omni-directional source is suspended in free space the directivity factor $Q = 1$ and no allowance should be made. This may be the case when considering boiler flues or engine exhaust pipes.

Normally however the source is located on a surface and the source will radiate noise in a hemispherical pattern. In this case the directivity factor $Q = 2$ and an allowance of +3 dB should be made re L_p from equation 11.1. If the source is located at the junction of two surfaces $Q = 4$ and an allowance of +6 dB has to be applied.

For sources located at the junction of three surfaces $Q = 8$ and an allowance of +9 dB is made.

Figure 11.3 provides an illustration of these surface directivity patterns.

Example 1. A chiller is to be located externally next to a rooftop plantroom wall. The octave band sound power levels have been supplied. The sound pressure level at 16 m is required.

	Octave Band Frequency (Hz)							
	63	125	250	500	1k	2k	4k	Hz
L_w	85	87	88	90	89	85	81	dB
$10\log_{10}(1/4\pi r^2)$	−35	−35	−35	−35	−35	−35	−35	dB
$10\log_{10} Q$	+6	+6	+6	+6	+6	+6	+6	dB
L_p	56	58	59	61	60	56	52	dB

For this example $r = 16$m and $Q = 4$.

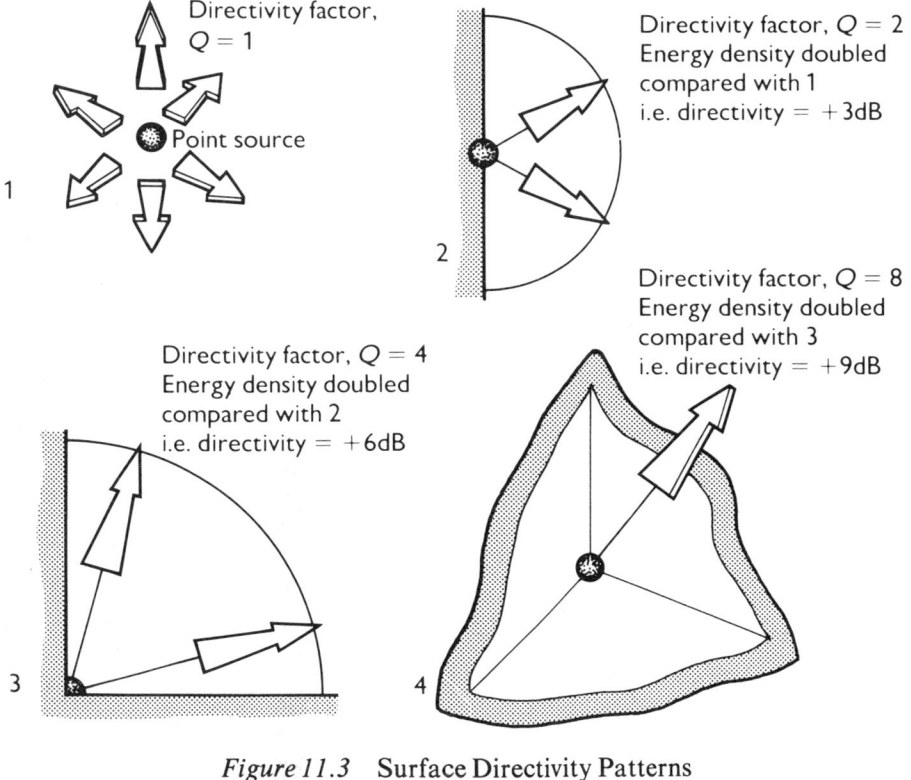

Figure 11.3   Surface Directivity Patterns

11.3 ACOUSTIC PROPAGATION

Once either the sound power level or the sound pressure level and the frequency content are known together with the directivity pattern associated with the source the noise level at the receiver's position is basically assessed from the following:

Distance from source to receiver and the dimensions of the source in relation to the propagation distance (11.3.1)
Whether screening exists between the source and receiver (11.3.2)
The existence of any reflections at the receiving position (11.3.3)

Other factors including atmospheric conditions and ground absorption which although discussed briefly later would not normally be included in a noise to exterior calculation unless the receiving position is at a great distance (11.3.4 and 11.3.5)

11.3.1 Attenuation with Distances

If the source can be considered to be a point source in relation to its distance from the receiving position the normal inverse square law of 6 dB per doubling of distance can be used.

$$L_p = L_w + 10\log_{10}\left(\frac{1}{4\pi r^2}\right) \qquad\qquad 11.2$$

This can be simplified as:

$$L_p = L_w - 20\log_{10}r - 11 \text{ dB} \qquad\qquad 11.3$$

where L_p = sound pressure level at receiver;
$\quad L_w$ = sound power level of source;
$\quad\quad r$ = distance between source and receiver, m; and
$\quad$ 11 dB = correction assuming spherical radiation = $10\log_{10}(1/4\pi r^2)$

However attenuation with distance may be taken directly from the chart shown in Figure 11.4 where these numbers are to be directly subtracted from the sound power level, L_w.

If the receiving position is located close to the source this inverse square law does not apply. The attenuation with distance for the three possible types of source, point, line and plane area, is shown in Figure 11.5.

For a line source, the reduction in L_p with distance up to the point a/π from the source, where a is the length of the source, is 3 dB per doubling of distance and thereafter the normal 6 dB per doubling of distance.

For a plane area source there is no "attenuation" with distance up to the point c/π, where c is the shorter dimension of the source area, the attenuation then becomes 3 dB per doubling of distance up to b/π, where b is the longer dimension of the source area, and thereafter 6 dB per doubling of distance.

Figure 11.6 enables an estimation of the total distance correction for various sizes of line or plane area source to be made, which may be subtracted directly from the sound power level, L_w.

Example 2. The Chiller described in Example 1 is 3 metres long by 3 metres high. What is the sound level at one metre and five metres from the chiller?

	Octave Band Frequency (Hz)							
	63	125	250	500	1k	2k	4k	
L_w	85	87	88	90	89	85	81	dB
$Q = 4$	+6	+6	+6	+6	+6	+6	+6	dB
to one metre	−11	−11	−11	−11	−11	−11	−11	dB
L_p at one metre	80	82	83	85	84	80	76	dB
to five metres	−25	−25	−25	−25	−25	−25	−25	dB
L_p at five metres	66	68	69	71	70	66	62	dB

11.3.2 Screening

If the source and receiver are not in direct line of sight then additional attenuation results from the screening effect that results.
This is illustrated in Figure 11.7.

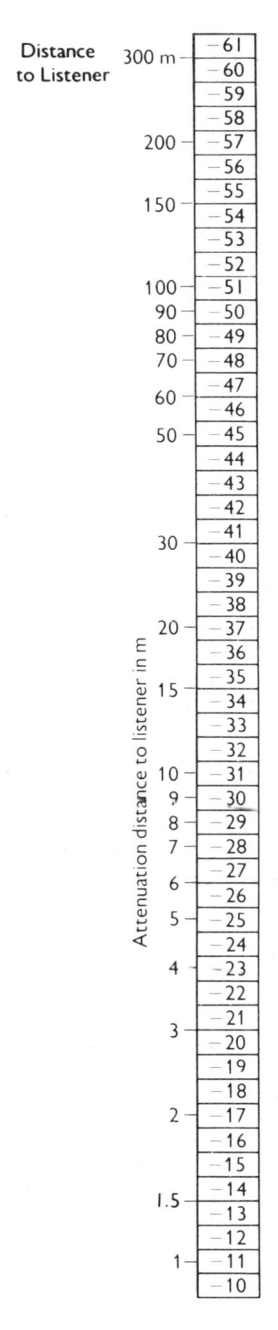

Figure 11.4   Attenuation with Distance from a Point Source

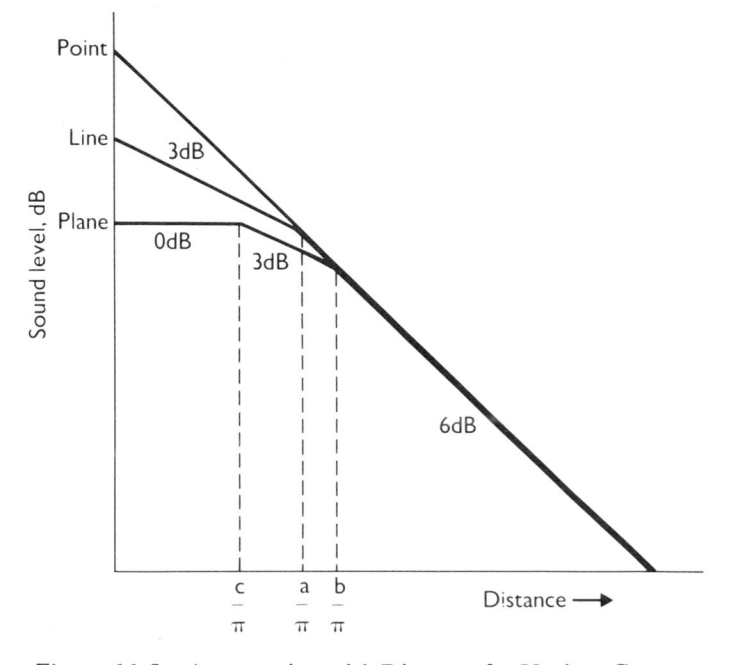

Figure 11.5   Attenuation with Distance for Various Sources

The screening effect varies with frequency because acoustic diffraction, responsible for the screening attenuation, is dependent on the wavelength of the sound in relation to the size of the screen. Screening is less effective at low frequencies than high. This is because the wavelength of low frequency sound is often equal to or greater than the dimensions of the screen and, as such, is not greatly affected by the presence of the screen. However the wavelength of high frequency sound is very small in comparison to the screen. High frequency sound is, therefore, more affected by the screen and a more pronounced cut-off will occur on the receiver side of the screen.

The attenuation provided by an infinitely long screen can be conveniently determined using Figure 11.8.

The examples shown are most frequently used in practice.

(a) δ is positive i.e., there is no direct line of sight between the source and receiver. An example of this case is the louvre in a roof top plantroom which is affected by shielding provided by a parapet at the edge of the building.

(b) δ is negative i.e., there is a line of sight between source and receiver but there is still an effect from a barrier adjacent to the source. An example of this is a cooling tower adjacent to a wall. Although there is unlikely to be any high frequency loss a useful degree of low frequency attenuation can be achieved.

(c) δ is positive but takes into account the effect of a thick barrier which could be an earth mound or a building.

Distance in metres (left axis); markers at 50, 30, 20, 10, 5, 3, 2, 1.

Distance	LINE SOURCE m				RECTANGULAR PLANE SOURCE m × m							
	3	10	30	100	1 × 3	1 × 10	3 × 3	3 × 10	3 × 30	10 × 10	10 × 30	30 × 30
50	45	45	45	45	45	45	45	45	45	45	45	45
	44	44	44	44	44	44	44	44	44	44	44	44
	43	43	43	43	43	43	43	43	43	43	43	43
	42	42	42	42	42	42	42	42	42	42	42	42
30	41	41	41	41	41	41	41	41	41	41	41	41
	40	40	40	40	40	40	40	40	40	40	40	40
	39	39	39	40	39	39	39	39	39	39	39	39
20	38	38	38	39	38	38	38	38	38	38	38	38
	37	37	37	39	37	37	37	37	37	37	37	37
	36	36	36	38	36	36	36	36	36	36	36	36
	35	35	35	38	35	35	35	35	35	35	35	35
	34	34	34	37	34	34	34	34	34	34	34	34
	33	33	33	37	33	33	33	33	33	33	33	33
10	32	32	32	36	32	32	32	32	32	32	32	32
	31	31	31	36	31	31	31	31	31	31	31	31
	30	30	30	35	30	30	30	30	30	30	30	30
	29	29	30	35	29	29	29	29	30	29	30	30
	28	28	29	34	28	28	28	28	29	28	29	30
	27	27	29	34	27	27	27	27	29	27	29	30
5	26	26	28	33	26	26	26	26	28	26	28	30
	25	25	28	33	25	25	25	25	28	25	28	30
	24	24	27	32	24	24	24	24	27	24	27	30
	23	23	27	32	23	23	23	23	27	23	27	30
	22	22	26	31	22	22	22	22	26	22	26	30
	21	21	26	31	21	21	21	21	26	21	26	30
3	20	20	25	30	20	20	20	20	25	20	25	30
	19	20	25	30	19	20	19	20	25	20	25	30
2	18	19	24	29	18	19	18	19	24	20	25	30
	17	19	24	29	17	19	17	19	24	20	25	30
	16	18	23	28	16	18	16	18	23	20	25	30
	15	18	23	28	15	18	15	18	23	20	25	30
	14	17	22	27	14	17	14	17	22	20	25	30
	13	17	22	27	13	17	13	17	22	20	25	30
1	12	16	21	26	12	16	12	16	21	20	25	30
	11	16	21	26	11	16	11	16	21	20	25	30
	10	15	20	25	10	15	10	15	20	20	25	30

Figure 11.6 Estimation of sound pressure level from Line and Plane Sources

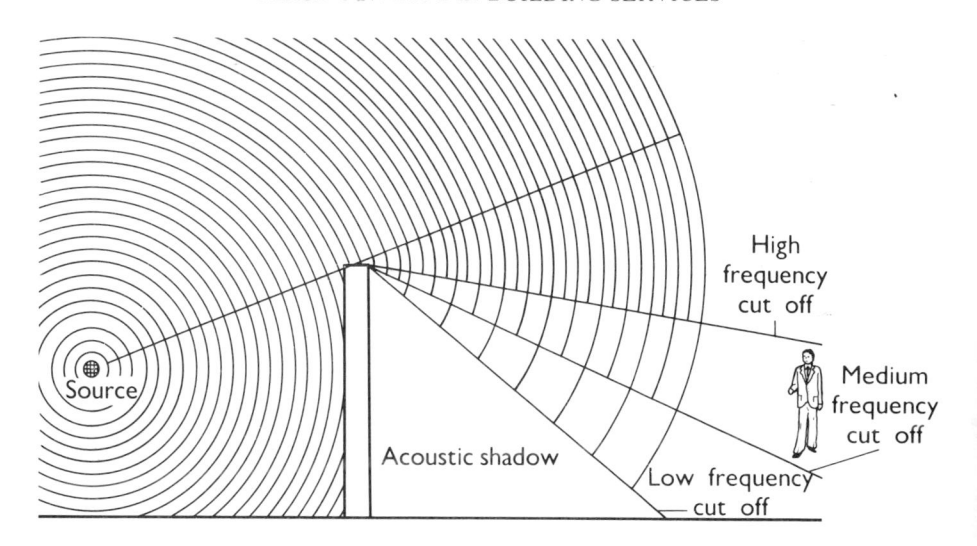

Figure 11.7   Acoustic Screening

Example 3. The Chiller described in Example 1 is to be located away from reflecting surfaces radiating hemispherically. The Chiller is located on steelwork 0.5 m high. The centre of the Chiller is therefore 2 metres above the roof. An infinitely long screen 4 metres high is to be located mid-way between the Chiller and a receiver positioned 16 metres away. What is the screening effect and the resultant sound pressure level. The screening is shown in Figure 11.9.

Length of path $A_1 + B_1 = 16.49$m

$\delta = 16.49 - 16 = 0.49$m

	Octave Band Centre Frequency							
	63	125	250	500	1k	2k	4k	Hz
L_w	85	87	88	90	89	85	82	dB
$10\log(1/4\pi r^2)$	-35	-35	-35	-35	-35	-35	-35	dB
Effective screening	-9	-10	-12	-15	-18	-20	-23	dB
L_p	44	45	44	43	36	30	24	dB

For this example the sound is approximated to emitting from the physical centre. It should be noted that in some cases the source should be take as some other location i.e. if there is a fan on top of the unit the source may be taken as being located on the top.

If the screen is of finite length, noise can be transmitted around the edges and the effect of the screen will be reduced. In this case the attenuation around the sides must

Additional Attenuation due to Screens
$\delta = A + B - d$

δ(m)	63	125	250	500	1k	2k	4k	Hz
−0.3	1	0	0	0	0	0	0	dB
−0.2	2	1	0	0	0	0	0	
−0.1	3	2	1	0	0	0	0	
−0.05	3	3	2	1	0	0	0	
−0.01	4	4	4	3	3	2	1	
0	5	5	5	5	5	5	5	
0.01	5	6	6	6	7	8	8	
0.05	7	7	8	9	10	12	13	
0.1	7	8	9	10	11	14	16	
0.2	8	9	10	11	14	16	19	
0.3	8	9	10	13	16	18	20	
0.4	9	10	12	14	17	20	22	
0.5	9	10	12	15	18	20	23	
1.0	11	12	14	18	20	23	25	
1.5	13	14	16	19	22	25	27	
2.0	14	15	18	20	24	27	29	
3.0	15	17	20	22	25	28	30	
4.0	16	18	20	24	26	30	31	
5.0	16	18	21	25	27	30	32	

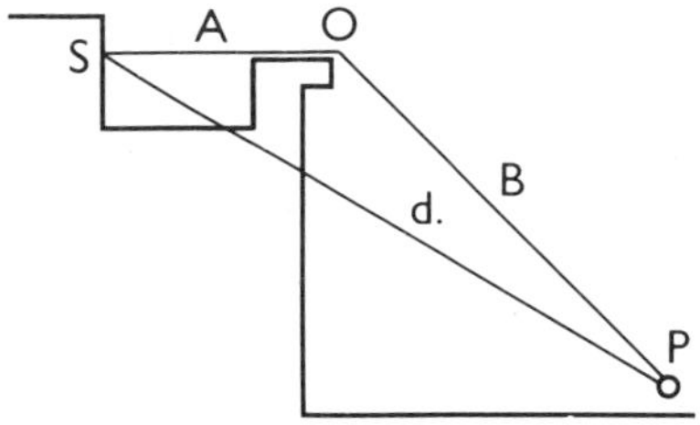

(a)

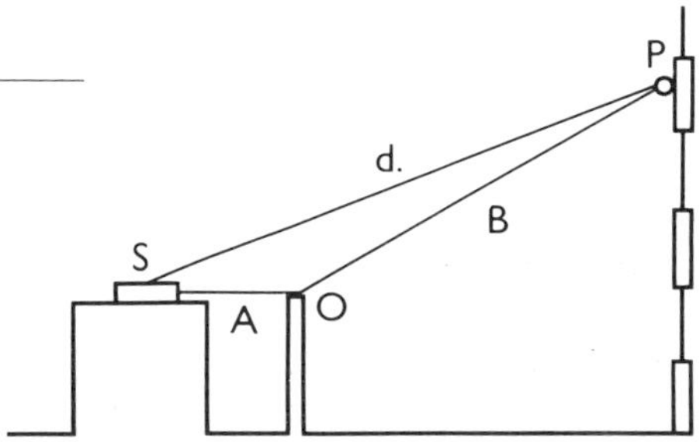

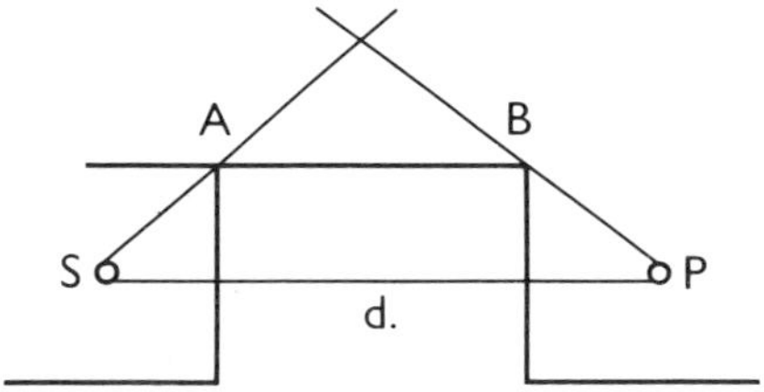

(c)

(b)

Figure 11.8 Attenuation Due to Screening

be determined in an exactly similar manner in order to establish the reduced screen attenuation.

If for example the attenuation over the top of the screen is calculated to be 10 dB and around the ends of the screen is 12 dB, the contribution from each path should be logarithmically added to yield an overall attenuation of 8 dB.

Example 4. The screen in Example 3 is only 10 metres long. What is the screening effect and the resultant sound pressure level. The screening is shown in Figure 11.9.

Length of path $A_1 + B_1 = 16.49$ m
Length of path $A_2 + B_2 = 17.88$ m
Length of path $A_3 + B_3 = 20$ m
$\delta_1 = 1.8$ m, $\delta_2 = 0.49$ m, $\delta_3 = 4$ m

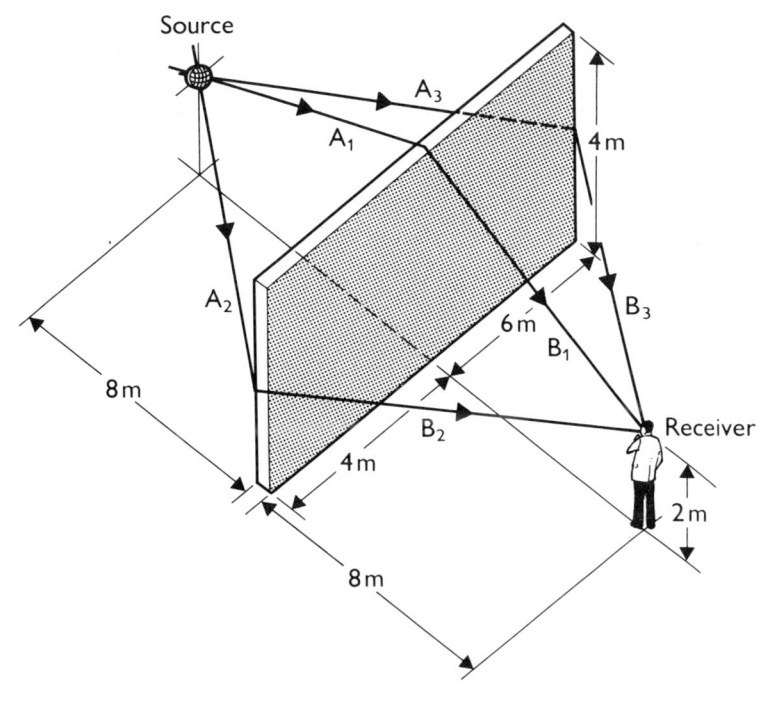

Figure 11.9 Screening Example

			Octave Band Centre Frequency				
	63	125	250	500	1k	2k	4k	Hz
Path 1	−14	−15	−17	−20	−23	−26	−28	dB
Path 2	−9	−10	−12	−15	−18	−20	−23	dB
Path 3	−16	−18	−20	−24	−26	−30	−31	dB
Effective attenuation	−7	−8	−10	−13	−16	−19	−21	dB

			Octave Band Centre Frequency				
	63	125	250	500	1k	2k	4k	Hz
L_W	85	87	88	90	89	85	81	dB
$Q = 2$	+3	+3	+3	+3	+3	+3	+3	dB
$10\log_{10} \dfrac{1}{4\pi r^2}$	−35	−35	−35	−35	−35	−35	−35	dB
Screening	−7	−8	−10	−13	−16	−19	−21	dB
L_p	46	47	46	45	41	34	28	dB

The construction of a screen need only be lightweight in most cases. Since the realistic limit for average screening is about 15 dB, the screening structure needs to provide an average Sound Reduction Index of no more than 20–25 dB, that is, have a superficial weight of approximately 6 kg/m². Acoustic louvres could be considered suitable screen material.

The structural and material requirements of the screen will often mean that more substantial screens than those required for acoustic reasons will be used. Although external screens do not have to be massive structures for acoustic reasons they must be impermeable, stable enough to withstand wind loads, durable enough to stand up to accidental damage and must resist weathering.

11.3.3 Reflection

In determining the sound level at a receiving position it may be necessary to establish the level at the facade of a building or similar. In this case for the broad band random noise considered here the reflection at this facade causes the sound level at positions close to the facade to increase by way of pressure doubling. If the receiving position is up to one metre from the facade a correction of +3 dB should be applied to take account of this reflection.

If for example the sound level calculated at a receiving position is 59 dB(A), but the receiving position is one metre from a large reflecting surface the resultant level will be 62 dB(A).

Pure tones and narrow bandwidth sounds create complex interference sound fields and generalised guidance cannot be given.

11.3.4 Atmospheric Attenuation

Weather conditions can affect sound propagation over long distances, however these effects are only of significance at distances of greater than 0.5 km.

The sound level at a receiving position at a long distance from a source will not generally follow the normal inverse square law with distance (attenuation at 6 dB per doubling of distance). Furthermore it will tend to fluctuate with time.

The factors affecting attenuation are:

(a) temperature and humidity — at temperatures greater than 15°C and humidities less than about 60% RH, the additional attenuation of sound over long distances becomes significant.

(b) wind gradients — wind effect on sound propagation is due to the wind gradient, that is, the increase in wind velocity with height above the ground, and not simply the wind "blowing noise along". The effect of wind gradient on sound is shown simplistically in Figure 11.10. The wind gradient causes the normally omnidirectional straight line sound rays to "bend upwards" on the upwind side of the source producing an acoustic shadow and downwards on the downward side to produce relatively higher levels. In most windy situations a wind gradient will be set up due to drag at ground level.

(c) temperature gradients — under normal conditions air temperature decreases with height — Figure 11.11a. During temperature inversions however the opposite occurs and symmetrical bending of sound rays takes place, as in Figure 11.11b.

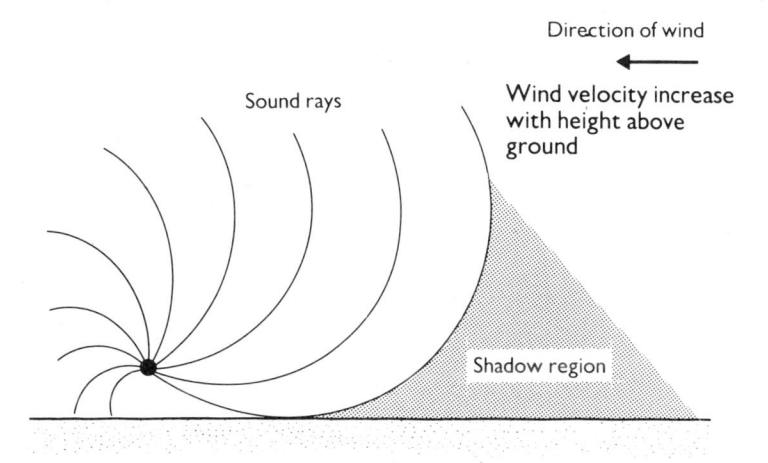

Figure 11.10   Effect of Wind Gradient

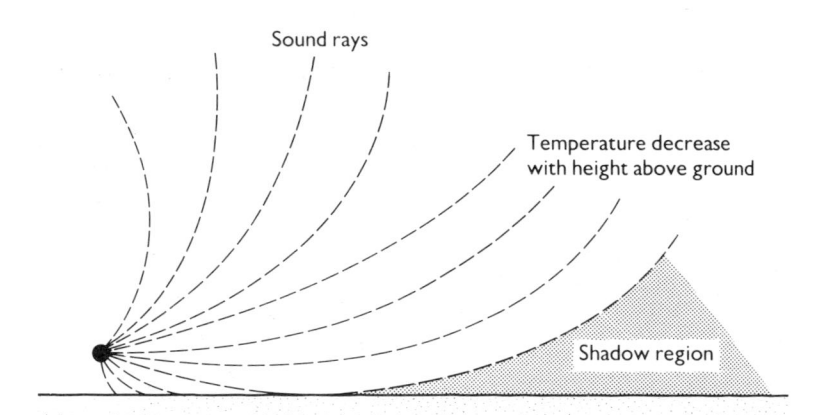

Figure 11.11a   Effect of Temperature Gradient — normal conditions

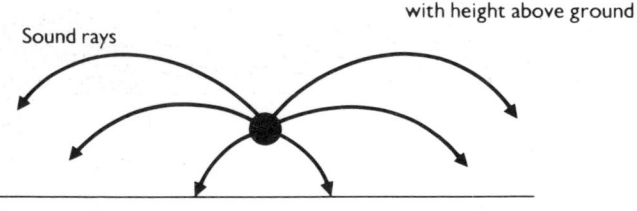

Figure 11.11b   Effect of Temperature Gradient — Temperature Inversion

(a) Propagation over hard ground: correction in dB(A) as a function of horizontal distance from edge of nearside carriageway d and height above ground h.

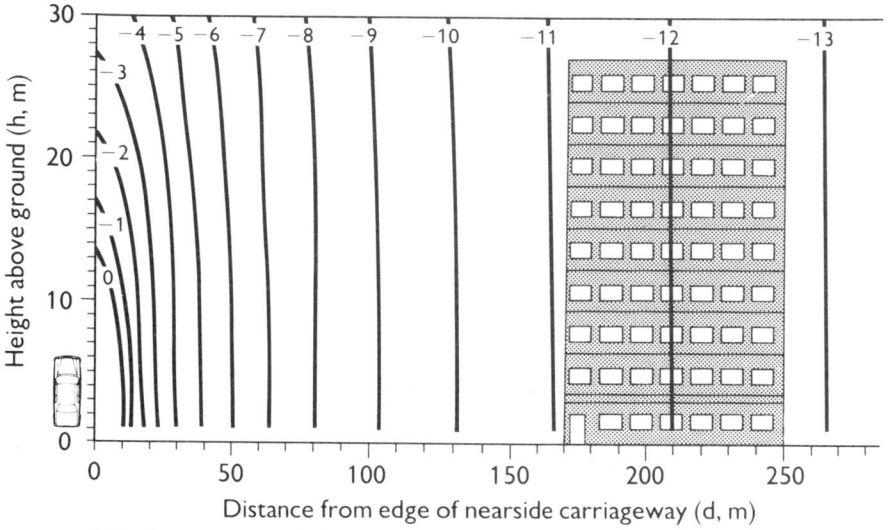

Effective source position
3.5 m in from kerb and 0.5 m high   Valid for distances greater than 4 m

(b) Propagation over grassland: correction in dB(A) as a function of horizontal distance from edge of nearside carriageway d and height above ground h.

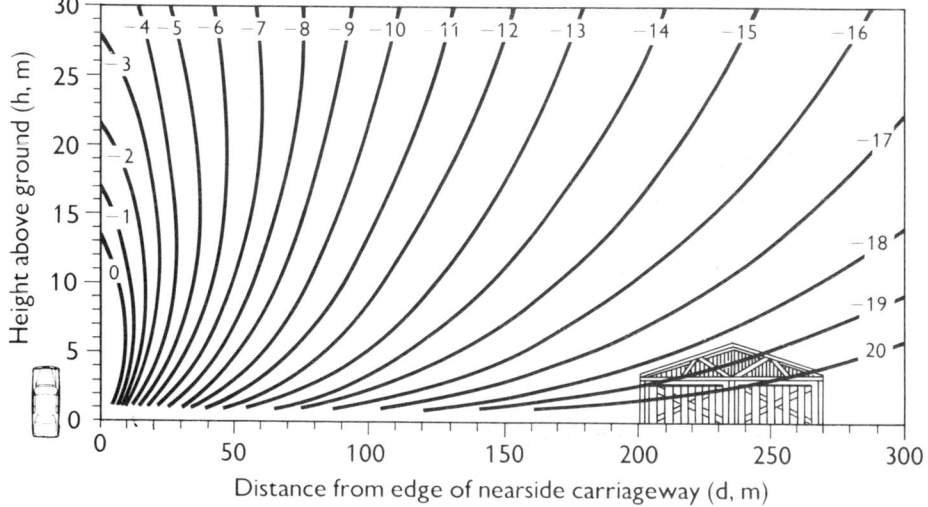

Effective source position
3.5 m in from kerb and 0.5 m high   Valid for distances greater than 4 m

Figure 11.12   BRE Traffic Attenuation over (a) hard surface, (b) soft surface

Fluctuations in sound levels at a receiving location at a long distance from a source have widely differing effects at different frequencies and can have considerable variations between peaks. The size of the fluctuations downwind increase with frequency and the distance from the source, whereas upwind the fluctuations are understandably greater near the shadow zone boundary.

In a stable atmosphere, such as weak winds and clear nights, the fluctuations typically range across 5 dB while in an unstable turbulent atmosphere, such as strong winds and clear sunny days, fluctuations from peak to peak are between 15 and 20 dB.

11.3.5 Attenuation Due to Ground Absorption

Ground covered with dense vegetation causes a greater attenuation than 6 dB per doubling of distance providing both source and receiver are relatively close to the ground. Dense woods also will provide additional attenuation but a single row of trees or shrubs will give no significant additional attenuation. Figures 12(a) and (b) shows how traffic noise in dBA decreases both over a hard surface and grassland.

11.4 CALCULATION OF SOUND TRANSMITTED TO INSIDE

Taking account of the factors discussed so far it is possible to calculate the sound level at an external receiving position due to the various types of building services noise propagating to the exterior. Figure 11.13 shows the steps that need to be taken into account.

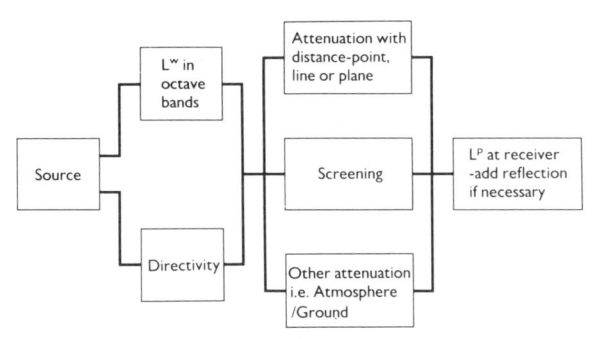

Figure 11.13   Block Diagram for Noise Control Calculation Problem

If the break-in of noise to a building from the exterior has to be assessed it is necessary to take the calculation a stage further. The resulting internal level will depend on the compound sound insulation provided by the building structure and the effect of the receiving room. In this case:

$$L_{pi} = L_{pe} - R + 10\log_{10}\frac{S}{A}$$ 11.4

where

L_{pi} = internal sound pressure level;
L_{pe} = external sound pressure level calculated or measured at the building facade;
R = Sound Reduction Index of the building structure;
S = surface area of building facade for this room, m^2; and
A = room absorption, m^2

If the room under consideration is an open plan space and hence supports a non-diffuse sound field, the internal position has to be selected at which the sound level should be calculated. For guidance the sound propagation in the open plan office will vary approximately as given in Figure 11.14. If a distance from the external wall of 2 metres is selected the correction is usually in the order of 3 dB i.e.,

$$L_{pi} = L_{pe} - R - 3\,dB \qquad\qquad 11.5$$

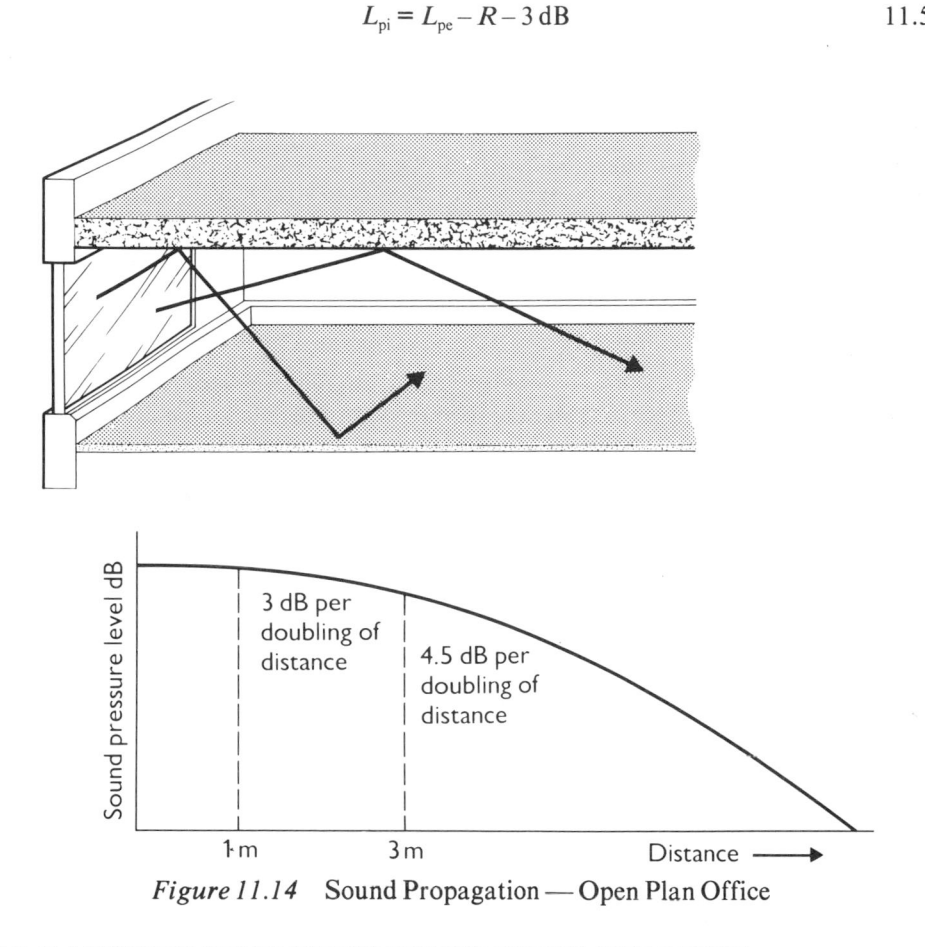

Figure 11.14   Sound Propagation — Open Plan Office

11.5 ASSESSMENT OF NOISE PROBLEMS TO THE EXTERIOR

In order to assess whether the sound transmitted to outside will be a problem, comparison can be made of either the predicted sound level or the measured sound level to the background noise level with the source switched off as was discussed in Chapter 4 with respect of BS4142. This is dependent on the time of operation of the source.

Example 5. The Chiller described in Example 4 is to be located in an area where the background sound level is 52 dBA during the day and 35 dBA at night. Will the screen provide sufficient noise control?

				Octave Band Centre Frequency				
	63	125	250	500	1k	2k	4k	Hz
L_p of Chiller	46	47	46	45	41	34	28	dB
Correction for dBA	−25	−16	−9	−3	0	+1	+1	dB
dBA Level	21	31	37	42	41	35	29	dB

Total dBA level = 46dBA.

Daytime background level is 52 dBA. Chiller is 46 dBA and is, therefore, unlikely to cause problems.

Night-time background level is 35 dBA. Chiller is 46 dBA and is, therefore, likely to lead to complaints (see Chapter 4, Design Criteria, for background noise criteria).

The screen is, therefore, sufficient providing the Chiller only operates during the day. If it runs through the night the screen will not provide sufficient noise control

In the case of breakin to the building served by the plant, noise criteria may have already been selected in order to define the preferred internal acoustic environment. In this case comparison can be made of either the predicted or measured sound level to the noise criteria to assess whether a noise problem exists or may exist.

11.6 NOISE CONTROL

The major methods of noise control are:

Relocation or reorientation of noise source — 11.6.1
Replacing the noise source with a quieter alternative or possibly limiting its period of operation — 11.6.2
Controlling the noise at source — 11.6.3
Screening — 11.6.4

11.6.1 Relocation

It is occasionally possible at design stage to relocate an item of plant with a potential noise problem to a non-critical situation. As an example a compressor planned to be sited external to a building could be relocated in a plantroom. Possibly a roof top plantroom may be located in the centre of large roof area in such a way as to increase the acoustic screening.

In addition it may be preferable to group all noisy sources together in order to reduce the amount of communal noise control necessary.

For a noise source with distinct directivity characteristics it may be possible to reorientate the source to effectively reduce the noise in the direction of the most sensitive receiving location. For example if an externally located plantroom may be redesigned so that ventilation openings and doors are positioned in the plantroom walls that do not face directly onto nearby residential property potential problems or costly noise control may be avoided.

11.6.2 Alternative

At the design stage it is sometimes possible to select a quieter type of plant to carry out the same operations or to select plant on which it is easier to apply noise control measures. In the case of a low speed reciprocating compressor an alternative could be a high speed screw compressor. Since the noise emanating from the screw type will be at a higher frequency it will be easier to control than the "diffusive" low frequency of the reciprocating type.

The period of operation of such plant as boilers or cooling towers may be limited so that during the night a potential noise problem is avoided. However in some cases, the normal operating noise is not a problem but the transient noise level at start up and shut down may be louder and as such cause a problem.

11.6.3 Noise Control at Source

In this section it has been assumed that the source under consideration is operating efficiently and that no reductions of noise are possible from the moving parts of the plant.

Noise control at source can be undertaken in the following ways:

installing attenuator units or acoustic louvres
constructing an enclosure around the plant

In plantrooms with atmospheric louvres providing fresh air intakes or exhaust outlets, it may be possible to provide a conventional absorptive attenuator unit in the duct between the fan and louvre. Alternatively if only a modest degree of attenuation is required then it could be provided by the installation of an acoustic louvre. Care must be taken in selection of an acoustic louvre to meet a specified attenuation, since the acoustic performance is sometimes given in the form of a noise reduction which includes an additional 6 dB attenuation above the simple insertion loss as normally stated for an attenuator.

Enclosing the sound source can sometimes be the most effective method of noise control. Enclosures can vary from lightweight materials such as plywood or plasterboard up to heavyweight materials such as dense concrete blocks. Proprietary acoustic enclosures such as 50 mm of glass fibre between 2 skins of steel sheet provide sound reduction somewhere between a lightweight and heavyweight material. To be most effective the source must be totally enclosed. It is, however, necessary with a lot of equipment to provide ventilation for cooling. Air cooled condensers and generators are typical examples. Unless only small sound reduction is required, any openings for ventilation must not degrade the performance of the enclosure and should, therefore, either contain attenuators or acoustic louvres. When only a small reduction is required a partial enclosure, which provides natural ventilation, may be sufficient. In this case the material used only needs to be lightweight as the sound reduction only requires a small weight.

11.6.4 Screening

The installation of screening between the source and receiver can provide a practical solution if only 10–15 dB(A) attenuation is required. The dimensional details of the screen have already been discussed and can be assessed by using Figure 11.8.

If a screen is designed as a method of noise control it may be necessary to consider the reflected noise from the screen back towards the source. This could be important for example if a screen is located close to a louvred opening in a plantroom below an office. This is shown in Figure 11.15.

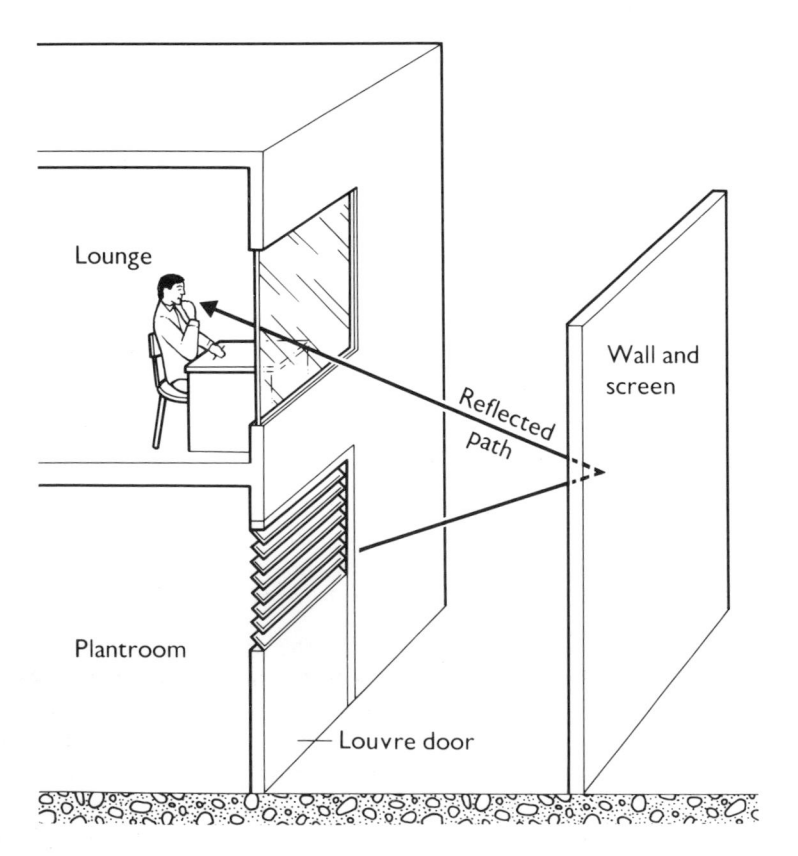

Figure 11.15 Reflections from Screens

In this case the surface of the screen on the source side should be faced with an acoustic absorbent material. Woodwool slabs are useful in this case since if rendered on only one side, the remaining side has a relatively high sound absorbent surface.

11.7 NOISE RADIATION FROM AN INTERNAL PLANTROOM

Air handling plant in an internal plantroom require to take in fresh air and/or to discharge used air. Other equipment such as chillers and pumps require ventilation to avoid overheating. Air intake and discharge associated with air handling plant will normally be through ductwork and grilles or louvres although in some cases one of these paths may be directly into the plantroom. Plantroom ventilation will either be mechanical through a fan system or naturally through louvred walls or louvred doors.

In both ducted and louvred intakes it is necessary to calculate the sound power level at the plantroom boundary. If there are several sources each has to be taken into account. As with externally located noise sources factors affecting the level of noise radiating from the source are:

Sound power level of the source
Frequency content of the sound
Directivity of the source dependent both on its dimensions and its location

The sound power level of a source should be obtained from manufacturers and as explained earlier it is given in terms of dB re 10^{-12} watts. In the case of moving systems the in-duct sound power level of the fan should be obtained. Any duct losses or losses through air handling unit components between the fan and atmosphere or the plantroom should be taken into account in calculating the sound transmitted to atmosphere/the plantroom. In the case of systems ducted to the plantroom wall this will give the sound power at the wall. In the case of non-ducted systems it is necessary to calculate the resultant sound pressure level in the plantroom as described in Chapter 3. The following correction can then be applied to this sound pressure level to give the effective sound power level at the opening L_w where

$$L_w = L_p + 10\log_{10}S - 6\,\text{dB} \qquad 11.6$$

where L_p = reverberant sound pressure level in the plantroom, dB; and
S = area of aperture, m².

In the case of other equipment it is necessary to obtain the sound power level of the equipment and calculate the resultant sound pressure level in the plantroom, as described in Chapter 3. In calculating the resultant sound pressure level it is necessary to take all items of equipment into account. If a manufacturer can only supply noise data in terms of sound pressure level it is necessary to obtain information on the distance at which the measurements were made and under what conditions ie free-field, semi-reverberant, and then estimate the sound power level of the source.

Once the sound power level at the plantroom wall is known the sound pressure level at distances from the plantroom can be calculated as discussed earlier. In calculating this sound level both distance and directivity must be taken into account.

Example 6. The Chiller described previously is to be located in a plantroom of 200 m³ with a reverberation time of 1.5 s. An acoustic louvre 2 m x 2 m is to be located in one wall. What is the sound pressure level at 16 m from the louvre?

			Octave Band Frequency (Hz)					
	63	125	250	500	1k	2k	4k	Hz
L_W	85	87	88	90	89	85	81	dB
Room volume — 200m³	−9	−9	−9	−9	−9	−9	−9	dB
Reverberation time — 1.5s	+2	+2	+2	+2	+2	+2	+2	dB
L_p in plantroom — reverberant	78	80	81	83	82	78	74	dB
Area of louvre — 4m²	+6	+6	+6	+6	+6	+6	+6	dB
−6	−6	−6	−6	−6	−6	−6	−6	dB
L_W at louvre	78	80	81	83	82	78	74	dB
Directivity — Q	+6	+7	+8	+9	+9	+9	+9	dB
Distance − 16m	−35	−35	−35	−35	−35	−35	−35	dB
L_p at 16m	49	52	54	57	56	52	48	dB

If acoustic louvres were used the sound reduction of the louvres could be subtracted from the above levels to obtain the sound level with acoustic louvres.

CHAPTER 12

CONSTRUCTION SITE NOISE AND VIBRATION

12.1 INTRODUCTION

Construction activities are often inherently noisy and can also produce high levels of structure borne or ground borne vibration. In this Chapter, we will consider just two aspects relating to construction site noise and vibration transfer, namely:

(a) transfer to the environment outside the boundary of the construction site, and
(b) transfer within an existing building during refurbishment works.

12.2 TRANSFER OF CONSTRUCTION SITE NOISE AND VIBRATION TO THE ENVIRONMENT OUTSIDE THE SITE BOUNDARY

12.2.1 Legislation

The transfer of noise and vibration from construction sites to the environment outside the site boundary is subject to the legislation contained in Sections 60 and 61 of the Control of Pollution Act, 1974. The Act, which came into force in 1975, does not set limits or advise on control techniques but does empower local authorities to act if they feel a nuisance exists, or is likely to occur.

There are two procedures that may be adopted in applying the Act and these have become known as the "notice" and "consent" procedures.

The "notice" procedure is described under Section 60 of the Act and places the onus for action on the local authority, whereby a notice would be served on a contractor specifying one or more of the following:

(a) Plant or machinery which is or is not to be used
(b) Hours during which works may be carried out
(c) Noise levels which may be emitted
(d) Provide for any change of circumstances

In acting under Section 60, the local authority must however have regard to:

(a) Relevant Codes of Practice issued under this Part of the Act.
(b) The need for ensuring that the best practicable means are employed to minimise noise.
(c) The interest of the recipient before specifying plant or machinery.
(d) The need to protect any persons from the effect of noise.

A person served with a notice may appeal against the notice to a magistrates' court within 21 days from the service of the notice.

The second approach, outlined in Section 61 of Act, places the onus on the contractor or other persons responsible to apply to the local authority for consent giving details of working methods and noise control procedures. The local authority may either grant such a consent or grant a consent subject to certain conditions.

It is a requirement of the Act that the local authority shall inform the applicant of its decision on its application within 28 days from receipt of the application.

If:

(a) the local authority does not give a consent within the 28 days period, or
(b) the local authority gives its consent within the 28 days period, but attaches any condition to the consent or limits or qualifies the consent in any way,

the applicant may appeal within 21 days from the end of that period.

12.2.2 British Standard BS5228

The control of noise and vibration transfer from construction works to the environment outside the site boundary poses special problems. The works are normally carried out in the open air and, although they may be of temporary duration, they can cause disturbance while they last, particularly in built up areas or areas containing sensitive buildings such as hospitals or schools.

Noise and vibration from construction sites can arise from a wide variety of activities and plant, and their intensity can vary markedly for different phases of the works. There is a requirement, therefore, for identifying effective methods of noise and vibration control to assist the responsible contractor in satisfying relevant environmental standards.

The degree and type of control required will vary from site to site and from activity to activity. In order to include the costs of any noise or vibration control in the tender price for the job, it is necessary to predict the noise or vibration levels from the proposed method of working with reasonable accuracy at the tender stage of a contract. Suitable techniques for noise prediction do exist, while reliable techniques to predict vibration are yet to be evolved.

There is often a need to monitor construction site noise and vibration to demonstrate that the site operator is working within the required environmental standards. For noise, such monitoring is often required over 12 hour working days, and this has led to consideration of sampling techniques to save man power and equipment costs.

All of the above aspects relating to noise are considered in BS5228: 1984 Noise Control on Construction and Open Sites. The standard consists of four parts, namely:

Part 1: Code of practice for basic information and procedures for noise control
Part 2: Guide to legislation for noise control applicable to construction and demolition, including road construction and maintenance
Part 3: Code of practice for noise control applicable to surface coal extraction by open cast methods
Part 4: Code of practice for noise control applicable to piling operations (in course of preparation).

The standard does not deal in any detail with vibration from construction operations.

Part 1 of the standard gives sufficient information and plant noise data to allow the A weighted equivalent continuous sound level from static and mobile plant to be calculated at the construction site boundary with reasonable accuracy. Where this level is above the standards set by the local authority, or is otherwise thought likely to be the cause of a nuisance, then consideration must be given to the various techniques of noise control. Part 1 of BS5228 also gives guidance on the monitoring of noise from sites for the purposes of assessing compliance with noise control targets.

12.2.3 Environmental Noise Standards

There are no set environmental noise standards applicable to construction operations and, since works are often of limited duration and sporadic in nature, the guidance available for more continuous and long term sources such as transportation and industrial processes are far from appropriate. It is the responsibility of the local authority to set noise limits appropriate to the area in which works are to be undertaken. However, several factors should be considered in setting standards including:

(a) ambient noise in the vicinity of the site prior to commencement of construction works.
(b) the time of day during which noise producing activities occur.
(c) the duration of the noise producing activities during the overall contract period.
(d) the character of the noise.
(e) the character of the area in the vicinity of the site.

Perhaps the most important consideration is the pre-existing ambient noise level. If this is high any increase which lasts over a long period of time is undesirable and such intrusion should, wherever possible, be avoided. However, certain construction operations are intrinsically noisy and silencing to bring levels down to desirable long term environmental standards would often preclude practicable progression of the works. Therefore, there is a strong case for some relaxation of environmental noise standards if works are to be of short duration and if best practicable means of noise control have been demonstrated. Noisy activities can often be successfully carried out with minimum disturbance if they are timed to coincide with periods of low noise sensitivity. For instance the majority of people are at work and away from residential areas during the daytime whilst it is during evenings, early mornings or weekends that office environments will be least affected.

Where pre-existing ambient noise levels are low, it is impossible to carry out any construction works without having some effect on the local environmental noise climate even if the most stringent noise control is adopted. Therefore, noise limits which prohibit any increase in ambient noise levels under these circumstances are totally restrictive to the contractor and consequently inappropriate. What needs to be considered is the type of work to be carried out and the quietest methods of achieving the desired result without the imposition of restrictive costs for noise control. Flexibility on behalf of the local authority and a responsible attitude on behalf of the contractor is, therefore, called for in these situations to define and implement the best practicable means of noise control.

12.2.4 Environmental Vibration Standards

Vibration standards fall mainly into two categories:

(a) Those appropriate to human comfort.
(b) Those appropriate to superficial or structural damage to buildings.

There is a wide disparity in the standards proposed in the various references on this subject.

It is generally accepted that humans are particularly sensitive to vibration stimuli and any perception tends to lead to concern, if not direct disturbance. However, a good deal of this concern is associated with fear of damage to property and much can be done to alleviate such fears through good public relations on the part of the contractor. Experience suggests that, provided measurement surveys are carried out to demonstrate vibrations are not excessive in terms of damage, the following vertical vibration peak particle velocities will be tolerated during the daytime:

Single event impulsive vibrations (blasting etc): 12 mm/s
Continuous impulsive or steady vibrations of
limited duration (piling etc): 2.5 mm/s

It is unreasonable to expect people to withstand such intrusion from more continuous sources either for extended periods or during the evening or night-time.

Vibration velocities of this magnitude are not likely to lead to even superficial damage to structures, although if they are continued over extended periods, say several years, fatigue failures may occur.

Experience suggests that the following vibration limits in terms of resultant peak

CONCRETE MIXER

The ubiquitous concrete mixer with its "chugging" engine, seen here in traditional but unfortunate mode with the engine cover up and, hence, losing the useful few dB's of engine casing noise screening.

GENERAL SITE SCENE

Amongst much general purpose site construction equipment, the tall tower cranes dominate, with the potential for high-up unscreened noise sources radiating in all directions. Tower cranes are now subject to European Economic Community noise control limits.

CIRCULAR SAW

A particularly irritating noise source, which is not at all easy to reduce or control whilst still maintaining its easy access and general usability.

ROAD DRILL

Probably the most striking and intense noise source, which often imposes itself close to everyday and otherwise peaceful locations — in this case, the village market square. However, the drills here are of an advanced design with not only a quieter conceptual hammer design, but also the inclusion of built-in acoustic jackets. Nevertheless, the working rattling chisel then remains the key noise source.

Courtesy: Compair-Hofman

particle velocities can be used to assess the likelihood of damage. These figures are relevant to continuous steady vibration lasting for relatively short periods, that is days rather than years:

Onset of architectural damage: 10 mm/s
Onset of structural damage: 25 mm/s

12.2.5 Major Sources of Noise on Construction Sites and Their Control

The major sources of noise on construction sites likely to cause environmental disturbance can be summarised as follows:

(a) Impulsive noise sources; hammer driven piling rigs, rock drills, concrete breakers etc.
(b) Earth moving plant; bulldozers, scrapers etc.
(c) Stationary plant; compressors, generators, pumps etc.

Methods of reducing noise from such plant are described in BS5228. They fall into two main categories:

silencing noise at source
controlling the spread of noise

A further source of noise is blasting operations which require special consideration.

12.2.5.1 Silencing Noise at Source. There are several general methods of silencing noise at source including:

(a) The use of more efficient exhaust silencers than those fitted to the manufacturers standard plant.
(b) The use of acoustic shrouds for plant such as concrete breakers.
(c) The use of damped tool bits, damped sheet piles and damped metal casings.
(d) The use of acoustic enclosures.
(e) The use of alternative or "super silenced" plant.
(f) Ensuring that manufacturers engine covers and access panels are kept closed.
(g) Good maintenance standards and care on site.

Hammer driven piling is one of the most intense noise sources encountered on construction sites and one of the most difficult to silence. Extensive conventional silencing methods will probably only yield between 5 and 10 dB(A) attenuation over the standard plant where 30 dB(A) would be more desirable. Alternative methods of piling, where appropriate can provide more beneficial reductions in sound levels. For example, bored piling rigs can be between 10 and 20 dB(A) quieter than a hammer driven rig and some of the proprietary silenced piling systems can be up to 30 dB(A) less noisy.

Pneumatic rock drills and concrete breakers also generate high noise levels and although manufacturers silencers can provide useful attenuation, it is often necessary to provide additional silencing by enclosing the machine in a suitably ventilated portable or fixed acoustic enclosure.

Alternatively, hydraulic or electric plant is available which is considerably quieter than the corresponding pneumatic plant.

Noise from earthmoving plant can often be effectively controlled by fitting engine covers and more efficient exhaust silencers. However, "super silenced" plant is now available from certain manufacturers which can be used in particularly sensitive situations.

Noise from stationary plant can be controlled by the use of suitably ventilated portable or fixed acoustic enclosures. To be effective such enclosures should cover the machines as fully as possible, be constructed of a material which has adequate sound insulation characteristics and be lined internally with an efficient sound absorbing material so that noise reflected internally does not readily transmit through openings.

Since openings are nearly always necessary for ventilation purposes, then a material exhibiting an average sound insulation value of 25 dB should normally be sufficient. Such materials as chipboard, woodwool cement slabs 50mm thick faced with 13mm plaster, brickwork, plasterboard etc., would be adequate. The resulting insulation is likely to be of the order of 20 dB which is a useful reduction for machinery noise. If, however, the noise source is predominatly low frequency in nature, then heavier materials should be used.

The internal acoustically absorbent lining should normally be 25 mm thick material such as mineral wool behind a perforated facing.

It should be emphasised that advice should be obtained from the manufacturer of the machinery to be enclosed to ensure that adequate ventilation and access to the machine is provided.

"Super silenced" petrol or diesel driven compressors, generators and pumps are also available. These are particularly useful since they are often required to run continuously for long periods both day and night and consequently are often subject to strict limitations on noise output.

Other simple noise control measures can provide useful reductions in overall site sound levels. These include the use of screws as opposed to nails, the use of sharp saws and drills and good practice when handling materials such as scaffold poles by avoiding dropping them from a height.

From 26th March 1986, all new compressors, welding generators, hand-held concrete breakers, tower cranes and power generators have all been subject to legislation as to their noise output. The legislation is in the form of Directives which, incorporated in a document entitled "The Construction Plant and Equipment

(Harmonisation of Noise Emission Standards) Regulations 1985", specify the permitted noise levels for such machinery when used as construction plant. All such new construction plant and equipment needs to be marked or "plated" to show it has been subjected to an EEC type examination and that a certificate of conformity is held by the supplier.

While this legislation will inevitably lead to quieter construction sites, compliance with the provisions of the legislation will not prove a defence against any action taken under the Control of Pollution Act, 1974. Also it will take time for the older equipment, not yet subject to these certifications, to become redundant and be replaced.

12.2.5.2 Controlling the Spread of Noise. There are basically two methods of controlling the spread of noise from construction sites. The first is to increase the distance between the noise source and the receiver and the second is to introduce noise reduction screens.

(a) *Distance Attenuation.* Increasing the distance between the noise source and the receiver can be a very effective method of noise control. Sound levels fall off between 3 dB(A) and 6 dB(A) for each doubling of distance from the source depending on the extent of the source. For stationary plant such as compressors and generators, a 6 dB(A) per doubling of distance can be assumed and consequently effective control can often be achieved by locating such plant away from any noise sensitive areas close to the site.

(b) *Screening.* Acoustic barriers in the form of carefully sited sand bags, stacks of bricks, top soil, woodwool slabs, plywood sheets fixed to scaffold support or even site buildings themselves can normally provide between 5 and 10 dB(A) attenuation of noise generated on site. This can be increased up to 15 dB(A) under certain circumstances.

To be effective, such barriers should be 5 times as long as they are high or, if this is not possible, the ends of the barrier should bend around the source. The minimum height of the barriers should be such that no part of the noise source is visible from the receiving point.

Further, the barrier should be placed as close as possible to either the noise source or the receiver position. No gaps should occur at joints between panels making up the barrier. The barrier material should have a weight in excess of 7 kg/m^2, although there is little point in making the barrier too heavy since the overall effectiveness will be controlled by sound passing round the ends and over the top of the barrier.

For major civil engineering works it may be possible to use substantial earth bunds as noise screens during construction. These bunds can often be landscaped to eventually become an attractive part of the environment once work is complete.

Care should be taken in siting barriers since it is possible for the barrier to reflect sound so that a problem is simply transferred from one location to another.

12.2.6 Major Sources of Vibration on Construction Sites and Their Control

The major sources of vibration on construction sites are blasting, piling and ground consolidation operations, and occasionally heavy earthmoving plant. If not adequately controlled, vibration from blasting operations can be severe enough to cause disturbance and occasionally superficial, or even structural, damage. The basic method of controlling blasting vibrations is to limit the charge weight per detonation to give acceptable levels at sensitive buildings.

Vibrations from piling will rarely be intense enough to cause structural damage and observations of even superficial damage are very limited. Disturbance, annoyance and even fear are the major concern over vibration from such sources. There is, however, little that can be done to control the magnitude of the vibrations from a given piling method apart from the following:

(a) For drop hammers, limit the distance through which the hammer drops. This leads to somewhat unpredictable results, although the general trend is for peak particle velocity to reduce as the distance of drop reduces.
(b) For drop hammers, reduce the weight of the hammer. This can have a significant effect; for example reduce the hammer weight from 2.75 to 1.00 tonne could reduce peak particle velocity in a structure from approximately 1.4 mm/s, which can be classified as distinctly perceptible, to 0.4 mm/s which can be classified as just perceptible.
(c) For single and double acting hammer systems, reduce the energy input into the pile.

If there is any doubt regarding the levels of vibration which are likely to occur from vibration generating operations, measurement surveys should be carried out either prior to the development commencing through site trials and/or during the critical stages of the works programme. Such an approach will show compliance with acceptable standards and alleviate the layman's justified fear of damage to his property which is often the root cause of complaints regarding vibration.

12.3 TRANSFER OF CONSTRUCTION SITE NOISE AND VIBRATION WITHIN AN EXISTING BUILDING DURING REFURBISHMENT WORK

It is often necessary to carry out construction or demolition operations during refurbishment work on a building while other areas of the same building or adjacent buildings are still in normal occupancy. It is, therefore, important to limit the transfer of noise and vibration from the area being worked on to these other areas which may be noise-sensitive.

BREAKING OUT

Whilst expected to be noisy in the immediate working vicinity, such activities often occur in partially occupied buildings, whereupon the diffusive structure-borne noise becomes a particularly obtrusive aggravation. This is especially true if the distant source involves a building modification of no concern to yourself.

MOBILE COMPRESSOR

The energy source for many construction site tools, and seen here in an acoustically treated format with cleanly designed covers in resin-bonded glass fibre. Apertures have been minimised.

Transfer of noise from one part of a building to another is via two mechanisms, namely:

airborne noise
structureborne noise

As discussed in Chapters 7 and 11 airborne noise transfer can normally be effectively controlled by the introduction of silenced or muffled plant and the erection of screens and/or enclosures as described above. Airborne noise transfer is also controlled to a large extent by the structure of the building itself.

The transfer of structureborne noise or vibration generated by construction operations is much more difficult to control. Sources of structureborne noise/vibration include breakers, cutters, drills, hammer drills, hammers, vibrators and other plant associated with refurbishment operations. Once vibration from such plant enters the building fabric, it can travel large distances with little attenuation to be:

(a) radiated as noise from the surfaces of rooms remote from the source and
(b) transferred to areas housing vibration sensitive equipment such as electron microscopes or X-ray machinery.

Each situation including potential vibration sources must be assessed separately, but some general guidelines regarding controlling transfer are given below:

(i) For "breaking out" works, wherever possible the part of the structure forming the mechanical link between the area being worked and the noise/vibration sensitive areas should be removed first, thus severing the noise/vibration transmission path.

(ii) Consideration should be given to using techniques of "breaking out" which do not produce high levels of continuous vibration and consequent noise regeneration. Such techniques include chemical and hydraulic bursting, thermic bursting, thermal lances, diamond core drills and saws and special equipment which breaks the concrete in bending.

The conventional method of breaking concrete is to use pneumatic or hydraulic breakers which are noisy and as they break the concrete by impact, they create high levels of vibration and structureborne noise. The methods mentioned above work on alternative principles and these are expanded below:

(a) *Chemical Bursting.* This technique relies on the expansive stress generated by the hydration of certain chemical compounds. Firstly, small holes are drilled in the concrete to be demolished. These are then filled with the chemical compound mixed with water. After some 10 to 20 hours the concrete is cracked enough to be removed by manhandling. Reinforcing bars can be cut using a thermic lance or oxyacetylene torch. This method does require the drilling of holes but such operations can be undertaken relatively quickly at non-noise-sensitive times. The impact of the drilling operations can be minimised by using diamond drills or thermic boring.

(b) *Hydraulic Bursting*. Hydraulic bursting can be used successfully with diamond drills or thermic boring to break out brick or concrete foundation structures.

The basic method of operation is to drill holes in the concrete into which the burster is inserted. The burster consists of a hydraulic cylinder with a number of pistons which operate radially such that when the cylinder is pressurised the force exerted by the pistons splits the concrete. This is done at several positions on the foundation being worked such that manageably-sized concrete slabs are formed. Reinforcing bars can be cut using a thermic lance or oxyacetylene torch. This method, as with chemical bursting, does require the drilling of holes but such operations can be undertaken relatively quickly at non noise sensitive times. The impact of these drilling operations can be minimised by using diamond drills or thermic boring.

(c) *Thermic Boring*. This method uses a thermic lance to burn holes into the concrete. The thermic lance consists of a steel tube packed with steel rods with one end connected to the lance holder. The other end of the lance holder is connected to an oxygen bottle via a hose and pressure regulator valve. The lance is lit by heating the free end and passing oxygen through it. Sections of concrete can be cut out by burning a series of holes close to each other and finally burning through the structure remaining between the holes.

(d) *Diamond Core Drills and Saws*. The degree of structureborne noise generated when cutting or drilling concrete with diamond saws or bits is generally considerably lower than that generated using percussive equipment, principally as a result of their sharp cutting edges. Consequently, whenever such operations are planned close to noise sensitive areas, such techniques should be considered.

(e) *Concrete Breakers Working in Bending*. The Building Research Establishment have developed a concrete breaking attachment for a hydraulic excavator known as the Nibbler. The breaker does not use impact, but rather snaps the concrete in tension by applying a bending movement. This again will considerably reduce the degree of structureborne noise generated during breaking operations.

MULTI-PURPOSE EXCAVATOR — JCB

As a result of current and pending European Economic Community legislation, machines of this type have noise emission and interior cab noise limitations which have forced the manufacturers to address the noise problems. Consequent to this, "noise" has become a selling feature, and developments continue to reduce levels, even below legal limits.

CEMENT MIXING LORRY

A multiplicity of noise sources which can be controlled down to appropriate construction site levels. However, it is interesting, but a corollary, that when travelling in reverse such "forward facing" vehicles need to emit a warning sound to be heard above the limited site noise levels. This could become an annoyance to nearby locations previously protected by site-controlled noise levels.

CHAPTER 13

LABORATORY TESTING

13.1 INTRODUCTION

The basic criterion for a good building service system is for the correct solution of not only the thermal performances but also the aerodynamic and acoustic requirements. It is not possible to achieve satisfactory results without adequate performance data, a prerequisite for a successful design. The "Design for Sound" of air conditioning systems is dependent upon the noise generation, flow generated noise or attenuation of its component parts. There are two ways in which information can be obtained. The performance of a piece of equipment can sometimes be obtained on site. However, it is not always possible to control all the variables involved in such a test, consequently interpretation of the results may be open to discussion. Alternatively, equipment or parts of air moving systems can be studied in a properly equipped laboratory where each variable can be individually controlled under known conditions. The design engineer needs to know the performance of each component part and its contribution to the whole in order to select noise control equipment or to reduce noise by re-design.

13.2 NOISE GENERATION

For a typical air supply the primary air mover, be it a centrifugal, axial or mixed flow fan is usually looked upon as the largest single noise source in the system and is generally provided with a primary attenuator. Between this primary attenuator and conditioned rooms, air passes through considerable lengths of ducting experiencing change of direction, change of duct cross sectional area and other flow regulating devices passing sometimes through a secondary attenuator before it discharges from a grille or diffuser. Each component can, in the course of moving or controlling the flow of air, generate noise. Equally well some components, whether intentionally or not, can attenuate some of the noise in the system by natural attenuation. It is important, if a specified room design criterion is to be met, to know how much noise and how much attenuation will be provided by each individual component of the system. Equally the interaction between components should be evaluated. All this has been discussed in the previous chapters but the information required and presented had to be confirmed by measurements.

351

The design engineer needs therefore to know for each component, how much noise does it add and how much noise will it remove? However, because of the realities of a system which involves interaction with many different pieces of equipment, the problem is further complicated by the need not only to know the answers to these two questions, but a third and more difficult question, namely: "will the unit yield the same performance data when operating near to other pieces of equipment in the particular system design?"

Product performance data is normally acquired by carrying out a series of measurements according to national or international test procedures. Manufacturers of fans, attenuators and other heating, ventilating and air conditioning equipment usually publish test information measured in accordance with the relevant British or International standard. It is easy therefore for direct comparisons between different manufacturers published performance data to be made. In the absence of the relevant test code it is necessary for a manufacturer to devise a series of tests which will provide useful information. However extra caution is required as different results can be obtained from the same piece of equipment subject to the conditions of the test.

For unusual applications, laboratory testing is vital to be successful in the acoustic design of a system, and laboratory "mock-ups" of the site situation become appropriate. Although the questions are endless typical examples would be:

How does the total radiated sound power level of an axial fan vary as inlet conditions are changed?
What is the effect upon the flow generated sound power level on an attenuator when operating in highly irregular flow patterns immediately downstream of a bend?
Is the sound power level generated by a constant flow rate regulator affected by unusual air distribution at the inlet to the regulator?
Do turning vanes, which might be of considerable use in minimising pressure loss of airflow round a bend, significantly contribute to the overall sound power level in the system?
What are the flow generated noise characteristics of T pieces, and other standard and non standard take-offs, abrupt transformations, right angle, mitre and radius bends more especially when variously combined in series?
How much flow generated noise occurs inside convoluted flexible spiral wound duct work?
Is there a limiting angle to which ductwork can be bent before the flow generated noise levels become unacceptably high?
How important are the flow conditions at a grille or diffuser if minimum flow generated sound power levels are to be achieved?

Having established why the design engineer needs to know the acoustic performance of each element in the system how does one obtain this information? Some is empirical data established over the years by researchers, such as the attenuation provided by ducts and bends, Figure 13.1 (See also Chapter 7).

Attenuation in dB/metre run

Straight Duct Circular/oval or Rigid Walled (unlined)	x (mm)	Octave Band Centre Frequency (Hz)						
		63	125	250	500	1k	2k	4k
	75–200	0.07	0.10	0.10	0.16	0.33	0.33	0.33
	200–400	0.07	0.10	0.10	0.16	0.23	0.23	0.23
	400–800	0.07	0.07	0.07	0.10	0.16	0.16	0.16
	800–1500	0.03	0.03	0.03	0.07	0.07	0.07	0.07
Straight Duct rectangular (unlined)	x (mm) side dimension	Octave Band Centre Frequency (Hz)						
		63	125	250	500	1k	2k	4k
	75–200	0.16	0.33	0.49	0.33	0.33	0.33	0.33
	200–400	0.49	0.66	0.49	0.33	0.23	0.23	0.23
	400–800	0.82	0.66	0.33	0.16	0.16	0.16	0.16
	800–1500	0.66	0.33	0.16	0.10	0.07	0.07	0.07

Figure 13.1a   Attenuation Per Metre of Unlined Duct

Duct Width/ Diameter D mm	Octave Band Centre Frequency							
	63	125	250	500	1k	2k	4k	Hz
150–250	–	–	–	–	1	2	3	dB
250–500	–	–	–	1	2	3	3	
500–1000	–	–	1	2	3	3	3	
1000–2000	--	1	2	3	3	3	3	

Figure 13.1b   Attenuation of Radius Bends or mitre bends with large turning vanes

This data was produced by experimental tests in a fully instrumented test laboratory such that only the attenuation of the duct was measured. Even with this information, on-site conditions can seriously affect the acoustic performance of a total system. Similarly data has been established on the performance of air moving devices such that the noise generation of a particular type of fan can be estimated from a series of graphs — Figure 13.2. However these graphs and tables are only suitable for approximate calculations, and guidance as described in Chapter 7. Precise data is required for such devices which may vary from design to design, manufacturer to manufacturer, and application to application.

To establish performance data of equipment, a controlled environment is essential. An acoustic test laboratory which is also able to provide an airflow facility will enable suitable tests to British or International Standards for almost all types of equipment. The test procedures and the facilities themselves must conform to a rigid standard in order to accurately provide this essential data. Just what can be provided, how tests are controlled and conducted are illustrated in the following paragraphs.

Generally the facilities must provide for measurement from 40 Hz to 16000 Hz. In addition they must be adequately isolated from background noise and vibration in order for these conditions not to affect the test measurements. Ideally both a room which reflects sound, known as a reverberation room and the opposite an echo free

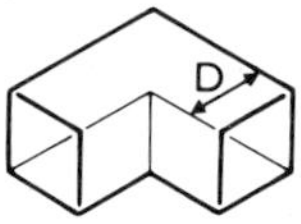

Duct Width	Octave Band Centre Frequency							
D mm	63	125	250	500	1k	2k	4k	Hz
75–100	–	–	–	–	1	7	7	dB
100–150	–	–	–	–	5	8	4	
150–200	–	–	–	1	7	7	4	
200–250	–	–	–	5	8	4	3	
250–300	–	–	1	7	7	4	3	
300–400	–	–	2	8	5	3	3	
400–500	–	–	5	8	4	3	3	
500–600	–	–	6	8	4	3	3	
600–700	–	1	7	7	4	3	3	
700–800	–	2	8	5	3	3	3	
800–900	–	3	8	5	3	3	3	
900–1000	–	5	8	4	3	3	3	
1000–1100	1	6	8	4	3	3	3	
1100–1200	1	7	7	4	3	3	3	
1200–1300	1	7	7	4	3	3	3	
1300–1400	2	8	7	3	3	3	3	
1400–1500	2	8	6	3	3	3	3	
1500–1600	3	8	5	3	3	3	3	
1600–1800	5	8	4	3	3	3	3	
1800–2000	6	8	4	3	3	3	3	

Duct Width	Octave Band Centre Frequency							
D mm	63	125	250	500	1k	2k	4k	Hz
75–100	–	–	–	–	2	13	18	dB
100–150	–	–	–	1	7	16	18	
150–200	–	–	–	4	13	18	18	
200–250	–	–	1	7	16	18	16	
250–300	–	–	2	11	18	18	17	
300–400	–	–	4	14	18	18	17	
400–500	–	1	5	16	18	16	17	
500–600	–	1	8	17	18	16	17	
600–700	–	2	13	18	18	17	18	
700–800	–	3	14	18	17	16	18	
800–900	–	4	15	18	18	17	18	
900–1000	–	5	16	18	17	17	18	
1000–1100	1	7	17	18	16	17	18	
1100–1200	1	8	17	18	16	17	18	
1200–1300	1	10	17	18	16	18	18	
1300–1400	2	11	18	18	16	18	18	
1400–1500	2	12	18	18	16	18	18	
1500–1600	3	14	18	18	17	18	18	
1600–1800	4	15	18	18	17	18	18	
1800–2000	5	16	18	17	17	18	18	

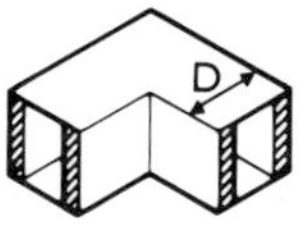

Lining thickness $= \dfrac{D}{10}$

Lining to extend distance 2D or greater

Figure *13.1c* Attenuation of Mitre (90°) Bends

room known as an anechoic room are desirable. In addition a linked pair of rooms forming a transmission suite are necessary to enable Sound Reduction Index tests to be carried out. To enable air conditioning systems and equipment to be tested a silenced air supply is required such that the noise measured is due to the device under test and is not affected by the air supply system.

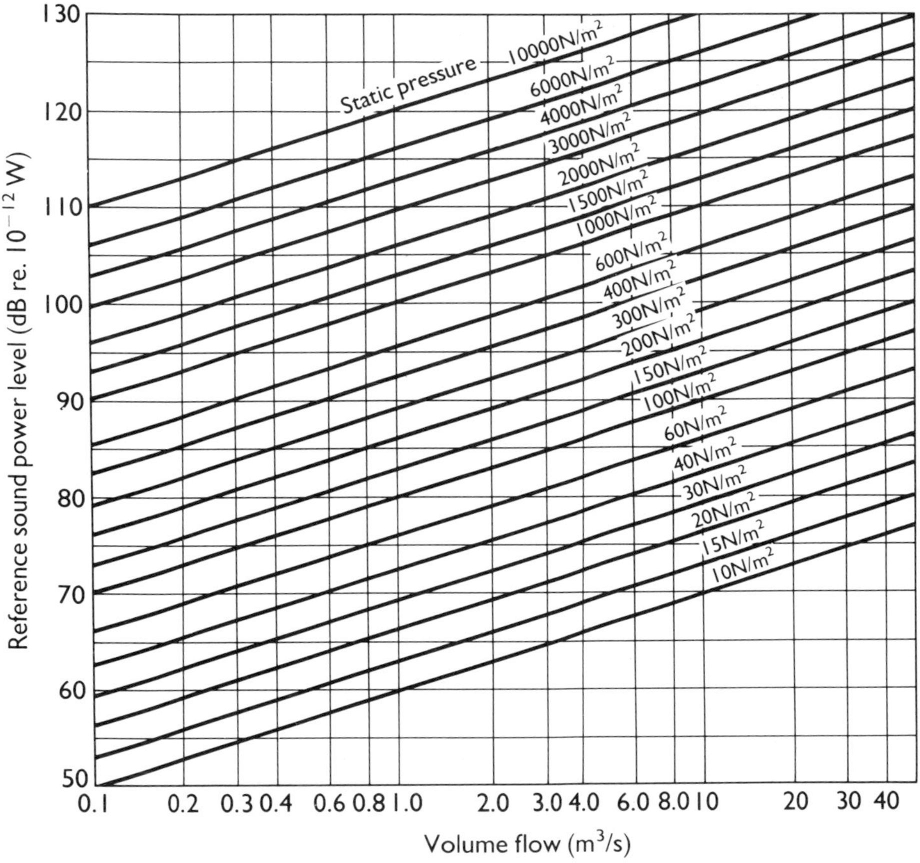

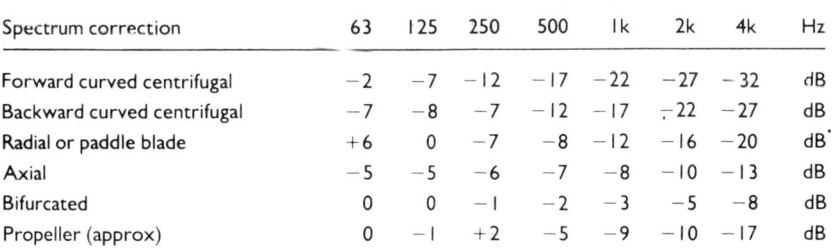

The chart shows: Reference sound power level (dB re. 10^{-12} W) on the y-axis (50 to 130) versus Volume flow (m³/s) on the x-axis (0.1 to 40). Lines labelled with static pressure values: 10000N/m², 6000N/m², 4000N/m², 3000N/m², 2000N/m², 1500N/m², 1000N/m², 600N/m², 400N/m², 300N/m², 200N/m², 150N/m², 100N/m², 60N/m², 40N/m², 30N/m², 20N/m², 15N/m², 10N/m².

Spectrum correction	Octave Band Centre Frequency							
	63	125	250	500	1k	2k	4k	Hz
Forward curved centrifugal	−2	−7	−12	−17	−22	−27	−32	dB
Backward curved centrifugal	−7	−8	−7	−12	−17	−22	−27	dB
Radial or paddle blade	+6	0	−7	−8	−12	−16	−20	dB
Axial	−5	−5	−6	−7	−8	−10	−13	dB
Bifurcated	0	0	−1	−2	−3	−5	−8	dB
Propeller (approx)	0	−1	+2	−5	−9	−10	−17	dB

Figure 13.2   Estimation of Fan Sound Power Level

13.3 REVERBERATION ROOMS

At its simplest this may be described as a room with walls, floor and ceiling that provide little sound absorption. The purpose of this condition is to cause multiple reflections of the sound waves such that sound arriving at any one measuring point has travelled a multiplicity of paths and comes from several different directions. With a correctly designed reverberation room the acoustic energy is equally distributed throughout the main volume of the room. In these conditions the acoustic field is said to be diffuse. To ensure there are no predominant standing waves in the room the walls are often built out of parallel. Alternatively fixed or slowly rotating reflective surfaces known as diffusers are included within the room to minimise measurement error. The physical size of the room determines, with other factors, the lower limiting frequency at which measurements can be made with a satisfactory degree of reliability. The larger the room the lower the limiting frequency. The layout of SRL's Laboratory complex is shown in Figure 13.3 with physical details in Appendix 13.A.

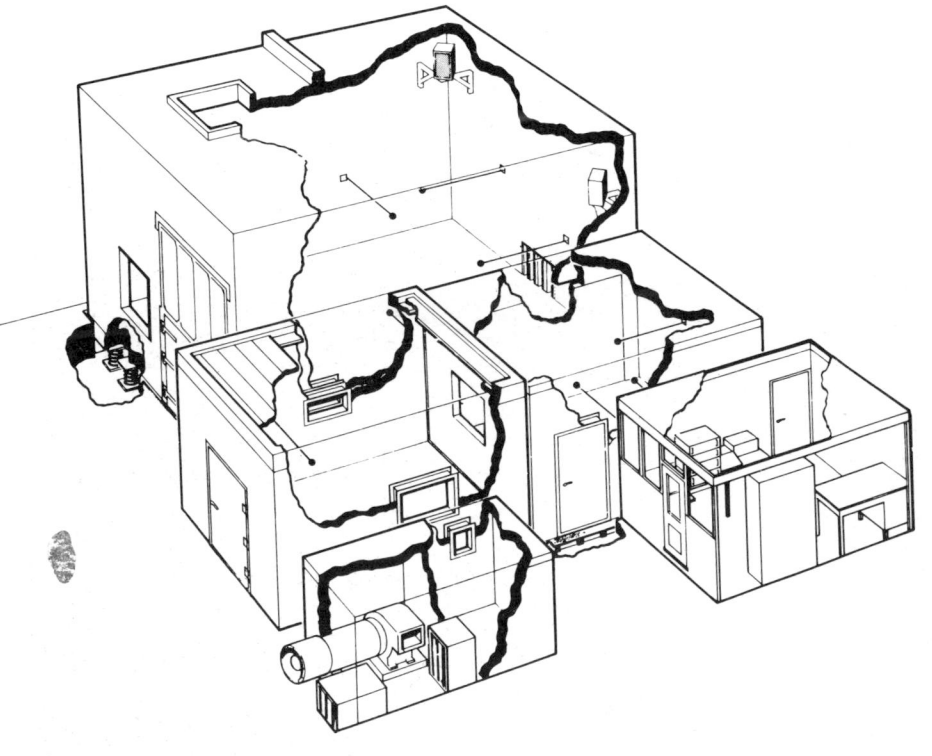

Figure 13.3   SRL's Acoustic and Airflow Test Laboratory

To minimise ground borne vibration and flanking transmission the room should be mounted on vibration isolators, preferably springs in order to adequately isolate at low frequencies and noise stop pads to limit high frequency transmission.

The openings in the room to provide for access and airflow must not be such as to cause a significant reduction in the acoustic integrity of the structure. Double sealed 45 dB acoustic access doors will satisfy these needs. Similarly any outlet duct for the air used in tests must be fitted with a high performance attenuator to prevent outside background airborne noise from affecting the measurements made within the room. Other apertures for various types of tests must also be adequately sealed with materials of sufficient mass for the area of opening in order to maintain the acoustic integrity of the structure.

13.4 ANECHOIC ROOMS

The antithesis of a reverberation room is a room with highly absorbent surfaces to minimise reflected sound. The surface treatment necessary to achieve this condition can be seen in Figure 13.4 — anechoic wedges. The shape, size and material composition of the anechoic wedge, as it is known, has been the subject of considerable research.

Figure 13.4   Anechoic Room

The longer the wedge the lower the limiting frequency at which the test cell can be operated and meet the high absorption requirements. Additionally an air gap at the base of the wedge between the wedge and the inner surface of the room may assist the absorptive effect. An anechoic room is designed to simulate free-field conditions and the inverse square law decay with distance will be obeyed. Consequently it permits investigations of the way in which acoustic energy is radiated from a specific source configuration - the source directivity. An anechoic room is not generally so convenient for measurement of equipment overall sound power levels. It is however the only laboratory facility by which directivity of noise may be studied. In common with a reverberation room, its isolation from surrounding structures is important. The efficiency of this isolation is a limiting factor in the extent to which meaningful measurements can be taken from very low sound power generating sources.

13.5 TRANSMISSION SUITE

A transmission suite is a facility of two adjacent reverberant rooms for measuring the airborne sound reduction index of panels, doors, windows and other pieces of equipment designed to minimise transfer of noise from one side to the other — normally to BS 2750. The suite consists of two rooms — a source room and a receiving room — the test panel being installed in an opening between the two rooms. The surfaces in the source room are generally hard and the room volume is large enough to ensure the test panel is subject to a measurable random incidence sound field. As the source room is normally very noisy, attempts to minimise structure borne vibrations are less critical. However the receiving room, as shown in Figure 13.3, must be vibration isolated from the surrounding structures, especially the noisy source room, to prevent structure and ground borne flanking transmission.

Again to assist in limiting flanking transmission and to ensure test results are of the item under test and are not influenced by the surrounding structure the test sample should totally fill the aperture between the chambers, that is a minimum surface area of 10 m² for most associated standards. Alternatively the surrounding structure must have a Sound Reduction Index at least 10 dB higher in each octave band than the device to be tested, taking into account the ratio of the respective areas, in order that the results are not influenced by sound energy radiating through the surrounding structure affecting the measurement. Figure 13.5 gives examples of this requirement.

For structures which cannot be tested in a vertical manner, such as floating floors, ceilings and so on, a "piggy back" transmission suite is necessary. This is where the source and receiving rooms are one above the other rather than side by side — Figure 13.3.

13.6 IN-DUCT TEST FACILITY FOR ATTENUATION

This is an important piece of equipment for measuring the insertion loss performance of devices. Being self contained, it does not require a reverberation room or anechoic

AREA RATIO	SOUND REDUCTION Index Difference
0.1:1	0.0 dB
0.2:1	3.0 dB
0.4:1	6.0 dB
0.6:1	8.0 dB
0.8:1	9.0 dB
1:1	10 dB
2:1	13 dB
4:1	16 dB
6:1	18 dB
8:1	19 dB
10:1	20 dB
20:1	23 dB
40:1	26 dB
60:1	28 dB
80:1	29 dB

Figure *13.5* Examples of Required Sound Reduction Differences Between Sample and Surrounding Structure Against Area Ratio.

room. As shown in Figure 13.6a the apparatus consists of loud speakers at one end of a duct system through which predetermined noise levels can be played, an inlet duct, a test section in which the item under test can be substituted, an outlet duct containing measuring microphones and an exponential expander leading to an anechoic termination — see Figure 13.6b. The anechoic termination is provided to ensure that any noise having passed through the test section and measured by the microphones is not reflected back to modify the level measured. The difference in sound pressure level registered at the fixed microphone sampling stations with and without the test item gives a direct measure of the insertion loss of any barrier such as an attenuator or sound resistant panel that might be placed in the test section.

$$\text{Insertion loss, I.L.} = L_1 - L_2 \qquad\qquad 13.1$$

where L_1 = average sound pressure level without sample and
L_2 = average sound pressure level with sample.

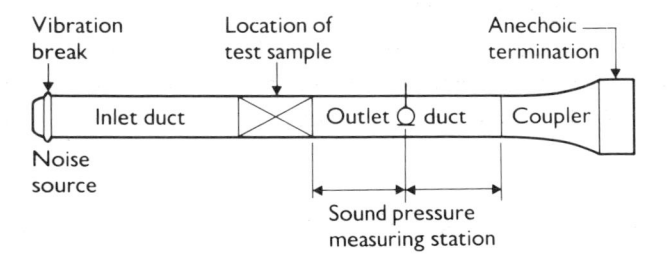

Figure 13.6a Progressive Wave Measuring Tube

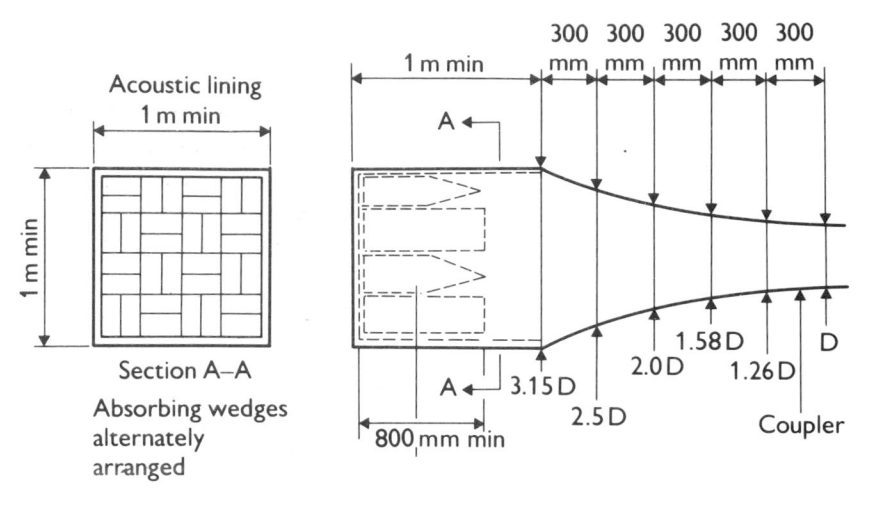

Figure 13.6b Anechoic Termination

Figure 13.6c Rectangular Attenuator being installed on to Source Inlet Duct for Insertion Loss Measurement

13.7 AIR MOVEMENT

Whilst the in-duct test facility can provide the static insertion loss of devices, it is essential to assess the dynamic performance of such elements by testing the effect on the performance of flow generated noise. A desirable feature of any acoustic testing laboratory designed to investigate mechanical services equipment is therefore an adequate air moving capability. For many years it has been a requirement to provide large volumes of air at pressures up to 1000 Pa for testing attenuators. Because of this additional requirement to measure the attenuator flow generated noise characteristics it is necessary that the noise of the air supply is relatively low.

The particular air supply chosen for SRL's Laboratory, shown in Figure 13.3 comprises a large double inlet centrifugal fan, driven by a hydraulic pump/motor system supplied by a V6 petrol engine. This gives great flexibility and control over the flow of air provided to the test rooms. The fan can deliver from 0.02 m^3/s to 14 m^3/s at pressures up to 2500 Pa. Because of the two stage flexibility and control offered by this engine/hydraulic system, low flow rates against low pressure losses can also be developed with ease. The noise from the fan and motor unit is attenuated by two banks of attenuating splitters and a 10 metre length of absorbent lined supply duct over 1 m in cross sectional area. This arrangement can give background noise levels in the measuring rooms of below NC15. A system of louvres in the fan room enables suction tests also to be made by extracting air from the test rooms through the unit on test.

13.7.1 Flow Measurement

The measurement of airflow during acoustic testing is as important as the acoustic measurement. Generally noise levels are proportional to the sixth power of airflow rate.

There are many methods of measuring the flow of air in ducts each with their own degree of accuracy and complexity of application. These generally fall into two types, those using pressure measurements and those using velocity measurements.

Figure 13.7 lists these with their applications and limitations.

The methods using pressure measurement achieve the flow rate by presenting a fixed area to the flow and measuring the pressure differential across the obstruction which can be directly related to flow. The following instruments can be calibrated directly in terms of volume flow.

13.7.1.1 Orifice Plates. The orifice exists in many different forms, round edge, square edge and sharp edge. These items refer to the particular form of the inside face of the orifice aperture. There are further variations in the pressure sensing locations before and after the orifice plate. The total volume flow in a duct must pass through the orifice plate aperture, thereby producing a pressure difference across the device

Measurement Means	Application	Range m/s	Precision	Limitations
Deflecting vane type anemometer (velometer)	Air velocities in rooms, at outlets, etc. directional	0.15 to 120	5%	Not well suited for duct readings needs periodic check calibration
Revolving vane type anemometer	Moderate air velocities in ducts and rooms, somewhat directional	0.5 to 15	5–20%	Extremely subject to error with variations in velocities with space or time, easily damaged, needs periodic calibration
Pitot Static tube	Standard Instrument for measurement of duct velocities via pressure	1.0 to 50 with micromanometer 50 and above with manometer	1–5%	Accuracy falls off at low end of range static tappings sensitive to flow direction
Hot wire anemometer	a) Low air velocities – directional and non-directional b) High air velocities c) Transient velocity and turbulance	0.005 to 5.0 up to 300	1–20% 1–10%	Requirres accurate calibration at frequent intervals Complex, costly
Orifice and manometer	Flow through pipes, ducts and plenums	Above Reynolds number of 5000	1–5%	Coefficent and accuracy influenced by approach conditions
Nozzle and manometer	Flow through pipes, ducts and plenums	Above Reynolds number of 5000	1–5%	Coefficent and accuracy influenced by approach conditions
Venturi tube and manometer	Same as orifice but used where permissible pressure drop is limited	Above Reynolds number of 5000	1–5%	Coefficent and accuracy influenced by approach conditions
Rotameters	Normally used for liquids	Any	1%	Must be calibrated for the liquid with which used

Figure 13.7   Airflow Measuring Devices and their precision

which is directly related to the volume flow rate. Such measuring instruments are used for laboratory standards but are rarely found on site. They produce a high pressure drop and require many hydraulic diameters of straight ductwork upstream to produce fully developed turbulent flow conditions before the constriction. Reference should be made to BS 1042 for the design of orifice plates — Figure 13.8.

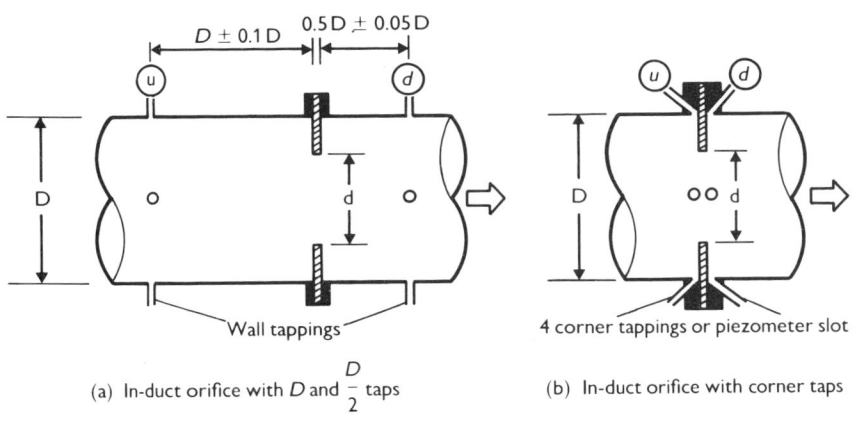

$D \pm 0.1\,D$ $0.5\,D \pm 0.05\,D$

Wall tappings

4 corner tappings or piezometer slot

(a) In-duct orifice with D and $\dfrac{D}{2}$ taps

(b) In-duct orifice with corner taps

Figure 13.8   Orifice Plates

13.7.1.2 Venturi Tube. The Venturi can be looked upon as an aerodynamically improved version of the orifice. A total pressure loss is caused across a constriction but it is not abrupt. Air in the duct is steadily accelerated to a maximum velocity and then decelerated under similar conditions. The length of ducting required for this device is greater than the orifice but much of the head loss is recovered as a result. Again BS 1042 is the reference for the design of Venturi Tubes — Figure 13.9.

13.7.1.3 Nozzles. Nozzles are a form of constriction which guide air from a plenum entry condition smoothly into the area of maximum constriction. At this point, however, the air discharges via a duct of the same cross-section and no further contraction or expansion is experienced — Figure 13.10a. The nozzle, therefore, has a lower resistance, compared to the orifice. They are frequently seen on site and used almost exclusively as the volume flow monitoring device in leakage testers. For a wide range of air flow rates a series of nozzles can be mounted in a panel within a plenum. Each nozzle has a removable plug allowing the correct size of nozzle required for the particular air flow rate to be measured accurately — Figure 13.10b.

13.7.1.4 Pitot-Static Tube. A pitot-static tube is the most accurate device for measuring velocities at a point in an airstream above 5 m/s. It is a double tube, one inside the other, with a 90° bend — Figure 13.11. The outer tube has a series of holes which detect the static pressure, while the facing hole is "pointed" into the airstream. In practice this produces a stagnation point at which the total pressure can be measured. The local velocity pressure is then obtained using a manometer as a differential pressure measuring device. Following a traversing and averaging technique together with knowing the area of the duct in which the measurements are taken, the volume flow rate can be accurately obtained. The total pressure is insensitive to yaw while the static is very sensitive. Hence the tube should be yawed to give the maximum differential pressure. The pitot-static tube is normally considered to be a standard device not requiring calibration.

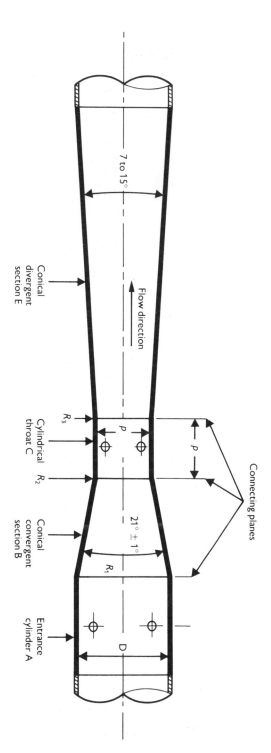

Figure 13.9 Classical Venturi Tube

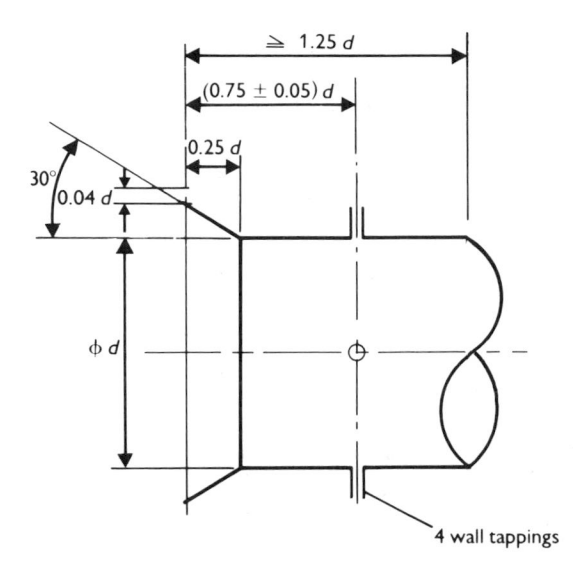

Figure 13.10a Conical Inlet Geometry

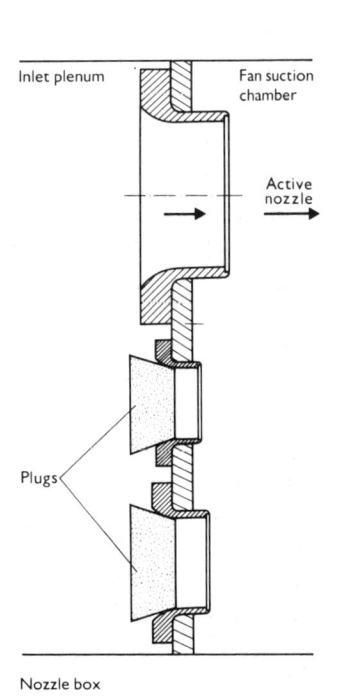

Figure 13.10b Nozzle Box

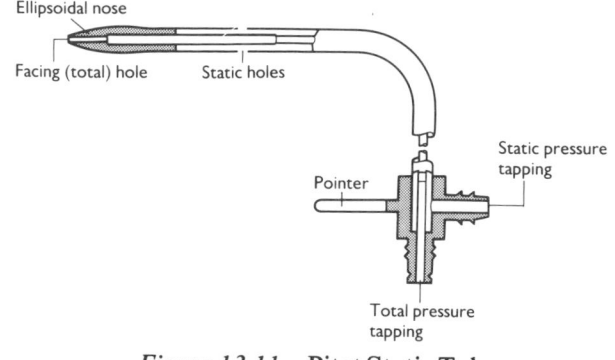

Ellipsoidal nose
Facing (total) hole
Static holes
Static pressure tapping
Pointer
Total pressure tapping

Figure 13.11 Pitot Static Tube

KEEP CLEAR

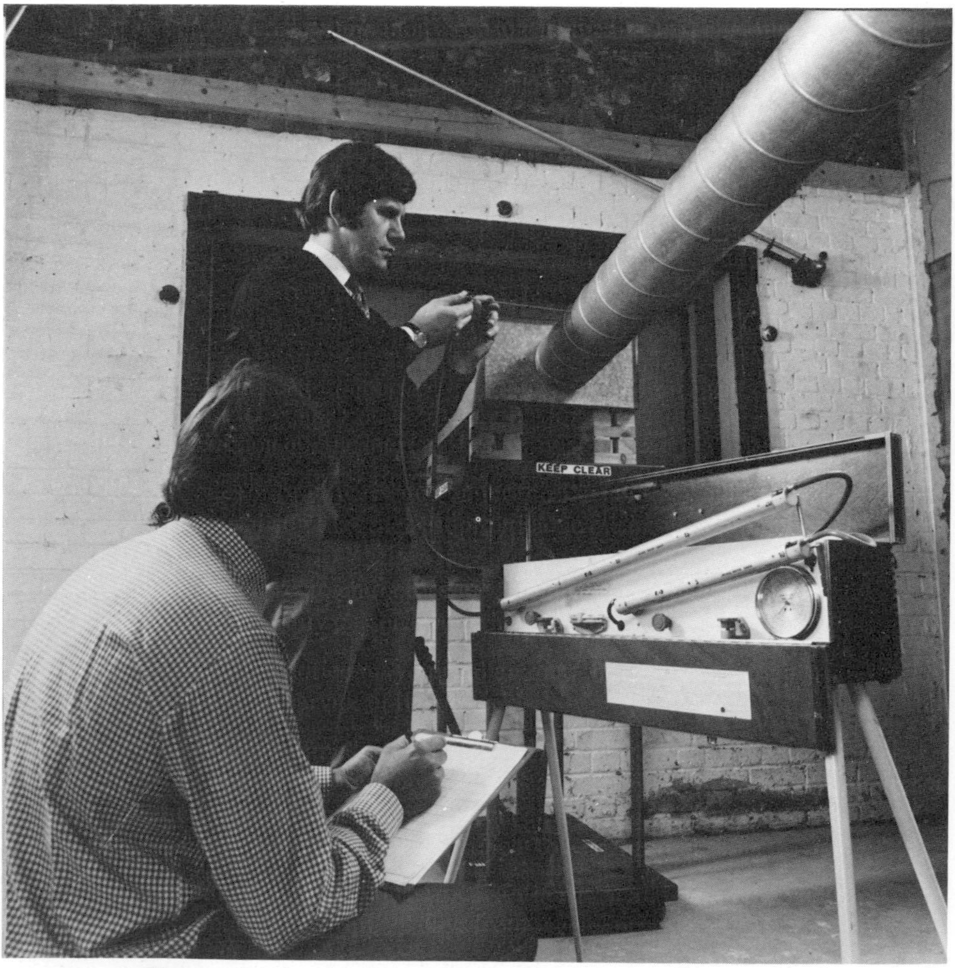

Figure 13.11b Air Flow Measurement using a Pitot Travers

The types of test capable of being carried out in a representative Laboratory complex are as follows.

Generated sound power measurements on:

Air handling units
Lighting systems
Cooling systems
Fans
Fan coil units

In addition to the simple single pitot-static tube requiring multiple measurements there are devices which purport to average the many readings necessary to obtain an accurate flow. These can be in the form of a grid system simulating the multiple point measurement of the pitot-static tube or a single tube which extends across the duct with several total head tappings and normally a single static head tapping either on the downstream face of the tube or at the base of the tube by the duct wall. Their degree of accuracy approximates to that of the pitot-static tube method but they do have to be calibrated against a known flow to obtain an instrument constant.

To measure the volume flow rate with devices measuring velocity at a point or over a comparatively small part of the flow area the appropriate area is needed for each velocity measurement. Even in an ideal system where the velocity is essentially uniform there will be a wide variation of velocity at the edge in the boundary layer which represents a surprisingly large proportion of the total area even in fully turbulent flows. In ducts it is normal procedure to arrange measuring positions to divide the area into specified zones — BS848. However, in most practical ventilation systems gross variations in velocity across the duct due to upstream fittings such as bends, make the division into equal rectangles a less accurate method. The following devices can be used to measure air velocity in ducts.

13.7.1.5 Airborne Tracer Techniques. Measurements are made by timing the rate of movement of a smoke tracer in an airflow. This is rarely used for quantitative methods but is an invaluable technique for the commissioning engineer when studying airflow movements. Titanium tetrachloride vapour, though unpleasant if inhaled, can be squirted into trouble areas using an injector and is especially useful for detecting equipment casing and duct leakage paths.

13.7.1.6 Deflecting Vane Anemometer. Frequently known as a velometer and consists of a balanced, damped, spring controlled pivoted vane which may be deflected by a very light current of air to carry a pointer over a scale — Figure 13.12. In large ducts the whole meter may be inserted in the airflow. For smaller ducts a special probe is inserted into the duct and linked to the meter using a pair of rubber tubes. Thus air from the duct is taken via the deflecting vane before being returned to the duct. Different scales are provided for use with different probes for measuring velocities at ventilation grilles, discharge openings or wall and ceiling diffusers.

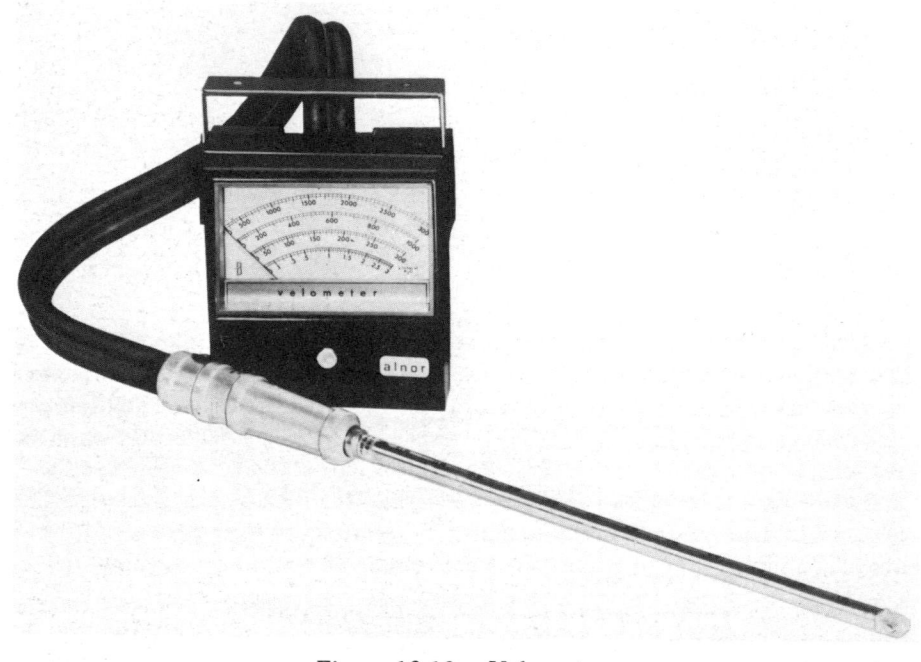

Figure 13.12 Velometer

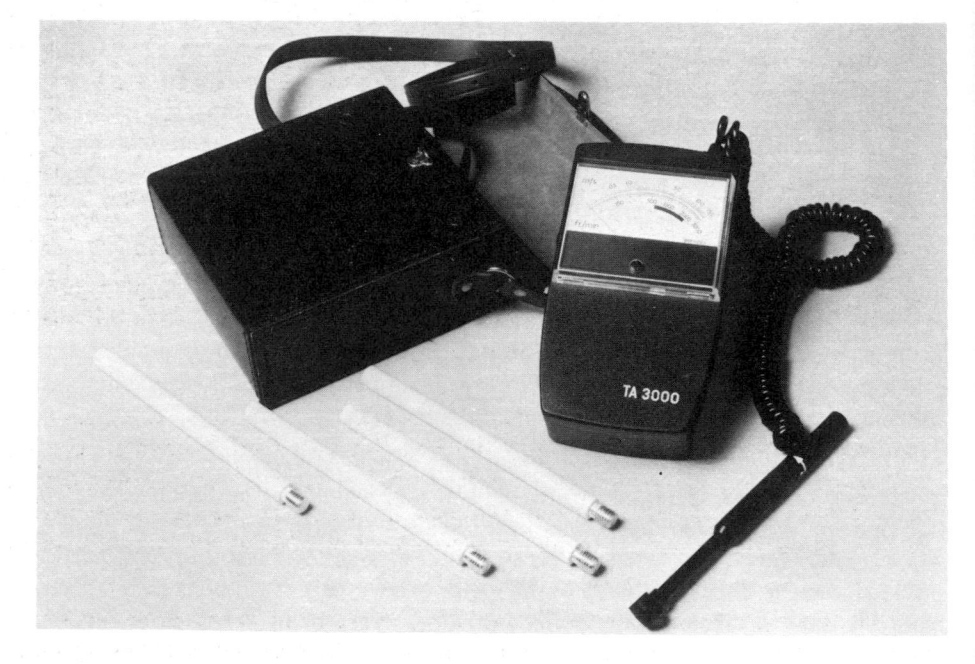

Figure 13.13 Hot Wire Anemometer

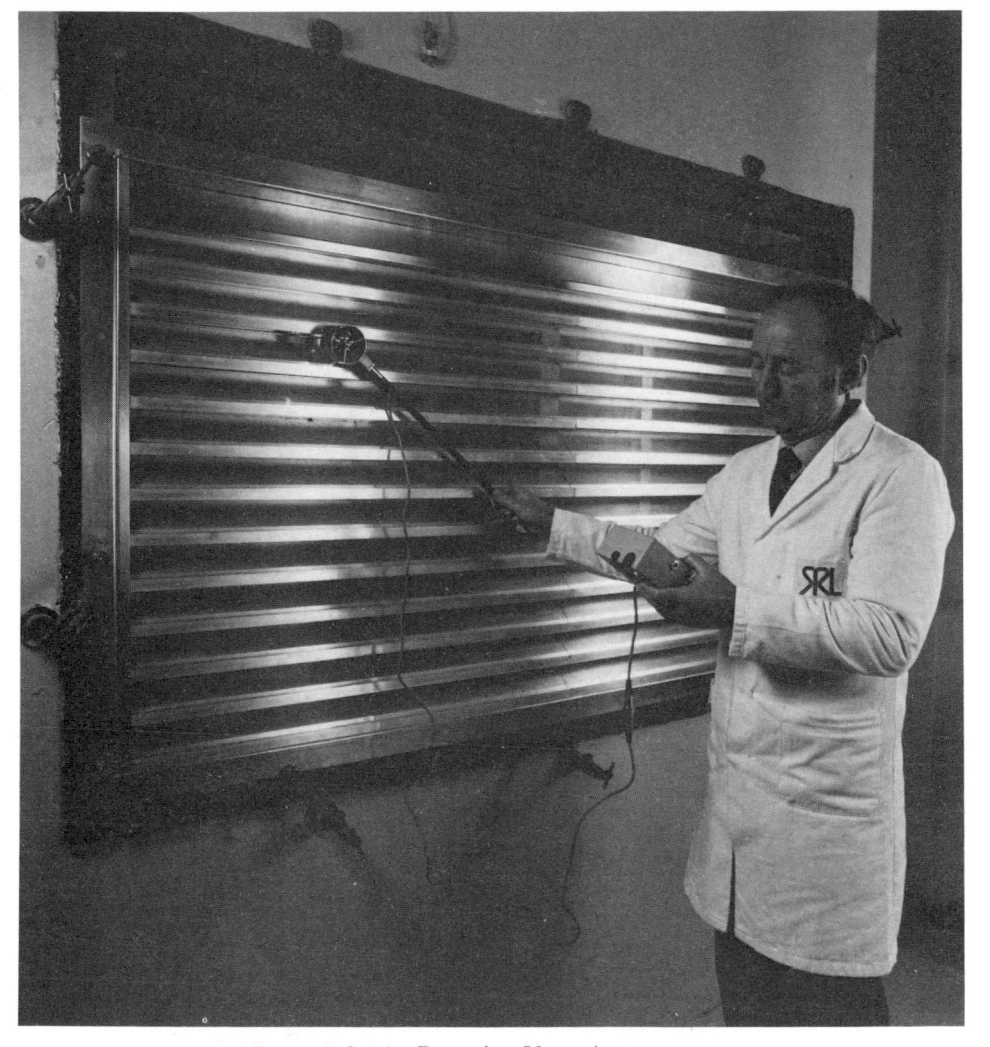

Figure 13.14 Rotating Vane Anemometer

13.7.1.7 Hot Wire Anemometer. This is a device which measures the cooling effect of airflow over a heated wire and usually compares this with a reference shielded wire, — Figure 13.13. The cooling effect is measured electrically and displayed as a reading on a voltmeter calibrated directly in speed. The device is particularly suitable for low speeds, it is simple to use, but gives no indication of flow direction. It is not an absolute device and requires calibrating frequently.

13.7.1.8 Rotating-Vane Anemometer. The rotating-vane anemometer measures velocity in terms of velocity head. The mode of operation is by means of a revolving vane, driven by the airflow to be measured — Figure 13.14. An electrical transducer counts

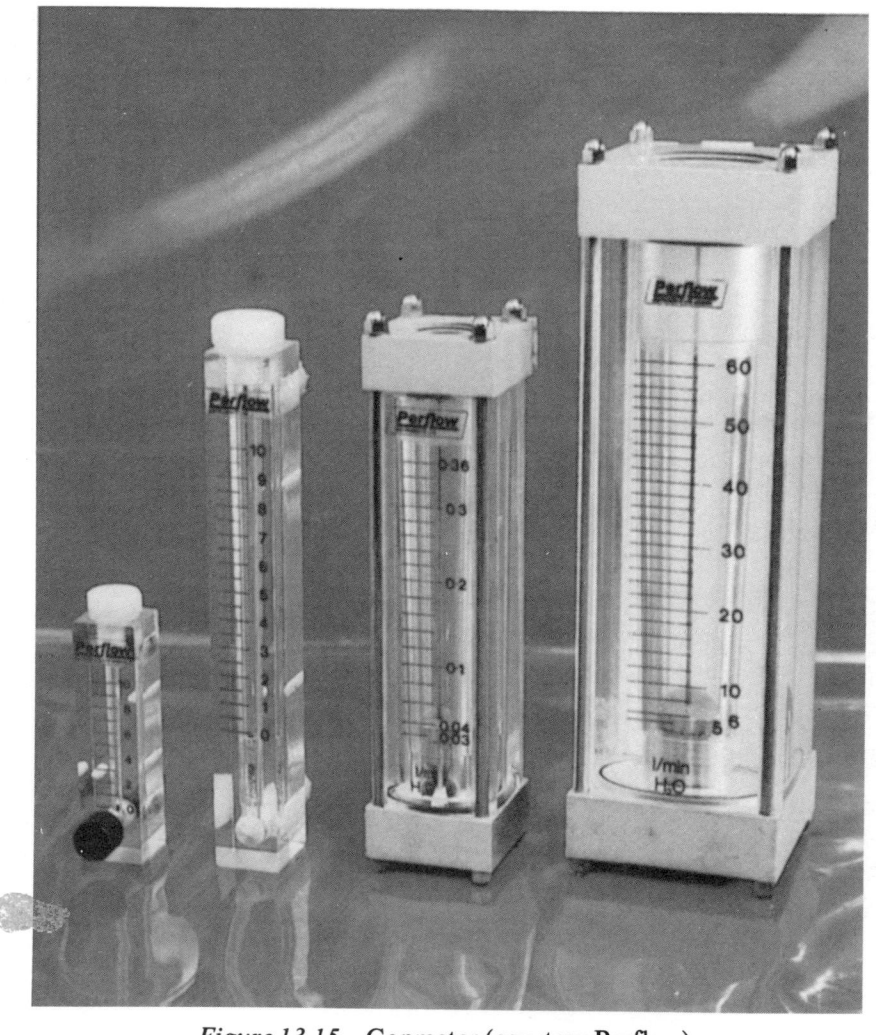

Figure 13.15 Gapmeter (courtesy Perflow)

the vane passage frequency and displays the equivalent velocity on the meter. Whereas a pitot-static tube or hot wire anemometer measures the velocity over a very small area, the rotating vane anemometer "averages" over a considerably larger area. Whether or not the spacial average produced is valid depends on the flow pattern. For uniform airflow it is a very convenient means of velocity measurement, but for narrow jets such as issue from grilles or diffusers its operation is very suspect in an airflow containing a spread. It is not a standardised instrument and due to its delicate nature also requires frequent calibration.

13.7.1.9 Gapmeter. Gapmeter devices are frequently used for liquid flow measurement, and occasionally for air. In this device the fluid flows up a tapered tube

past a "float" which lifts until its weight is balanced by the pressure drop across it —
Figure 13.15. Since the weight of the float is constant, the pressure drop and velocity
past the float are also constant and the flow-rate can be read off from the position of
the float against a scale on the side of the device.

13.8 TYPES OF TEST

Given an acoustics and airflow test laboratory with the facilities mentioned there are
many types of tests which can be carried out on many different types of equipment as
illustrated in the following list. However a well equipped laboratory also provides an
excellent location in which total systems or part of such systems may be assembled
and studied in great detail. There are significant advantages in using such system
mock-ups. In the absence of the normal restrictions of the buildings it is easier to gain
access to almost any part of the system and make measurements at that point. It is
also possible with equipment available in the laboratory to investigate the affect of
changing, adding or removing individual components of that system in order to
achieve the desired criterion.

Induction units
Lawn mowers
Domestic Appliances
Roof units
Room conditioners

Airflow generated sound power measurements on:

Attenuators
Ceiling systems
Diffusers
Grilles
Louvres
Terminal units
Ventilators

Sound reduction properties of:

Bricks and blocks
Ceiling systems
Ceiling tiles
Curtain walling systems
Doors
Cross talk attenuators
Acoustic louvres
Panels
Partitions
Windows

Sound absorption properties of:

Building blocks
Carpets
Ceiling tiles
Flooring systems
Foam
Mineral wools
Screens

Airflow characteristics of:

Air handling units
Attenuators
Ceiling systems
Curtain walling systems
Diffusers
Grilles
Fans
Induction Units
Louvres
Terminal units
Ventilators
Windows

Impact sound reduction on:

Floor systems
Carpets

13.9 TEST METHODS

The demand for reliable test data has resulted in an increased need for internationally recognised test procedures. Test information is of little value unless one knows how the test information was obtained. It can also prevent a potential customer making meaningful comparisons between different manufacturers products. Bodies such as the International Organisation for Standardisation (ISO), British Standards Institute (BSI), the European Heating and Ventilation Society (Eurovent) and the American Society of Heating Refrigeration and Air Conditioning Engineers (ASHRAE) have produced test codes for most pieces of equipment commonly encountered by the building services engineer. A high degree of similarity exists between these organisations standards, occasional differences arising where conditions might be unique to one country.

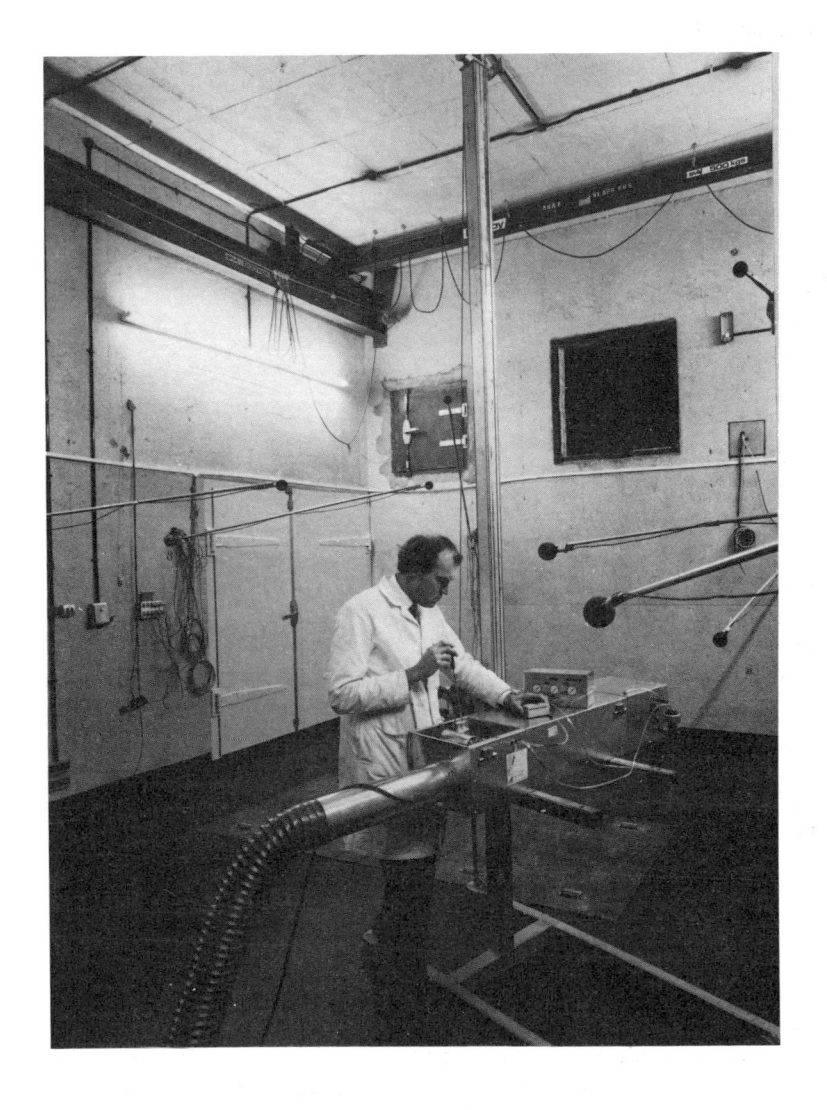

TERMINAL UNIT

Unit being set up for its very important aerodynamic set point before the acoustic reverberation room measurements of total combined outlet and casing radiated sound power level, which in this case included breakout noise from the flexible feeder duct.

13.9.1 Generated Sound Power Level Measurements

This is the largest area of testing, offering a wide variety of codes each applicable to a certain group of equipment. BS 848 Part 2 1966 "Methods of Testing Fans" was one of the earlier attempts to formalise the measurement of fan noise in this country. This was extensively revised in 1985. The code provides three methods by which fan sound power levels can be measured. The in-duct, free-field, and reverberant field methods are detailed.

Free-field testing in an anechoic room is a comparatively tedious and time consuming operation of measuring equipment sound power levels, but as mentioned earlier this is partly compensated for by the additional information concerning the directivity of the noise source and the improved accuracy. When this is not required then the reverberant field method of test is both quicker and more convenient. It is significant that the test codes listed below are all written around the reverberant field method of testing providing an alternative method of in-duct tests when air movement is not involved or where the device to be tested is normally mounted in a ducted system.

13.9.1.1 Fan Testing. The acoustic testing of fans is carried out in accordance with BS 848:Part 2:1985. Only two classifications of test are necessary when measuring fans to determine sound power levels. If the fan has a duct on the inlet and/or outlet side then the sound power levels on the sides which are ducted have to be determined by an in-duct method. If the fan is unducted on the inlet and/or outlet side then the sound power levels on the sides which are unducted have to be determined by a method where sound pressure levels are measured on a measurement surface or at discrete points enveloping the fan. The second classification can use an anechoic or semi-anechoic room or a reverberation room provided the fan size permits qualification of the room. When the fan size is sufficiently large such that the anechoic or semi-anechoic or reverberation room cannot be used then the semi-reverberant method may be applied. It should be noted that this method may be used even for the case where the fan size is small.

Figures 13.16 and 13.17 illustrate recommended test assemblies for an in-duct and reverberant field sound power testing.

The standard covers four installation types representative of the actual fan operating conditions in service. In the previous 1966 code it was possible to obtain an induct sound power level from an open sound power level, and vice versa, by using an end reflection correction. Except for small fans, under 150 mm diameter, this is no longer possible and the fan must be tested in its typical in-service arrangement.

Figure 13.16 Induct Simultaneous Measurement of Inlet and Outlet Noise

13.9.1.2 In Duct Method. All connections between the fan and the ducts shall be rigid, unless a vibration isolating coupling is used as an inherent part of the fan system. An anechoic termination is fitted to the ductwork to limit the reflection of sound from the termination of the duct which could affect the measurements. An example of an anechoic termination is given in Figure 13.18.

A microphone is located within the test ductwork at a specific point as determined from the standard and measurements made at three positions 120° from each other. To eliminate the effect of the air flow over the microphone either a sampling tube, nose cone or foam ball is fitted to the end of the microphone. These windshields may affect the sensitivity of the microphone and correction factors need to be applied as well as corrections for the frequency responce of the microphone, the flow velocity correction of the sampling tube (if fitted), the modal correction for the windshield (if

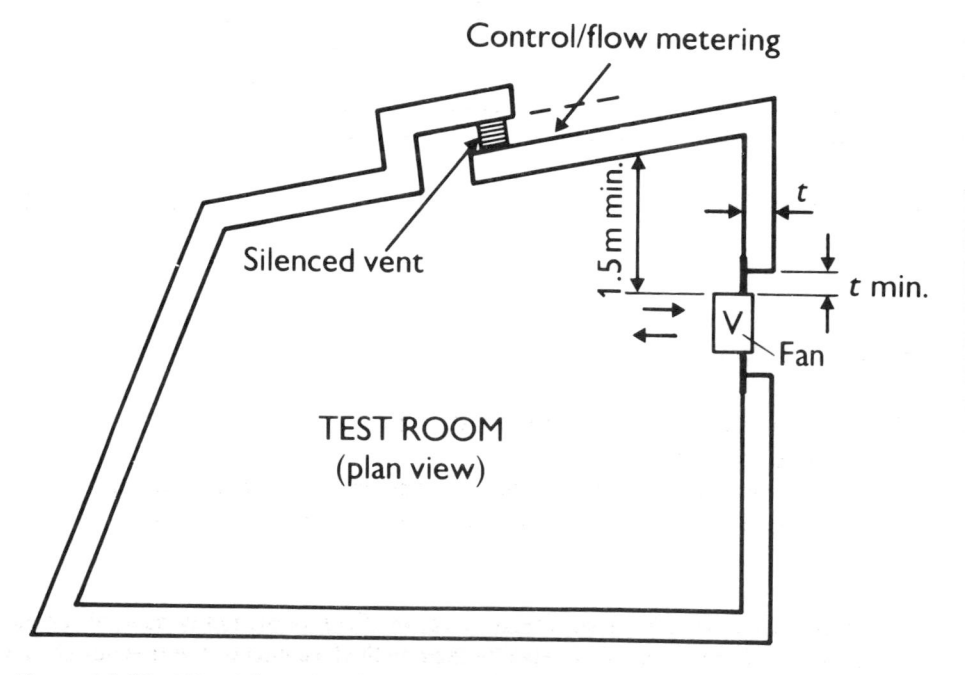

Figure 13.17a Reverberation Room Test For Free Inlet/Free Outlet Configuration

fitted) and a correction for the self generated noise of the windshield. From the sound pressure level measured by the microphone the fan sound power can be calculated by:

$$\text{Sound Power Level, } L_w = L_p + 10\log_{10}\left(\frac{A}{A_0}\right) + B + C \qquad 13.2$$

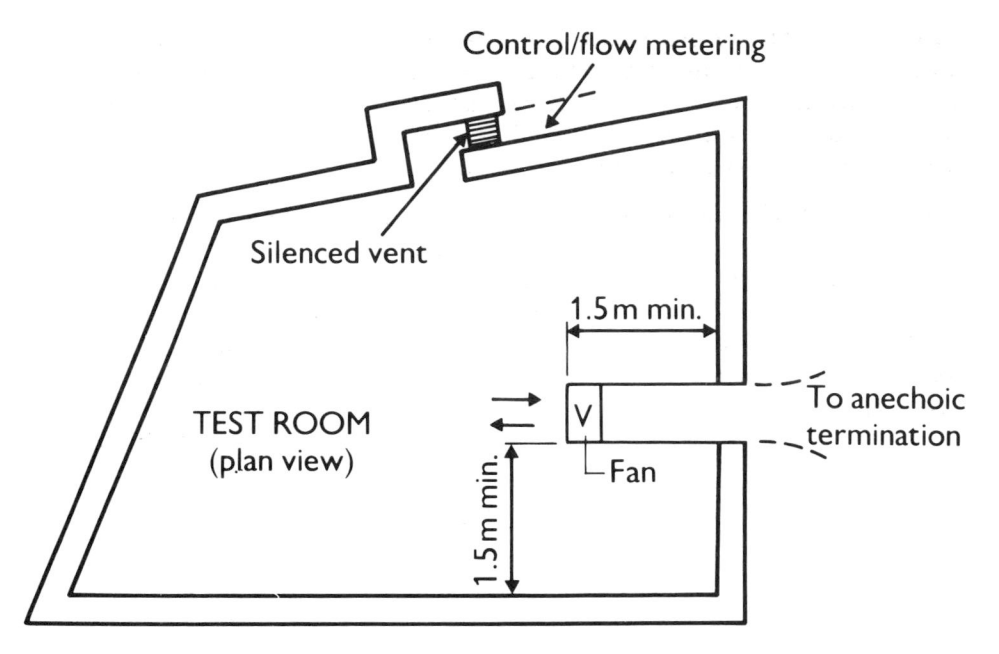

Figure 13.17b Reverberation Room Test For Free Inlet/Outlet, Ducted Outlet/Inlet Configuration

where L_w = sound power level;
$\quad L_p$ = average sound pressure level;
$\quad A$ = cross sectional area of the test duct, m²;
$\quad A_0$ = reference area = 1 m²;
$\quad B$ = correction factor for temperature and pressure, for air = 0; and
$\quad C$ = compound correction factor for microphone and windshield.

13.9.1.3 Reverberant Field Method. The reverberant field method may be used for the determination of the sound power level of fans on the free inlet or the free outlet side or for fans too small (less than 150 mm in diameter) to be tested by the in-duct method. The average sound pressure level in the reverberation room is measured by time averaged sampling at several different microphone locations. The sound power level can be obtained by:

$$\text{Sound Power Level, } L_w = L_p - 10\log_{10}\left(\frac{T}{T_0}\right) + 10\log_{10}\left(\frac{V}{V_0}\right) - 14 - B \qquad 13.3$$

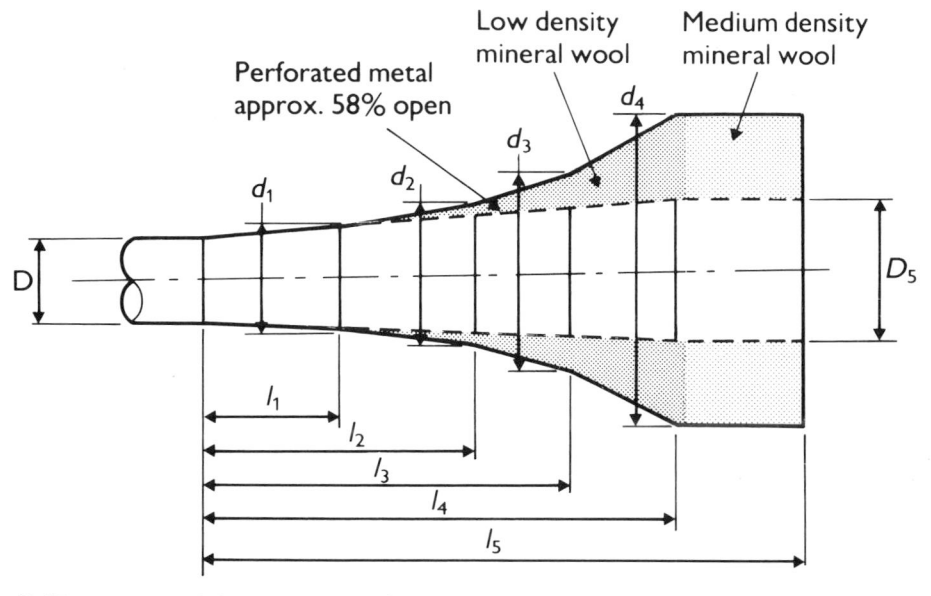

D (Duct internal diameter)   d = external diameter   l = duct length

$$d_1 = 1.15\,D\text{·}\qquad l_1 = 1.44\,D$$
$$d_2 = 1{,}64\,D\qquad l_2 = 2.89\,D$$
$$d_3 = 2.25\,D\qquad l_3 = 3.89\,D$$
$$d_4 = 3.44\,D\qquad l_4 = 5.11\,D$$
$$D_5 = 1.67\,D\qquad l_5 = 6.44\,D$$

Figure 13.18   Anechoic Termination Allowing Airflow

where L_w = sound power level;
  L_p = average sound pressure level;
  T = reverberation time of test room, s;
  V = volume of the test room;
  B = correction factor for temperature and pressure, for air = 0;
  T_0 = 1.0s; and
  V_0 = 1 m³.

The induct sound power levels for small fans (i.e., less than 150mm diameter) can be obtained by adding a correction for each octave band end reflection — Figure 13.19 to that calculated from the above formulae.

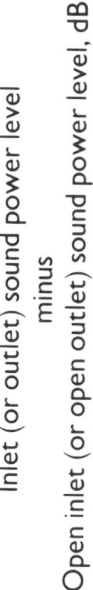

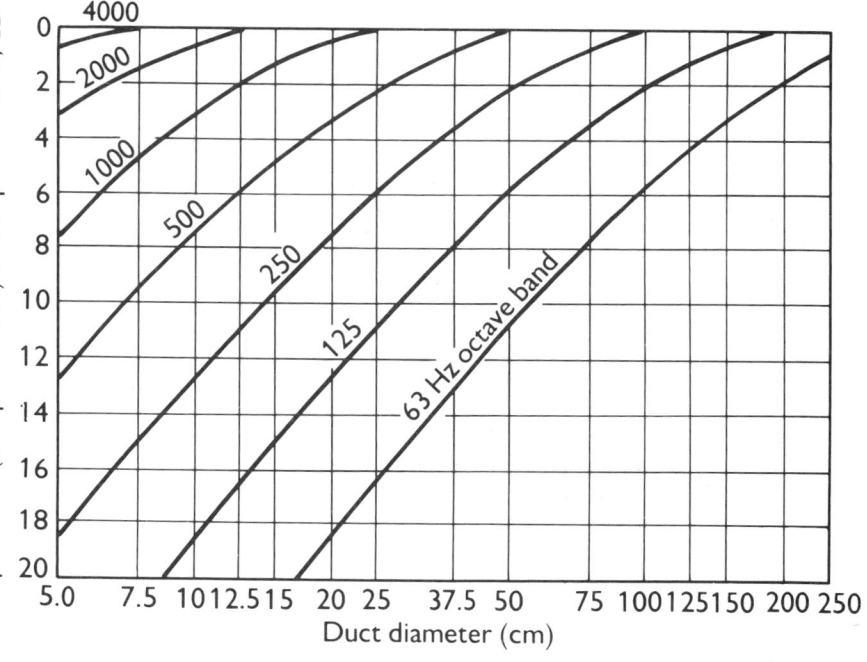

NOTE. For rectangular ducts, the effective diameter may be taken as $1.13\sqrt{A}$ where A is the cross-sectional area in cm².

Figure 13.19   End Reflection Correction

13.9.1.4 Anechoic or Semi Anechoic Method. The test environments that are suitable for measurements according to this method include:

(a) an anechoic room which provides a spherical free field
(b) an anechoic room which provides a hemispherical free field over a reflecting plane
(c) a flat outdoor area complying with certain requirements
(d) a room in which the contributions of the reverberant field to the sound pressures on the measurement surface are small compared with those of the direct field of source.

Conditions described under (d) are met in very large rooms as well as in smaller rooms with sufficient sound absorptive materials on their walls and ceilings.

The test room should be large enough and as free as possible from reflecting objects with the exception of the reflecting plane. The measurement surface within a test room should be within a sound field which is sufficiently free of undesirable sound reflections from the room boundaries and outside the near field of the fan under test.

Measurements are made on an hypothetical surface enveloping the fan, normally in the shape of a hemisphere. Depending upon the type of fan installation under test up to 20 microphone measuring positions may be required. The sound power level can be calculated by:

$$\text{Sound Power Level, } L_w = L_p + 10\log_{10}\left(\frac{S}{S_0}\right) - K - B \qquad 13.4$$

where L_w = sound power level;
L_p = average sound pressure level;
S = area of measuring surface m^2;
S_0 = reference area = 1 m^2;
K = correction factor for test environment; and
B = correction factor for temperature and pressure, for air = 0.

Measurements for all types of test methods have to be made in third octave band widths from 50 Hz to 10 kHz and reported in the respective one third octave band widths. This enables assessment of narrow band noise such as blade passing frequencies. The aerodynamic and acoustic tests to BS 848 cannot be carried out at the same time as the flow staightener for the former must be removed for the latter to reduce flow generated noise from the source. The new code is therefore much more complicated than the old. It is hoped that the increase in knowledge of fan performance and the improvement in the data collection over the intervening 20 years will enable much more accurate data to be obtained which could lead to the reduction in attenuator sizes for room noise control which could pay back the considerable increased cost of fan testing to the new standard.

13.9.1.5 Roof Units. For testing exhaust roof and wall ventilators, consideration must be given to their operating position during the test. Some have fans having anti-vibration mounts, others have gravity operated shutters. They must be operated therefore in a representative site position. HEVAC have published a test code No.SFB (1961)(S7) which makes recommendations about the method of testing of such fans. Figure 13.20 shows the recommended test position. Apart from the operational procedure, resultant sound pressure levels in the reverberation room are measured in exactly the same way as for other fans as described earlier.

13.9.1.2 Flow Generated Sound Power Level Measurements. *Air Terminal Devices* include high velocity single or dual duct systems, with constant or variable volume flow rate regulators and sound attenuating sections, terminal reheat devices, diffuser and

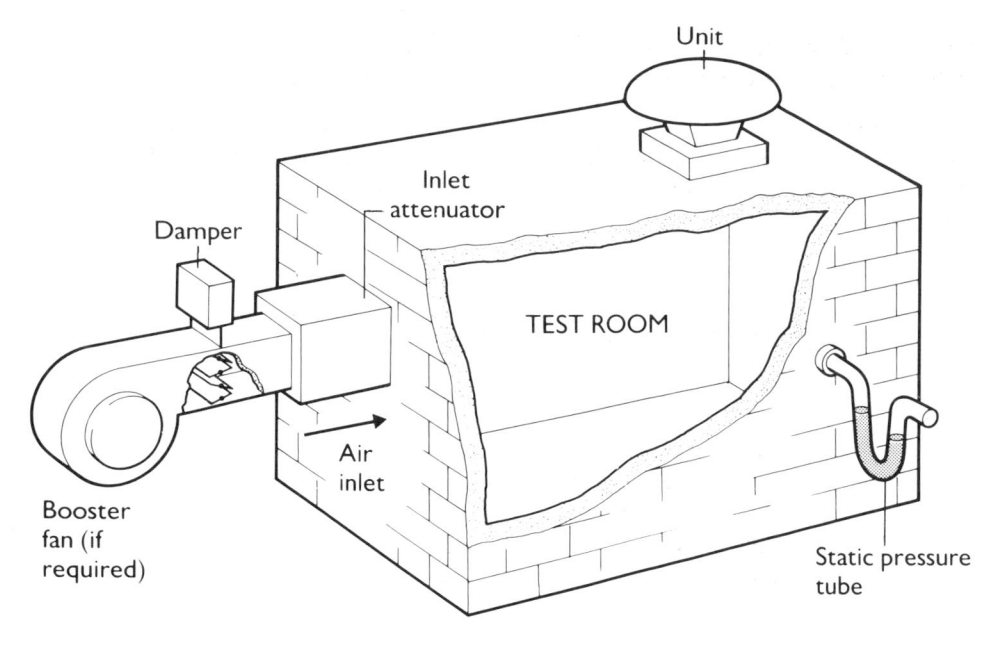

Figure 13.20   Roof Extract Fan Test Method

octopus outlets. The British Standards BS.4718; BS.4773; BS.4856; BS4857 and BS 4979 and other equivalent international standards make recommendations for measurement of airborne sound power, casing radiated sound power, and insertion loss of the device. Figure 13.21 shows the recommended locations of different terminal devices in a reverberation room. Although the principle of measuring sound power in the reverberation room is similar to that described for fan testing, most codes include precautions to ensure measurements are generated entirely by the unit on test. For example, the air supply must be sufficiently quiet to ensure no contribution is added to the unit sound power level. This can represent a problem with continuous flow monitoring unless a suitably quiet method of airflow generation is selected. Noise from the air supply system can be checked by inserting an attenuator, known as a noise check attenuator, having a performance of at least 5 dB in the octaves of interest upstream of the device under test. If the noise level produced does not fall by more than 1 dB the test facility is considered to be acceptable and the test is continued with this noise check attenuator still in place.

In addition most standards for testing air terminal devices also allow the use of a reference sound source for calibrating the reverberation room. This usually takes the form of some aerodynamic source of known sound power output which can be operated in the test chamber, located as shown in Figure 13.21 at Position C, so that the mean time averaged sound pressures can be measured. The correction between the mean sound pressure level and the known sound power level of the sound source is then calculated.

For measuring casing radiated sound power level it is necessary that the unit under investigation is operated entirely within the reverberant room, Figure 13.21 position E. It is important that the discharge airborne sound power level is sufficiently attenuated so as not to contribute to the casing radiated sound power levels. Likewise the air supply system must be adequately contained to prevent radiation of noise through the supply ducting.

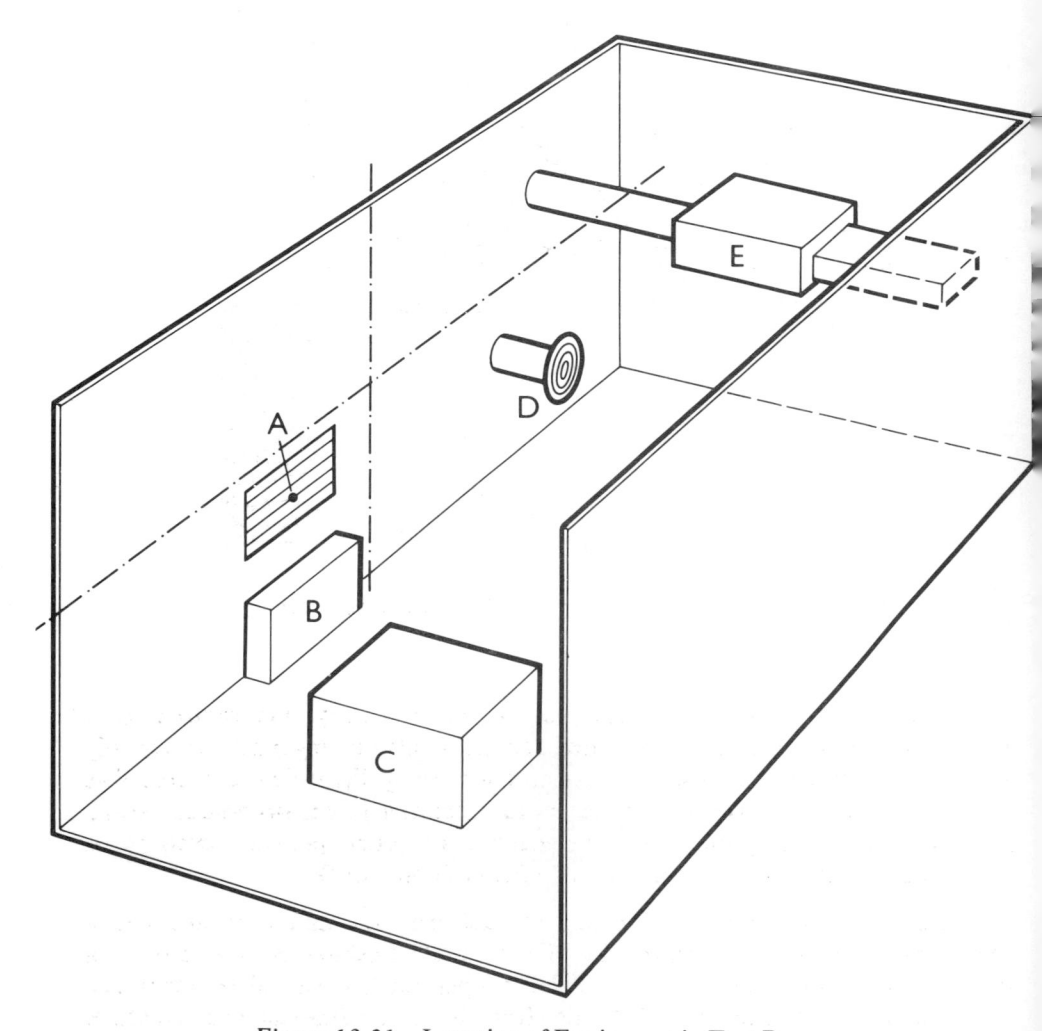

Figure 13.21   Location of Equipment in Test Room

Figure 13.21 (cont.) Air Terminal Device in test location, D.

Grille and Diffuser noise is generally a function of face velocity and pressure drop. It is essential that a silent air supply is used and the supply ducting must be internally smooth so as not to create flow generated air noise. For grilles which very often have very low noise levels, it is important that the reverberation rooms airborne sound and vibration isolation is good enough to enable such sound pressure levels to be measured. Test location shown in Figure 13.21, Position A or D.

Induction and Fan Coil Unit Testing. In each case the unit is mounted in a reverberation chamber as shown in Figure 13.21, Position B. The mounting conditions of the unit under test is particularly important. This should be so arranged as to minimise the transmission of vibration from the unit into the test room structure. Assuming that the operating conditions, such as temperature, humidity, air volume, motor speed and static pressure drop, have been selected and suitably maintained during the test, the total sound power radiated from the unit into the room may be measured.

For induction units, noise levels of the primary air supply must be checked using a noise check attenuator as described above to ensure they are not adding to the total sound power level measured.

In the case of units with ducted fresh air inlets, measurement of sound power radiated from the unit through this ducting may be carried out by locating the unit on the outside of the reverberation chamber and ducting the noise into the room to obtain the units natural insertion loss.

Units such as "through the wall" room air conditioners require additional airborne sound reduction index test, to assess the sound reduction index when placed in a building envelope. This is measured in a transmission suite, in the same way as for other airborne Sound Reduction Index tests according to BS.2750 detailed later in this chapter.

13.9.2 Attenuator Testing

BS4718:1971 concerns the performance measurement of unit attenuators and covers static insertion loss of broad band airborne sound, airflow generated noise and aerodynamic pressure loss. Figure 13.22 is the recommended attenuator airflow generated noise test rig. The fan-attenuator combination provides quiet air flow which must be maintained to within 2%.

The resultant sound pressure levels are measured in the reverberation room and the generated sound power level is calculated from:

$$\text{Sound Power Level, } L_w = L_p + 10\log_{10}V - 10\log_{10}T - 14 + X_R \qquad 13.5$$

where L_w = sound power level;
$\quad L_p$ = average sound pressure level;
$\quad T$ = reverberation time, s;
$\quad V$ = room volume, m^3; and
$\quad X_R$ = end reflection factor.

Insertion loss is tested using either the in-duct test facility in Figure 13.6a or a reverberation room. In each case the insertion loss is obtained from the difference in mean sound pressure levels at the measuring positions firstly with an empty duct and secondly with the attenuator in position.

Background noise levels must be at least 6 dB and preferably 10 dB below the measured values particularly with the attenuator in test. Similarly care must be taken to ensure that breakout from the duct does not flank the attenuator and break back into the ductwork on the quiet side of the attenuator. Casing breakout noise can be measured as shown in Figure 13.21, Position E.

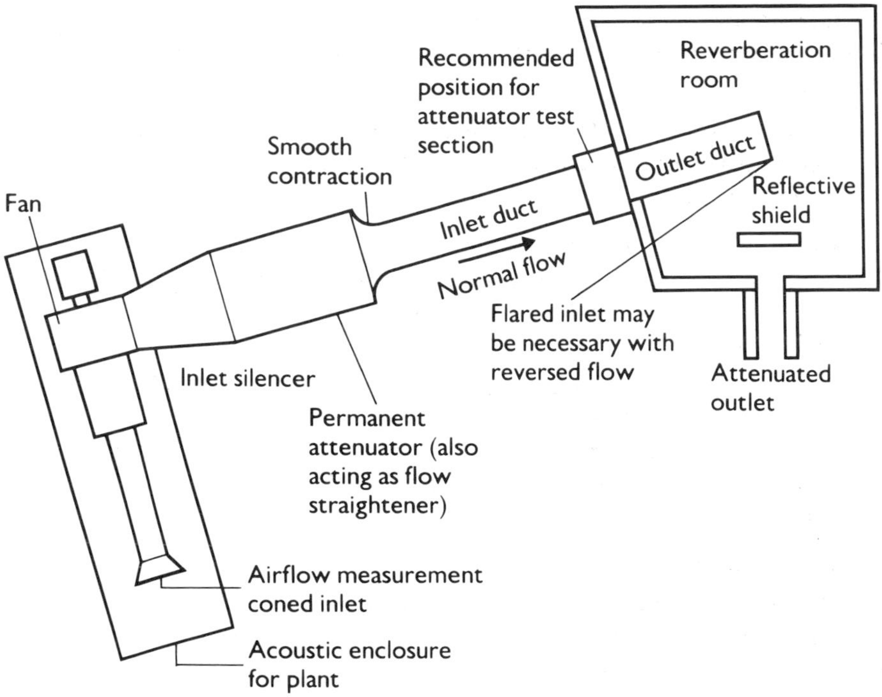

Figure 13.22 diagram labels:

Recommended
position for
attenuator test
section

Reverberation
room

Smooth
contraction

Outlet duct

Fan

Reflective
shield

Inlet duct

Normal flow

Inlet silencer

Flared inlet may
be necessary with
reversed flow

Attenuated
outlet

Permanent
attenuator (also
acting as flow
straightener)

Airflow measurement
coned inlet

Acoustic enclosure
for plant

Minimum duct lengths:
Inlet duct 5 D or 4 m whichever is greater
Outlet duct 5 D or 4 m whichever is greater

Figure 13.22   Attenuator Flow Generated Noise Test Rig

13.9.3 Airborne Sound Reduction Index

BS2750 and ISO 140 contains recommendations for laboratory measurements of airborne and impact sound reduction index. Figure 13.23 is a diagrammatic representation of the test facilities required. Subject to certain conditions being met, sound is generated in the transmitting room and the average sound pressure level measurements are recorded on both sides of the test sample.

Using airborne sound:

$$\text{Sound Reduction Index: } R = L_1 - L_2 + 10\log_{10}\left(\frac{S}{A}\right) \qquad 13.6$$

where L_1 = average sound pressure level in source room;
L_2 = average sound pressure level in receiving room;
S = area of partition under test, m^2;

$$A_2 = \text{receiving room absorption} = 0.163\ \frac{V_2}{T_2}$$

where V_2 = receiving room volume, m^3 and T_2 = receiving room reverberation time, s.

Performance data is computed usually at frequencies from 100 Hz to 3150 Hz in third octaves and presented as the numerical average of these values.

An alternative and increasingly used value is derived from BS.5821. This gives a "Method of rating airborne sound insulation in buildings and of interior building elements" and is depicted as "R_w", and known as the Weighted Sound Reduction Index. This is obtained by comparing the results of a Sound Reduction Index test with a reference curve allowing a maximum of 2.0 dB for the mean unfavourable deviations.

The American equivalent Sound Transmission Class STC, makes a similar calculation between the 125 Hz and 4000 Hz third octave bands allowing a maximum deviation of 8 dB in any one third octave band in addition to the maximum of 2.0 dB for the mean unfavourable deviations.

With the increasing interest in accreditation there comes a tendency to report data as simply and concisely as possible with the minimum of conditions and calculations.

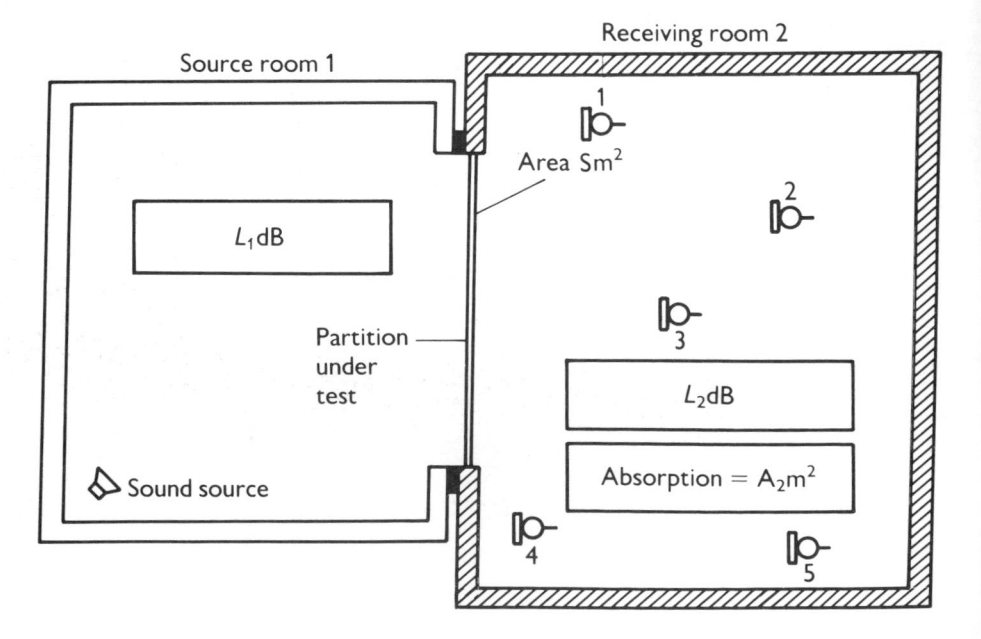

Figure 13.23   Measurement of Sound Reduction Index

Figure 13.23 (cont.)   Sliding Patio Doors under test to evaluate their Sound
Reduction Index

Whilst there are limitations to the over implementation of this philosophy one correction of importance concerns the calculation for compound partitions when a high transmission loss door, 50 dB, is being tested in a good wall, of say 55 dB, when one considers that the door typically occupies only 20% of the total test partition. Since the actual wall is not usually tested and a $\pm$ 2 dB margin of error may exist at each predicted third octave band, the correction leaves the final "accredited" door-alone performance with an even larger tolerance on the 95% certainty value. Hence there is a tendency to fully report the test construction of wall plus door and report only the performance of the compound combination. Others with expert experience may then manipulate this accredited value to best effect.

The requirements of the Building Regulations 1985 section E — Airborne and Impact sound — also refer to sound insulation tests on walls and floors carried out in accordance with BS2750. However these refer to a different interpretation of tests which are normally carried out in the field. Laboratory tests are still valid but are reported in terms of Weighted Standardised Level Difference (D_nT,w) established using BS5821 from Level Differences determined by BS2750 where the Standardised Level Difference (D_nT) is established by:

$$\text{Standardised Level Difference, } D_nT = L_1 - L_2 + 10\log_{10}\left(\frac{T}{T_0}\right) \qquad 13.7$$

where D_nT = Standardised Level Difference;
L_1 = average sound pressure level in source room;
L_2 = average sound pressure level in receiving room;
T = reverberation time in receiving room, s; and
T_0 = reference reverberation time = 0.5s.

Sound Reduction Index for a number of common materials are given in Appendix 13.B.

13.9.4 Impact Sound Transmission Loss

The test facilities are similar to those required for airborne transmission loss in accordance with the BS2750. For impact transmission loss a reference tapping machine is operated on the test floor constructed as a roof and sound pressure measurements recorded in the reverberant receiving room below.

$$\text{Normalised impact sound pressure, } L_N = L_2 - 10\log_{10}\left(\frac{A}{A_0}\right) \qquad 13.8$$

where L_2 = average sound pressure level in receiving room;
A = receiving room equivalent absorption area, m^2; and
A_0 = reference absorption area, 10 m^2.

Care must be taken to ensure that the airborne noise radiating from the tapping machine does not flank the floor structure, raising the measured levels.

In the Building Regulations 1985 section E — Airborne and Impact Sound — the requirements are set out in terms of Weighted Standardised Sound Pressure Level ($L_nT_{,w}$) using BS5281. These figures are derived from the standardised sound pressure levels obtained in accordance with BS2750 from:

$$\text{Standardised sound pressure level, } L_nT = L_1 - 10\log_{10}\left(\frac{T}{T_0}\right) \qquad 13.9$$

where L_1 = average sound pressure level in the receiving room;
T_1 = reverberation time of the receiving room; and
T_0 = reference reverberation time = 0.5 s.

13.9.5 Sound Absorption Coefficients

These are usually measured in the presence of a random incidence or a normal incidence sound field.

Random Incidence Method. BS3638 and ISO354 refer to methods for the measurement of sound absorption coefficients in a reverberation room. Subject to certain conditions being met with respect to the test room and dimensions and mounting conditions of the test sample, which must be between 10 and 12 m^2, the absorption coefficients for the material may be calculated by measuring the change of reverberation time in the test room with and without a sample in place. Figure 13.24 shows a typical test taking place. The absorption coefficient for each third octave band typically between 100 Hz and 5 kHz are found from:

$$\text{Absorption coefficient, } \alpha = \left(\frac{V}{S}\right)\left(\frac{55.3}{c}\right)\left(\frac{1}{T_2} - \frac{1}{T_1}\right) \qquad 13.10$$

where V = test room volume, m^3;
S = sample surface area, m^2;
c = velocity of sound in air, m/s;
T_1 = reverberation time of the empty room, s; and
T_2 = reverberation time of the room containing sample, s.

Absorption coefficients for common materials are given in Appendix 13.C.

Normal Incidence Method. This method is usually reserved for testing porous and perforated panel absorbents and uses the standing wave tube — Figure 13.25. The American Society for Testing Materials is one of the few organisations who have compiled a test code using this particular piece of equipment. Generally, the value of normal incidence absorption coefficients is less than the random incidence data. There are graphs available for the conversion of normal incidence absorption coefficients to the more useful random incidence absorption coefficients. However,

Figure 13.24   Typical Reverberation Room Test

the standing wave tube method and the resultant normal incidence absorption coefficient are of more use for comparisons between materials than for acoustic calculations since in real life random incidence is normally encountered.

13.10 AERODYNAMIC AND ACOUSTIC MEASURING INSTRUMENTS

A wide range of equipment exists for use in air flow and noise measurements. The following gives some idea of what is currently available and used in laboratory tests.

Figure 13.25   Standing Wave Tube

13.10.1 Noise Generating Equipment

A loudspeaker is simply a piece of equipment for converting electrical signals into mechanical vibrations which in turn radiate in the form of airborne sound waves. Some loudspeakers work better in a preferred range of frequencies and in order to cover the entire frequency spectrum required for a test it may be necessary to use two or more loudspeakers. Cross over networks are electrical devices designed to channel the electrical signals to the loudspeaker best suited for their reproduction. Audio amplifiers increase the low voltage signal to a power level suitable for driving the loudspeakers.

Signal sources are many and varied but all produce a fluctuating periodic or random electrical signal with a selected frequency content. Pre-recorded tapes are often used which provide a repeatable signal source and are convenient for field testing. Alternatively, there are signal generators for sine waves, frequency modulated sine waves, bands of random noise, white noise and pink noise and selective filters for spectrum shaping of the noise generated to simulate particular conditions.

13.10.2 Noise Measuring Equipment

All these operate by measuring the amplitude of an electrical signal. It is essential therefore to convert the airborne sound pressure into an electrical analogy. This is performed by a microphone. For vibrations, an accelerometer is used.

Having obtained the electrical signal, its processing depends very much on what information content is required. Most frequency analysers measure not only the level of the signal but the spectral distribution. Analysis is normally carried out in octave or third-octave band widths but narrow band analysers are also available for detailed information of a spectrum content.

The information can be displayed in several different ways. Nearly all analysis equipment is provided with a meter from which a direct reading can be obtained. When a permanent record is required the analyser output is fed to a high speed pen recorder. More recent developments have given us the ability to analyse data in real time. Sophisticated equipment of this type coupled to a computer —Figure 13.26 can provide accurate and speedy procurement and analysis of test results.

Figure 13.26 Computer Controlled Analysis Equipment

13.10.3 Calibration Equipment

The reliability and absolute accuracy of laboratory tests are only as good as the calibration of the test instrumentation. Microphones and their analysis instruments are usually calibrated using a reference sound source such as a pistonphone or an electronic calibrator. In each case the calibrator is fitted over the microphone creating a fully enclosed void in which the reference sound pressure level is generated. The pistonphone produces this reference level by driving two small pistons on a cam, while the electronic calibrator may be thought of as a small loudspeaker driven by a sine wave signal generator.

13.10.4 Airflow Measuring Equipment

As already outlined there are many ways in which volume flow or velocity can be measured. In an acoustic testing laboratory, when noise generation may be undesirable, some of these methods are less suitable than others. Orifice plates, vane anemometers and to a lesser extent venturies and Dahl tubes are all potential noise sources and great care must be exercised in their use.

In quiet conditions a pitot-static tube, a laminar flow device or a hot wire anemometer would be the least likely to contribute to the existing sound level. Figure 13.27 provides a useful reference table detailing not only the accuracy of each method but an indication of its inherent pressure loss and noise generation characteristics.

Device	Measuring Principle	Airflow Resistance	Ease of Continuous Measurements	Suitability for use in acoustic tests
Pitot tube	Senses total and static head	Negligible	Unsuitable	Almost silent
Venturi	Produces pressure drop related to flow through it	Medium	Good	Quiet. May need extra attenuation for critical tests
Orifice plate	Similar to Venturi	High	Good	Noisy. Will require attenuation
Conical inlet nozzle	Similar to Venturi but with plenum inlet	Average	Good	Usually used on fan tests, relatively quiet.
Hot wire anemometer	Depends upon the cooling effect of air flow on a heated wire.	Negligible	Unsuitable	Almost silent
Vane anemometer	Windmill speed related to air flow through it	Low	Poor except under special circumstances	Noisy. Will require extra attenuation
Laminar flow device	Produces pressure drop linearly related to flow through its honeycombs	Average	Good	Quiet. Attenuation only needed for critical tests
Flow grid	Average pitot device	Low	Good	Quiet. Attenuation only needed for critical tests
Annubar	Averageing pitot device	Negligible	Good	Almost silent

Figure 13.27   Airflow Measuring Equipment

13.11 ACOUSTIC MEASUREMENT TECHNIQUES

Many acoustic tests are available to measure the sound power levels of equipment. In view of the fact that sound power is not a parameter than can be measured directly, it is necessary to measure a physical manifestation of this, such as sound pressure level or sound intensity and subsequently, convert it to sound power level. This can be performed with a knowledge of the room characteristics in which the sound pressure levels are measured.

13.11.1 Reverberation Room Sound Pressure Levels

A reverberant room produces a diffuse sound field and it is essential to measure the mean sound pressure level of this field. In practice, the sound pressures vary from place to place, so that some method of sampling this sound field is used. This is normally achieved by taking the average result of a number of randomly selected time averaged readings. The greater the number of measurements the greater the accuracy of the result. Care must be taken in the selection of these positions and no microphone measuring position should be closer than half a wavelength to a reflecting surface or another measuring position where the wavelength chosen must be that of the centre frequency of the lowest band of interest. Such rooms are usually restricted to octave and one third band analysis and are not suitable for pure tone power measurements.

13.11.2 Anechoic Room Measurements

If free field conditions can be closely approximated it is possible to determine the sound power level and directivity of a source by measuring sound pressure levels at a number of points in an anechoic room. Additionally narrow bandwidths and pure tones may be measured. The measuring positions should be equi-distant from the source and evenly distributed about it. The points may be considered to be on the surface of an imaginary sphere centred on the source and the radius of which is at least three times the largest dimension of the source and, at least equal to one wavelength for the lowest frequencies to be measured. To obtain the overall sound power level, and if the total range of sound pressure levels is less than 6 dB, an arithmetic average is taken and substituted in the equation:

$$\text{Sound Power Level, } L_w = \overline{L}_p + 20\log_{10}r + 11 \text{ dB} \qquad 13.11$$

where $\overline{L}_p$ = average sound pressure level; and
$\quad\quad r$ = sphere radius, m.

For the directivity factor in a particular direction, the sound pressure level in the relevant direction is measured. The difference between this and the average is called the directivity index:

$$\text{Directivity Index} = L_{pi} - \overline{L}_p \, \text{dB} \qquad\qquad 13.12$$

where L_{pi} = sound pressure level in direction of interest
and $\overline{L}_p$ = average of all sound pressure levels.

The directivity factor, Q, is obtained by converting the directivity index to a power ratio thus:

$$\text{Directivity factor, } Q = \text{antilog}_{10} \left(\frac{DI}{10} \right) \qquad\qquad 13.13$$

where DI = directivity index, dB, from equation 13.12.

13.11.3 Limitations of Background Noise

Ideally, sound pressure measurements of equipment should be independent of contributions from other sound sources. Hence the need for measuring rooms to be isolated from extraneous noise and vibration. If the increase in the sound pressure level in any given frequency interval, with the sound source operating, compared to the ambient sound pressure level is 10 dB or more, then the total level due to source and ambient level is effectively the sound pressure due to the source. If the difference is less than 10 dB Figure 13.28 shows the corrections to be made to the source level.

Difference	Subtract from total
10+	0
6 to 9	1
4 or 5	2
3	3
2	(4 or 5)

Figure 13.28   Decibel Subtraction

13.12 DISCUSSION

This chapter has concentrated on the reasons for and how laboratory testing of materials and devices take place. Laboratory data is an essential and integral part of the Building Services Engineer's stock in trade. Whilst it may be true that the controlled conditions in an acoustic laboratory may not relate directly with the site situation, the information obtained, if properly applied will make a significant contribution to the overall outcome of every project. A following example will make this point clear.

13.12.1 Attenuators

Laboratory tests may show that a particular attenuator has a specific capacity for reducing the level of sound which arrives at the unit as a plane wave. It does not follow in a site situation that the only incidence on the attenuating section will be in the form of a plane wave. Indeed, the nearer the sound source is to the attenuator, the more the sound incidence becomes random. In this case the attenuator may have a better insertion loss performance and Figure 13.29 shows the improvement to be expected.

When high attenuations are expected the implication is that the in-duct sound power levels downstream of the attenuator will be low. If however, the outlet duct is in a region of high noise level then the in-duct sound power levels may be increased as a result of "break in" thus reducing the nett attenuator performance unless preventive action is taken.

The published data for attenuator pressure loss and regeneration indicates the performance to be expected with ideal flow conditions. Site conditions are frequently far from ideal. Not only will the pressure loss be higher, but the uneven flow through the attenuator will increase the regenerated sound power levels.

The nature of the test code BS4718:1971 ensures that the inlet ductwork is free from vibration: the insertion loss results therefore do not show any reduction resulting from

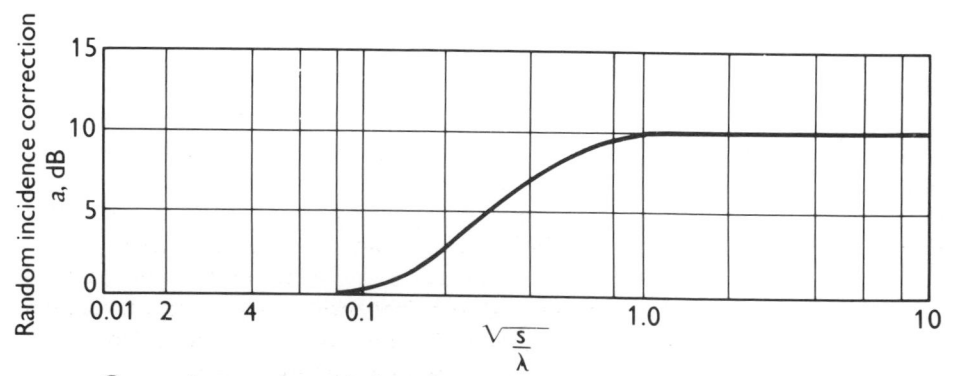

Correction a to be added to plane-wave incidence insertion loss of a duct to obtain the insertion loss for a random incidence of sound S = open area of duct; λ = wavelength of sound in consistent units.

Figure 13.29   Attenuator correction for random incidence

such problems. For site applications on plant with high vibration levels, the quoted attenuator performance may only be realised after adequate measures have been taken to isolate the attenuator from these vibrations.

13.12.2 Fans

The effect of inlet and discharge conditions on a fan are important to the generated noise levels. Laboratory data is obtained when the flow through the fan is ideal. It is frequently necessary on site to isolate the fan from its supply duct. The practice is to use a flexible collar which can cause an airway constriction close to the fan and possible increase the total radiated sound power level (Figure 13.30) as already mentioned in Chapter 7. This can be minimised by putting the collar as far away from the fan as is practically possible in a position where the air flow will be less turbulent. Often axial flow fans are supplied with close coupled matching cylindrical attenuators, in which case the flexible should be at the ends of these attenuators (Figure 13.31).

13.12.3 Vibration Isolation of Room Conditioners

In the mounting of unitary room conditioners it is necessary to keep to a minimum the transmission of vibration (due to out of balance forces within the fan and mains induced "hum") to the surroundings via walls etc. In the laboratory test the unit is adequately mounted on anti-vibration devices and its noise levels measured accordingly. If similar site mountings are not achieved, then the unit can cause a whole floor or wall area to re-radiate acoustic energy resulting in higher noise levels than the acoustic design and test predicts.

13.12.4 Terminal Unit Noise Breakout

Assuming a quiet air supply, the major noise source in a high pressure unit is the variable or constant volume flow rate regulator or controller. Laboratory data establishes how much noise comes from the air outlet of the unit and how much radiates from the casing. What may not be realised is that the regulator noise will propagate upstream as well as downstream. Depending on the nature of the supply duct wall the noise breakout in certain frequency bands can be higher than the sound power levels radiated by the casing. In critical areas where the designer is hoping to achieve low noise levels this source must be taken into account. Upstream attenuators may be used to reduce these levels. Alternatively if the feed ducting passes from a non critical area, the ductwork in the critical area may be suitably lagged.

13.12.5 Flow noise in Supply Ducting

As a general rule wherever pressure is lost in a typical duct run some of this energy is converted to sound. Ductwork with smooth rigid internal faces is much less likely to

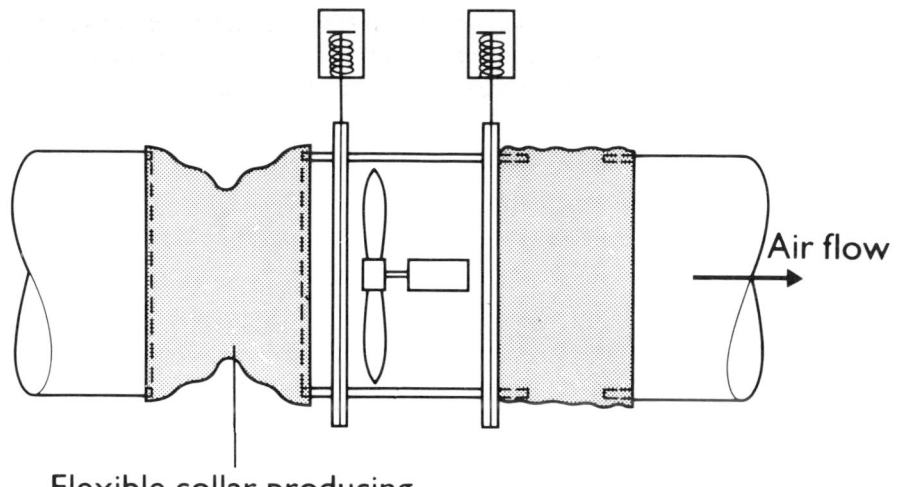

**Flexible collar producing
turbulent inlet conditions
at the fan**

Figure 13.30   Increased Fan Noise Generation Resulting from Poor Inlet
Conditions

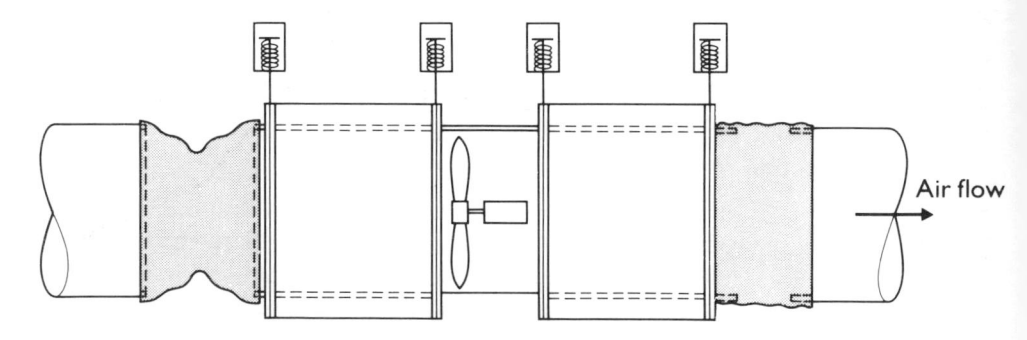

Figure 13.31   Close Coupled Attenuators on Axial Flow Fan

cause flow generated noise than spiral wound flexible ductwork. If a site situation demands the use of flexible ducting then the effect of regeneration can be reduced by keeping the air velocities as low as possible, preferably less than 10 m/s.

A secondary effect of regeneration which may be significant in some conditioned areas is that of break out.

A typical increase in sound power levels which result from replacing a 600 mm length of 2 mm steel ducting with flexible fabric ductwork is shown in Figure 13.32. This result demonstrates a high frequency weakness in the flexible ductwork.

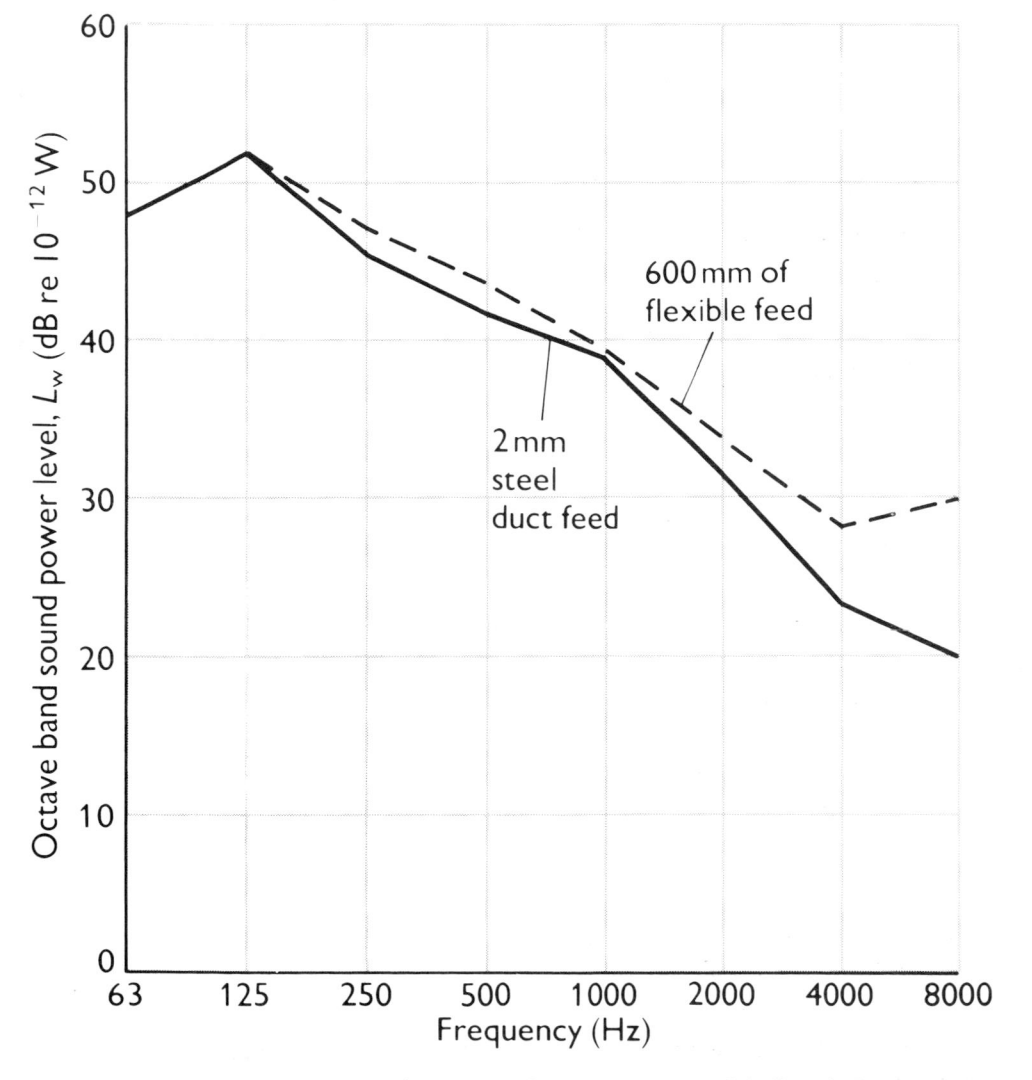

Figure 13.32   Increased Breakout from 600mm Flexible Supply Duct

Laboratory acoustic performance for grille and diffuser air outlets is obtained in the knowledge that the air approaching the unit is quiet and any use of flexible feeds on site can lead to a modification of this condition. Hence the grilles may appear to be noisier than the manufacturer's claim. The problem is then the isolation of the seperate noise sources and the possible inclusion of secondary attenuators before the grille.

13.12.6 Inlet Conditions

Because of site space restrictions it is frequently found that inlet conditions to a unit are very different from those under which its noise levels were measured in the laboratory. Due allowance should be made for this when estimates are made. The use of inlet cones can help to minimise regeneration noise where air flows are required to expand or contract. Flow straightners will reduce turbulence but care is necessary to ensure that their own self generated mid and high frequency noise does not worsen the situation.

Additionally, misalignment of the inlet flow, especially offset joggles, on to the regulator can increase the flow generated noise of this regulator, and hence all dependent noise outputs, by as much as 6 dB.

13.12.7 Transmission Barriers

All laboratory tests for Sound Reduction Index are conducted in a wall which has a high transmission loss compared to the test unit. The barrier may be mounted on site in a wall of much lower performance. There is no point in using a door with a Sound Reduction Index greater than the wall into which it is to be mounted — Figure 13.33 and if this is done it would be incorrect to expect the performance of the combinations to be as high as the laboratory test for the door alone.

% of wall occupied by door	30 dB Door			38 dB Door		
	45 dB	50 dB	55 dB	45 dB	50 dB	55 dB Wall
100	30	30	30	38	38	38
80	31	31	31	39	39	39
60	32	32	32	40	40	40
50	33	33	33	40	41	41
40	34	34	34	41	42	42
30	35	35	35	42	43	43
20	36	37	37	42	44	45
10	39	40	40	44	46	47
5	41	42	43	44	47	50
4	41	43	43	44	48	50
3	42	44	45	45	48	51
2	42	45	46	45	49	52
1	43	47	48	45	49	53

Figure 13.33 Variation of Combined Sound Reduction Index for Door/Wall Barriers

13.12.8 System Mock-Up in a Laboratory

Some of the numerous problems encountered when applying laboratory data to site situations have already been outlined. Whilst recognising that tests on different units must be conducted according to a standardised and repeatable set of conditions, it is equally important for the design engineer to have a knowledge of the extent to which each and every variation from the idealised test will influence the applied acoustic performance of the unit. It would be impracticable for any system to be installed unless there was sufficient evidence to indicate that when commissioned all areas served by this system will meet their design criteria. Clearly a considerable wealth of data is now available to permit a designer to cope with most site problems. Yet there

Figure 13.34   Floor being assembled for Impact Sound Transmission evaluation

Figure 13.35   Acoustic Ceiling Tiles being installed for Sound Absorption test

are still occasions when new combinations of circumstance arise which have not previously been encountered. It is at this point that the importance of the laboratory and its flexibility becomes indispensable.

Given a laboratory with the facilities and equipment already discussed it provides an excellent location in which total systems or parts of such systems may be assembled and studied in great detail. There are significant advantages in using system mock-ups. In the absence of the normal confines of a building it is easier to gain access to almost any part of a system and make measurements at that point. It is also possible with the equipment available in the laboratory to investigate the effect of changing, adding or removing individual components of that mock up.

Laboratory testing can therefore provide for accurate data, for comparison of products, for the design of simple systems and for the development of more complex systems by virtue of "mock-ups". Adequate data from tests in controlled conditions can therefore save time and money if correctly applied.

13.13 CONCLUSIONS

This chapter has illustrated some of the areas where the building services engineer can encounter noise problems; the things that may go wrong in the absence of reliable product performance data, or when it is incorrectly applied. A laboratory can be a powerful aid to manufacturers or designers. Without it one can endeavour to succeed by a mixture of intuition, experience and plain guess work, but for how long? With it data measured in accordance with the relevant codes can be produced and the suitability of new ideas can be investigated before undertaking the enormous expense of installation.

Equally well one can see the importance of such a facility in achieving the most economic solution of site problems when they are encountered. The flexibility of a laboratory and the ability to control each variable make it an ideal place for the simulation and investigation of such problems.

APPENDIX 13A SOUND RESEARCH LABORATORIES LABORATORY DIMENSIONS

Reverberation Room

Length 8.3 m
Width 6.7 m
Height 5.4 m
Nominal volume 300 m^3
Roof unit test aperture 1.425 m x 1.425 m
Fan inlet test aperture 1.425 m x 1.425 m
"Transmission Loss" aperture 2.0 m long x 1.2 m high

Lower North Transmission Room

Length 4.9 m
Width 3.9 m
Height 3.1 m
Nominal volume 60.0 m^3

South Transmission Room

Length 4.5 m
Width 3.9 m
Height 3.1 m
Nominal volume 54.5 m³

Upper North Transmission Room

Length 4.6 m
Width 3.9 m
Height 2.9 m
Nominal volume 52.0 m³

APPENDIX 13B SOUND REDUCTION INDICES

	Thickness mm	Weight kg/m²	\multicolumn{7}{c}{Octave Band Frequencies Hz}
			63	125	250	500	1k	2k	4k
3 mm lead sheet	3	34	25	30	31	27	38	44	33
18 g galvanised sheet steel	1.2	10	8	13	20	24	29	33	39
18 g fluted steel panels stiffened at edges, joints sealed	1.2	39	25	30	20	22	30	28	31
Chipboard sheets on wood framework	19	11	14	17	18	25	30	26	32
Fibreboard sheets on wood framework	12	4	10	12	16	20	24	30	31
Plasterboard sheets on wood framework	9	7	9	15	20	24	29	32	35
Plywood sheets on wood framework	6	3.5	6	9	13	16	21	27	29
Woodwool slabs									
unplastered	25	19	0	0	2	6	6	8	8
plastered (12 mm on each face)	50	75	18	23	27	30	32	36	39
Walls									
Single leaf brick, plastered both sides	125	240	30	36	37	40	46	54	57
Solid breeze blocks, plastered (12 mm both sides)	125	145	20	27	33	40	50	57	56
Solid breeze blocks, unplastered	75	85	12	17	18	20	24	30	38
Hollow cinder concrete blocks									
painted	100	75	22	30	34	40	50	50	52
unpainted	100	75	22	27	32	37	40	41	45
"Thermalite" blocks	100	125	20	27	31	39	45	53	38

	Thickness mm	Weight kg/m^2	Octave Band Frequencies Hz						
			63	125	250	500	1k	2k	4k
280 mm brick, 56 mm cavity, strip ties, outer faces plastered 12 mm	300	380	28	34	34	40	56	73	76
50 mm × 100 mm studs, 9 mm plaster board and 12 mm plaster coat both sides	142	60	20	25	28	34	47	39	50
Windows									
Single glass in heavy frame	6	15	17	11	24	28	32	27	35
	8	20	17	18	25	31	32	28	36
9 mm panes in separate frames, 50 mm cavity	62	34	18	25	29	34	41	45	53
6 mm panes in separate frames, 100 mm cavity	112	34	20	28	30	38	45	45	53
6 mm panes in separate frames, 188 mm cavity	200	34	25	30	35	41	48	50	56
6 mm panes in separate frames, 188 mm cavity with absorbent blanket in reveals	200	34	26	33	39	42	48	50	57
6 mm panes in same frame, 12 mm cavity	25	30	18	21	25	29	34	28	35
Doors									
Flush panel, hollow core, normal cracks as usually hung	43	9	9	12	13	14	16	18	24
Solid hardwood, normal cracks as usually hung	43	28	13	17	21	26	29	31	34
Typical proprietary "acoustic" door	100	—	37	36	39	44	49	54	57
Floors									
T & G boards, joints sealed	21	13	17	21	18	22	24	30	33
T & G boards, 12 mm plasterboard ceiling under, with 3 mm plaster skim coat	235	31	15	18	25	37	39	45	45
T & G boards "floating" on glass wool mat	240	35	20	25	33	38	45	56	61
Concrete, reinforced	100	230	32	37	36	45	52	59	62
	200	460	36	42	41	50	57	60	65
	300	690	37	40	45	52	59	63	67
126 mm concrete with "floating" screed	190	420	35	38	43	48	54	61	63

APPENDIX 13C ABSORPTION COEFFICIENTS OF COMMON MATERIALS

—compiled from Sound Research Laboratories tests and other published data

N.B. The following data should be considered only as a guide since many factors will influence the actual absorption (e.g. method of mounting, decorative treatment, thickness, properties of surrounding structure)

Material	\multicolumn{6}{c}{Octave Band Frequencies, Hz.}
	125	250	500	1k	2k	4k
Sprayed Acoustic Plaster	0.30	0.35	0.5	0.7	0.7	0.7
Board on Joist Floor	0.15	0.2	0.1	0.1	0.1	0.1
Breeze Block	0.2	0.3	0.6	0.6	0.5	0.5
Brickwork—plain/painted	0.05	0.04	0.02	0.04	0.05	0.05
Concrete, Tooled Stone, Granolithic	0.02	0.02	0.02	0.04	0.05	0.05
Cork, 25 mm—solid backing	0.05	0.1	0.2	0.55	0.6	0.55
Fibreboard						
—solid backing	0.05	0.1	0.15	0.25	0.3	0.3
—25 mm air space	0.3	0.3	0.3	0.3	0.3	0.3
Glass, 3–4 mm	0.2	0.15	0.1	0.07	0.05	0.05
> 4 mm	0.1	0.07	0.04	0.03	0.02	0.02
Plaster, lime or gypsum						
—solid backing	0.03	0.03	0.02	0.03	0.04	0.05
—on lath/studs, air space	0.3	0.15	0.1	0.05	0.04	0.05
Plywood/Hardboard, air space	0.32	0.43	0.12	0.07	0.07	0.11
Wood Blocks/Lino/Rubber Flooring	0.02	0.04	0.05	0.05	0.1	0.05
Wood Panelling, 12 mm on 25 mm battens	0.31	0.33	0.14	0.1	0.1	0.12
Carpet, Haircord on felt	0.1	0.15	0.25	0.3	0.3	0.3
Pile and thick felt	0.07	0.25	0.5	0.5	0.6	0.65
Air Absorption (Relative Humidity = 50%) − (x) (per m³)	0.0	0.0	0.0	0.003	0.007	0.02
Audience per person	0.33	0.40	0.44	0.45	0.45	0.45
Upholstered seat	0.45	0.60	0.73	0.80	0.75	0.64
Proprietary Ceiling Tile (typical values) A. Fixed to Solid backing Mineral Wool/Fibre	0.10	0.25	0.70	0.85	0.70	0.60
Perforated Metal, 31 mm thick (absorbent infill)	0.10	0.30	0.65	0.75	0.65	0.45
B. On Battens—25 mm to 50 mm thick Mineral Wool/Fibre	0.15	0.35	0.65	0.80	0.75	0.70
Perforated Metal, 31 mm thick (absorbent infill)	0.2	0.55	0.80	0.80	0.80	0.75

Material	Octave Band Frequencies, Hz.					
	125	250	500	1k	2k	4k
C. Suspended						
Mineral Wool/Fibre	0.50	0.60	0.65	0.75	0.80	0.75
Perforated Metal, 31 mm thick (absorbent infill)	0.25	0.55	0.85	0.85	0.75	0.75

CHAPTER 14

OTHER BUILDING SERVICES PLANT

14.1 INTRODUCTION

In the previous Chapters we have built up to and considered the major noise and vibration features of building services air distribution in a fairly consistent and inter-related manner.

On the way a few diversions were allowed, more especially in the single Chapter devoted to vibration control, but a number of product application areas remain and these will be considered in this Chapter.

We shall consider, Lifts, Escalators, Hotels, Wall-mounted equipment, Floating floors, Cooling towers, Condenser units, Standby diesel generator sets, Standby gas turbine generators, Boilers.

14.2 LIFTS

For the noise control engineer the most appropriate lift is the now comparatively rare use of hydraulic elevators. These units use a telescopic hydraulic ram sunk into a basement location, activated most usually by oil and occasionally by water — as still is the case in many districts of central London. The plunger of this cylinder acts directly on the elevator base and, as such, one immediate advantage is the elimination of a penthouse or overhead support for machinery — Figure 14.1. Generally, they are only applied to low rise buildings up to around five storeys high and operate at quite modest speeds in the region of one metre per second. However, exceptions prove the rule and going back into the early 1900s, buildings as high as thirty storeys were silently served by such lifts and achieved speeds of nearer to three metres per second. Nevertheless, the mechanical services engineer does still have a minor noise problem to address inasmuch as the power unit will necessarily contain a pressure generating pump, and this does need noise and vibration control. It lends itself to remote location, most conveniently in the plantroom already benefitting from noise and vibration control procedures.

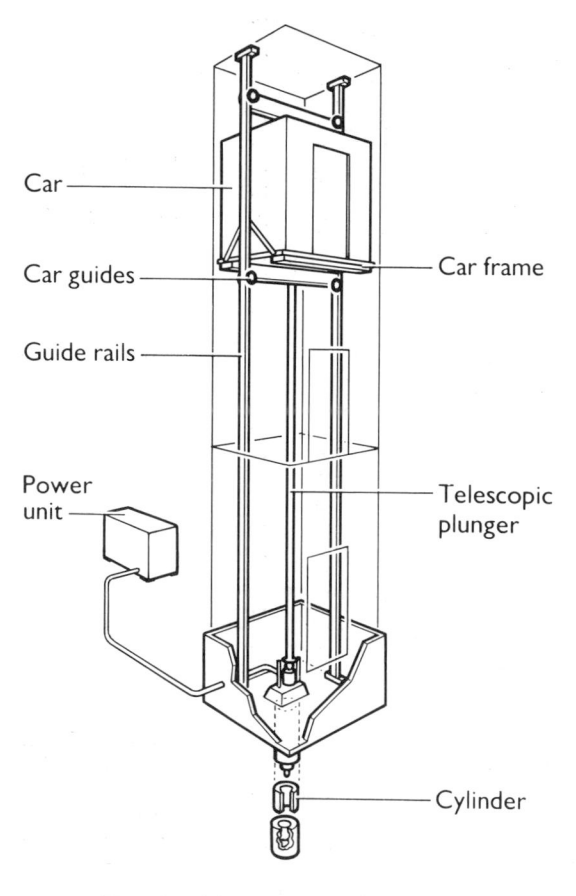

Car

Car guides ———————————————— Car frame

Guide rails

Power
unit ——————————————————— Telescopic
                                plunger

—————————————————— Cylinder

Figure 14.1   Hydraulic Elevator

Most traditional lifts consist of an electric traction engine pulling on wire ropes attached to the top of the lift car — Figure 14.2. The motor unit is often located on top of the building in a dedicated penthouse lift room, and necessarily is usually associated with a large inertia block supported on resilient mounts with a static deflection rating of some 10-20 mm — Figure 14.3. This not only isolates motor and gear noise, but will considerably alleviate the start-up "thud" of some systems. There is no reason why the motor unit should not be basement mounted and perhaps this is an under-used variation in building services. Such a location is usually a much less noise sensitive area fixed down to a comparatively massive basement slab via resilience and does not call for the mainly inconvenient penthouse lift motor room.

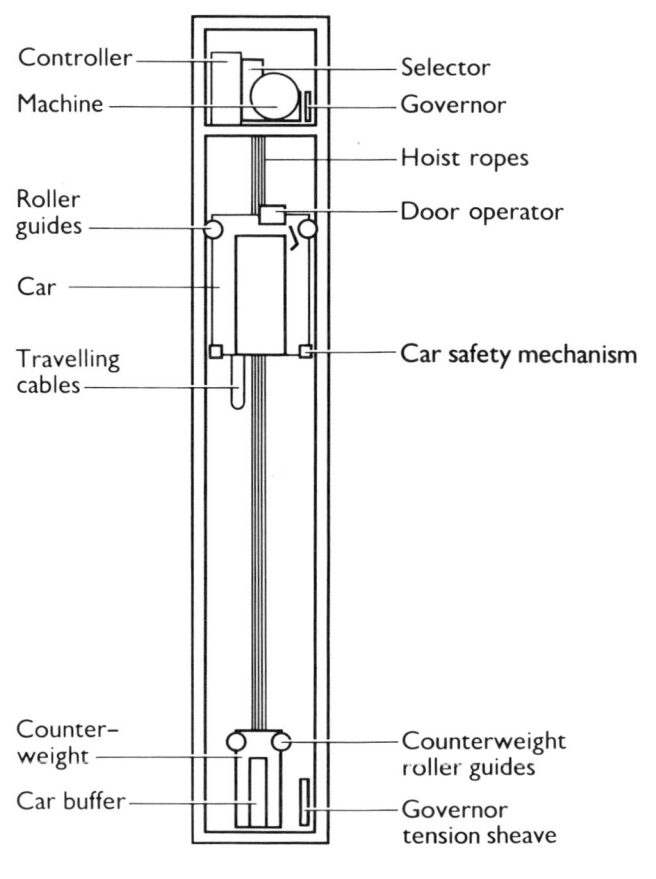

Controller — Selector
Machine — Governor
— Hoist ropes
Roller guides — Door operator
Car
Travelling cables — Car safety mechanism
Counter-weight — Counterweight roller guides
Car buffer — Governor tension sheave

Figure 14.2 Wire Rope Lift

However, the length of the hoist rope is doubled as a result of the up and over action, which aggravates the springiness inherent in this rope. Due consideration should be addressed to this location for the traction unit.

With gearless traction lifts, speeds as high as 8 metres per second can be reached, which gives a new noise control problem in the region of the passenger door access to the lift shaft. The high speed lifts push in front of them a pressure wave which acts quite strongly on the shaft access door as it passes by. Hence, these doors, most usually of the sliding variety, must be adequately designed and retained in order not to vibrate and rattle. This becomes especially relevant to the increasing use of prestigious reception areas directly outside in the lift lobby.

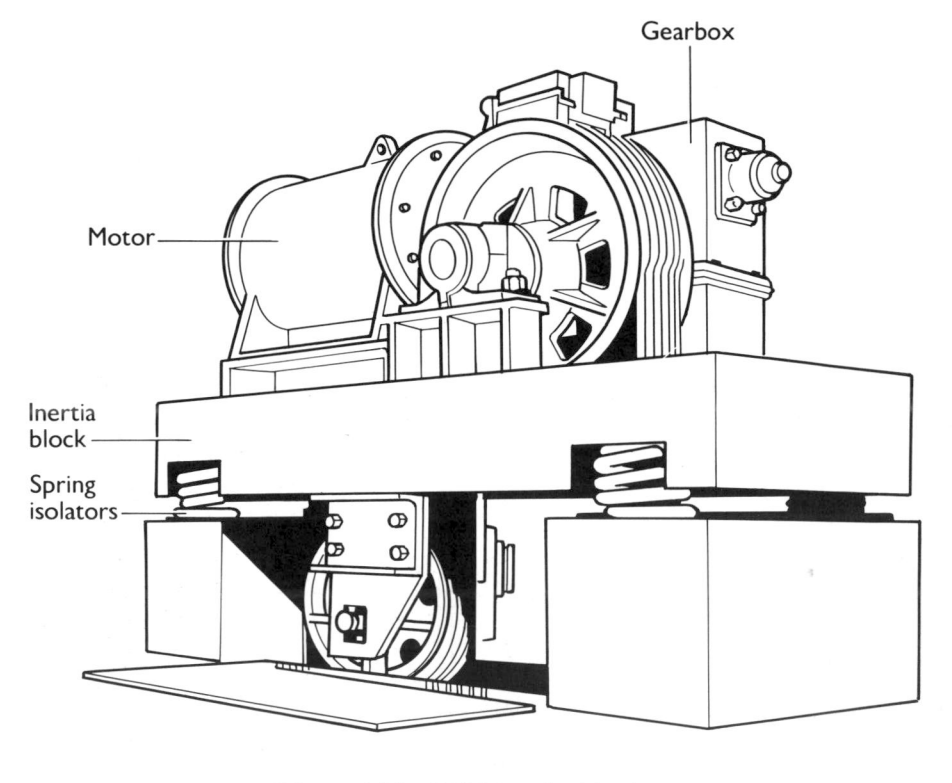

Figure 14.3   Lift Traction Engine

14.3 ESCALATORS

The basic escalator system is shown in Figure 14.4 and as can be seen is assembled on a substantial structural steel frame supported between the two floor levels. The opportunity for straightforward mechanical noise in the chain drives and step runners is the paramount concern of the noise control engineer and becomes aggravated as the wear and tear builds up, even whilst remaining well within the working maintenance acceptabilities. The motor drive is invariably of the worm gear type and puts a certain amount of vibration into this substantial sub-frame.

The vibration and radiation of noise from the sub-frame and working features will be the basic noise radiation from the functional unit. To some extent, this can and will be contained within decorative and dirt excluding cover arrangements such as the inspection area for the motor unit and the ceiling to be seen from the floor below. To avoid coupling of these inevitable vibrations into the structure, resilient mounting of

LIFT
Whilst over the past decade lifts have become very much an architectural feature, they must remain smooth and quiet in their operation for their full impact to be appreciated, as here in the Metrocentre.

Courtesy: O & K Escalators Ltd.

ESCALATORS

Elegance and quietness is the ambience created by this water stream installation, which also clearly shows the boxed-in underside of the split stage two. Such covers will supply breakout noise control, but if not well damped and vibration isolated, may also become their own source of structure-borne noise.

Courtesy: O & K Escalators Ltd

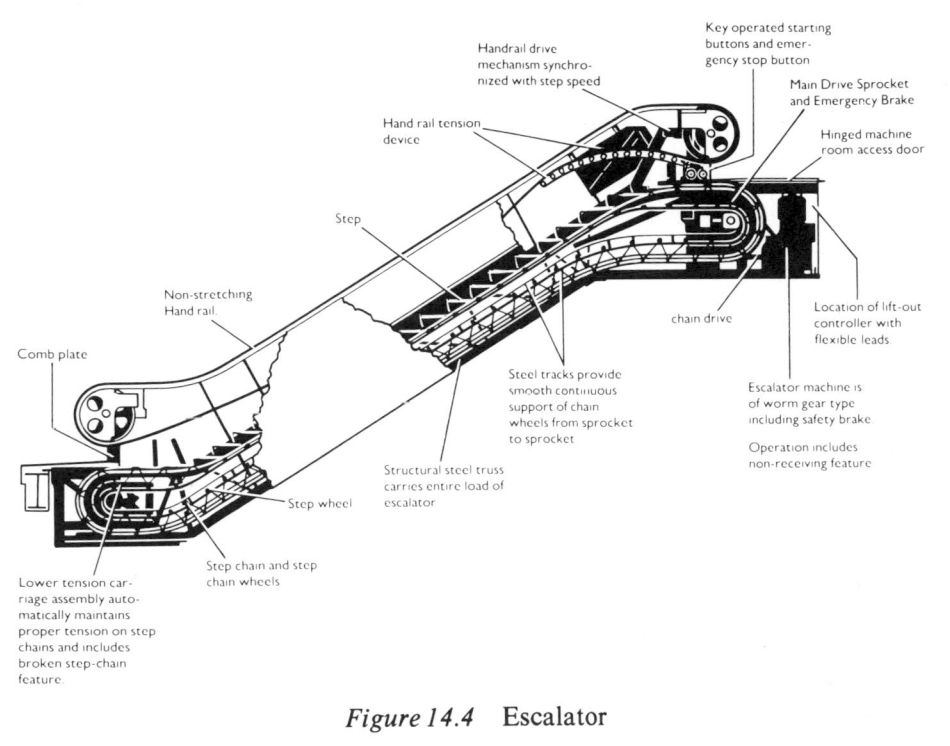

Key operated starting buttons and emergency stop button

Handrail drive mechanism synchronized with step speed

Main Drive Sprocket and Emergency Brake

Hand rail tension device

Hinged machine room access door

Step

Non-stretching Hand rail

chain drive

Location of lift-out controller with flexible leads

Comb plate

Steel tracks provide smooth continuous support of chain wheels from sprocket to sprocket

Escalator machine is of worm gear type including safety brake

Operation includes non-receiving feature

Structural steel truss carries entire load of escalator

Step wheel

Step chain and step chain wheels

Lower tension carriage assembly automatically maintains proper tension on step chains and includes broken step-chain feature

Figure 14.4   Escalator

this substantial sub-frame should be considered if low noise levels below 50 dBA are required in nearby areas. Additional decorative panelling added around the balustrades beneath the moving handrail can act as efficient noise radiators if rigidly linked to the support frame, and an extra noise radiation problem is created. The use of damping materials on such panelling constitutes a worthwhile noise control procedure, more especially for remedial situations. The moving handrail is most traditionally of a non-stretch rubber-like composite construction, and if not correctly maintained can lead to annoying squeak and squeal.

It should be noted that the speed of operation will be a most sensitive feature to noise generation, as the generated sound power levels will approximately follow a 12 dBA per doubling of speed rule.

14.3.1 Travelators

For noise control purposes these installations may be considered as "horizontal elevators", although the sub frame is often less substantial as continuous building structure support is often available. Hence vibration isolation cannot just be concentrated at the two ends, as with inclined elevators, and many resilient pads or even continuous resilient mat will be necessary.

14.4 HOTEL BEDROOMS

This is a unique situation where occupancy leads to contrasting demands on the smaller living units, usually placed compactly together. The most obvious is the requirement for sleep at less than NC25 next to rooms with televisions on, baths running, wardrobe doors sliding and switches clicking, plus the additional need for impact noise control from floors above.

To truly isolate these inconsistent situations, studio construction techniques aimed at the same problem need to be adopted with the walls of each room isolated and constructed on independent floating floors, as indicated in Figure 14.5. However, in reality a compromise is required and attention to detail and juxtaposition of items assists.

Bathrooms should be constructed back to back, and similarly headboards back to back. Switches and plugs and sockets should be avoided on the bedhead wall. Where possible, switches of a variety with minimal click should be employed. Where switches, sockets and cableways do exist on opposite sides of a common wall, then the convenience of mounting units back to back should be avoided and their locations staggered, as in Figure 14.6.

Carpets will remain the main form of impact noise control and should, where appropriate, be followed through into the bathroom areas.

However, certain stylists will insist on tiled floors for bathroom/toilet areas, whereupon full impact noise control can only be achieved through employment of a resilient floor construction. Modest specifications not necessarily up to that of the airborne noise control floating floors will usually suffice.

Water flush cisterns can sometimes be incorporated into clever double wall arrangements at the bathroom interface. They should not be rigidly coupled to the structure, and should preferably be chosen as "close coupled" where the cistern is supported off the pan itself and a resilient wall fixing is only required to supply stability and not load carrying capacity. The use of syphonic discharge cisterns is often recommended for their advantageous noise control features, but they are not a favoured option by the plumbing fraternity.

The general noise of water flow in pipes is not helped by a move towards smaller bore piping and will certainly be aggravated by the recently approved (UK) introduction of high pressure systems. Generally, the adoption of larger bore pipes will solve most water flow noise problems simply associated with the resulting lower flow speeds. Nevertheless, some valve and tap noise will remain in the system. This must not be coupled into the structure and requires the use of resilient pipe clamps, even of the simplistic rubber lined variety. Recently developed silent cistern float valves, such as the Torbek valve from Ideal-Standard, are a substantial improvement of some 15 dBA and may also be applied to any water tank refill system as frequently experienced in the domestic environment.

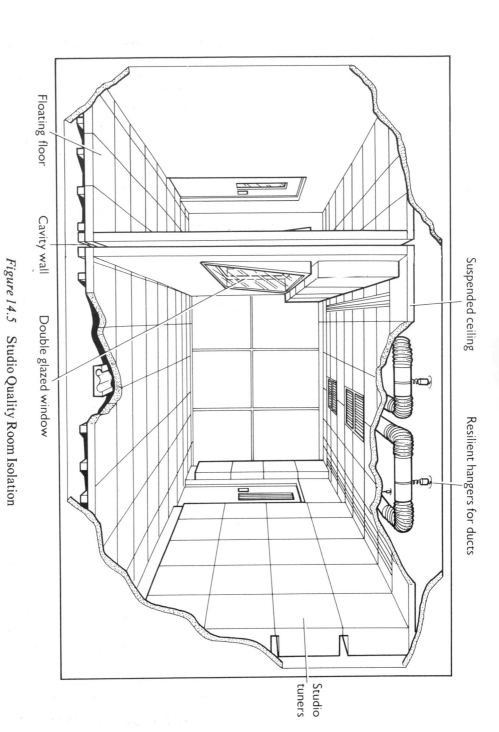

Floating floor

Cavity wall

Double glazed window

Figure 14.5 Studio Quality Room Isolation

Suspended ceiling

Resilient hangers for ducts

Studio tuner's

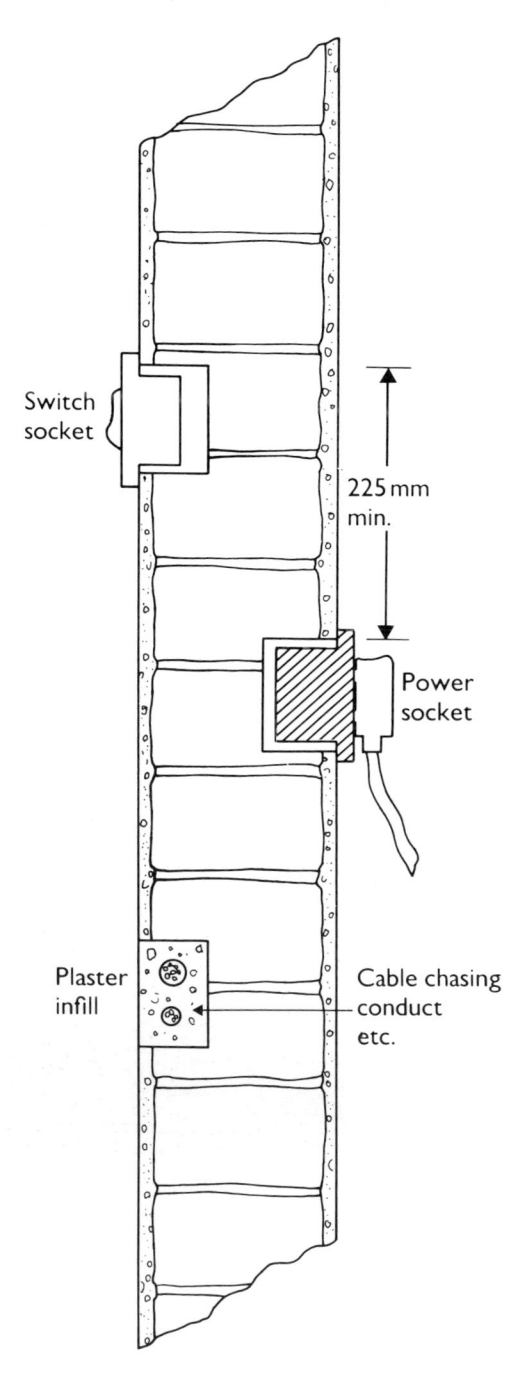

Figure 14.6   Socket and Switch Locations

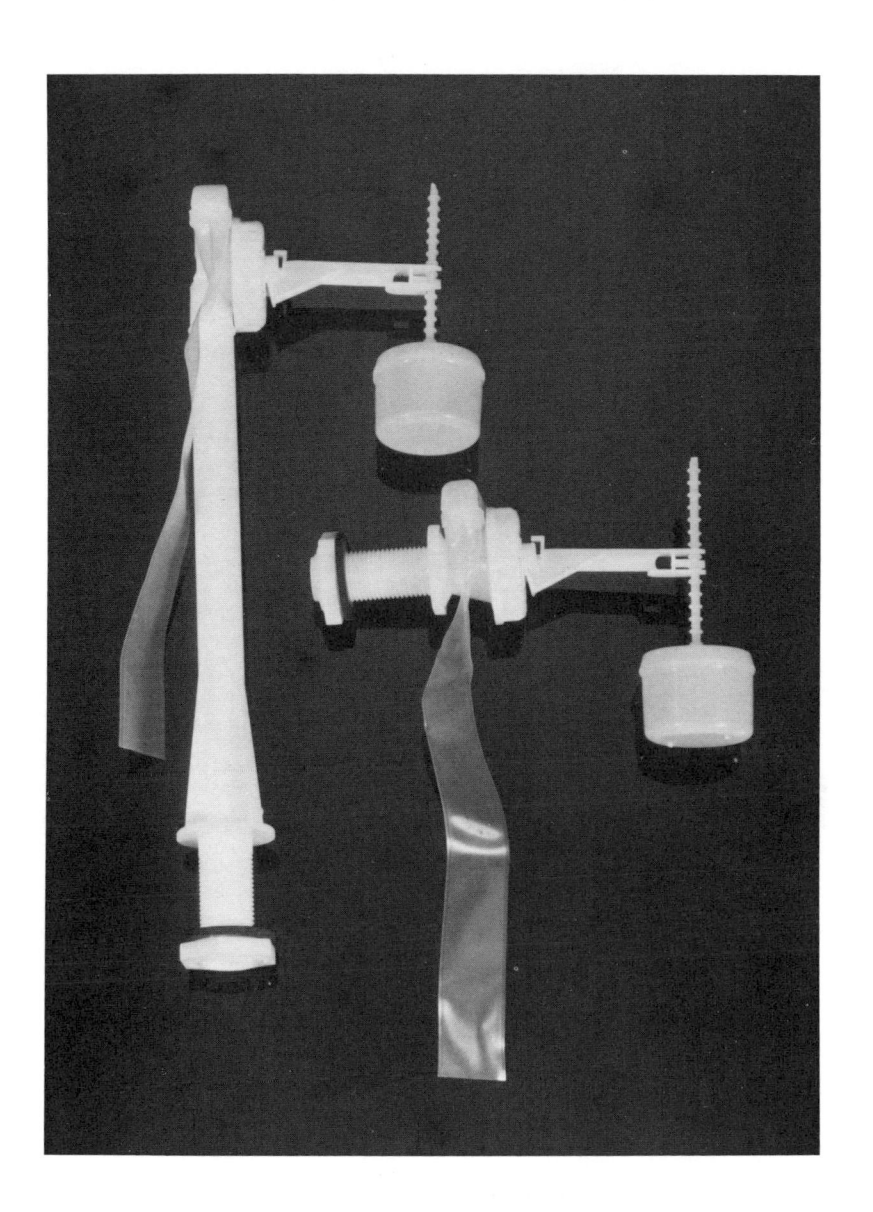

TORBEK CISTERN VALVES
Not silent, but a great improvement on the traditional ballcock valve.

Whilst hinged wardrobe doors can quite readily be provided with slamproof noise closures, the space efficient use of sliding doors is favoured, often, being of a substantial nature and sometimes incorporating heavy glass mirrors. The runner arrangement needs careful attention. Generally, a suspension runner will be quieter than a bottom runner system, and remarkably quiet designs incorporating resilient castors can be found. The rail arrangement may also lend itself to resilient mounting off from the structure. Shock absorbing bump stops will be required at both ends of the doors travel.

Airborne noise from the corridor is most appropriately controlled by the incorporation of a double door arrangement each end of the corridor, naturally formed at the entrance to the room alongside the bathroom. Hence, the doors will be separated by around three metres, and each door can now be of quite modest acoustic performance each with an average Sound Reduction Index of 30 dB. The combination with this large spacing will approach a sound reduction of 60 dB. Such a high insulation would be impossible to achieve with a single practical door arrangement.

Whilst the above recommendations have been presented with very much the simplistic stereotyped hotel bedroom in mind, the principles can readily be applied to more adventurous and lavish hotel suites.

14.5 WALL MOUNTED EQUIPMENT

Many odd pieces of noisy equipment are designed for wall mounting, such as hand driers, telephones, hair driers and toilet extract units. When screwed directly to a surface, not only do they make a little more noise in the source room but can become decidedly audible in the "room next door". This is a result of the audible structural vibration excited into the supporting panel.

The solution lies in resilient mountings and a few devices do come with sufficiently soft mountings for the task. However, generally for good isolation a supplementary panel will need to be pre-resiliently mounted to the structural surface. Since this is difficult to do in anything but an obvious manner, it needs to be executed as a separate architectural feature which may benefit visually if applied to a prepared countersunk location — Figure 14.7.

14.6 FLOATING FLOORS

This really is one of the more interesting aspects of noise control by way of inventiveness that has been introduced over the last 20 years. It's application has allowed the juxta-position of noisy and quiet areas into adjacent locations. To this end it must have been a great advantage to the architect in his task of creating a building suitable for the purpose to hand. It is now possible, even if only just possible, to have plantrooms with 110 dBA next to NC20 listening environments — studios, audiology suites, concert halls, theatres, lecture rooms.

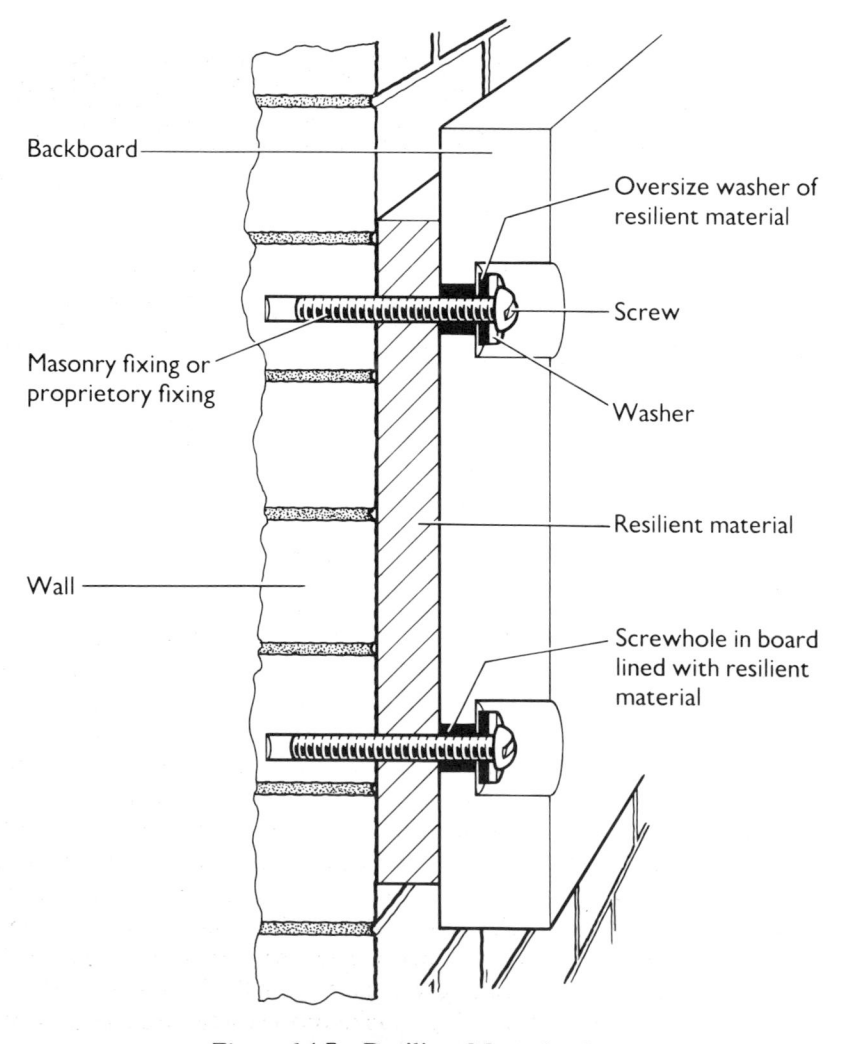

Figure 14.7 Resilient Mounting Panel

The acoustic principle involved is exactly the same as that of double glazing expounded in Chapter 5 as a method of beating the mass law. This feature of beating the mass law is the key to engineering and architectural acceptance in that a comparatively lightweight compound structure results with very high Sound Reduction Indices. The principle is illustrated in Figure 14.8 as the creation of two horizontal slabs of concrete, the lower one being the structural floor potentially with load bearing beams incorporated, the upper being a non-structural concrete raft, which is traditionally in the region of 100 mm thick reinforced concrete. However,

Upper slab

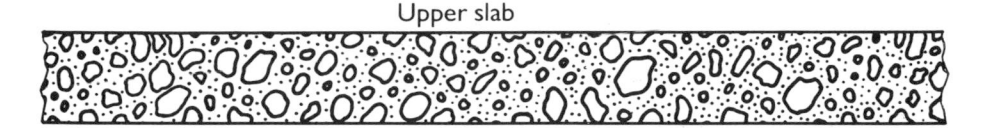

Air gap 50 mm

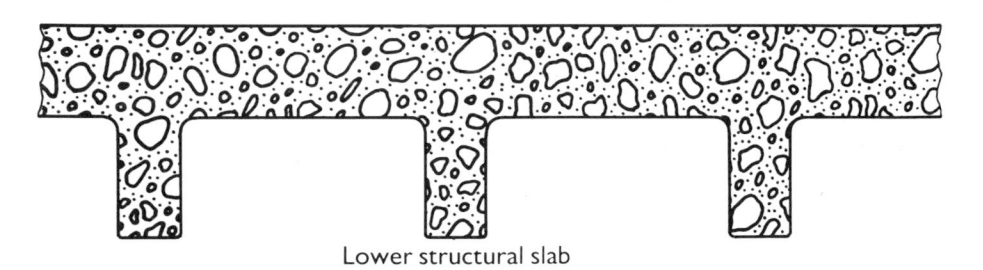

Lower structural slab

Figure 14.8 Double Floor Slab Principle

this needs supporting up from the structural floor, if it is to be even moderately load bearing, transferring its weight and loading through to the structural support members. This must be achieved in a resilient manner without rigid links, in an attempt to keep the two heavy membranes as a double dynamic panel system. This is easy for the double glazing situation where the glass panels are vertical and notionally non-load bearing. So in Figure 14.9 we see the introduction of suitable resilient supports which have taken on various forms over the years — rubber pads, glass fibre pads, coil springs, folded steel springs. Each have their own merits from a technical, engineering, architectural and customer acceptance point of view, but little will be discussed on these points here. The height allowed for such resilience is again traditionally in the region of 50 mm and, hence, oddly enough the innocent looking trapped air space becomes the controlling resilient feature linking the two slabs. For the dimensions illustrated so far, the spring mass resonance of this trapped air will be in the region of 15 Hz.

This then immediately gives some guidance to the resilient requirements of the previously mentioned support members, on which there is always the additional constraint of wishing to be as rigid as reasonable to minimise the stress transfer requirements on the upper floated slab, reinforced as it may be. Hence, the mass resonant frequency of the slab mass on this resilient pad need only be placed between 15 Hz and 20 Hz to combine satisfactorily and equitably with the trapped air stiffness. Whilst it might be tempting to suppose the inclusion of spring mounts set at a resonant frequency of 5 Hz would be an advantage, possibly alleviating the need for primary vibration isolation of the machine, this would be a false assumption due to the stiffness of the trapped air spring. Larger air spaces and heavier upper floor masses can be introduced in an attempt to justify this concept, but this procedure has not been followed up as a general practice for the application of mass effective floating floors. Structural engineering considerations lead to the separation of these resilient mounts being in the region of 300-500 mm, although even greater spacings are sometimes permissible on pedestrian areas.

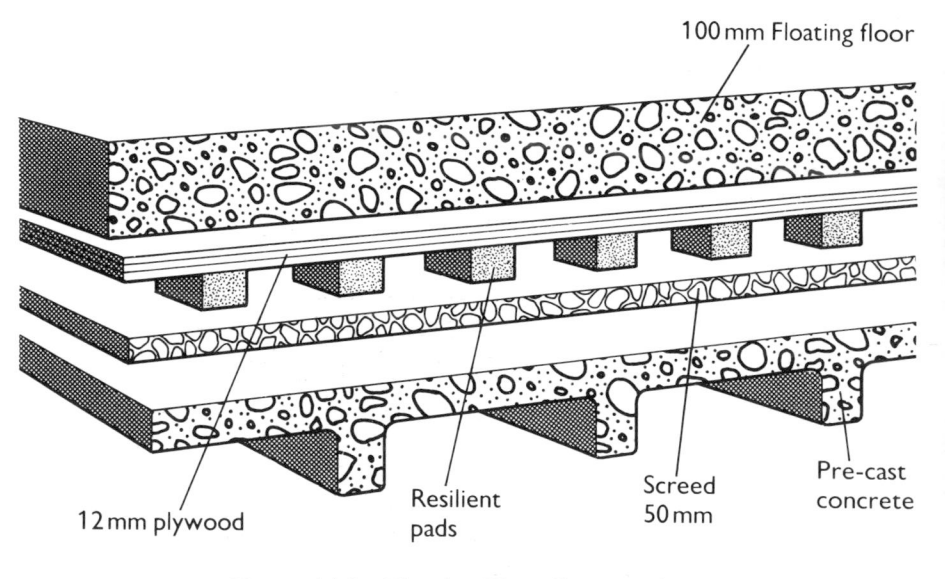

Figure 14.9   Floating Floor Construction

Thus, the floor does not constitute a vibration isolation situation at low frequencies, but only begins to offer effective isolation at frequencies above 30 Hz. So we see that it is to be considered as a normal isolating construction concerned with the audible frequencies at 30 Hz and above. At these frequencies the resilient connection (air spring plus resilient supports) leaves the two slabs essentially decoupled to commence action in a similar manner to double glazing, beating the mass law with a Sound Reduction Index increasing at 10 dB per octave, in contrast to the single simple panels' 5 dB per octave — Figure 14.10. Therefore, we see from Figure 14.10 the very high Sound Reduction Indices which may be achieved from these floors, more especially at frequencies of 125 Hz and above.

Now all this can only be achieved with very thorough flanking control, as whilst we may have created a substantial barrier beneath a noisy plantroom through the floor construction, flanking through the side walls will leave the net sound reduction only some 8 dB better than that of the wall. Often this 8 dB increase is sufficient and the high insulation values available at, say, 1 kHz are not necessary and the addition of

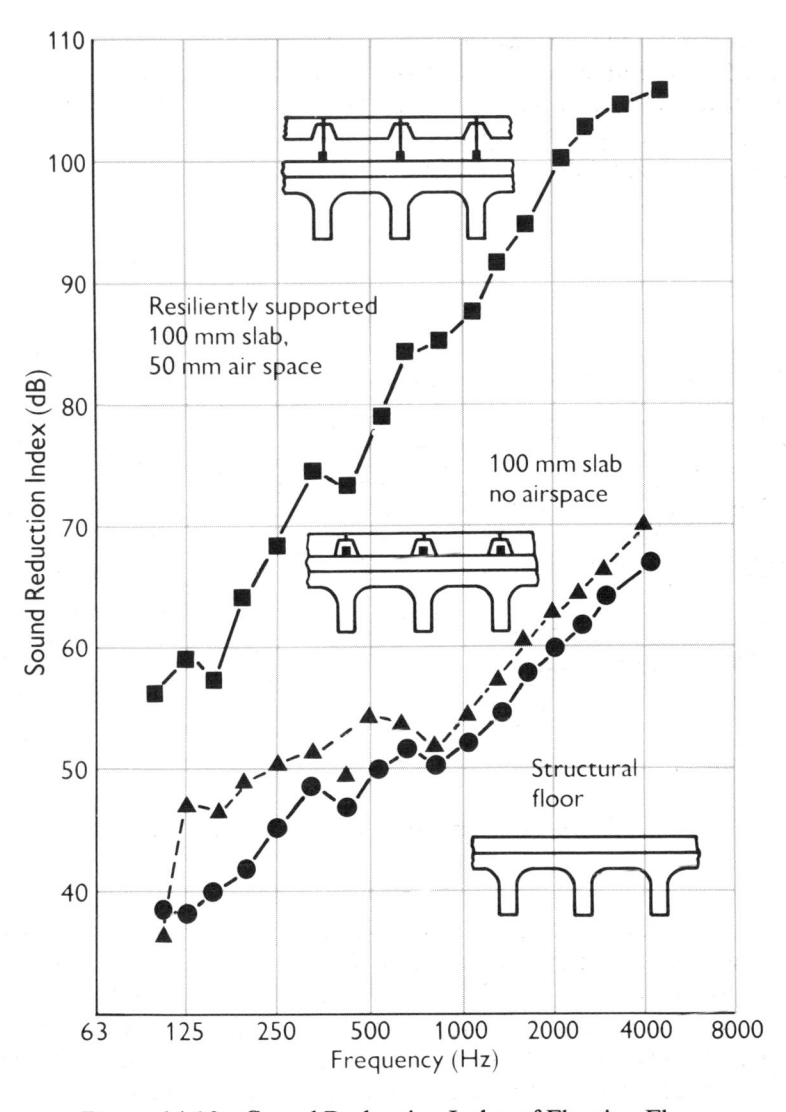

Figure 14.10   Sound Reduction Index of Floating Floor

the flanked floating floor alone adds that little bit more, 8 dB, to the initial insulation without adding too much mass. Should the full Sound Reduction Index potential be required, then it will be necessary to also float the walls either off from the floating floor, or independently resiliently floated from the main structural slab. Sometimes it is convenient to create within the plantroom a large proprietary metal acoustic enclosure on the floating floor, which will then supply most of the flanking control.

These are the features which need total incorporation for the isolation in, for instance, broadcasting and studio environments where noisy plantrooms are located adjacent to NC20 and even lower areas.

When we turn to the total engineering application of such a floor, bearing in mind we were initially talking about noisy plantrooms, we have to realise that the point loads available from the primary spring isolators of the equipment bear heavily over very restricted areas. These point loads need to be distributed through this non-structural upper floated slab to the resilient mounts outlined above. It is mainly at this stage that the relative merits of the resilient mounts come into their own and the total engineering of the applied system requires various resiliences over the areas of the plantroom floor to pick up and support these high point loads without sacrificing the continuity and homogeneity of the floating slab. To this end the non-linearity offered by the glass fibre system has so far proved the best engineering concept when appropriately required for plantrooms. Penetrations and upstands through the floated slab to locally support high loads is not a preferred technique, which also adds quite considerable to the structural and contractual complexity of the applied procedures.

As was warned in Chapter 6 on vibration isolation, this resilient floating floor is not a substitute for the primary requirements of low frequency vibration isolation. All that was said before applies and because we have deliberately introduced a resiliently mounted floor slab with a natural resonant frequency around 20 Hz, we have introduced a comparatively low frequency resonant system into the supporting structure. This situation would even apply if the floating floor had been incorporated in a basement situation for flanking noise control, as could just be the case for a studio complex. Consequently, most equipment on the floating floor must be isolated from this resonant system with resilience having a static deflection of at least 10mm, and this is most usually achieved in building services by the use of open coil steel springs. However, please note this is notionally a minimum deflection requirement and other floor span and machine type parameters may demand even higher deflection springs, as recommended in Chapter 6.

14.7 COOLING TOWERS

Discussion will be restricted to the application where the water to be cooled runs down a series of surface area creating slats over which air passes naturally with convective assistance, or more appropriately for this Chapter, with fan assistance. Whilst the water flow and dripping can be a noise problem, it is the fan assistance which usually calls up noise control procedures in building services. Two key locations occur; firstly and most commonly the rooftop situation, and secondly the light-well situation.

The feature most commonly leading to noise control application restrictions is the airflow duty of the circulation fan, often of simple propeller type, which is extremely sensitive to back pressure reducing the volume throughput and cooling air. To this

Figure 14.11   Acoustic Louvres (courtesy Sound Attenuators Ltd)

end the simple application of duct attenuator techniques is too restrictive on the airflow. If sufficient space is available, simple acoustic screening by wooden, masonry or acoustic panelling spaced reasonably clear of the tower will solve rooftop application problems with respect to sensitive locations at lower levels. However, for protection at higher levels, and similarly for the light-well situations, it is necessary to incorporate a roof, possessing noise stopping properties. For both these situations the acoustic louvre shown in Figure 14.11 is proving the most appropriate solution, where it is formed as a complete surround wall to the tower, which with its comparatively large open area adds little back pressure to the system. The roof may then be a suitable panelling system, or if even more circulation is demanded this can also be an acoustic louvre arrangement.

If the comparatively modest acoustic performance from these acoustic louvres is not sufficient, then it will prove necessary to change from the simple propeller fan to an axial or even centrifugal fan arrangement able to deal with the back pressures of splitter type attenuators controlling the inlet and outlet airborne noise.

Large cooling towers are often designed knowing full well that acoustic treatment will be necessary and correspondingly more powerful fans are incorporated.

14.8 CONDENSER UNITS

Condenser units imply the use of fan assisted air cooling over a dry coil and fan noise control remains the key problem. The procedures spelt out for cooling towers are appropriate for this facet of the unit.

However, a compressor is also incorporated which will require separate noise and vibration isolation considerations. The unitary nature of the product does not usually facilitate local noise and vibration control on the compressor, such as acoustic enclosures and improved resilient isolators; sometimes the resilient isolators are already appropriate for all but the most critical applications. If additional noise and vibration control is required, the whole condenser unit will need to be treated as the culprit even though the source of the problem may be the inaccessible compressor.

14.9 STANDBY DIESEL GENERATOR SETS

These units, or their gas turbine equivalents, have become very much a standard part of building services plantroom packages, although quite often they find themselves away from the main plantroom in more ideal and isolated situations. In our experience the two main locations contrast from the extremes of basement locations to rooftop locations. In both situations there are two obvious noise sources and controls, namely, the containment of the engine casing noise, most familiar by its tappit-like sound, and silencing of the exhaust system.

In parallel with this airborne noise control comes the most obvious need for structure-borne vibration control, which must also encompass the low frequency rotation generated motions, together with the vibration excitations up into the audible range. The vibration isolation approaches have been covered in Chapter 6 and for diesel generating sets come down fairly universally to the need to incorporate an inertia block weighing approximately the same as the diesel generator set. This is mounted on open coil springs with deflections varying from 20 mm to 150 mm depending on the precise location, the main aim being to avoid the excitation of structural resonances, the first and lowest ones of which are usually the loaded floor constructions. However, to meet the structure-borne noise control demands these open coil springs must incorporate noise stop pads, more familiarly of an elastomeric nature such as neoprene rubber.

In this application, the inertia block serves several purposes:

1. Rigidity between the diesel engine and alternator set
2. Added mass to reduce the vibration amplitudes of the mounted engine, acting favourably towards the long-life of the machine
3. Adds stability to the resiliently mounted assembly by lowering the centre of gravity and minimising the excitation of rocking modes, more especially during start up and run down

This technique is much preferred to the alternative of direct soft spring mounting of the diesel generating set feet, to which are added bump stops to control the considerable lurch at start up. These bump stops tend to introduce a thud into the structure, the alarming nature of which is best avoided.

Having resiliently isolated the machine from below, all connections from the machine to the structure must either be sufficiently resilient in their own rights, such as electric cables, or require a degree of resilient support from the structure. Such an item will be the exhaust system itself, which will generally be pipework of a substantial nature which may or may not include a "flexible connector" to the engine. These flexible connectors often become very rigid with age and are of unpredictable performance and at any rate do not isolate the high levels of acoustic noise level content within the pipe — that is the later task of the exhaust silencer. Hence, the exhaust pipe needs supporting on resilient hangers, the first of which nearest to the engine should have a similar static deflection to that of the main inertia block springs. Supports at the distance of more than 50 times the exhaust pipe diameter from the engine may be of a simple acoustic resilient nature, most conventionally based on engineering aerocore glass fibre or elastomeric compounds.

The noise levels from the carcass of the engine are of high intensity and should the installation require to be inaudible in surrounding rooms, then dedicated detailed acoustic enclosure arrangements will be required. Simply putting the engine "in the old basement laundry room" with a new door will rarely satisfy surrounding occupants. For basement situations, the surrounding walls, ceiling and floors require treatment, whilst for rooftop situations a floating floor and acoustic canopy arrangement will control the noise intrusion. Such an acoustic canopy is usually of a proprietary format with the necessary doors and penetrations based on 1–2 mm steel skins, with between 50 and 100 mm of acoustically absorbent lining.

For the basement location the situation can become very complex, as flanking control into the walls must also be guarded against. Again, it may be most cost effective to assemble, within the room, a convenient canopy arrangement built up from unitary sections, based as previously, on 1–2 mm steel with 100 mm of acoustic internal absorbent lining. In both cases, access doors must be of suitable acoustic grade, which generally necessitates airtightness.

So immediately a new problem occurs for this acoustic canopy, of supplying sufficient cooling/ventilation for the engine. Water cooling is often available and this can sometimes be fed to remote cooling coils. More traditionally a radiator fan is close-coupled to the engine, and this will need its input carefully attenuated keeping a close eye on the pressure development abilities of such fans. With electricity generation at hand and to spare, it is often wise to devise a system free of this shaft coupled radiator fan and incorporate an electrically driven exhaust or inlet fan more able to deal with the back pressures of this necessary attenuation for the cooling air inlet and exhaust. The main silencer or silencers of the engine exhaust system will be proprietary products, the science of which, as for the duct silencers, is beyond the scope of this book.

14.10 STANDBY GAS TURBINE GENERATORS

These compact units historically based on the same principles as aircraft jet propulsion systems, are surprisingly not very noisy in their basic format. The power unit driving the electric generator consists of two parts, these being a gas generator and a power turbine. However, these two units are so tightly coupled together that they become considered as one. Consequently, three regions of noise control become apparent — Figure 14.12 and are:

1. Fresh air inlet noise
2. Casing radiation
3. Outlet exhaust noise.

14.10.1 Fresh Air Inlet Noise

This spectrum — Figure 14.13, is extremely rich in high frequency noise to the extent that due attention must be paid to its precise attenuation, unlike many other aspects in building services noise control where the noise control procedures will automatically take care of the high frequency much as if an "also ran" requirement. Not so here for the inlet noise control, but, supplied with the relevant sound power or sound pressure information the treatments are very much in keeping with everything that has gone before.

A fresh air inlet is ducted from atmosphere and usually contains air filtering apparatus, to which must be added a noise control duct attenuator. With the need for good high frequency noise control, this will incorporate comparatively narrow splitter airways. Usually the coned inlet nozzle of the air intake to the gas generator compressor demands a short plenum from which to draw.

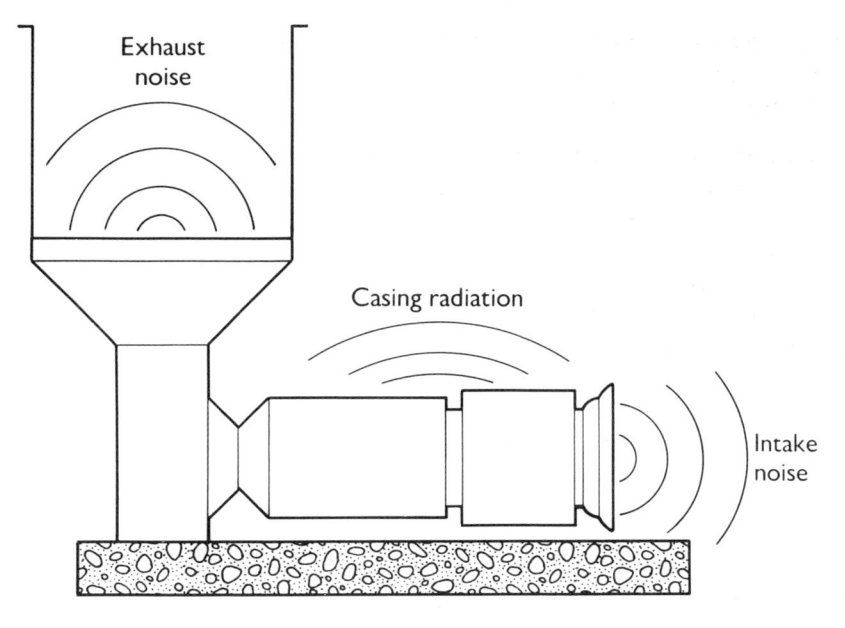

Figure 14.12 Gas Turbine Noise Sources

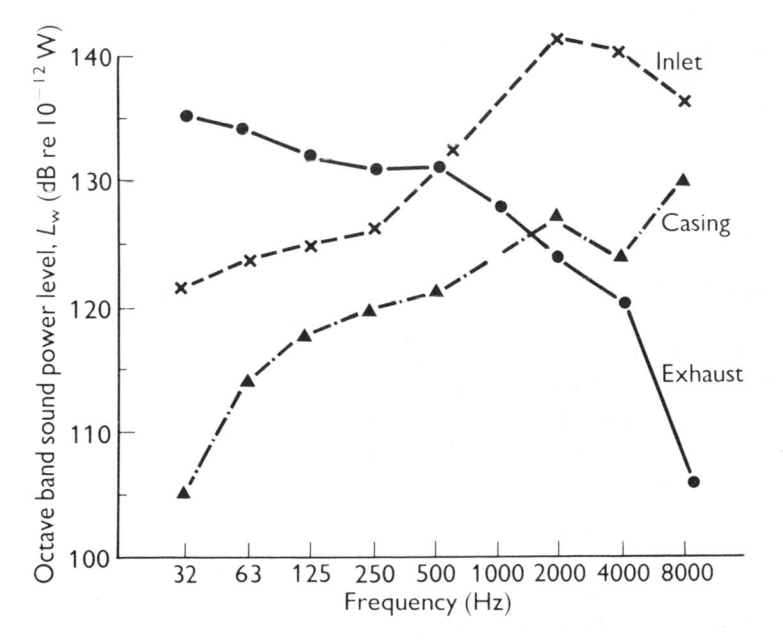

Figure 14.13 Gas Turbine Noise Spectra

14.10.2 Casing Radiation

The casing radiation spectrum — Figure 14.13, whilst being somewhat rich in high frequency sounds lends itself very well to the rising Sound Reduction Index of airborne noise control panels. The enclosure arrangements are generally quite straightforward and naturally incorporate internal sound absorbing linings based on protective faced mineral fibres to hold down any potential reverberant sound field growth. Such an enclosure would generally form the first line of attack in a building services project, the enclosed unit being in a plantroom.

14.10.3 Outlet Exhaust Noise

This is the challenging noise source, as not only is it rich in low frequencies but it is also at elevated temperatures in the region of 300–500°C. This hot exhaust gas will be ducted to atmosphere and the necessary specialist duct attenuator will be incorporated. The design is obviously of a specialist nature, but will generally include thicker splitters than average to deal with the low frequencies, and appropriate construction to deal with the temperatures and differential expansions involved. Stainless steel is a popular material for the surfaces of the hot gas exhaust attenuator in contact with the gases, but standard steels of appropriate gauge are probably more cost effective, especially bearing in mind the standby nature of the plant.

14.9.4 Package

The specialist nature of these three noise control procedures often leads to the standby gas turbine generator being supplied as a packaged item within its own metal acoustic enclosure with full inlet and exhaust gas noise control attenuators in place. Such a unit can then be directly installed into the plantroom layout, always remembering the thermal caution and lagging required on the now attenuated exhaust discharge.

Generally, the alternator itself does not constitute much of a noise problem and may be enclosed within this unitary enclosure or left outside. However, as with the diesel generator sets, rigidity of the coupling between power turbine and the alternator is of equally paramount importance and some form of inertia block is usually supplied. But very much unlike the diesel generator set, high deflection low frequency spring isolators are not usually required and good quality noise control vibration isolators based on synthetic rubber or engineering glass fibre will suffice for the necessary vibration control.

14.11 BOILERS

Here it is necessary to consider the boiler size range from large domestic units as used in town office refurbishments, to the vast units applied to multi-storey office blocks and industrial service energy centres.

In general terms, three distinct noise sources exist:

the fan or blower
flame or combustion noise
subsidiary flue system combustion oscillation

Staying initially with the smaller domestic range, which is located closely amongst the occupied space, the blower may well be the dominant plantroom noise source via its air inlet ports, whilst a well-established combustion flame within the water jacket remains well subdued. Here a local screen or hood can provide the 10 dBA of reduction typically demanded and yet retain easy maintenance access. In contrast, the flame roar may dominate through an over-generous but flimsy inspection cover, more especially when the combustion zone is being pushed hard and concentrated over a small volume. A burner of large flame size, or better still a multiplicity of small flames, will be quieter for a given heat output than concentrated, but possibly very efficient, combustion zones. Nevertheless, faced with a very fixed noisy boiler, the options to become involved in combustion noise research are out of the question and the manufacturers limited range of alternative burner units must be investigated, whilst maintaining an acceptable heat output. This is usually a plausible option, but most probably with a decrease in the thermal efficiency rather than the absolute output.

Wall flame boilers are particularly quiet but usually have an annoying "buzz" on start-up but which, being of a mid-frequency nature, will be easily insulated by even lightweight plant room walls.

Start-up is probably the most annoying and startling feature of most on-off blown flame boilers as a result of the explosive and impulsive like "whoomph". However for the lighter duty boilers the modulated flame is once more available and this tends to minimise or even completely avoid this start-up by modulating up and down the gas supply and largely avoiding the "off" state altogether.

Turning now to the larger central plant boiler units, they must be analysed in the straightforward ways described in earlier chapters, in conjunction with reliable sound output power or even pressure levels for the specific duty and burner units proposed. Noise power guidance related to thermal duty ratings is fairly unreliable except by ensuring a prediction technique very much on the high side. This is then grossly unfair on the quieter units which can downrate on this clumsy guidance by over 10 dB.

It is with these larger units, and their larger bore flues, that unstable combustion of a low frequency nature can occur as a result of the particular geometry chosen much in the same way as an organ pipe resonates. Such conditions are rarely predicted and become the subject of site remedial problems. Stainless steel flue attenuators can sometimes help by damping the resonance down rather than by duct borne energy absorption. However, simple and convenient changes in the flue layout can also be effective such as

length change - longer or shorter
addition of an outlet deflector
cutting the outlet at a generous angle rather than straight across.

The vibration output of these large boilers must not be forgotten or underestimated. Manufacturers' levels are not likely to be available and as with much other building services plant — Chapter 6 guidance is the best available advice. A rigid inertia base or frame under the boiler should be supported on nominally 20 mm deflection open coil springs which incorporate adjustable limit stops. Following on from this, the flue must not be too rigidly coupled to any supports and a high temperature flexible connector may be necessary into any substantial rigid chimney.

SUGGESTED FURTHER READING

Books

Controlling Noise in Building Services, Sound Research Laboratories Ltd (1987).
Practical Building Acoustics, Sound Research Laboratories Ltd (1976).
Noise Control in Industry, Sound Research Laboratories Ltd (1978).
Instrumentation for Practical Noise Measurement, Sound Research Laboratories Ltd (1978).
Airflow and Acoustic Commissioning of H & V Systems, Sound Research Laboratories Ltd (1985).
Basic Vibration Control, Sound Research Laboratories Ltd (1977).
Acoustics, Noise and Buildings, P. H. Parking, H. R. Humphreys and J. R. Cowell (1979).
Reduction of Machinery Noise, M. J. Crocker (1974).
Woods Practical Guide to Fan Engineering, B. B. Daly (1985).
Handbook of Noise and Vibration Control, R. H. Warring (1983).
Fan Engineering, Buffalo Forge Co. (1970).
Acoustics, L. L. Beranek (1986).
Community Noise Rating, T. J. Schultz (1982).
The Noise Handbook, W. Tempest (1985).
The Effects of Noise on Man, K. D. Kryter (1985).
Noise and Man, W. Burns (1968).
Vibration and Sound, P. M. Morse (1986).
Principles and Applications of Room Acoustics, L. Cremer, H. A. Muller and T. J. Schultz (1982).
Noise Control, C. M. Harris (1979).
Room Acoustics, H. Kuttruff (1979).
Noise and Vibration Control, L. L. Beranek (1971).
Dynamic Vibration Absorbers, J. B. Hunt (1979).
Noise Control—Principles and Practice, Bruel and Kjaer.
Machinery Health Monitoring, Bruel and Kjaer.
Measuring Sound, Bruel and Kjaer.
Measurements in Building Acoustics, Bruel and Kjaer.
Vibration Testing, Bruel and Kjaer.
Environmental Noise Measurement, Bruel and Kjaer.
Sound Intensity, Bruel and Kjaer.
Vibration Problems in Engineering, Timoshenko, Young and Weaver (1974).
Noise, Buildings and People, D. J. Croome (1977).

Papers

Generalised Theory for Computing Noise from Turbulence in Aerodynamic Systems—ASHRAE Journal, H. C. Hardy (1963).

Noise in Ventilating Systems — H & V Journal, G. Porges (1966).

Investigations of Aerodynamic Noise in Duct System Components for High Velocity Plant—BRS, H. Brockmeyer (1969).

In High Velocity Systems Duct Parts Create Sound — ASHRAE Journal, W. F. Kerka (1960).

The Problem of Duct Generated Noise and its Prediction — ASHRAE, C. G. Gordon (1968).

American Society of Heating, Refrigeration and Air Conditioning Engineers, Inc — ASHRAE Handbook — HVAC Systems, (1987).

A Study to Determine the Noise Generation and Noise Attenuation of Lined and Unlined Duct Fittings — ASHRAE RP–265, (1983).

Air Flow Generated Noise: Part I. Grilles and Dampers — HVRA Report 75, M. J. Holmes (1973).

Air Flow Generated Noise: Part II, Bends with Turning Vanes — HVRA Report 78, M. J. Holmes (1973).

Aerodynamic Noise Generation by Duct Elements — ASHRAE Paper 2/47, C. E. Bullock (1970).

The Problem of Duct-Generated Noise and Its Prediction — ASHRAE Paper 2070, C. G. Gordon (1968).

Spoiler Generated Flow Noise, Results — JASA 45, C. G. Gordon (1969).

Flow-Generated Noise Research Report — VDI–Z 7 (20), M. Hecki (1969).

INDEX